T0310088

OXIDATION OF AMINO ACIDS, PEPTIDES, AND PROTEINS

Wiley Series of Reactive Intermediates in Chemistry and Biology

Steven E. Rokita, Series Editor

Quinone Methides
Edited by Steven E. Rokita

Radical and Radical Ion Reactivity in Nucleic Acid Chemistry
Edited by Marc Greenberg

Carbon-Centered Free Radicals and Radical Cations
Edited by Malcolm D. E. Forbes

Copper-Oxygen Chemistry
Edited by Kenneth D. Karlin and Shinobu Ihoh

Oxidation of Amino Acids, Peptides, and Proteins: Kinetics and Mechanism
By Virender K. Sharma

OXIDATION OF AMINO ACIDS, PEPTIDES, AND PROTEINS

Kinetics and Mechanism

VIRENDER K. SHARMA, PhD

A JOHN WILEY & SONS, INC., PUBLICATION

Published by John Wiley & Sons, Inc., Hoboken, New Jersey
Published simultaneously in Canada

For general information on our other products and services or for technical support, please contact our Customer Care Department within the United States at (800) 762-2974, outside the United States at (317) 572-3993 or fax (317) 572-4002.

Wiley also publishes its books in a variety of electronic formats. Some content that appears in print may not be available in electronic formats. For more information about Wiley products, visit our web site at www.wiley.com.

Library of Congress Cataloging-in-Publication Data:
Sharma, Virender K.
Oxidation of amino acids, peptides, and proteins : kinetics and mechanism / Virender K. Sharma.
p. ; cm. – (Wiley series on reactive intermediates in chemistry and biology)
Includes bibliographical references and index.
ISBN 978-0-470-62776-1 (cloth)
I. Title. II. Series: Wiley series on reactive intermediates in chemistry and biology.
[DNLM: 1. Oxidation-Reduction. 2. Amino Acids–metabolism. 3. Environmental Pollutants–chemistry. 4. Peptides–metabolism. 5. Proteins–metabolism. QU 125]
547'.23–dc23

2012040268

Printed in the United States of America

10 9 8 7 6 5 4 3 2 1

I dedicate this book to my late mother
Krishna Devi Sharma
who has been my inspiration and strength

CONTENTS

PREFACE TO SERIES

Most stable compounds and functional groups have benefited from numerous monographs and series devoted to their unique chemistry, and most biological materials and processes have received similar attention. Chemical and biological mechanisms have also been the subject of individual reviews and compilations. When reactive intermediates are given center stage, presentations often focus on the details and approaches of one discipline despite their common prominence in the primary literature of physical, theoretical, organic, inorganic, and biological disciplines. The *Wiley Series on Reactive Intermediates in Chemistry and Biology* is designed to supply a complementary perspective from current publications by focusing each volume on a specific reactive intermediate or target and endowing it with the broadest possible context and outlook. Individual volumes may serve to supplement an advanced course, sustain a special topics course, and provide a ready resource for the research community. Readers should feel equally reassured by reviews in their speciality, inspired by helpful updates in allied areas, and intrigued by topics not yet familiar.

This series revels in the diversity of its perspectives and expertise. Where some books draw strength from their focused details, this series draws strength from the breadth of its presentations. The goal is to illustrate the widest possible range of literature that covers the subject of each volume. When appropriate, topics may span theoretical approaches for predicting reactivity, physical methods of analysis, strategies for generating intermediates, utility for chemical synthesis, applications in biochemistry and medicine, impact on the environment, occurrence in biology, and more. Experimental systems used to explore these topics may be equally broad and range from simple models to complex arrays and mixtures such as those found in the final frontiers of cells, organisms, earth, and space.

Advances in chemistry and biology gain from a mutual synergy. As new methods are developed for one field, they are often rapidly adapted for application in the other. Biological transformations and pathways often inspire analogous development of new procedures in chemical synthesis, and likewise, chemical characterization and identification of transient intermediates often provide the foundation for understanding the biosynthesis and reactivity of many new biological materials. While individual chapters may draw from a single expertise, the range of contributions contained within each volume should collectively offer readers with a multidisciplinary analysis and exposure to the full range of activities in the field. As this series grows, individualized compilations may also be created through electronic access to highlight a particular approach or application across many volumes that, together, cover a variety of different reactive intermediates.

Interest in starting this series came easily, but the creation of each volume of this series required vision, hard work, enthusiasm, and persistence. I thank all of the contributors and editors who graciously accepted the challenge.

John Hopkins University STEVEN E. ROKITA

INTRODUCTION

I dedicate this book to my mother, Mrs. Krishna Devi Sharma. Like all children, I believed that my mother was divine and a living goddess on earth. It would be a failing if I didn't tell you her story because I feel so blessed to have been born to her. She came from a small village named Polia Prohitan in Himachal Pradesh, India, and while she did not have the opportunity to be formally educated, she was a highly evolved, religious, wise, and unusually intelligent person. Quiet and soft-spoken, she was the center of our entire household. Her love for the family was the binding force that provided each of us with the help and motivation we needed to go forward in difficult and testing times. She instilled a sense of strict discipline tempered with strong values, reminding us always: "Good things bring good results."

All of a sudden, she began to experience weakness. It took the doctors some time to figure out why. Those were tough times for our family. I could not bear the thought of losing her as she became sicker with each passing day. It was during that period I had the thought that, as a chemist, I might be able to do something to unravel the mystery of her illness and save her. When she passed away, I decided to continue to pursue the idea of bringing pure science and medical science together in the study of proteins, hoping to contribute to social and humanitarian needs in a humble way. My mother continues to be my beacon as I explore the study of protein oxidation. In the fall of 2009, I met Ms. Anita Lekhwani at an American Chemical Society meeting and told her about my wish to write this book in honor of my mother. She encouraged me to take on the task.

A major cause of protein degradation results from its oxidation by reactive species, generated in the body. The deleterious reactions are involved in oxidative stress and in several diseases including cancer, aging, vascular problems, and Alzheimer's and Parkinson's diseases. Chapter 1 of the book describes reactive species (radical and nonradical) associated with various diseases. Protein structures are important in understanding their reactions with reactive species and hence are reviewed in Chapter 1. Because of my background in environmental chemistry, the significance of reactive species in the atmosphere, disinfection processes, and remediate pollutants is briefly given in Chapter 1 as well. Examples of environmental processes continue to be part of subsequent chapters of the book. Emphasis on thermodynamics (or speciation) of amino acids and reactive species as well as the effect of metal–ligand binding in oxidation chemistry is based on my PhD training under the supervision of Professor Frank J. Millero and is presented separately in Chapter 2. My introduction to protein chemistry began when I observed studies on the interactions of proteins with hydroxyapatite/octacalcium phosphate, conducted by senior researcher Dr. Mats Johnson under the guidance of Professor George H. Nancollas. Mats is very supportive of this book. My initial exposure to radical chemistry was at the laboratory of Dr. Benon H.J. Bielski at Brookhaven National Laboratory (BNL), where I was Research Associate for two years. The time spent at BNL was very helpful in writing this book.

In the biological environment, protein damage is governed by the concentrations of reactants under localized environments and reaction rates of amino acid side chains and backbone (or peptide) of proteins with the reactive species. Chapters 3–6 present the kinetics and mechanisms of reactive halogen, oxygen, nitrogen, carbon, sulfur, and phosphate species as well as of reactive high-valent Cr, Mn, and Fe species. The generation of these species in the laboratory for performing kinetics studies is initially presented. The importance of fundamental properties of the reactive species (equilibrium, pH, and redox potentials) is stressed in each of the chapters. Because amino acids, peptides, and proteins are involved in the formation of disinfection by-products in the treatment of water, the oxidation of these molecules by chlorine, chlorine dioxide, and ozone is also discussed in Chapters 3 and 4. Chapter 5 deals mainly with the recently studied oxidative nitrogen species, peroxynitrite, on which several oxidative studies have been performed in the last few years. Reactive species of C, S, and P are also covered in Chapter 5, which also demonstrates their importance in the remediation of pollutants in aquatic environments. High-valent Cr species, which are carcinogenic and toxic, are discussed with respect to their reactions with biomolecules (Chapter 6). These reactions are reviewed in Chapter 6. Finally, Chapter 6 covers the oxidation of amino acids, peptides, and proteins by permanganate, ferryl, and ferrates species.

It is my intention that the literature presented in the book is beneficial to chemists, biochemists, and environmental chemists and engineers. While writing the book, I came across hundreds of references for each of the chapters, and I attempted to cite as many of those sources as possible. I apologize in advance to those researchers whose work I unintentionally overlooked.

Wiley found the book to be an excellent fit for the *Reactive Intermediates in Chemistry and Biology* book series. I sincerely thank the series editor,

Professor Steve Rokita, for his guidance. Steve constantly reminded me about the theme of the book in order to focus me on the subject matter. I thank nears and dears, Puja, Narinder, Devinder, Kiran, and Renu, as well as my high school chemistry teacher, M.P. Khurana, for their support. I am very thankful to my friend and colleague Professor Mary Sohn, who provided corrections for some of chapters. Rachel Gilman is acknowledged for her help in the editing of the text. I thank Erik Casbeer for making some figures for this book.

1

REACTIVE SPECIES

Reactive intermediates and oxidative damage of proteins are important in biomedical research due to their roles in pathologies and aging [1–5]. Reactive species are also associated with important mediators in a wide range of biological processes such as signaling, for proper synaptic plasticity, and normal memory [3, 6–9]. Additionally, nitroxidative species contribute to pain and central sensitization [10, 11]. The amounts of reactive species during neurodegenerative diseases and aging increase to higher levels than the antioxidants present in a cell can handle. The reactive species that participate in a large number of reactions in diseases [4, 12–17] include both free radicals and nonradical species (Table 1.1) [18–20]. Reactive oxygen species (ROS) include superoxide anion ($O_2^{-\bullet}$), hydroperoxyl ($HO_2^{\bullet}$), alkoxyl ($RO^{\bullet}$), peroxyl ($ROO^{\bullet}$), hydroxyl radical ($^{\bullet}OH$), hydrogen peroxide (H_2O_2), ozone (O_3), singlet oxygen ($^1O_2$), and hypochlorous acid (HOCl). The ROS initiate many reactions, for example, the primary mitochondrial ROS, $O_2^{-\bullet}$, reacts with superoxide dismutase (SOD) to form H_2O_2, which then reacts further with metal ions or their complexes (Fenton and Fenton-like reactions) to produce $^{\bullet}OH$. Other intermediates are reactive nitrogen species (RNS), which include nitric oxide ($NO^{\bullet}$), nitrogen dioxide radical ($NO_2^{\bullet}$), peroxynitrite ($OONO^-$), peroxynitrous acid (OONOH), alkylperoxynitrite (ROONO), and nitrosyl (NO^+).

ROS and RNS are interconnected and cause protein damage in biological processes. $O_2^{-\bullet}$, $NO^{\bullet}$, and $ONOO^-$ are associated with neuroimmune activation,

Oxidation of Amino Acids, Peptides, and Proteins: Kinetics and Mechanism, First Edition.
Virender K. Sharma.

TABLE 1.1. Various Reactive Species

Free Radicals	Nonradicals
Reactive oxygen species (ROS)	
Superoxide, $O_2^{-\bullet}$	Hydrogen peroxide, H_2O_2
Hydroxyl, $^\bullet OH$	Hypobromous acid, HOBr[a]
Hydroperoxyl, $HO_2^\bullet$ (protonated superoxide)	Hypochlorous acid, HOCl[b]
Carbonate, $CO_3^{-\bullet}$	Ozone, O_3[c]
Peroxyl, $RO_2^\bullet$	Singlet $O_2^1\Delta g$
Alkoxyl, $RO^\bullet$	Organic peroxides, ROOH
Carbon dioxide radical, $CO_2^{-\bullet}$	Peroxynitrite, $ONOO^-$[d]
Singlet $O_2^1\Sigma_g^+$	Peroxynitrate, O_2NOO^-[d]
	Peroxynitrous acid, ONOOH
	Peroxomonocarbonate, $HOOCO_2^-$
	Nitrosoperoxycarbonate, $ONOOCO_2$
Reactive nitrogen species (RNS)	
Nitric oxide, $NO^\bullet$	Nitrous acid, HNO_2
Nitrogen dioxide, NO_2[c]	Nitrosyl cation, NO^+
Nitrate, $NO_3^\bullet$	Nitroxyl anion, NO^-
	Dinitrogen tetroxide, N_2O_4
	Dinitrogen trioxide, N_2O_3
	Peroxynitrite, $ONOO^-$[d]
	Peroxynitrate, O_2NOO^-
	Peroxynitrous acid, ONOOH[d]
	Nitronium cation, NO_2^+
	Alkyl peroxynitrites, ROONO
	Alkyl peroxynitrates, RO_2ONO
	Nitryl chloride, NO_2Cl
	Peroxyacetyl nitrate, $CH_3C(O)OONO_2^-$[c]
Reactive chlorine species (RCS)	
Atomic chlorine, $Cl^\bullet$	Hypochlorous acid, HOCl[b]
	Nitryl chloride, NO_2Cl[e]
	Chloramines
	Chlorine gas (Cl_2)
	Bromine chloride (BrCl)[a]
	Chlorine dioxide (ClO_2)
Reactive bromine species (RBS)	
Atomic bromine, $Br^\bullet$	Hypobromous acid (HOBr)
	Bromine gas (Br_2)
	Bromine chloride (BrCl)

"ROS" is a collective term that includes both oxygen radicals and certain nonradicals that are oxidizing agents and/or are easily converted into radicals (HOCl, HOBr, O_3, $ONOO^-$, $^1O_2$, H_2O_2).
All oxygen radicals are ROS, but not all ROS are oxygen radicals. Peroxynitrite and H_2O_2 are frequently erroneously described in the literature as free radicals, for example. "RNS" is a similar collective term that includes NO and NO_2 as well as nonradicals such as HNO_2 and N_2O_4.
"Reactive" is not always an appropriate term: H_2O_2, $NO^\bullet$, and $O_2^{\bullet-}$ react fast with few molecules, whereas $^\bullet OH$ reacts fast with almost everything. Species such as $RO_2^\bullet$, $NO_3^\bullet$, $RO^\bullet$, HOCl, HOBr, $CO_3^\bullet$, $CO_2^\bullet$, $NO_2^\bullet$, $ONOO^-$, NO_2^+, and O_3 have intermediate reactivities.

[a] HOBr and BrCl could also be regarded as RBS.

[b] HOCl and HOBr are often included as ROS, although HOCl is also an RCS.

[c] Oxidizing species formed in polluted air that are toxic to plants and animals. $NO_2^\bullet$ is also produced *in vivo* by myeloperoxidase and from $ONOO^-$ [19]. Ozone might also be produced *in vivo*, although the chemistry involved is unclear [20].

[d] $ONOO^-$, O_2NOO^-, and ONOOH are often included as ROS but are also classifiable as RNS.

[e] NO_2Cl can also be regarded as a RNS.

Adapted from Halliwell [18] with the permission of the International Society of Neurochemistry.

supraspinal descending facilitation, and nitroxidative stress [21]. The $O_2^{-\bullet}$ species is produced from mitochondria and NADPH oxidase, while NOS enzymes synthesize $NO^{\bullet}$ through enhanced nociception and the activation of the *N*-methyl-D-aspartate receptor. Both $O_2^{-\bullet}$ and $NO^{\bullet}$ form $ONOO^-$, which inactivates the glutamate transporter, manganese superoxide dismutase (MnSOD), and glutamate synthase, which increases the production of additional nitroxidative species [10]. In addition to ROS and RNS, other reactive species also involved in various biological activities include the carbonate radical ($CO_3^{-\bullet}$) and the organic radical, $R^{\bullet}$ (thiyl and protein radicals). Metals such as Cr, Mn, and Fe in their high-valent sates are also involved in reactions with molecules of biological importance. Reactive intermediates may also be produced by UV radiation in the presence of oxygen [22].

1.1 DISEASES

1.1.1 Neurodegenerative Diseases

Generally, there are four common features in neurodegenerative diseases, which are interrelated with one another [23–25]. These include (1) both ROS and RNS working together to cause damage in the degenerative disease and also to create a vicious cycle by stimulating proinflammatory gene transcription in glia; (2) participation of redox-active (e.g., Cu and Fe) and redox-inactive (e.g., Zn) metal ions; (3) abnormal functioning of mitochondria; and (4) accumulation of misfolded or unfolded proteins in brain cells, which leads to Alzheimer's disease (AD), Parkinson's disease (PD), Huntington's disease (HD), frontotemporal labor degeneration (FTLD), multiple sclerosis, and amyotrophic lateral sclerosis (ALS) (Table 1.2) [26]. A recent study demonstrated the role of RNS in protein misfolding, mitochondrial dysfunction, and synaptic injury [27]. Most of the folded proteins display toxicity toward cultured neuronal cells *in vitro* and, hence, may be related to the degeneration and loss of nerve cells *in vivo*. The molecular mechanism of toxic effects is not fully understood, but changes in membrane permeability, influx of Ca^{2+}, and oxidative damage induction, followed by apoptosis have been suggested [26].

The chemistry of neurotoxicity is presented in Figure 1.1 [23]. The $O_2^{-\bullet}$ species is produced from mitochondrial proteins and mutationally altered or damaged proteins, which subsequently generate H_2O_2. The $^{\bullet}OH$ species is generated through reactions of H_2O_2 with $O_2^{-\bullet}$ (Haber–Weiss reaction) and transition metal ion (generally Cu(I) and Fe(II)) (Fenton reaction). The resulting oxidized metal ions (Cu(II) and Fe(III)) can be reduced by cellular reductants such as thiols, vitamin C, and vitamin E. SOD also transforms $O_2^{-\bullet}$ to O_2 and H_2O_2, while catalase (CAT) removes H_2O_2. Oxidation of protein side chains generate hydroxylated and carbonyl products (Chapters 4 and 5). Oxidation also occurs through halogenated species (e.g., HOCl) (Chapter 3). The

TABLE 1.2. Misfolded Proteins and Their Associated Neurodegenerative Diseases

Protein(s)	Disease(s)	Lesion(s)
Aβ and tau protein	AD, Down's syndrome	Senile plaques, NFTs
α-Synuclein	Parkinson's disease, dementia with Lewy bodies	Lewy bodies and Lewy neurites
Tau protein	FTLD	Tau inclusions
PrP	Transmissible prion disease	Amyloid plaques and prion rods
Huntington	HD	Intranuclear inclusions
ABri/ADan and tau protein	British and Danish familial dementias	Amyloid plaques and NFTs
SOD-1	Motor neuron disease	SOD-1 inclusions
TDP-43	FTLD, motor neuron disease	Intracellular inclusions

Mutations in the genes encoding these proteins invariably give rise to familial forms of neurodegenerative disease.

Aβ, amyloid β-peptide; ADan, Danish dementia peptide; SOD-1, superoxide dismutase 1; TDP-43, TAR DNA-binding protein 43; NFT, neurofibrillary tangle.

Adapted from Allsop et al. [26] with the permission of the Biochemical Society.

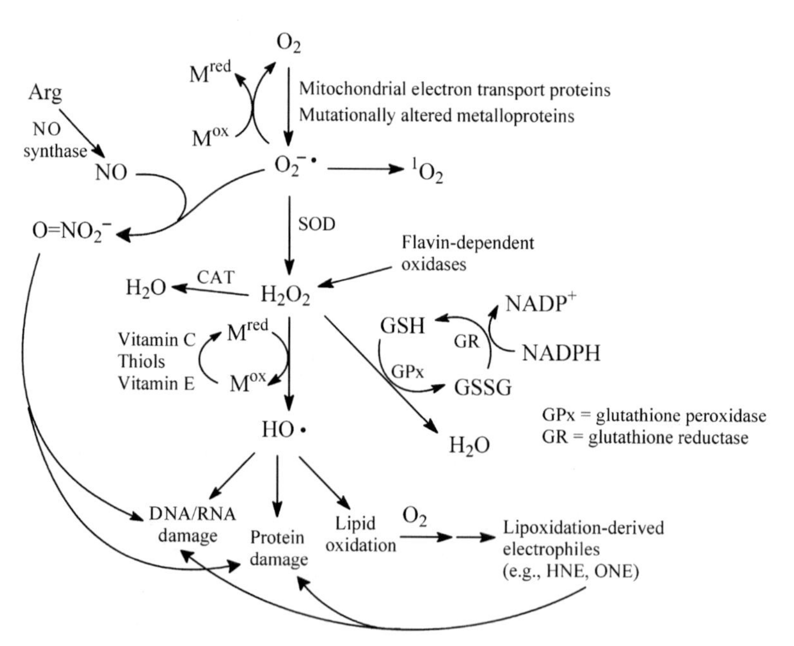

Figure 1.1. Scheme of oxidative stress biochemistry and neurotoxicity. HNE, 4-hydroxy-2-nonenal; ONE, 4-oxo-2-nonenal (adapted from Sayre et al. [23] with the permission of the American Chemical Society).

roles of reactive species in common neurodegenerative diseases, Alzheimers disease (AD) and Parkinson's disease (PD), as well as in aging, are summarized below.

1.1.1.1 Alzheimers Disease. AD was first discovered in 1907 and is an irreversible disease, which causes unusual behavior, memory loss, personality changes, and a decline in the ability to concentrate [28]. AD directly affects ~10% of humans by age 65 and ~50% by age 85 [28, 29]. Two major hypotheses have been proposed to explain AD. The first suggests that an abnormal processing of the amyloid precursor protein (APP) takes place during the neurodegeneration, which results in generation, aggregation, and deposition of the Aβ peptide [30]. This process may facilitate neurofibrillary tangle (NFT) peptide formation and, consequently, cell death (see Table 1.2), and is classified as an amyloid cascade hypothesis. This hypothesis is reasonably supported by genetic studies [31–33]. APP encodes the Aβ peptide, while the protein genes, PS1 and PS2, encode transmembrane proteins. The second hypothesis suggests cytoskeletal changes take place during neurodegeneration, in which hyperphosphorylation and aggregation of tau processes contribute to the activation of cell death [34]. The amyloid cascade hypothesis has been investigated extensively [28, 35–38]. Recently, mass spectrometry (MS) has been applied *in vitro* and *in vivo* to learn the role of the Aβ peptide in AD [39]. Studies include elucidating the structure of the Aβ peptide and its interaction with metals (e.g., Cu(II)) [28]. Copper, iron, and zinc have been determined in amyloid plaques from the brains of those with AD [40].

The size of the Aβ peptide varies from 39 to 43 amino acids, produced from the sequential β- and γ-secretase processing of APP [39]. Cu(II) binds to the Aβ peptide through tyrosine (Tyr10) and histidine (His13, His14, and His6) [41]. The Aβ complex of Cu(II) has a high positive reduction potential. The neurotoxicity of the Aβ peptide has been suggested to relate to the production of $^{\bullet}OH$ in the copper-mediated oxidation of ascorbate ($AScH^-$) in the presence of oxygen and H_2O_2 (Eqs. 1.1–1.4) [42]:

$$A\beta\text{-Cu(II)} + AScH^- \rightarrow A\beta\text{-Cu(I)} + ASc^{\bullet-} + H^+ \tag{1.1}$$

$$A\beta\text{-Cu(II)} + \cdot + ASc^{\bullet-} \rightarrow A\beta\text{-Cu(I)} + ASc \tag{1.2}$$

$$A\beta\text{-Cu(I)} + H_2O_2 \rightarrow A\beta\text{-Cu(II)} + {}^{\bullet}OH + OH^- \tag{1.3}$$

$$A\beta\text{-Cu(I)} + O_2 \rightarrow A\beta\text{-Cu(II)} + O_2^{\bullet-}. \tag{1.4}$$

1.1.1.2 Parkinson's Disease. PD is a neurodegenerative movement disorder in which pathophysiological features include the accumulation of intracellular inclusions and degeneration and death of dopaminergic neurons of substantia nigra (SN), a part of the midbrain [43]. Mitochondrial dysfunction, oxidative stress, abnormal protein accumulation, and protein phosphorylation are important molecular mechanisms that compromise dopamine neuron function

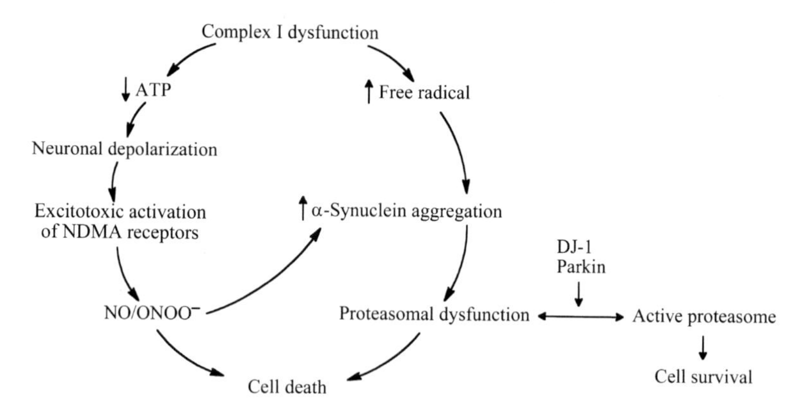

Figure 1.2. Complex I deficiency may be central to sporadic PD (adapted from Dawson and Dawson [43] with the permission of the American Association for the Advancement of Science).

and survival in the pathogenesis of both sporadic and familial PD [44]. The role of ROS in sporadic PD is shown in Figure 1.2 [43]. Derangements in complex I (NADH CoQ10 reductase) and oxidative stress lead to aggregation and accumulation of α-synuclein [43]. Dysfunction of complex I results in the formation of free radicals and a decrease in the level of the formation of ATP, which ultimately result in neuronal depolarization and excitotoxic neuronal injury. The involvement of RNS also increases oxidative stress and injury (see Fig. 1.2) [45].

The role and significance of glutathione (GSH) in PD has been studied in detail [36, 37, 46]. A cell has a redox equilibrium, which involves GSH and GSH disulfide [47]. Under oxidative stress, the equilibrium moves toward disulfide and, hence, depletes GSH. A significant decrease in levels of GSH in SN has been observed in PD, which lowers the activity of the mitochondrial complex I [48–50]. This process accelerates the buildup of defective proteins, which impairs the ubiquitin–proteasome pathway during protein degradation, resulting in cell death of the SN dopaminergic neurons. The presence of abnormal protein aggregates is a characteristic of neurodegeneration during PD. The role of iron in PD has been reported [49, 51]. Iron transporters, divalent metal transporter 1 (DMT1) and ferroportin, may be involved in dopaminergic brain areas where the production of H_2O_2 favors the Fenton reaction. The chelation of iron may be effective in preventing or delaying the progression of PD [52].

1.1.2 Metals in Human Diseases

In biological systems, redox metals such as Cu, Fe, Cr, and Co are involved in redox cycle reactions and generate reactive intermediates like $O_2^{\bullet-}$ and $NO^{\bullet}$, which cause the disruption of homeostasis. This creates an excess amount of

ROS and RNS over available antioxidant biomolecules that subsequently induce modification of proteins, peroxidation of lipid, and damage of DNA resulting in diseases such as cancer, cardiovascular disease, diabetes, and atherosclerosis [35]. The significance of Cu and Fe in AD and PD is described in the previous subsection. Elevated levels of copper have been detected in tissues of serum and tumors of cancer patients in comparison to healthy persons. The ratios of Cu:(Fe, Zn, and Se) have also been detected at higher levels in cancer patients than in normal subjects [53]. Copper can catalyze the formation of ROS and also decrease the levels of GSH [54, 55]. Increased levels of copper could play an important role in the development of various cancers. The interaction between Cu and homocysteine may be involved in atherosclerosis [56]. This interaction generates free radicals, which oxidize low-density lipoprotein (LDL), and oxidized products have been found in atherosclerosis plaques.

Excess levels of iron can be toxic due to its role in producing hydroxyl radicals (Fenton reaction, Fig. 1.3), which are involved in oxidizing DNA. More than 100 oxidized products, both carcinogenic and mutagenic, have been identified. The most common DNA oxidized product is 8-hydroxyguanine (Fig. 1.3) [35]. The lipid peroxidation process, involved in coronary artery disease, is catalyzed by iron [57] (Fig. 1.3). The initial formation of the peroxyl radical ($ROO^•$) rearranges into endoperoxides through a cyclization reaction, which leads to the formation of malondialadehyde (MDA). The other major product of the lipid peroxidation process is 4-hydroxy-2-nonenal. Reactions between MDA and DNA bases produce adducts (Fig. 1.3) such as M_1C, M_1A, and M_1G formed from DNA bases cytosine, adenosine, and guanine, respectively [58]. A significant increase in the levels of 8-OH-G, 2-hydroxy-adenine, and 8-hydroxy-adenine adducts have been detected in rectum and colon biopsies [59].

Chromium(VI) at high levels is considered to be a risk to human health. Some compounds of Cr(VI) may cause skin cancer when they come in contact with skin. Further details on the toxic effects of Cr and related mechanisms are discussed in Chapter 6. Vitamin B_{12} is 4% cobalt by mass, which contributes to toxic and carcinogenic effects. Cobalt in the presence of O_2 generates Co(I)-$OO^•$, which, through catalysis by SOD, forms Co(I) and H_2O_2 [60, 61]. The reaction between Co(I) and H_2O_2 results in $^•OH$, which induces damage to DNA and inhibits DNA repair [62].

The redox inactive metals (Zn, Cd, and Pb) generate toxic effects when undergoing bond formation with sulfhydryl groups of proteins and depletion of GSH. The exposure of Pb to adolescents represents a serious health threat worldwide [63]. The toxic effects of Pb include hypertension, cognitive impairments, and neurological disorders [64, 65]. Both ROS and RNS have been implicated in hypertension due to their exposure to humans [66]. A number of studies have shown arsenic is carcinogenic [67]. Adverse health effects of arsenic from contaminated water include anemia, skin lesions, peripheral

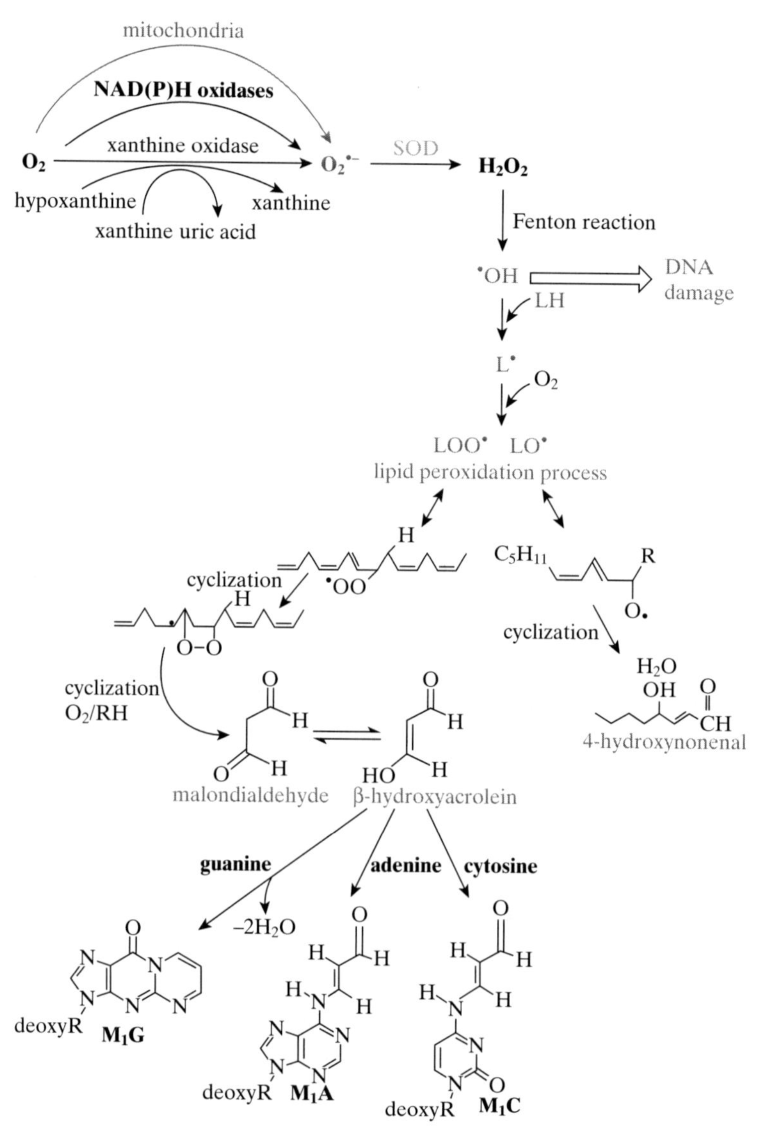

Figure 1.3. ROS formation and the lipid peroxidation process (adapted from Jomova and Valko [35] with the permission of the Elsevier Inc.). See color insert.

neuropathy, and tumors [68]. Evidence of both ROS and RNS involvement has been detected in the metabolism of arsenic [60, 69, 70].

In recent years, the use of metal nanoparticles in various applications has increased greatly and further studies are in progress to evaluate their toxicity [71, 72]. For example, silver nanoparticles and Ag^{+} ions have been shown to generate ROS [73, 74, 74–77]. It has been suggested that ROS, produced by silver nanoparticles, may disrupt the production of ATP and damage cell membranes. The specific type of DNA damage or chromosomal aberrations needs to be investigated. DNA damage response in silver nanoparticle-treated cells by the upregulation of damage response proteins may be identified by conducting protein expression analysis [78]. Recently, there has been more emphasis on protein–nanoparticle interactions for the development of functional and safe nanoparticles [79]. Significantly, a role of modified fullerenes and cerium oxide nanoparticles has been demonstrated in protecting mammalian cells against damage, which can possibly be caused by ROS and RNS [80].

The biological consequences of various reactive species, mentioned above, are determined by rates of their formation and decay in different intra- and extracellular environments. Steady-state concentrations of reactive species can be estimated using their reaction rates with various constituents of environments. Proteins are made of ~70% of the mass of organic constituents related to living matter and are an important target of reactive species. The modifications of proteins by reactive species are governed by factors such as structure, redox properties, and acid–base chemistry of proteins [81]. At a molecular level, assessing the protein structure is critical to understand the modifications induced by oxidation processes and thus the function of proteins [82, 83]. The next section describes the progress that has been made in revealing structures of proteins. The role of redox properties of proteins in oxidative reactions is explained for thiols as an example in Section 1.3.5. The influence of pK_a and the speciation of reactive amino acids and their moieties in proteins are presented in Chapter 2. Redox potentials, reaction rates, and oxidative mechanisms of reactive species are given in Chapters 3–6. The involvement of reactive species in environmental processes (e.g., disinfection and remediation) is also presented in Section 1.4 and in Chapters 3–6.

1.2 PROTEIN STRUCTURE

The details of protein structure are very important in understanding biochemical functions such as energy conversion, transport, enzyme catalysis, and host defense [84, 85]. Studies on protein structure are also imperative to understanding various *in vivo* processes, such as cell–cell communication, ligand binding, folding, and transport of proteins across membranes [86–88]. A number of techniques have been applied to determine the structure of proteins at all structural levels [87, 89–93]. Conformational changes in proteins in response to alternations in the environment of a solvent can be observed by using calorimetric and optical methods [94]. Optical methods include UV–vis, infrared, and fluorescence spectroscopy, and circular dichroism (CD), which have been applied in measuring protein thermodynamic stabilities based on the denaturant-induced unfolding transition [95]. X-ray crystallography has been used extensively to obtain high-resolution structural information on proteins [96]. This technique has limitations because partially unfolded proteins do not crystallize, and hence, it is not always clear whether conformations in the solid state are identical to the bulk solution phase structures [97]. Time-resolved X-ray techniques may shed more light on protein dynamics [96]. Nuclear magnetic resonance (NMR) spectroscopy can provide insight on the structure and dynamics of proteins in solution [98–100]. Modern instruments have eliminated the limitations of high concentrations of proteins using traditional NMR spectroscopy when collecting data. However, determining structural information for proteins larger than 40–50 kDa remains a challenge for NMR spectroscopy. Progress in the mapping of proteins is summarized below, which includes the application of oxidative labeling of proteins.

1.2.1 Oxidative Labeling

1.2.1.1 Carbonyl Labeling. Proteins containing carbonyls are produced by reacting them with 2,4-dinitrophenylhydrazine (DNPH), followed by separation using gel electrophoresis [90]. The identification and quantification of derivatized protein carbonyls can also be carried out by ratiometric Raman spectroscopy [101]. Another approach of protein carbonyl labeling is by the reaction with biotin hydrazide under mild pH conditions [102]. One drawback of labeling is that both approaches cannot distinguish primary and secondary carbonyls as well as carbonyls from glycation [103, 104]. It is also possible that both reactants may not react with all types of oxidized amino acids [105].

1.2.1.2 Cysteine Residue Labeling. Labeling of the free SH group by a reagent based on iodoacetamide, maleimide, and 5,5′-dithiobis-(2-nitrobenzoic acid) allows indirect assessment of the oxidation of Cys [106–108]. Increased Cys oxidation causes a decrease in the amount of labeling. One other approach uses the blocking of reduced thiols present in solution before the reduction of oxidized thiols with dithiothreitol (DTT), followed by labeling with iodoacetamidofluorescein for separation using two-dimensional gel electrophoresis [109]. Thiol-specific biotin-HPDP (N-[6-(Biotinamido)hexyl]-3′-(2′-pyridyldithio)propionamide) labeling has also been demonstrated for proteins containing reversibly oxidized cysteines [110].

1.2.1.3 Reactive Species Labeling. Oxidized forms of chlorine, bromine, and iodine readily react with Cys, Met, and aromatic amino acids to modify the side chains of proteins (Fig. 1.4) [111, 112]. Aside from the side chains shown in Figure 1.4, cystine also reacts to yield *N*-dischlorocystine. The stoichiometric amounts of oxidizing agents used in the reaction mixtures determine the extent of oxidation of Cys, Met, and cystine. The reaction between halide anions and oxidizers such as H_2O_2 and $^{\bullet}OH$ may also form oxidized halide intermediates. Significantly, intermediates may react with water to form

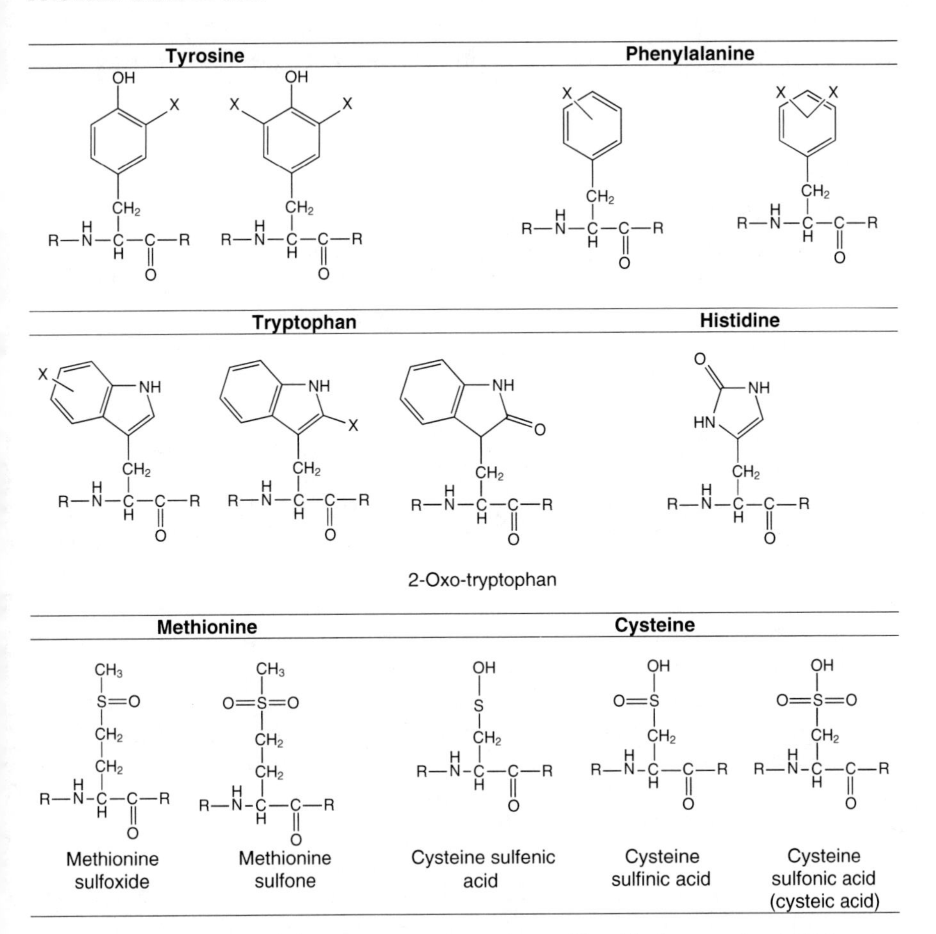

Figure 1.4. Structures of the most common amino acid oxidation products. X denotes either a halogen or a hydroxyl group (adapted from Roeser et al. [112] with the permission of Springer Americas).

hydroxylated products as well as products of adjacent peptide bond cleavage. Examples include simultaneous oxidative halogenation and cleavage of C-terminal peptide bonds of Trp and Tyr. The size of reactants, pH, and redox potential of either Trp or Tyr determine the selectivity of the cleavage reaction [113]. HOCl and hypobromous acid (HOBr) form mostly halogenated Trp, Tyr, and 2-oxo Trp [114–116]. RNS preferentially oxidize Met, Cys, His, Tyr, Phe, and Trp. The reaction of NO in oxygenated Cys resulted in nitrosation of the thiol group [117]. RNS are selective for site-directed labeling of peptides and proteins. For example, the reduction of nitrated Trp or Tyr to their corresponding aromatic amines is a suitable approach for site-directed labeling [118, 119].

Most of the labeling of peptides and proteins has been performed using ROS, particularly hydroxyl radicals [87]. The detailed chemistry of reactive halogen, nitrogen, and oxygen species-mediated oxidation of side chains is described in Chapters 3–5.

1.2.1.4 Hydroxyl Radical Labeling. The probing of the solvent-accessible surface area (SASA) of proteins using reactions with $^{\bullet}OH$ radical and subsequent identification of oxidized sites by MS has been widely used for the oxidative labeling of proteins [86, 87, 120–129]. Because of the low specificity of the $^{\bullet}OH$ radical, a number of target sites in the protein are modified, resulting in an incorporation of oxygen atoms into amino acid side chains. This yields characteristic +16 Da adducts in the mass spectrum. The oxidative labeling of the target sites in proteins is influenced by intrinsic reactivity and solvent accessibility [87, 130]. Hydroxyl radicals preferentially react with sulfur-containing heterocyclic and aromatic amino acids with rate constants varying from 5×10^{-9} to 1×10^{10}/M/s [1, 12, 131–134]. However, α-carbon hydrogen atoms, such as those found in Gly and Ala, have been found to react slower with the $^{\bullet}OH$ radical ($k \sim 10^{9}$/M/s) [131]. More is discussed in Chapter 4. The reactivity of amino acid residues with $^{\bullet}OH$ radicals and the ability of MS to detect oxidized products under aerobic conditions have the potential to probe the structure of proteins. The primary oxidation products for the amino acid side chains are presented in Table 1.3 [87, 135, 136]. Various methods of $^{\bullet}OH$ generation have been described, including Fenton chemistry, electrochemistry, coronal discharge methods, photochemistry, radiolysis, and pulse electron beams (Chapter 4) [87, 87, 137, 138].

A scheme of $^{\bullet}OH$ radical probing of the protein is presented in Figure 1.5 [87]. In this approach, the side chains of the protein and its complexes are modified using $^{\bullet}OH$ radicals, followed by X-ray exposure and digestion with proteases [82]. Quantification by MS determines the extent of modification. A tandem MS technique is applied to assess particular modified sites. This quantitation provides information about the solvent accessibility of each peptide in both the isolated and complexed states of the protein. The slower rate of binding proteins compared to the free protein indicates the influence of the binding process on the peptide containing the reactive side chains. Furthermore, allosteric changes in the conformation during binding may also produce an increase in reactivity [87]. This phenomenon can be observed in dose–response experiments. The use of a freeze-drying technique for the removal of H_2O_2 to determine the conformation of a protein was demonstrated to be unsuitable due to the oxidation of proteins under cold conditions [139].

There has been recent focus on the duration of either electron or laser pulses to the protein in order to prevent structural equilibrium during the timescale of exposure [140–143]. The use of timescales in submicroseconds has been determined for oxidative protein surface mapping [140, 142, 143]. A continuous-flow capillary setup has also been employed with an objective of exposing individual molecules of a protein to only a single $^{\bullet}OH$ pulse [123,

TABLE 1.3. Primary Products and Corresponding Mass Changes for Various Amino Acid Side Chains Subjected to Radiolytic Modification and Detectible by Mass Spectrometry

Side Chain	Side-Chain Modification and Mass Changes
Cys	Sulfonic acid (+48), sulfinic acid (+32), hydroxy (–16)
Cystine	Sulfonic acid, sulfinic acid
Met	Sulfoxide (+16), sulfone (+32), aldehyde (–32)
Trp	Hydroxy- (+16, +32, +48, etc.), pyrrol ring-open (+32, etc.)
Tyr	Hydroxy- (+16, +32, etc.)
Phe	Hydroxy- (+16, +32, etc.)
His	Oxo- (+16), ring-open (–22, –10, +5)
Leu[a]	Hydroxy- (+16), carbonyl (+14)
Ile[a]	Hydroxy- (+16), carbonyl (+14)
Val[a]	Hydroxy- (+16), carbonyl (+14)
Pro	Hydroxy- (+16), carbonyl (+14)
Arg	Deguanidination (–43), hydroxy- (+16), carbonyl (+14)
Lys	Hydroxy- (+16), carbonyl (+14)
Glu	Decarboxylation (–30), hydroxy- (+16), carbonyl (+14)
Gln	Hydroxy- (+16), carbonyl (+14)
Asp	Decarboxylation (–30), hydroxy- (+16)
Asn	Hydroxy- (+16)
Scr[b]	Hydroxy- (+16), carbonyl (–2-or +16-H_2O)
Thr[b]	Hydroxy- (+16), carbonyl (–2-or +16-H_2O)
Ala	Hydroxy- (+16)

[a] For aliphatic side chains, +14 Da products are normally much less than +16 Da products.
[b] For Ser and Thr, only trivial amounts of +16 and –2 Da products were found.
Adapted from Xu and Chance [87] with the permission of the American Chemical Society.

137, 144]. Different oxidation products detected in model proteins, ubiquitin and apomyoglobin, using a nanosecond laser-induced photochemical oxidation method are presented in Table 1.4 and Table 1.5 [141]. A single shot of the laser resulted in 70% and 90% oxidation of ubiquitin and apomyoglobin, respectively, and different surface amino acids residues were oxidized. The liquid chromatography (LC)/MS/MS analysis of trypsin digested apomyoglobin showed 99.3% sequence coverage. Extensive protein surface modifications of ubiquitin and β-lactoglobulin were seen in one submicrosecond electron beam pulse [142]. Fast photochemical oxidation of protein (FPOP) utilizing •OH radicals has also been used to characterize the epitope of the serine protease thrombin [145].

1.2.2 Other Techniques

In the last two decades, the approach of side-chain modification coupled with proteolysis and MS has been applied extensively to analyze protein

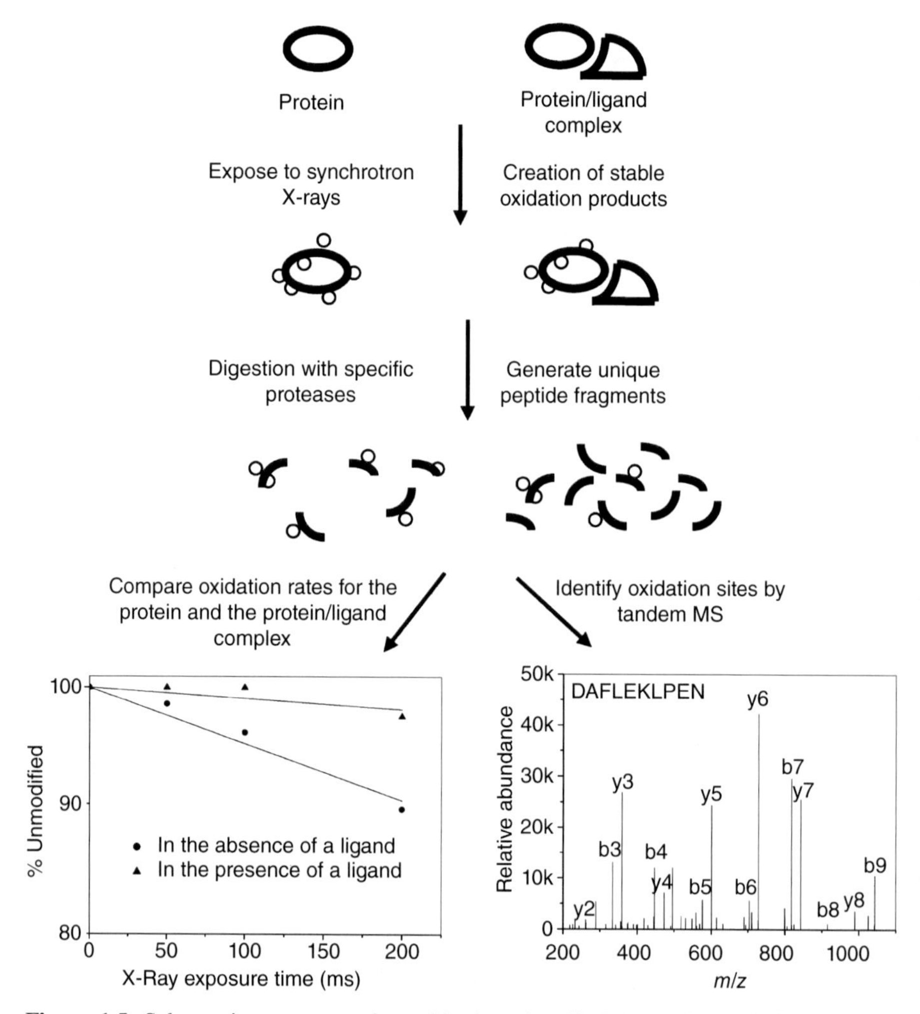

Figure 1.5. Schematic representation of hydroxyl radical footprinting (adapted from Xu and Chance [87] with the permission of the American Chemical Society).

conformations. Protein cleavage can be performed either enzymatically or chemically for MS analysis. Common proteases used are trypsin, chymotrypsin, pepsin, savinase, proteinase K, endoproteinase Asp-N, endoproteinase Lys-C, and metalloendopeptidase Lys-N [90, 92, 146]. A mixture of several proteases has been used to achieve cleavage at a wide range of sites in proteins [147]. However, protease is large and may not have access to the structural backbone of protein for cleavage in order to probe the structure precisely. A microwave method has also been applied to protein digestion [148]. Cyanogen bromide

TABLE 1.4. Number of Oxidations Detected by LC/FT-MS and Reactive Amino Acid Residues Identified by LC/MS/MS for Ubiquitin

Peptide Residues	Measure Mass (Da) by LC/FT-MS[a]	Oxidation Detected by LC/FT-MS[a]	Sites of Oxidation by LC/MS/MS[a]	Solvent-Accessible Surface Area by NMR (A^2)
1–6	780.4201	M + O	M1	23.57
			Q2	69.87
12–27	1802.9145	M + O	T12	39.63
			P19	65.70
			S20	70.83
			N25	61.76
34–42	1070.5102	M + 2O	I36	32.28
			P37	67.15
			P38	36.52
43–47	699.3591	M + O		
48–63	1794.8742	M + O		
64–72	1082.6081	M + O		

[a] Six oxidized peptides were found in LC/FT-MS analysis, whereas three oxidized peptides were detected in LC/MS/MS analysis. The solvent-accessible surface areas were calculated using GETAREA 1.1 [337] using NMR structure of bovine ubiquitin [338].

Adapted from Aye et al. [141] with the permission of the American Chemical Society.

(CNBr) probes the C-terminal side of Met if it is not oxidized (MetO); however, CNBr is highly toxic [90].

There have been several reviews on the role of MS in the field of proteomics [89–91, 149–154]. Analysis of MS is based on the mass-to-charge ratio (m/z) of gaseous ions or their fragments and corresponding signal intensities. The m/z of the resulting gas phase ions can be correlated to structural properties of proteins in solution. Electrospray ionization (ESI)-MS and matrix-assisted laser desorption/ionization (MALDI)-MS techniques have been applied to determine the structure of peptides and proteins [155–167]. ESI produces multiple charged peptides, while MALDI generates singly charged peptides, and their fragmentation has been carried out by postsource decay (PSD) [168, 169]. High-quality MS/MS spectra of peptides are produced by high- and low-energy collision-induced dissociation (CID) on time-of-flight (TOF)/TOF instruments [170]. MALDI-MS analysis has also been enhanced by combining it with the ion trap, Fourier transform mass spectrometer, and triple quadrupole (Q_qQ_{LIT}).

Figure 1.6 shows the fragmented ions that may result from high- and low-energy CID. Generally, high-energy CID results in all types of fragmented ions, while only **b**, **y** **a**, and **z** fragments are observed in low-energy CID spectra

TABLE 1.5. Number of Oxidations Detected by LC/FT-MS and Reactive Amino Acid Residues Identified by LC/MS/MS for Apomyoglobin

Peptide Residues	Measure Mass (Da) by LC/FT-MS[a]	Oxidation Detected by LC/FT-MS[a]	Site of Oxidation Detected by LC/MS/MS[a]	Solvent-Accessible Surface Area by NMR (A^2)
1–16	1830.8891	M + O	S3	58.7
			E6	47.5
			W7	19.0
			Q8	83.7
			L11	42.6
			K16	47.8
17–31	1637.8371	M + 1O	Q26	19.3
	1653.8321	M + 2O	E27	54.1
32–42	1286.6506	M + 1O		
	1302.6455	M + 2O		
43–47	699.3591	M + O		
51–62	1366.6286	M + O	M55	7.3
64–77	1521.9242	M + 1O	H64	33.0
	1537.9191	M + 2O	T66	35.7
88–98	1248.6620	M + 2O	P88	61.6
			L89	48.6
			S92	23.9
			H93	58.5
			H96	55.9
119–133	1517.6568	M + 1O	M131	0.2
	1533.6517	M + 2O	T132	52.3
134–139	763.4228	M + O		
140–145	646.3286	M + O		
148–153	665.3020	M + O		

[a] Eleven peptides are detected as oxidized peptides in LC/FT-MS analysis, although six oxidized peptides are observed in LC/MS/MS. The solvent-accessible surface areas were calculated using GETAREA 1.1 [337] using NMR structure of horse heart myoglobin [339].

Adapted from Aye et al. [141] with the permission of the American Chemical Society.

[157]. Loss of H_2O, CO, and ammonia from the sequence fragmented ions can usually be observed by comparing the two high- and low-energy CID spectra [154]. The fragmentation efficiency of the chemical modification of a peptide can significantly improve in low-energy CID. A protonated basic amino acid residue or positively charged derivative located at the N-terminus of a peptide promotes the fragmentation of b-ion fragments (see Fig. 1.6). Quaternary ammonium and phosphonium groups have been shown to enhance fragmentation [171, 172]. The sulfonation of peptides by reagents such as chlorosulfonylacetyl chloride, sulfobenzoic acid cyclic anhydride, 4-sulfophenyl isothiocyanate, and 3-sulfopropionic acid *N*-hydroxysuccinimide ester display

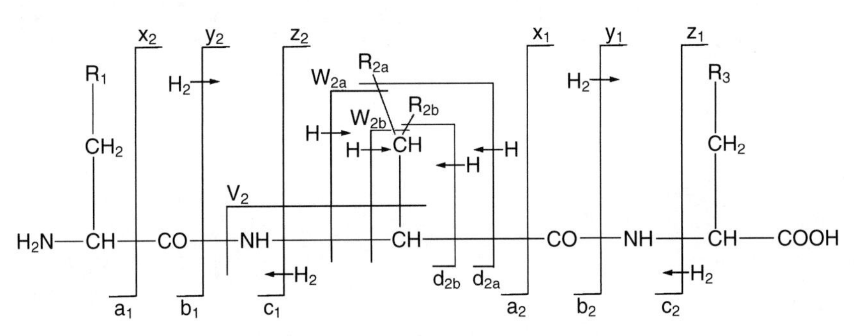

Figure 1.6. Fragment ions of peptide (adapted from Soltwisch and Dreisewerd [167] with the permission of the American Chemical Society).

an enhancement of fragmentation along the peptide chain [161, 173–176]. The interpretation of spectra was simplified because only y-ions were detected.

The amino group of the lysine residue in peptides was selectively modified by reactions with peroxycarbonate [167]. The lysine peroxycarbamates formed undergo hemolytic fragmentation under low-energy CID in ESI and MALDI-MS. The fragmentation of deuterated analogues of modified lysine resulted in the formation of a-, c-, or z-types of ions with MS, which suggested the involvement of a free radical mechanism in the fragmentation. The results in MALDI-MS are shown in Table 1.6 [157]. Mostly, the fragments of a- and z-ions were observed at the Lys site with one Lys or multiple Lys in the sequence of peptides. Entry 4 showed the disulfide bond remains intact; however, the peptide bond near the Lys was cleaved. Entry 1 in Table 1.6 showed the loss of a side chain of Val at the N-terminus in the fragment with *m/z* 961.45. Side-chain losses of Lys, Tyr, and Ser were also observed (entries 2, 3, and 5 in Table 1.6).

Other dissociation and ionization techniques besides CID have also been applied in the MS-based structural determination of peptides and proteins. These include ultraviolet photodissocation, nanoelectrospray ionization (nanoESI), easy ambient sonic spray ionization (EASI), easy ambient sonic spray ionization-membrane interface mass spectrometry (EASI-MIMS), laser ablation electrospray ionization (LAESI), desorption electrospray ionization (DESI), electrospray-assisted laser desorption/ionization (ELDI), electron capture dissociation (ECD), electron transfer dissociation (ETD), infrared multiphoton dissociation (IRMPD), atmospheric pressure chemical ionization (APCI), and ambient liquid mass spectrometry (ALMS) [162, 163, 165–167, 177–189]. CID and IRMPD normally produce heterolytic cleavage of the amide bonds in the polypeptide chain and resulting b and y fragments having N- and C-termini, respectively [190]. ECD cleaves the N–C_α bonds giving mainly fragmented ions of the N-terminus c' and C-terminus $z^\bullet$. A recent study compared top-down CID, CID, and IRMPD of nitrated proteins: hen egg-white lysozyme (HEWL), myoglobin, and cytochrome *c* [191]. The total number of fragments in ECD was greater than those in IRMPD or CID, but

TABLE 1.6. Fragments of Selectively Modified Peptides Studies by MALDI-MS

Parent Peptide	Fragments			
VQGEESNDK	a_8 Calc: m/z 831.35 Exp. 831.39	NH=CH-CO-QGEESNDK[a] Calc: m/z 961.47 Exp. 961.45		
KKALRRQETVDAL	$z^{\bullet}_{12}$ Calc: m/z 1382.78 Exp. 1382.91	NH=CH-CO-KALRRQETVDAL[a] Calc: m/z 1455.89 Exp. 1455.82		
YEVHHQKLVFF	$[a_7 + Na]^{+a}$ Calc: m/z 916.46 Exp. 916.52	NH=CH-CO-EVHHQKLVFF[a] Calc: m/z 1339.88 Exp. 1339.82		
CGNLSTCMLGTYTEDFNKFH TFPQTAIGVGAP-NH_2	$a^{\bullet}_{18}$ Calc: m/z 1947.83 Exp. 1947.82	$a^{\bullet}_{15}$ Calc: m/z 1552.82 Exp. 1552.79		
SYSMEHFRWGKPVGKKR	a_{11} Calc: m/z 1381.65 Exp. 1381.61	a_{15} Calc: m/z 1762.85 Exp. 1762.82	a_{16} Calc: m/z 1890.98 Exp. 1890.90	NH=CH-CO-YSMEHFRWGKPVGKKR[a] Calc: m/z 2062.09 Exp. 2061.99
Ac-SYSMEHFRWGKPVGKKR	a_{11} Calc: m/z 1423.66 Exp. 1423.62	a_{15} Calc: m/z 1804.90 Exp. 1804.83	a_{16} Calc: m/z 1932.99 Exp. 1932.92	
SYSMEHFRWGKPVGKKRRPVKVYP	a_{11} Calc: m/z 1381.65 Exp. 1381.58	a_{15} Calc: m/z 1762.85 Exp. 1762.78	a_{16} Calc: m/z 1890.98 Exp. 1890.86	a_{21} Calc: m/z 2527.40

[a] Loss of side chain of N-terminus.

Adapted from Yin et al. [157] with the permission of the American Society of Mass Spectrometry.

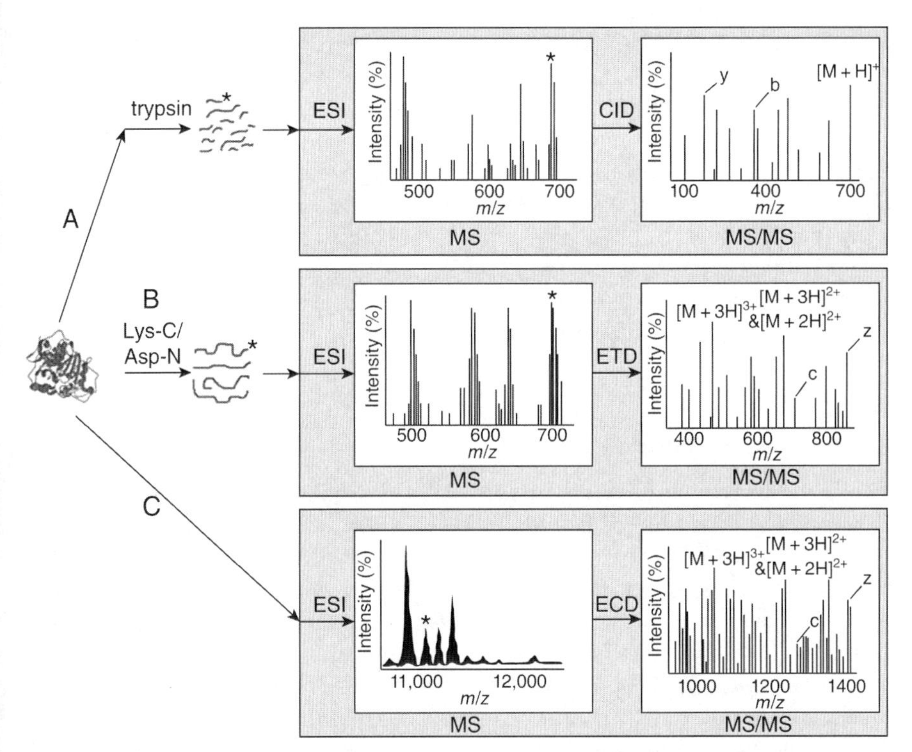

Figure 1.7. MS-based approaches A, B, and C. The gray boxes symbolize what takes place inside the MS equipment; the stars denote a certain peptide or protein (from an oxidized sample), chosen for CID, ETD, or ECD (adapted from Törnvall [90] with the permission of the Royal Society of Chemistry). See color insert.

more cleavages were generated in the vicinity of the sites of nitration in CID than IRMPD and ECD. ELDI is suitable for MS and top-down MS/MS of large proteins up to 29 kDa [164, 165]. The use of LC ESI tandem mass spectrometry (MS/MS) to sequence peptides generated from a digest of proteome sample has also been demonstrated [158].

MS-based approaches are generalized in Figure 1.7 [90]. In the bottom-up approach (A), tryptic proteolysis is followed by MS and CID-MS/MS analysis of relatively small +1 and +2 charged species, while further species proteolysis is followed by MS and ETD-MS/MS of larger +2, +3, and +4 charged species in the middle-down approach. A top-down approach involves ESI-MS on the intact protein to determine the total protein mass followed by ECD-MS/MS of a selected protein species with the modification of interest. The bottom-up approach is the most commonly used method for peptide mass analysis. Several labeling and separation methods are applied in the analysis of complex protein mixtures.

1.3 REACTIVE SPECIES

Amino acids play many roles in the metabolic activities of humans [192]. For example, amino acids are involved in the efficiency of utilizing dietary proteins, the cell-specific metabolism of nutrients, cell signaling, and oxidative stress [193, 194]. Reactive intermediates produced in the body influence metabolic activities by altering side chains of amino acids and modifying protein backbones (polypeptide chains) [195]. This section summarizes the involvement of reactive intermediates in biological systems. The generation and reactivity of the intermediates are described in Chapters 3–6.

1.3.1 Halogen Species

The heme enzyme myeloperoxidase (MPO) is linked to atherosclerosis [196]. Reactions between halide/pseudohalide (Cl^-, Br^-, and SCN^-) ions and hydrogen peroxide (H_2O_2) are catalyzed by MPO to produce HOCl, HOBr, and hypothiocyanous acid (HOSCN) [197]. The concentrations of Cl^-, Br^-, and SCN^- in plasma are in the ranges of 100–140 mM; 20–100 μM, and 20–250 μM, respectively. HOCl and HOBr are powerful oxidants and react at multiple targets (Chapter 3) [198]. These oxidants can cause tissue damage at excessive levels. HOCl and HOBr have been implicated in kidney disease, inflammatory bowel disease, cardiovascular disease, asthma, cystic fibrosis, rheumatoid arthritis, and neurodegenerative conditions [199, 200]. SCN^- reacts much faster with H_2O_2 (730-fold) than with Cl^-; hence, most of H_2O_2 is transferred into HOSCN. A high level of HOSCN has been observed in smokers due to their elevated levels of SCN^- [197]. HOSCN is less reactive compared to HOCl and HOBr, but is highly selective. Reactions of HOSCN are predominately with thiols (e.g., GSH) (Chapter 3). The ratio of [GSH]/[oxidized glutathione (GSSG)] may be useful to detect early atherosclerosis [201]. Overall, the concentration of SCN^- determines the damage induced by MPO. The chemistry of HOCl, HOBr, and HOSCN and their oxidation of amino acids including thiols are discussed in Chapter 3. Halamines (e.g., RNHCl and RHBr) can be produced as intermediates in reactions of HOCl and HOBr and may participate in reactions of biological systems. Hydrolysis, halogen transfer, and reduction of halamines are presented in Chapter 3.

Chlorine dioxide (ClO_2) is an important biocide and bleach [202]. Generation of ClO_2 using different methods including that from chlorite ion at neutral pH under mild conditions in the presence of water-soluble, manganese porphyrins and porphyrazines [203] is discussed in Chapter 3. Mechanisms of decomposition of ClO_2 under neutral and alkaline conditions are also given in Chapter 3. Finally, oxidation of inorganic and organic compounds (e.g., amino acids, peptides, and proteins) by ClO_2 is summarized in Chapter 3.

1.3.2 Oxygen Species

ROS play fundamental roles in maintaining health and in developing diseases, and estimation of their generation in isolated mitochondria has been evaluated as 0.1–2.0% of all consumed oxygen [204]. Mitochondria can produce $O_2^{-\bullet}$ at complex I (NADH-coenzyme Q reductase) and complex III (ubiquinone–cytochrome *c* oxireductase) [13, 205]. The generation of ROS in mitochondria is monitored by $NO^\bullet$, which is transferred to other RNS (see Fig. 1.1). Thus, a low concentration of $NO^\bullet$ can increase the production of $O_2^{-\bullet}$ and H_2O_2, while the formation of $OONO^-$ results in from the reaction of $O_2^{-\bullet}$ and $NO^\bullet$ [206]. Other major cellular sources of ROS are lipoxygenases, Nox family of enzymes, peroxisomers, uncoupled nitric oxide synthase (NOS), cyclooxygenases, xanthine oxireductase, and cytochrome P450 family proteins [207]. Levels of ROS were shown to increase in mouse hepatic cells in culture and in mouse liver by cytochrome P450 enzyme-mediated processes [208].

Significantly, the basic biology of cells and tissues is affected by SOD, $O_2^{-\bullet}$, and H_2O_2 [209]. Due to the importance of SOD, several studies have been performed on their structures and their involvement in the dismutation of $O_2^{-\bullet}$ [210–216]. MnSOD is an essential enzyme affecting levels of ROS in mitochondria and the protection of cells from damage by ROS [210, 216]. Nickel superoxide dismutase (NiSOD) has also been studied to elucidate its role in the disproportionation of $O_2^{-\bullet}$ [217–219]. Part of the $O_2^{-\bullet}$ produced in the mitochondria may be converted to H_2O_2 by Cu,Zn-SOD. The mechanistic aspects of the reactions of SOD with $O_2^{-\bullet}$ are described in Chapter 4. Under unregulated conditions, ROS accumulate and result in oxidative damage to cellular proteins, lipids, and nucleic acids [5, 7, 220, 221]. However, SOD, CAT, and thiol-based redox couples in molecular systems neutralize threats from ROS to cells. Chapter 4 discusses the oxidation of functional groups on proteins by ROS ($O_2^{-\bullet}$, $^1O_2$, O_3, and $^\bullet OH$). Mechanisms of reactions are also given in Chapter 4.

Carbonate radical ($CO_3^{\bullet -}$) may also be considered as ROS due to its selectivity. The reactivity of $CO_3^{\bullet -}$ with amino acids is less than that of $^\bullet OH$, but it is more selective and may also be a mediator of protein modification in cellular environments under conditions of oxidative stress. Carbonate radicals may also play a role in the decomposition of peroxynitrite in the presence of bicarbonate. Reactions of $CO_3^{\bullet -}$ are discussed in Chapter 5. Modifications of proteins by peroxymonocarbonate (HCO_4^-) and carboxy radical ($CO_2^{\bullet -}$) are briefly presented in Chapter 4. Other species that may also be considered ROS are peroxyl ($ROO^\bullet$) and alkoxyl ($RO^\bullet$) radicals, which are formed from the reaction of carbon-centered radicals with oxygen. Peroxidation has been demonstrated to go through chain reactions and to generate these radicals. The breakdown of these radicals into carbonyl hydrperoxides, alcohols, and carbonyl groups suggests their importance in biological systems.

Sulfite and sulfate radicals ($^\bullet SO_3^-$ and $SO_4^{\bullet -}$) may be other ROS because of their possible reactions with amino acids and peptides (Chapter 5). $SO_4^{\bullet -}$ is a strong oxidant and its redox potential is similar to that of $^{\bullet}OH$ [222]. Both $SO_4^{\bullet -}$ and $^{\bullet}OH$ can possibly react with phosphate ions to produce phosphate radicals (Chapter 5). Reactions of phosphate radicals with aromatic amino acids and their peptides are discussed in Chapter 5.

1.3.3 Nitrogen Species

Nitric oxide ($^{\bullet}NO$) has been studied extensively due to its importance in biological chemistry. A low level of $^{\bullet}NO$ production is involved in several processes such as immune system control, blood pressure modulation, signal transduction, smooth muscle relaxation, memory, and inhibition of platelet aggregation [21, 223]. However, high levels of $^{\bullet}NO$ can result in injury to tissues [224]. Recent studies hypothesized that dietary nitrite may form $^{\bullet}NO$ in the human stomach [225, 226]. One of the reasons of $^{\bullet}NO$ toxicity may be its reaction with $O_2^{\bullet -}$ to form peroxynitrite. Peroxynitrite can oxidize as well as nitrate a number of biological molecules including proteins. Peroxynitrite can either directly promote one- and two-electron oxidations in molecules or decompose to secondary radicals such as $^{\bullet}OH$ and $^{\bullet}NO_2$ (Chapter 5). $^{\bullet}NO_2$ is a mild and selective oxidant (Chapter 5). The secondary radicals thus result in nitration and lipid peroxidation [227]. In the occurrence of high levels of bicarbonate in interstitial (30 mM) and intracellular (12 mM) fluids, the formation of a short-lived adduct between peroxynitrite and CO_2 occurs. This adduct decomposes into oxidizing intermediates, $CO_3^{\bullet -}$ and $^{\bullet}NO_2$. The kinetics of nitrogen species and nitration products of free amino acids and proteins are presented in Chapter 5.

1.3.4 Sulfur Species

Thiol proteins play key roles in diverse physiological processes [228, 229]. Thiol proteins are also important molecules by which reactive species involve themselves in cellular transduction pathways [230]. During oxidative stress, oxidation of protein thiols takes place. The reactions of protein thiols (Pr-SH) can occur through one-electron and two-electron pathways (Fig. 1.8) [230]. In two-electron pathways, sulfenic acid (Pr-SOH) is initially formed, which can undergo several secondary reactions. Sulfenic acid can also form mixed disulfides (Pr-SS-G) by reacting with GSH. The formation of sulfenic acid is a prominent feature in several acute and chronic diseases. Significantly, more than 200 different cellular proteins have been shown to undergo cysteine oxidation [231].

One-electron oxidants such as radicals and transition metal ions produce the corresponding thiyl radical (Pr-S$^{\bullet}$). This radical may react with either other biomolecules or may participate in further reactions with thiol molecules and oxygen. The reaction with oxygen ultimately forms superoxide and hydrogen peroxide. A recent study on the oxidation of methionine-lysine peptide by $^{\bullet}OH$ in the presence of CAT resulted in the formation of methionine

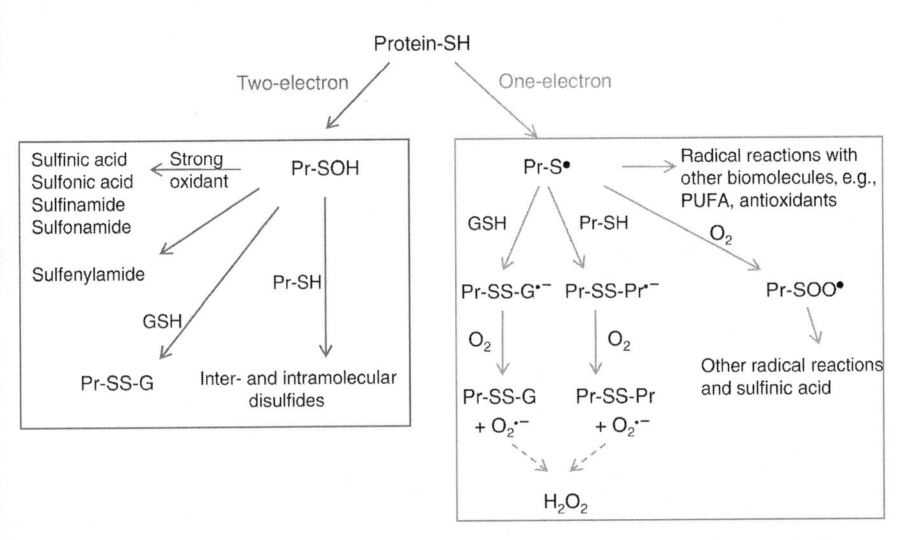

Figure 1.8. Oxidation of protein thiols by one-electron and two-electron pathways. PUFA, polyunsaturated fatty acids (adapted from Winterbourn and Hampton [230] with the permission of Elsevier Inc.) See color insert.

sulfoxide through a one-electron transfer [232]. Recently, the reactivity of simple thiols and protein thiols with H_2O_2 has been explained in detail [228]. Generally, the nucleophilic attack of thiolate on an electrophile occurs and rate constants vary by seven orders of magnitude (k(cysteine)) = 2.6×10^1/M/s; k(peroxiredoxin 2) = 1.0×10^8/M/s) [233, 234]. Thiyl radicals are involved in enzymatic and detoxifying pathways, and in recent years, the oxidizing power of the thiyl radicals produced by one-electron oxidations has been studied. In a recent work, cysteine thiyl radicals were shown to be involved in reversible intramolecular hydrogen transfer reactions with amino acid residues in peptides and proteins [235].

The reduction potential is the key parameter to express oxidation power and is pH dependent. The reduction potential of a one-electron couple, expressed as a midpoint electrode potential, $E_m(GS^•, H^+/GSH)$ for GSH (γ-glutamylcysteineglycine) in water as a function of pH has been determined (Fig. 1.9) [236]. E_m is an experimentally useful midpoint potential where the sum of concentrations of oxidants is equal to the sum of reductants [237]. In Figure 1.9, the solid line is for a simple thiol in which only thiol/thiolate ionization was considered with $pK_a = 9.2$ and assuming $E^o(RS^•/RS^-) = 0.8$V. However, the dashed line represents the considered four macroscopic pK_a values of the reductant (GSH, pK_{r1-4}), consisting of 2.05, 3.40, 8.72, and 9.49, and three pK_a values of the oxidant ($GS^•$, pK_{o1-3}), consisting of 1.8, 3.2, and 9.08, and also assuming $E^o(RS^•/RS^-) = 0.8$V for the fully deprotonated GSH [236]. The value of $E_m(GS^•, H^+/GSH) = 0.92$ at pH 7.4 was determined and compared to other biologically relevant free radicals (Table 1.7). The value of

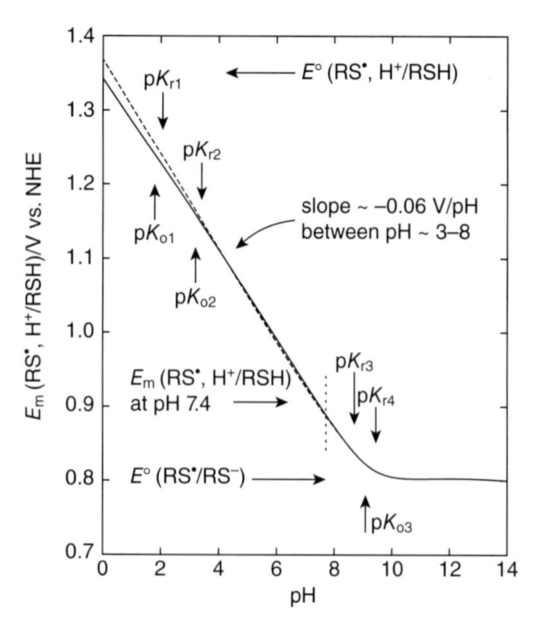

Figure 1.9. Expected form of the dependence upon pH of the midpoint electrode potential of the thiyl radical/thiol couple $E_m(RS^•, H^+/RSH)$ (adapted from Madej and Wardman [236] with the permission of Elsevier Inc.).

E_m at pH 7 for the thiyl radical of hydrogen sulfide is similar to the value of $GS^•$ at pH 7.4. This is an indication of the moderate oxidizing power of $GS^•$. However, $GS^•$ has a higher oxidizing power than do radicals of 8-oxo-7,8-dihydro-2′-deoxyguanosine, urate, and ascorbate [238–240]. The order of E_m in Table 1.7 suggests $GS^•$ would oxidize 8-oxo-7,8-dihydro-2′-deoxyguanosine but not guanosine [240]. The order also suggests $NO_2^•$ can oxidize GSH [238].

The role of thiyl radicals can be studied through analysis in biological systems [241]. However, thiyl radicals are highly reactive with half-lives on the order of microseconds; therefore, their direct detection is challenging in biological systems. Rate constants for hydrogen transfer reactions of the thiyl radical of cysteamine with amino acids and peptides are up to 10^5/M/s [242–245]. The reactions of thiyl radicals from GSH, cysteine, and pencillamine with $NO^•$ are in the range of (2–3) × 10^9/M/s [246]. A recent analysis of thiyl radicals suggests a spin trapping electron paramagnetic resonance technique using 5,5-dimethyl-1-pyrrolone *N*-oxide as a trapping agent is a promising approach for the analysis of thiyl radicals in biosystems [241].

In addition to thiyl radicals, the reactions of protein radicals with ascorbate have also been studied to determine if the loss of ascorbate in a living organism can be related to these reactions [247]. Protein radicals were generated on tryptophan and tyrosine residues of insulin, β-lactoglobulin, pepsin, chymotrypsin, and bovine serum albumin (k = 2.9–19) × 10^9/M/s. The tryptophanyl

TABLE 1.7. Oxidizing Power of Some Biological Free Radicals (Midpoint Electrode [Reduction] Potentials, E_m of Radicals, pH 7.0–7.4 vs. Norman Hydrogen Electrode [NHE]) (Values of the E_m of the Couple Were Taken from Madej and Wardman [236] with the Permission of Elsevier Inc.)

Radical Obtained from Oxidation of	Couple[a]	E_m
(Bi)carbonate	$CO_3^{\bullet-}$, H^+/HCO_3^-	1.74
Adenosine	$AdO^{\bullet}$/Ado	1.42
Methionine	$Met^{\bullet+}$/Met[b]	~1.2–1.5
Guanosine	$G(-H)^{\bullet}$, H^+/G	1.29
Tryptophan	$Trp^{\bullet+}$, H^+/TrpH	~1.05
Nitrite	$NO_2^{\bullet}/NO_2^-$	~1.04
Glutathione	$GS^{\bullet}$, H^+/GSH	0.92
Hydrogen sulfide	$S^{\bullet-}$, H^+/HS^-	0.92
Tyrosine	$TyrO^{\bullet}$, H^+/TyrOH	~0.90
8-Oxo-7,8-dihydro-2′-deoxyguanosine	8-OHdG$(-H)^{\bullet}$, H^+/8-OHdG	0.74
5-Hydroxylindole	5-Indoxyl$^{\bullet}$/5-hydroxyindole	0.64
Urate	UH^{-}, H^+/UH_2^-	0.59
Ascorbate	$Asc^{\bullet-}$, $H^+/AscH^-$	0.30

[a] Predominant forms of radical/reductant couples at pH 7–7.4 are shown.
[b] Irreversible process with rapid deprotonation from the radical cation.

radicals in chymotrypsin, pepsin, and β-lactoglobulin reacted with monohydrogen ascorbate with rate constants of 1.6×10^8/M/s, 1.8×10^8/M/s, and 2.2×10^7/M/s, respectively. Comparatively, the corresponding reaction of protein tyrosyl radicals had rate constants an order of magnitude slower. The rate constants of the reactions were not significantly affected by the location of radicals in the protein chains [247]. A separate study estimated a rate constant of 2.2×10^6/M/s for the tyrosyl radical with GSH, which is about two orders of magnitude slower than the reaction of the tyrosyl radical with ascorbate [248]. This indicates GSH plays a minor role in radical "repair," while the loss of ascorbate may be due to its reactions with protein radicals under physiological conditions.

Overall, the chemistry of reactive sulfur species is complex. In addition to cysteine and GSH, a biological role for H_2S is also being studied (Chapter 5) [249–251]. Physiological levels of H_2S are 50–160 μM and 10–100 μM in mammalian brain tissues and human plasma, respectively [252]. The comparison of sulfahydryl radicals derived from H_2S is compared with other thiyl radicals in Chapter 5.

1.3.5 High-Valent Cr, Mn, and Fe Species

Transition metal ions play important roles in a number of biological processes [35, 36, 253, 254]. Metal oxo and metal hydroxo moieties have been recognized in different biological oxidation and electron transfer reactions such as hydroxylation, epoxidations, sulfoxodation, dehalogentation, and deamination [255, 256]. The interactions of high-valent metal ions with proteins and DNA may lead to oxidative deterioration of biological molecules. For example, high-valent compounds of Cr pose serious danger to biological systems and are considered both toxic and carcinogenic (Chapter 6). The properties and reactivity of Cr species are discussed in Chapter 6. Chapter 6 also highlights the role of high-valent Cr species in carcinogenicity, genotoxicty, and cytotoxicity.

High-valent species of Mn and Fe have been reported as active intermediates in oxidation events of metalloenzymes [35, 257, 258]. Various complexes of Mn and Fe have been synthesized to mimic biologically important processes (Chapter 6). High-valent oxo complexes of Mn have been prepared to elucidate the role of Mn^{V}-oxo species in water oxidation at the oxygen-evolving center in photosystem II [259]. The aqueous chemistry of oxo compounds of Mn is presented in Chapter 6. The rates of reactions between Mn(VII) and amino acids are influenced by colloidal MnO_2 and phosphate species in solution (Chapter 6). Pathways of oxidation reactions carried out by Mn(VII) with amino acids and aminopolycarboxylates are described in Chapter 6.

High-valent iron complexes as intermediates have been invoked in reactions of heme and nonheme enzymes [260–264]. In recent years, several complexes of iron(IV) and iron(V) have been synthesized as models to mimic catalytic centers of enzymes [253, 265–270]. Properties of these complexes are presented in Chapter 6. Rate studies have revealed that metal ions influence the electron transfer from reductants to iron(IV) complexes (Chapter 6). A summary of the recent work on the kinetics and products of reactions of iron(IV) complexes with amino acids and peptides is presented in Chapter 6.

1.4 REACTIVE SPECIES IN ENVIRONMENTAL PROCESSES

1.4.1 Atmospheric Environment

Free radical species play a critical role in combustion, plasma environments, interstellar clouds, and atmospheric chemistry [271]. Reactions of radical species have been shown to have an adverse effect on human health and vegetation [272]. Species such as chlorine atoms, O_3, $^{\bullet}OH$, and $NO_3^{\bullet}$ can react with inorganic and volatile organic compounds of the atmosphere [273–276]. For example, the reaction of CO with $^{\bullet}OH$ is important in combustion reactions [271]. ROS are also formed in the reaction of O_3 with aerosol particles [277]. In recent years, reactions of $NO_3^{\bullet}$ have also been studied. $NO_3^{\bullet}$ radicals are formed through the reaction of $NO_2^{\bullet}$ with O_3 (Eq. 1.5):

$$O_3 + NO_2^{\bullet} \rightarrow NO_3^{\bullet} + O_2. \tag{1.5}$$

Kinetic studies on the reactions of $^{\bullet}OH$ and $NO_3^{\bullet}$ radicals with a number of 1-alkenes [CH_2=CHR] and 2-methyl-1-alkenes [CH_2=C(CH_3)R], where

$R = (CH_2)_nCH_3$ have demonstrated the carbon number of substituent group, n, increases the rate constants up to $n < 7$, followed by no further increase at $n \geq 7$. Significantly, substituent group carbon number n had no affect on the reaction of O_3 with the same alkenes. Steric and ring strain effects appear to control the rate constants of the reactions of these alkenes with $^{\bullet}OH$, $NO_3^{\bullet}$, and O_3 [278].

$NO_3^{\bullet}$ is a nighttime atmospheric oxidant and its reactivity with most organic molecules is many orders of magnitude higher than of O_3 or $NO_2^{\bullet}$ [279]. Reaction mechanisms of reactions of $NO_3^{\bullet}$ with molecules include electron transfer, hydrogen abstraction, and addition to the π system [279]. The reaction between $NO_3^{\bullet}$ and $NO_2^{\bullet}$ can take place to yield N_2O_5 in the absence of organic reactants, and N_2O_5 acts as a reservoir for $NO_3^{\bullet}$ [279]. Recently, the reactions of $NO_3^{\bullet}$ with *N*- and *C*-protected aromatic amino acids in the presence of O_2, O_3, $NO_2^{\bullet}$, and N_2O_4 have been studied [280, 281]. Consistent with the chemistry of $NO_3^{\bullet}$, the initial electron transfer step at the aromatic ring occurred, followed by multisteps, which finally yield nitroaromatic compounds [280]. The formation of products was not influenced by the presence of O_2 when reactions of $NO_3^{\bullet}$ with tyrosine and phenylalanine were studied. In the case of tryptophan, an intramolecular oxidizing cyclization involving the amide moiety leads to the formation of tricyclic products. Results suggest $NO_3^{\bullet}$ may cause damage to the peptides lining the respiratory tract, ultimately causing pollution-driven diseases [281].

1.4.2 Disinfection By-Products (DBP)

Dissolved organic nitrogen (DON) in the aquatic environment is of interest due to the bioavailability and connections between the carbon and nitrogen biogeochemical cycles [282]. The concentrations of nitrogen-containing substances in aquatic environments have increased due to inputs from agricultural runoff, deposition of NO_x, and wastewater effluents [283–285]. Dissolved free amino acids are one of the constituents of DON [286]. Amino acids have been found in a number of water resources including drinking water, lakes, rivers, marshes, and groundwaters [287–289]. Amino acids may also be found in industrial effluents because of their many uses and applications. Proteins have been detected in wastewater treatment effluents [290].

A number of disinfectants/oxidants are used in the treatment of water. Redox potentials of oxidants are provided in Table 1.8 [291–294]. Under acidic conditions, the redox potential of $^{\bullet}OH$ is one of the highest of any oxidant (Table 1.8). However, the potential is low under basic conditions. Under acidic conditions, the order of redox potentials of oxidants without chlorine and transition metal species are $^{\bullet}OH > SO_4^{-\bullet} > NO_3^{\bullet} > O_3 > H_2O_2 > CO_3^{-\bullet} > HSO_5 > O_2$. Redox potentials of some oxidants ($SO_4^{-\bullet}$, $NO_3^{\bullet}$, $CO_3^{-\bullet}$, HSO_5, ClO_2, and Cl_2) have no pH dependence. Ozone has a high redox potential in alkaline medium. Among the high-valent iron and manganese species, redox potentials of Fe(VI) (Fe^{6+}/Fe^{3+}) and Mn(VI) (Mn^{6+}/MnO_2) are similar and higher than Mn(VII).

TABLE 1.8. Redox Potentials for the Oxidants/Disinfectants Used in Water Treatment [291–294]

Oxidant	Reaction	E^o (V/NHE)
Hydroxyl radical	$^{\bullet}OH + H^+ + e^- \rightleftharpoons H_2O$	2.80
	$^{\bullet}OH + e^- \rightleftharpoons OH^-$	1.89
Sulfate radical	$SO_4^{-\bullet} + e^- \rightleftharpoons SO_4^{2-}$	2.43
Nitrate radical	$NO_3^{\bullet} + e^- \rightleftharpoons NO_3^-$	2.30
Manganate	$MnO_4^{2-} + 4H^+ + 2e^- \rightleftharpoons MnO_2 + 2H_2O$	2.26
	$MnO_4^{2-} + 8H^+ + 4e^- \rightleftharpoons Mn^{2+} + 4H_2O$	1.74
	$MnO_4^{2-} + 2H_2O + 2e^- \rightleftharpoons MnO_2 + 4OH^-$	0.60
Ferrate(VI)	$FeO_4^{2-} + 8H^+ + 3e^- \rightleftharpoons Fe^{3+} + 4H_2O$	2.20
	$FeO_4^{2-} + 4H_2O + 3e^- \rightleftharpoons Fe(OH)_3 + 5OH^-$	0.70
Ozone	$O_3 + 2H^+ + 2\,e^- \rightleftharpoons O_2 + H_2O$	2.08
	$O_3 + H_2O + 2\,e^- \rightleftharpoons O_2 + 2OH^-$	1.24
Hydrogen peroxide	$H_2O_2 + 2H^+ + 2\,e^- \rightleftharpoons 2H_2O$	1.78
	$H_2O_2 + 2\,e^- \rightleftharpoons 2OH^-$	0.88
Carbonate radical	$CO_3^{-\bullet} + e^- \rightleftharpoons CO_3^{2-}$	1.59
Permanganate	$MnO_4^- + 8H^+ + 5e^- \rightleftharpoons Mn^{2+} + 4H_2O$	1.51
	$MnO_4^- + 2H_2O + 3e^- \rightleftharpoons MnO_2 + 4OH^-$	0.59
Hypochlorite	$HClO + H^+ + 2\,e^- \rightleftharpoons Cl^- + H_2O$	1.48
	$ClO^- + H_2O + 2\,e^- \rightleftharpoons Cl^- + 2OH^-$	0.84
Peroxysulfate	$HSO_5 + e^- \rightleftharpoons HSO_5^-$	1.40
Perchlorate	$ClO_4^- + 8H^+ + 8e^- \rightleftharpoons Cl^- + 4H_2O$	1.39
Chlorine	$Cl_2 + 2\,e^- \rightleftharpoons 2Cl^-$	1.36
Dissolved oxygen	$O_2 + 4H^+ + 4\,e^- \rightleftharpoons 2H_2O$	1.23
	$O_2 + 2H_2O + 4\,e^- \rightleftharpoons 4OH^-$	0.40
Chlorine dioxide	$ClO_2(aq) + e^- \rightleftharpoons ClO_2^-$	0.95

This suggests both Fe(VI) and Mn(VI) have strong oxidizing power, particularly under acidic conditions.

Chlorine is a commonly used disinfectant in drinking water because it is readily available and effective. However, chlorine can react with amino acids and peptides to form DBPs, which have negative health effects [295–299]. The kinetics and mechanisms of the formation of DBPs are presented in Chapter 3. The chlorination products of amino acids include cyanogen chloride, chloroform, dichlroacetonitrile, monochloroamines, aldehydes, and nitriles [300–303]. Chlorinated products of peptides are more stable than chlorinated derivatives of amino acids; therefore, peptides in water are also of concern [300]. The chlorination of waters containing bromide and iodide ions may produce hypobromous (HOBr) and hypoiodous (HOI) acids [304, 305]. Thus, both HOBr and HOI may also result in DPBs [304, 305].

Monochloroamine (NH_2Cl) as a chlorine alternative is applied to reduce the formation of DPBs. During chloramination, lower concentrations of DBPs, such as trihalomethane and haloacetic acids, are produced than in chlorine-treated water [300]. Similar to chlorination, chloramination also produces haloacetonitriles and haloketones, which include dichloroacetonitrile, trichloroacetonitrile, and 1,1,1,-trichloro-2-propanone [300]. Other disinfectants, ClO_2 and O_3, are also applied as water treatments. Both disinfectants have high reactivities with amino acids and proteins (Chapters 3 and 4). The use of ClO_2 is beneficial in minimizing the formation of trihalomethanes, but ClO_2 itself reduces to ClO_2^- and ClO_3^-, which may cause hemolytic anemia and other health effects. Ozone is an efficient disinfectant, but it may react with Br^- in water to produce the carcinogenic bromate ion. Mn(VII) and Fe(VI) are other alternate disinfectants [306] and their reactions with amino acids, peptides, and proteins are presented in Chapter 6.

N-Nitrosodimethylamine (NDMA) is also of considerable concern as an environmental contaminant and has been detected in air, beverages, food products, and water [307–309]. NDMA is classified as a "probable human carcinogen" by the International Agency for Research on Cancer (IARC). A number of studies have shown that both chlorination and chloramination produce NDMA [310–312]. Ozonation has also been shown to result in the production of NDMA [313, 314]. The formation of NDMA in water usually occurs from the reaction of disinfectants (ClO_2, O_3, $^{\bullet}OH$, Mn(VII), and Fe(VI)) with nitrogen-containing precursors such as dimethylamine, tertiary amines, amine-containing polymers, and dimethylsulfamide [315]. The destruction of NDMA can be accomplished by oxidation, which include electrochemical, photolytic, photocatalytic, and chemical methods [315].

1.4.3 Oxidation Processes for Purifying Water

Oxidation processes using H_2O_2, ozone, the Fenton reaction, electron beam radiation, and ultrasound have been applied to degrade recalcitrant and emerging contaminants in water [316–318]. Generally, oxidation processes involve the formation of $^{\bullet}OH$, which reacts nonselectively with organics (see Chapter 4) [319, 320]. The reactions of O_3 with organics are selective (see Chapter 4). Generation of $^{\bullet}OH$ to oxidize compounds can also be achieved by applying UV/TiO_2, UV/H_2O_2, TiO_2-photocatalyzed, photoassisted Fenton, and electro-Fenton systems [291, 321, 322]. In recent years, studies have focused on photocatalysts under visible light to produce $^{\bullet}OH$ [322–324]. Sulfate radicals have also received attention in oxidation processes to destroy refractory organic contaminants, pharmaceutical and personal care products [222, 325–327]. More details on $^{\bullet}OH$ and $SO_4^{\bullet-}$ are presented in Chapter 4.

Among the high-valent metals, ferrate(VI) (Fe(VI), $Fe^{VI}O_4^{2-}$) has been shown to oxidize a number of inorganic and organic compounds in water [328–330]. Oxidations carried out by Fe(VI) are completed in shorter time periods than oxidations performed by Mn(VII) and Cr(VI) [331]. More details of the chemistry of high-valent compounds of iron, manganese, and chromium and their role in oxidizing organic compounds including amino acids and peptides are described in Chapter 6. Unlike chromium and manganese, processes using Fe(VI) are considered environmentally benign. Other oxidation states of iron, ferrate(V) (Fe(V)) and ferrate(IV) (Fe(IV)), have shown much higher reactivity than that of Fe(VI) and also have potential uses in oxidation processes [300, 332–334]. Synthetic methods of ferrates are summarized in Chapter 6. The reactivity of ferrates with molecules is highlighted for amino acids and peptides (Chapter 6). The generation of Fe(V) species in the Fe(VI)-TiO_2-UV system has also been suggested as a method to degrade pollutants and organisms [335]. The iron-tetraamidomacrocyclic ligand (Fe-TAML), a catalyst in trace amounts, activates H_2O_2 to produce reactive species, Fe^{IV}=O and Fe^{V}=O, which oxidize numerous compounds of environmental interest [336].

REFERENCES

1 Stadtman, E.R. Protein oxidation and aging. *Free Radic. Res.* 2006, *40*, 1250–1258.

2 Winterbourn, C.C. Reconciling the chemistry and biology of reactive oxygen species. *Nature Chem. Biol.* 2008, *4*, 278–286.

3 Murphy, M.P., Holmgren, A., Larsson, N.G., Halliwell, B., Chang, C.J., Kalyanaraman, B., Rhee, S.G., Thornalley, P.J., Partridge, L., Gems, D., Nyström, T., Belousov, V., Schumacker, P.T., and Winterbourn, C.C. Unraveling the biological roles of reactive oxygen species. *Cell Metab.* 2011, *13*, 361–366.

4 Halliwell, B. The wanderings of a free radical. *Free Radic. Biol. Med.* 2009, *46*, 531–542.

5 Baraibar, M.A., Hyzewicz, J., Rogowska-Wrzesinska, A., Ladouce, R., Roepstorff, P., Mouly, V., and Friguet, B. Oxidative stress-induced proteome alterations target different cellular pathways in human myoblasts. *Free Radic. Biol. Med.* 2011, *51*, 1522–1532.

6 Spickett, C.M. and Pitt, A.R. Protein oxidation: role in signalling and detection by mass spectrometry. *Amino Acids* 2012, *42*, 5–21.

7 Dickinson, B.C. and Chang, C.J. Chemistry and biology of reactive oxygen species in signaling or stress responses. *Nat. Chem. Biol.* 2011, *7*, 504–511.

8 Mailloux, R.J. and Harper, M.E. Uncoupling proteins and the control of mitochondrial reactive oxygen species production. *Free Radic. Biol. Med.* 2011, *51*, 1106–1115.

9 Gullotta, F., di Masi, A., Coletta, M., and Ascenzi, P. CO metabolism, sensing, and signaling. *Biofactors* 2012, *38*, 1–13.

10 Little, J.W., Doyle, T., and Salvemini, D. Reactive nitroxidative species and nociceptive processing: determining the roles for nitric oxide, superoxide, and peroxynitrite in pain. *Amino Acids* 2012, *42*, 75–94.

11 Salvemini, D., Little, J.W., Doyle, T., and Neumann, W.L. Roles of reactive oxygen and nitrogen species in pain. *Free Radic. Biol. Med.* 2011, *51*, 951–966.

12 Davies, M.J. The oxidative environment and protein damage. *Biochem. Biophys. Acta—Proteins Proteomics* 2005, *1703*, 93–109.

13 Paradies, G., Petrosillo, G., Paradies, V., and Ruggiero, F.M. Oxidative stress, mitochondrial bioenergetics, and cardiolipin in aging. *Free Radic. Biol. Med.* 2010, *48*, 1286–1295.

14 Paradies, G., Petrosillo, G., Paradies, V., and Ruggiero, F.M. Mitochondrial dysfunction in brain aging: role of oxidative stress and cardiolipin. *Neurochem. Int.* 2011, *58*, 447–457.

15 Monti, D., Ottolina, G., Carrea, G., and Riva, S. Redox reactions catalyzed by isolated enzymes. *Chem. Rev.* 2011, *111*, 4111–4140.

16 Madian, A.G., Diaz-Maldonado, N., Gao, Q., and Regnier, F.E. Oxidative stress induced carbonylation in human plasma. *J. Proteomics* 2011, *74*, 2395–2416.

17 Madian, A.G., Myracle, A.D., Diaz-Maldonado, N., Rochelle, N.S., Janle, E.M., and Regnier, F.E. Determining the effects of antioxidants on oxidative stress induced carbonylation of proteins. *Anal. Chem.* 2011, *83*, 9328–9336.

18 Halliwell, B. Oxidative stress and neurodegeneration: where are we now? *J. Neurochem.* 2006, *97*, 1634–1658.

19 Augusto, O., Bonini, M.G., Amanso, A.M., Linares, E., Santos, C.C.X., and De Menezes, S.L. Nitrogen dioxide and carbonate radical anion: two emerging radicals in biology. *Free Radic. Biol. Med.* 2002, *32*, 841–859.

20 Wentworth, P., Jr., Nieva, J., Takeuchi, C., Galve, R., Wentworth, A.D., Dilley, R.B., DeLaria, G.A., Saven, A., Babior, B.M., Janda, K.D., Eschenmoser, A., and Lerner, R.A. Evidence for ozone formation in human atherosclerotic arteries. *Science* 2003, *302*, 1053–1056.

21 Mel, A., Murad, F., and Seifalian, A.M. Nitric oxide: a guardian for vascular grafts. *Chem. Rev.* 2011, *111*, 5742–5767.

22 Grosvenor, A.J., Morton, J.D., and Dyer, J.M. Profiling of residue-level photo-oxidative damage in peptides. *Amino Acids* 2010, *39*, 285–296.

23 Sayre, L.M., Perry, G., and Smith, M.A. Oxidative stress and neurotoxicity. *Chem. Res. Toxicol.* 2008, *21*, 172–188.

24 Adly, A.A.M. Oxidative stress and disease: an updated review. *Res. J. Immunol.* 2010, *3*, 129–145.

25 Glade, M.J. Oxidative stress and cognitive longevity. *Nutrition* 2010, *26*, 595–603.

26 Allsop, D., Mayes, J., Moore, S., Masad, A., and Tabner, B.J. Metal-dependent generation of reactive oxygen species from amyloid proteins implicated in neurodegenerative disease. *Biochem. Soc. Trans.* 2008, *36*, 1293–1298.

27 Nakamura, T. and Lipton, S.A. Redox modulation by S-nitrosylation contributes to protein misfolding, mitochondrial dynamics, and neuronal synaptic damage in neurodegenerative diseases. *Cell Death Differ.* 2011, *18*, 1478–1486.

28 Gaggelli, E., Kozlowski, H., Valensin, D., and Valensin, G. Copper homeostasis and neurodegenerative disorders (Alzheimer's, prion, and Parkinson's diseases and amyotrophic lateral sclerosis). *Chem. Rev.* 2006, *106*, 1995–2044.

29 Fillit, H.M., O'Connell, A.W., Brown, W.M., Altstiel, L.D., Anand, R., Collins, K., Ferris, S.H., Khachaturian, Z.S., Kinoshita, J., Van Eldik, L., and Dewey, C.F. Barriers to drug discovery and development for Alzheimer disease. *Alzheimer Dis. Assoc. Disord.* 2002, *16*, S1–S8.

30 Hardy, J. and Selkoe, D.J. The amyloid hypothesis of Alzheimer's disease: progress and problems on the road to therapeutics. *Science* 2002, *297*, 353–356.

31 Goate, A., Chartier-Harlin, M.C., Mullan, M., Brown, J., Crawford, F., Fidani, L., Giuffra, L., Haynes, A., Irving, N., James, L., Mant, R., Newton, P., Rooke, K., Roques, P., Talbot, C., Pericak-Vance, M., Roses, A., Williamson, R., and Hardy, J. Segregation of a missense mutation in the amyloid precursor protein gene with familial Alzheimer's disease. *Nature* 1991, *349*, 704–706.

32 Levy-Lahad, E., Tsuang, D., and Bird, T.D. Recent advances in the genetics of Alzheimer's disease. *J. Geriatr. Psychiatry Neurol.* 1998, *11*, 42–54.

33 Sherrington, R., Froelich, S., Sorbi, S., Campion, D., Chi, H., Rogaeva, E.A., Levesque, G., Rogaev, E.I., Lin, C., Liang, Y., Ikeda, M., Mar, L., Brice, A., Agid, Y., Percy, M.E., Clerget-Darpoux, F., Piacentini, S., Marcon, G., Nacmias, B., Amaducci, L., Frebourg, T., Lannfelt, L., Rommens, J.M., and St George-Hyslop, P.H. Alzheimer's disease associated with mutations in presenilin 2 is rare and variably penetrant. *Hum. Mol. Genet.* 1996, *5*, 985–988.

34 De Ferrari, G.V. and Inestrosa, N.C. Wnt signaling function in Alzheimer's disease. *Brain Res. Rev.* 2000, *33*, 1–12.

35 Jomova, K. and Valko, M. Advances in metal-induced oxidative stress and human disease. *Toxicology* 2011, *283*, 65–87.

36 Jomova, K., Vondrakova, D., Lawson, M., and Valko, M. Metals, oxidative stress and neurodegenerative disorders. *Mol. Cell. Biochem.* 2010, *345*, 91–104.

37 Rivera-Mancía, S., Pérez-Neri, I., Ríos, C., Tristán-López, L., Rivera-Espinosa, L., and Montes, S. The transition metals copper and iron in neurodegenerative diseases. *Chem. Biol. Interact.* 2010, *186*, 184–199.

38 Sesti, F., Liu, S., and Cai, S.Q. Oxidation of potassium channels by ROS: a general mechanism of aging and neurodegeneration? *Trends Cell Biol.* 2010, *20*, 45–51.

39 Grasso, G. The use of mass spectrometry to study amyloid-β peptides. *Mass Spectrom. Rev.* 2011, *30*, 347–365.

40 Lovell, M.A., Robertson, J.D., Teesdale, W.J., Campbell, J.L., and Markesbery, W.R. Copper, iron and zinc in Alzheimer's disease senile plaques. *J. Neurol. Sci.* 1998, *158*, 47–52.

41 Hung, Y.H., Bush, A.I., and Cherny, R.A. Copper in the brain and Alzheimer's disease. *J. Biol. Inorg. Chem.* 2010, *15*, 61–76.

42 Dikalov, S.I., Vitek, M.P., and Mason, R.P. Cupric-amyloid β peptide complex stimulates oxidation of ascorbate and generation of hydroxyl radical. *Free Radic. Biol. Med.* 2004, *36*, 340–347.

43 Dawson, T.M. and Dawson, V.L. Molecular pathways of neurodegeneration in Parkinson's disease. *Science* 2003, *302*, 819–822.

44 Thomas, B. and Beal, M.F. Parkinson's disease. *Hum. Mol. Genet.* 2007, *16*, *Spec No. 2,* R183–R194.

45 Danielson, S.R. and Andersen, J.K. Oxidative and nitrative protein modifications in Parkinson's disease. *Free Radic. Biol. Med.* 2008, *44*, 1787–1794.

46 Bharath, S., Hsu, M., Kaur, D., Rajagopalan, S., and Andersen, J.K. Glutathione, iron and Parkinson's disease. *Biochem. Pharmacol.* 2002, *64*, 1037–1048.

47 Martin, H.L. and Teismann, P. Glutathione—a review on its role and significance in Parkinson's disease. *FASEB J.* 2009, *23*, 3263–3272.

48 Chinta, S.J. and Andersen, J.K. Nitrosylation and nitration of mitochondrial complex I in Parkinson's disease. *Free Radic. Res.* 2011, *45*, 53–58.

49 Chinta, S.J. and Andersen, J.K. Redox imbalance in Parkinson's disease. *Biochem. Biophys. Acta* 2008, *1780*, 1362–1367.

50 Zeevalk, G.D., Razmpour, R., and Bernard, L.P. Glutathione and Parkinson's disease: is this the elephant in the room? *Biomed. Pharmacother.* 2008, *62*, 236–249.

51 Sian-Hülsmann, J., Mandel, S., Youdim, M.B.H., and Riederer, P. The relevance of iron in the pathogenesis of Parkinson's disease. *J. Neurochem.* 2011, *118*, 939–957.

52 Jomova, K. and Valko, M. Importance of iron chelation in free radical-induced oxidative stress and human disease. *Curr. Pharm. Des.* 2011, *17*, 3460–3473.

53 Gupte, A. and Mumper, R.J. Elevated copper and oxidative stress in cancer cells as a target for cancer treatment. *Cancer Treat. Rev.* 2009, *35*, 32–46.

54 Prousek, J. Fenton chemistry in biology and medicine. *Pure Appl. Chem.* 2007, *79*, 2325–2338.

55 Speisky, H., Gómez, M., Burgos-Bravo, F., López-Alarcón, C., Jullian, C., Olea-Azar, C., and Aliaga, M.E. Generation of superoxide radicals by copper-glutathione complexes: redox-consequences associated with their interaction with reduced glutathione. *Bioorg. Med. Chem.* 2009, *17*, 1803–1810.

56 Haidari, M., Javadi, E., Kadkhodaee, M., and Sanati, A. Enhanced susceptibility to oxidation and diminished vitamin E content of LDL from patients with stable coronary artery disease. *Clin. Chem.* 2001, *47*, 1234–1240.

57 Touyz, R.M. and Schiffrin, E.L. Reactive oxygen species in vascular biology: implications in hypertension. *Histochem. Cell Biol.* 2004, *122*, 339–352.

58 Marnett, L.J. Lipid peroxidation—DNA damage by malondialdehyde. *Mutat. Res. Fundam. Mol. Mech. Mutagen.* 1999, *424*, 83–95.

59 Skrzydlewska, E., Sulkowski, S., Koda, M., Zalewski, B., Kanczuga-Koda, L., and Sulkowska, M. Lipid peroxidation and antioxidant status in colorectal cancer. *World J. Gastroenterol.* 2005, *11*, 403–406.

60 Valko, M., Morris, H., and Cronin, M.T.D. Metals, toxicity and oxidative stress. *Curr. Med. Chem.* 2005, *12*, 1161–1208.

61 Leonard, S., Gannett, P.M., Rojanasakul, Y., Schwegler-Berry, D., Castranova, V., Vallyathan, V., and Shi, X. Cobalt-mediated generation of reactive oxygen species and its possible mechanism. *J. Inorg. Biochem.* 1998, *70*, 239–244.

62 Galanis, A., Karapetsas, A., and Sandaltzopoulos, R. Metal-induced carcinogenesis, oxidative stress and hypoxia signalling. *Mutat. Res.—Genet. Toxicol. Environ. Mutagen.* 2009, *674*, 31–35.

63 Kumar, A. and Scott Clark, C. Lead loadings in household dust in Delhi, India. *Indoor Air* 2009, *19*, 414–420.

64 Patrick, L. Lead toxicity, a review of the literature. Part I: exposure, evaluation, and treatment. *Altern. Med. Rev.* 2006, *11*, 2–22.

65 Patrick, L. Lead toxicity part II: the role of free radical damage and the use of antioxidants in the pathology and treatment of lead toxicity. *Altern. Med. Rev.* 2006, *11*, 114–127.

66 Valko, M., Leibfritz, D., Moncol, J., Cronin, M.T.D., Mazur, M., and Telser, J. Free radicals and antioxidants in normal physiological functions and human disease. *Int. J. Biochem. Cell Biol.* 2007, *39*, 44–84.

67 Waalkes, M.P., Liu, J., Ward, J.M., and Diwan, B.A. Mechanisms underlying arsenic carcinogenesis: hypersensitivity of mice exposed to inorganic arsenic during gestation. *Toxicology* 2004, *198*, 31–38.

68 Jomova, K., Jenisova, Z., Feszterova, M., Baros, S., Liska, J., Hudecova, D., Rhodes, C.J., and Valko, M. Arsenic: toxicity, oxidative stress and human disease. *J. Appl. Toxicol.* 2011, *31*, 95–107.

69 Sharma, V.K. and Sohn, M. Aquatic arsenic: toxicity, speciation, transformations, and remediation. *Environ. Int.* 2009, *35*, 743–759.

70 Roy, A., Manna, P., and Sil, P.C. Prophylactic role of taurine on arsenic mediated oxidative renal dysfunction via MAPKs/NF-B and mitochondria dependent pathways. *Free Radic. Res.* 2009, *43*, 995–1007.

71 Teow, Y., Asharani, P.V., Hande, M.P., and Valiyaveettil, S. Health impact and safety of engineered nanomaterials. *Chem. Commun.* 2011, *47*, 7025–7038.

72 Barrena, R., Casals, E., Colón, J., Font, X., Sánchez, A., and Puntes, V. Evaluation of the ecotoxicity of model nanoparticles. *Chemosphere* 2009, *75*, 850–857.

73 Sharma, V.K., Yngard, R.A., and Lin, Y. Silver nanoparticles: green synthesis and their antimicrobial activities. *Adv. Colloid Interface Sci.* 2009, *145*, 83–96.

74 Smetana, A.B., Klabunde, K.J., Marchin, G.R., and Sorensen, C.M. Biocidal activity of nanocrystalline silver powders and particles. *Langmuir* 2008, *24*, 7457–7464.

75 Dallas, P., Sharma, V.K., and Zboril, R. Silver polymeric nanocomposites as advanced antimicrobial agents: classification, synthetic paths, applications, and perspectives. *Adv. Colloid Interface Sci.* 2011, *166*, 119–135.

76 He, D., Jones, A.M., Garg, S., Pham, A.N., and Waite, T.D. Silver nanoparticle-reactive oxygen species interactions: application of a charging-discharging model. *J. Phys. Chem. C* 2011, *115*, 5461–5468.

77 Hwang, E.T., Lee, J.H., Chae, Y.J., Kim, Y.S., Kim, B.C., Sang, B.I., and Gu, M.B. Analysis of the toxic mode of action of silver nanoparticles using stress-specific bioluminescent bacteria. *Small* 2008, *4*, 746–750.

78 Ahamed, M., Karns, M., Goodson, M., Rowe, J., Hussain, S.M., Schlager, J.J., and Hong, Y. DNA damage response to different surface chemistry of silver nanoparticles in mammalian cells. *Toxicol. Appl. Pharmacol.* 2008, *233*, 404–410.

79 Mahmoudi, M., Lynch, I., Ejtehadi, M.R., Monopoli, M.P., Bombelli, F.B., and Laurent, S. Protein-nanoparticles interactions: opportunities and challenges. *Chem. Rev.* 2011, *111*, 5610–5637.

80 Karakoti, A., Singh, S., Dowding, J.M., Seal, S., and Self, W.T. Redox-active radical scavenging nanomaterials. *Chem. Soc. Rev.* 2010, *39*, 4422–4432.

81 Marshall, N.M., Garner, D.K., Wilson, T.D., Gao, Y.G., Robinson, H., Nilges, M.J., and Lu, Y. Rationally tuning the reduction potential of a single cupredoxin beyond the natural range. *Nature* 2009, *462*, 113–116.

82 Wang, L. and Chance, M.R. Structural mass spectrometry of proteins using hydroxyl radical based protein footprinting. *Anal. Chem.* 2011, *83*, 7234–7241.

83 Shi, W. and Chance, M.R. Metalloproteomics: forward and reverse approaches in metalloprotein structural and functional characterization. *Curr. Opin. Chem. Biol.* 2011, *15*, 144–148.

84 Anfinsen, C.B. Principles that govern the folding of protein chains. *Science* 1973, *181*, 223–230.

85 Antoniou, D. and Schwartz, S.D. Protein dynamics and enzymatic chemical barrier passage. *J. Phys. Chem. B* 2011, *115*, 15147–15158.

86 Konermann, L., Tong, X., and Pan, Y. Protein structure and dynamics studied by mass spectrometry: H/D exchange, hydroxyl radical labeling, and related approaches. *J. Mass Spectrom.* 2008, *43*, 1021–1036.

87 Xu, G. and Chance, M.R. Hydroxyl radical-mediated modification of proteins as probes for structural proteomics. *Chem. Rev.* 2007, *107*, 3514–3543.

88 Chung, H.S., McHale, K., Louis, J.M., and Eaton, W.A. Single-molecule fluorescence experiments determine protein folding transition path times. *Science* 2012, *335*, 981–984.

89 Konermann, L., Pan, J., and Liu, Y.H. Hydrogen exchange mass spectrometry for studying protein structure and dynamics. *Chem. Soc. Rev.* 2011, *40*, 1224–1234.

90 Törnvall, U. Pinpointing oxidative modifications in proteins—recent advances in analytical methods. *Anal. Methods* 2010, *2*, 1638–1650.

91 Engen, J.R. Analysis of protein conformation and dynamics by hydrogen/deuterium exchange MS. *Anal. Chem.* 2009, *81*, 7870–7875.

92 Capelo, J.L., Carreira, R., Diniz, M., Fernandes, L., Galesio, M., Lodeiro, C., Santos, H.M., and Vale, G. Overview on modern approaches to speed up protein identification workflows relying on enzymatic cleavage and mass spectrometry-based techniques. *Anal. Chim. Acta* 2009, *650*, 151–159.

93 Roeser, J., Permentier, H.P., Bruins, A.P., and Bischoff, R. Electrochemical oxidation and cleavage of tyrosine- and tryptophan-containing tripeptides. *Anal. Chem.* 2010, *82*, 7556–7565.

94 Pain, R.H. *Mechanisms of Protein Folding*. Oxford University Press, Oxford, UK, 2000.

95 Fersht, A.R. *Structure and Mechanism in Protein Science*. W.H. Freeman & Co., New York, 1999.

96 Schotte, F., Lim, M., Jackson, T.A., Smirnov, A.V., Soman, J., Olson, J.S., Phillips, G.N., Jr., Wulff, M., and Anfinrud, P.A. Watching a protein as it functions with 150-ps time-resolved X-ray crystallography. *Science* 2003, *300*, 1944–1947.

97 Rhodes, G. *Crystallography Made Crystal Clear*. Academic Press, San Diego, CA, 2000.

98 Dempsey, A.C., Walsh, M.P., and Shaw, G.S. Unmasking the annexin I interaction from the structure of apo-S100A11. *Structure* 2003, *11*, 887–897.

99 Mittermaier, A. and Kay, L.E. New tools provide new insights in NMR studies of protein dynamics. *Science* 2006, *312*, 224–228.

100 Yi, S., Boys, B.L., Brickenden, A., Konermann, L., and Choy, W.Y. Effects of zinc binding on the structure and dynamics of the intrinsically disordered protein prothymosin α: evidence for metalation as an entropic switch. *Biochemistry* 2007, *46*, 13120–13130.

101 Zhang, D., Jiang, D., Yanney, M., Zou, S., and Sygula, A. Ratiometric Raman spectroscopy for quantification of protein oxidative damage. *Anal. Biochem.* 2009, *391*, 121–126.

102 Chung, W.G., Miranda, C.L., and Maier, C.S. Detection of carbonyl-modified proteins in interfibrillar rat mitochondria using N′-aminooxymethylcarbonyl hydrazino-D-biotin as an aldehyde/keto-reactive probe in combination with Western blot analysis and tandem mass spectrometry. *Electrophoresis* 2008, *29*, 1317–1324.

103 Shacter, E. Quantification and significance of protein oxidation in biological samples. *Drug Metab. Rev.* 2000, *32*, 307–326.

104 Levine, R.L., Williams, J.A., Stadtman, E.R., and Shacter, E. Carbonyl assays for determination of oxidatively modified proteins. *Methods Enzymol.* 1994, *233*, 346–357.

105 Sweetlove, L.J. and Møller, I.M. Chapter 1: oxidation of proteins in plants-mechanisms and consequences. *Adv. Bot. Res.* 2009, *52*, 1–23.

106 Cuddihy, S.L., Baty, J.W., Brown, K.K., Winterbourn, C.C., and Hampton, M.B. Proteomic detection of oxidized and reduced thiol proteins in cultured cells. *Methods Mol. Biol.* 2009, *519*, 363–375.

107 Eaton, P. Protein thiol oxidation in health and disease: techniques for measuring disulfides and related modifications in complex protein mixtures. *Free Radic. Biol. Med.* 2006, *40*, 1889–1899.

108 Fabisiak, J.P., Sedlov, A., and Kagan, V.E. Quantification of oxidative/nitrosative modification of CYS34 in human serum albumin using a fluorescence-based SDS-PAGE assay. *Antioxid. Redox Signal.* 2002, *4*, 855–865.

109 Baty, J.W., Hampton, M.B., and Winterbourn, C.C. Detection of oxidant sensitive thiol proteins by fluorescence labeling and two-dimensional electrophoresis. *Proteomics* 2002, *2*, 1261–1266.

110 McDonagh, B., Ogueta, S., Lasarte, G., Padilla, C.A., and Bárcena, J.A. Shotgun redox proteomics identifies specifically modified cysteines in key metabolic enzymes under oxidative stress in *Saccharomyces cerevisiae*. *J. Proteomics* 2009, *72*, 677–689.

111 Nagy, P. and Ashby, M.T. Reactive sulfur species: kinetics and mechanisms of the oxidation of cysteine by hypohalous acid to give cysteine sulfenic acid. *J. Am. Chem. Soc.* 2007, *129*, 14082–14091.

112 Roeser, J., Bischoff, R., Bruins, A.P., and Permentier, H.P. Oxidative protein labeling in mass-spectrometry-based proteomics. *Anal. Bioanal. Chem.* 2010, *397*, 3441–3455.

113 Harriman, A. Further comments on the redox potentials of tryptophan and tyrosine. *J. Phys. Chem.* 1987, *91*, 6102–6104.

114 Pattison, D.I. and Davies, M.J. Absolute rate constants for the reaction of hypochlorous acid with protein side chains and peptide bonds. *Chem. Res. Toxicol.* 2001, *14*, 1453–1464.

115 Fu, X., Wang, Y., Kao, J., Irwin, A., D'Avignon, A., Mecham, R.P., Parks, W.C., and Heinecke, J.W. Specific sequence motifs direct the oxygenation and chlorination of tryptophan by myeloperoxidase. *Biochemistry* 2006, *45*, 3961–3971.

116 Pattison, D.I. and Davies, M.J. Kinetic analysis of the reactions of hypobromous acid with protein components: implications for cellular damage and use of 3-bromotyrosine as a marker of oxidative stress. *Biochemistry* 2004, *43*, 4799–4809.

117 Di Simplicio, P., Franconi, F., Frosalí, S., and Di Giuseppe, D. Thiolation and nitrosation of cysteines in biological fluids and cells. *Amino Acids* 2003, *25*, 323–339.

118 Abello, N., Barroso, B., Kerstjens, H.A.M., Postma, D.S., and Bischoff, R. Chemical labeling and enrichment of nitrotyrosine-containing peptides. *Talanta* 2010, *80*, 1503–1512.

119 Zhang, Q., Qian, W.J., Knyushko, T.V., Clauss, T.R.W., Purvine, S.O., Moore, R.J., Sacksteder, C.A., Chin, M.H., Smith, D.J., Camp, D.G., II, Bigelow, D.J., and Smith, R.D. A method for selective enrichment and analysis of nitrotyrosine-containing peptides in complex proteome samples. *J. Proteome Res.* 2007, *6*, 2257–2268.

120 Maleknia, S.D., Wong, J.W., and Downard, K.M. Photochemical and electrophysical production of radicals on millisecond timescales to probe the structure, dynamics and interactions of proteins. *Photochem. Photobiol. Sci.* 2004, *3*, 741–748.

121 Nukuna, B.N., Sun, G., and Anderson, V.E. Hydroxyl radical oxidation of cytochrome c by aerobic radiolysis. *Free Radic. Biol. Med.* 2004, *37*, 1203–1213.

122 Sharp, J.S., Becker, J.M., and Hettich, R.L. Analysis of protein solvent accessible surfaces by photochemical oxidation and mass spectrometry. *Anal. Chem.* 2004, *76*, 672–683.

123 Hambly, D.M. and Gross, M.L. Laser flash photolysis of hydrogen peroxide to oxidize protein solvent-accessible residues on the microsecond timescale. *J. Am. Soc. Mass Spectrom.* 2005, *16*, 2057–2063.

124 Bridgewater, J.D., Lim, J., and Vachet, R.W. Transition metal-peptide binding studied by metal-catalyzed oxidation reactions and mass spectrometry. *Anal. Chem.* 2006, *78*, 2432–2438.

125 Sharp, J.S., Sullivan, D.M., Cavanagh, J., and Tomer, K.B. Measurement of multisite oxidation kinetics reveals an active site conformational change in Spo0F as a result of protein oxidation. *Biochemistry* 2006, *45*, 6260–6266.

126 Charvátová, O., Foley, B.L., Bern, M.W., Sharp, J.S., Orlando, R., and Woods, R.J. Quantifying protein interface footprinting by hydroxyl radical oxidation and molecular dynamics simulation: application to galectin-1. *J. Am. Soc. Mass Spectrom.* 2008, *19*, 1692–1705.

127 McClintock, C., Kertesz, V., and Hettich, R.L. Development of an electrochemical oxidation method for probing higher order protein structure with mass spectrometry. *Anal. Chem.* 2008, *80*, 3304–3317.

128 West, G.M., Tang, L., and Fitzgerald, M.C. Thermodynamic analysis of protein stability and ligand binding using a chemical modification- and mass spectrometry-based strategy. *Anal. Chem.* 2008, *80*, 4175–4185.

129 Kiselar, J.G., Datt, M., Chance, M.R., and Weiss, M.A. Structural analysis of proinsulin hexamer assembly by hydroxyl radical footprinting and computational modeling. *J. Biol. Chem.* 2011, *286*, 43710–43716.

130 Tong, X., Wren, J.C., and Konermann, L. γ-Ray-mediated oxidative labeling for detecting protein conformational changes by electrospray mass spectrometry. *Anal. Chem.* 2008, *80*, 2222–2231.

131 Garrison, W.M. Reaction mechanisms in the radiolysis of peptides, polypeptides, and proteins. *Chem. Rev.* 1987, *87*, 381–398.

132 Maleknia, S.D., Brenowitz, M., and Chance, M.R. Millisecond radiolytic modification of peptides by synchrotron X-rays identified by mass spectrometry. *Anal. Chem.* 1999, *71*, 3965–3973.

133 Maleknia, S.D., Ralston, C.Y., Brenowitz, M.D., Downard, K.M., and Chance, M.R. Determination of macromolecular folding and structure by synchrotron X-ray radiolysis techniques. *Anal. Biochem.* 2001, *289*, 103–115.

134 Houée-Levin, C. and Bobrowski, K. Pulse radiolysis of free radical processes in peptides and proteins. In *Radiation Chemistry: from Basic to Applications in Material and Life Sciences*, M. Spotheim-Maurizot, M. Mostafavi, and T.D. Jacquline, eds. EDP Sciences, L' Editeur, France, 2008, pp. 233–247.

135 Xu, G. and Chance, M.R. Radiolytic modification and reactivity of amino acid residues serving as structural probes for protein footprinting. *Anal. Chem.* 2005, *77*, 4549–4555.

136 Shcherbakova, I., Mitra, S., Beer, R.H., and Brenowitz, M. Fast Fenton footprinting: a laboratory-based method for the time-resolved analysis of DNA, RNA and proteins. *Nucleic Acids Res.* 2006, *34*, document no. e48.

137 Konermann, L., Stocks, B.B., and Czarny, T. Laminar flow effects during laser-induced oxidative labeling for protein structural studies by mass spectrometry. *Anal. Chem.* 2010, *82*, 6667–6674.

138 Konermann, L., Stocks, B.B., Pan, Y., and Tong, X. Mass spectrometry combined with oxidative labeling for exploring protein structure and folding. *Mass Spectrom. Rev.* 2010, *29*, 651–667.

139 Hambly, D.M. and Gross, M.L. Cold chemical oxidation of proteins. *Anal. Chem.* 2009, *81*, 7235–7242.

140 Gau, B.C., Sharp, J.S., Rempel, D.L., and Gross, M.L. Fast photochemical oxidation of protein footprints faster than protein unfolding. *Anal. Chem.* 2009, *81*, 6563–6571.

141 Aye, T.T., Low, T.Y., and Sze, S.K. Nanosecond laser-induced photochemical oxidation method for protein surface mapping with mass spectrometry. *Anal. Chem.* 2005, *77*, 5814–5822.

142 Watson, C., Janik, I., Zhuang, T., Charvátová, O., Woods, R.J., and Sharp, J.S. Pulsed electron beam water radiolysis for submicrosecond hydroxyl radical protein footprinting. *Anal. Chem.* 2009, *81*, 2496–2505.

143 Zhang, H., Gau, B.C., Jones, L.M., Vidavsky, I., and Gross, M.L. Fast photochemical oxidation of proteins for comparing structures of protein-ligand complexes: the calmodulin-peptide model system. *Anal. Chem.* 2011, *83*, 311–318.

144 Hambly, D.M. and Gross, M.L. Chapter 7: microsecond time-scale hydroxyl radical profiling of solvent-accessible protein residues. *Compr. Anal. Chem.* 2008, *52*, 151–177.

145 Jones, L.M., Sperry, J.B., Carroll, J.A., and Gross, M.L. Fast photochemical oxidation of proteins for epitope mapping. *Anal. Chem.* 2011, *83*, 7657–7661.

146 Yu, X.C., Joe, K., Zhang, Y., Adriano, A., Wang, Y., Santoro, H.G., Keck, R.G., Deperalta, G., and Ling, V. Accurate determination of succinimide degradation products using high fidelity trypsin digestion peptide map analysis. *Anal. Chem.* 2011, *83*, 5912–5919.

147 Zhong, M., Lin, L., and Kallenbach, N.R. A method for probing the topography and interactions of proteins: footprinting of myoglobin. *Proc. Natl Acad. Sci. U.S.A.* 1995, *92*, 2111–2115.

148 Hauser, N.J., Han, H., McLuckey, S.A., and Basile, F. Electron transfer dissociation of peptides generated by microwave D-cleavage digestion of proteins. *J. Proteome Res.* 2008, *7*, 1867–1872.

149 Aebersold, R. and Goodlett, D.R. Mass spectrometry in proteomics. *Chem. Rev.* 2001, *101*, 269–295.

150 Domon, B. and Aebersold, R. Mass spectrometry and protein analysis. *Science* 2006, *312*, 212–217.

151 Ong, S.E. and Mann, M. Mass spectrometry-based proteomics turns quantitative. *Nature Chem. Biol.* 2005, *1*, 252–262.

152 Ly, T. and Julian, R.R. Elucidating the tertiary structure of protein ions in vacuo with site specific photoinitiated radical reactions. *J. Am. Chem. Soc.* 2010, *132*, 8602–8609.

153 Ly, T. and Julian, R.R. Ultraviolet photodissociation: developments towards applications for mass-spectrometry-based proteomics. *Angew. Chem. Int. Ed.* 2009, *48*, 7130–7137.

154 Papayannopoulos, I.A. The interpretation of collision-induced dissociation tandem mass spectra of peptides. *Mass Spectrom. Rev.* 1995, *14*, 49–73.

155 Tanaka, K. The origin of macromolecule ionization by laser irradiation (Nobel lecture). *Angew. Chem. Int. Ed.* 2003, *42*, 3861–3870.

156 Fenn, J.B. Electrospray wings for molecular elephants (Nobel lecture). *Angew. Chem. Int. Ed.* 2003, *42*, 3871–3894.

157 Yin, H., Chacon, A., Porter, N.A., and Masterson, D.S. Free radical-induced site-specific peptide cleavage in the gas phase: low-energy collision-induced dissociation in ESI- and MALDI mass spectrometry. *J. Am. Soc. Mass Spectrom.* 2007, *18*, 807–816.

158 Wang, N. and Li, L. Exploring the precursor ion exclusion feature of liquid chromatography-electrospray ionization quadrupole time-of-flight mass spectrometry for improving protein identification in shotgun proteome analysis. *Anal. Chem.* 2008, *80*, 4696–4710.

159 Wang, N., Xie, C., Young, J.B., and Li, L. Off-line two-dimensional liquid chromatography with maximized sample loading to reversed-phase liquid chromatography-electrospray ionization tandem mass spectrometry for shotgun proteome analysis. *Anal. Chem.* 2009, *81*, 1049–1060.

160 Wang, N., Xu, M., Wang, P., and Li, L. Development of mass spectrometry-based shotgun method for proteome analysis of 500 to 5000 cancer cells. *Anal. Chem.* 2010, *82*, 2262–2271.

161 Lesur, A., Varesio, E., and Hopfgartner, G. Protein quantification by MALDI-selected reaction monitoring mass spectrometry using sulfonate derivatized peptides. *Anal. Chem.* 2010, *82*, 5227–5237.

162 Wang, H., Ouyang, Z., and Xia, Y. Peptide fragmentation during nanoelectrospray ionization. *Anal. Chem.* 2010, *82*, 6534–6541.

163 Shen, Y., Hixson, K.K., Tolić, N., Camp, D.G., Purvine, S.O., Moore, R.J., and Smith, R.D. Mass spectrometry analysis of proteome-wide proteolytic post-translational degradation of proteins. *Anal. Chem.* 2008, *80*, 5819–5828.

164 Peng, I.X., Ogorzalek Loo, R.R., Shiea, J., and Loo, J.A. Reactive-electrospray-assisted laser desorption/ionization for characterization of peptides and proteins. *Anal. Chem.* 2008, *80*, 6995–7003.

165 Peng, I.X., Ogorzalek Loo, R.R., Margalith, E., Little, M.W., and Loo, J.A. Electrospray-assisted laser desorption ionization mass spectrometry (ELDI-MS) with an infrared laser for characterizing peptides and proteins. *Analyst* 2010, *135*, 767–772.

166 Abzalimov, R.R. and Kaltashov, I.A. Electrospray ionization mass spectrometry of highly heterogeneous protein systems: protein ion charge state assignment via incomplete charge reduction. *Anal. Chem.* 2010, *82*, 7523–7526.

167 Soltwisch, J. and Dreisewerd, K. Discrimination of isobaric leucine and isoleucine residues and analysis of post-translational modifications in peptides by MALDI in-source decay mass spectrometry combined with collisional cooling. *Anal. Chem.* 2010, *82*, 5628–5635.

168 Spengler, B., Kirsch, D., Kaufmann, R., and Jaeger, E. Peptide sequencing by matrix-assisted laser-desorption mass spectrometry. *Rapid Commun. Mass Spectrom.* 1992, *6*, 105–108.

169 Gevaert, K., Demol, H., Martens, L., Hoorelbeke, B., Puype, M., Goethals, M., Van Damme, J., De Boeck, S., and Vandekerckhove, J. Protein identification based on matrix assisted laser desorption/ionization-post source decay-mass spectrometry. *Electrophoresis* 2001, *22*, 1645–1651.

170 Pittenauer, E. and Allmaier, G. High-Energy collision induced dissociation of biomolecules: MALDITOF/RTOF mass spectrometry in Comparison to tandem sector mass spectrometry. *Comb. Chem. High Throughput Screen.* 2009, *12*, 137–155.

171 Huang, Z.H., Wu, J., Roth, K.D.W., Yang, Y., Gage, D.A., and Watson, J.T. A picomole-scale method for charge derivatization of peptides for sequence analysis by mass spectrometry. *Anal. Chem.* 1997, *69*, 137–144.

172 Spengler, B., Luetzenkirchen, F., Metzger, S., Chaurand, P., Kaufmann, R., Jeffery, W., Bartlet-Jones, M., and Pappin, D.J.C. Peptide sequencing of charged derivatives by postsource decay MALDI mass spectrometry. *Int. J. Mass Spectrom. Ion Process.* 1997, *169–170*, 127–140.

173 Keough, T., Lacey, M.P., Fieno, A.M., Grant, R.A., Sun, Y., Bauer, M.D., and Begley, K.B. Tandem mass spectrometry methods for definitive protein identification in proteomics research. *Electrophoresis* 2000, *21*, 2252–2265.

174 Keough, T., Lacey, M.P., and Youngquist, R.S. Solid-phase derivatization of tryptic peptides for rapid protein identification by matrix-assisted laser desorption/ionization mass spectrometry. *Rapid Commun. Mass Spectrom.* 2002, *16*, 1003–1015.

175 Keough, T., Youngquist, R.S., and Lacey, M.P. Sulfonic acid derivatives for peptide sequencing by MALDI MS. *Anal. Chem.* 2003, *75*, 156A–165A.

176 Keough, T., Youngquist, R.S., and Lacey, M.P. A method for high-sensitivity peptide sequencing using postsource decay matrix-assisted laser desorption ionization mass spectrometry. *Proc. Natl Acad. Sci. U.S.A.* 1999, *96*, 7131–7136.

177 Sargaeva, N.P., Lin, C., and O'Connor, P.B. Identification of aspartic and isoaspartic acid residues in amyloid β peptides, including Aβ1-42, using electron-ion reactions. *Anal. Chem.* 2009, *81*, 9778–9786.

178 Woodin, R.L., Bomse, D.S., and Beauchamp, J.L. Multiphoton dissociation of molecules with low power continuous wave infrared laser radiation [23]. *J. Am. Chem. Soc.* 1978, *100*, 3248–3250.

179 Little, D.P., Speir, J.P., Senko, M.W., O'Connor, P.B., and McLafferty, F.W. Infrared multiphoton dissociation of large multiply charged ions for biomolecule sequencing. *Anal. Chem.* 1994, *66*, 2809–2815.

180 Takáts, Z., Wiseman, J.M., Gologan, B., and Cooks, R.G. Mass spectrometry sampling under ambient conditions with desorption electrospray ionization. *Science* 2004, *306*, 471–473.

181 Shiea, J., Huang, M.Z., Hsu, H.J., Lee, C.Y., Yuan, C.H., Beech, I., and Sunner, J. Electrospray-assisted laser desorption/ionization mass spectrometry for direct ambient analysis of solids. *Rapid Commun. Mass Spectrom.* 2005, *19*, 3701–3704.

182 Huang, M.Z., Yuan, C.H., Cheng, S.C., Cho, Y.T., and Shiea, J. Ambient ionization mass spectrometry. *Annu. Rev. Anal. Chem.* 2010, *3*, 43–65.

183 Sampson, J.S. and Muddiman, D.C. Atmospheric pressure infrared (10.6 µm) laser desorption electrospray ionization (IR-LDESI) coupled to a LTQ Fourier transform ion cyclotron resonance mass spectrometer. *Rapid Commun. Mass Spectrom.* 2009, *23*, 1989–1992.

184 Nemes, P., Goyal, S., and Vertes, A. Conformational and noncovalent complexation changes in proteins during electrospray ionization. *Anal. Chem.* 2008, *80*, 387–395.

185 McEwen, C.N., McKay, R.G., and Larsen, B.S. Analysis of solids, liquids, and biological tissues using solids probe introduction at atmospheric pressure on commercial LC/MS instruments. *Anal. Chem.* 2005, *77*, 7826–7831.

186 Haddad, R., Sparrapan, R., and Eberlin, M.N. Desorption sonic spray ionization for (high) voltage-free ambient mass spectrometry. *Rapid Commun. Mass Spectrom.* 2006, *20*, 2901–2905.

187 Haddad, R., Sparrapan, R., Kotiaho, T., and Eberlin, M.N. Easy ambient sonic-spray ionization-membrane interface mass spectrometry for direct analysis of solution constituents. *Anal. Chem.* 2008, *80*, 898–903.

188 Cody, R.B., Larame'E, J.A., and Durst, H.D. Versatile new ion source for the analysis of materials in open air under ambient conditions. *Anal. Chem.* 2005, *77*, 2297–2302.

189 Williams, J.P., Patel, V.J., Holland, R., and Scrivens, J.H. The use of recently described ionisation techniques for the rapid analysis of some common drugs and samples of biological origin. *Rapid Commun. Mass Spectrom.* 2006, *20*, 1447–1456.

190 Roepstorff, P. and Fohlman, J. Proposal for a common nomenclature for sequence ions in mass spectra of peptides. *Biomed. Mass Spectrom.* 1984, *11*, 601.

191 Mikhailov, V.A., Iniesta, J., and Cooper, H.J. Top-down mass analysis of protein tyrosine nitration: comparison of electron capture dissociation with "slow-heating" tandem mass spectrometry methods. *Anal. Chem.* 2010, *82*, 7283–7292.

192 Friedman, M. and Levin, C.E. Nutritional and medicinal aspects of D-amino acids. *Amino Acids* 2012, *42*, 1553–1582.

193 Wu, G. Amino acids: metabolism, functions, and nutrition. *Amino Acids* 2009, *37*, 1–17.

194 Jong, C.J., Azuma, J., and Schaffer, S. Mechanism underlying the antioxidant activity of taurine: prevention of mitochondrial oxidant production. *Amino Acids* 2012, *42*, 2223–2232.

195 Grimm, S., Höhn, A., and Grune, T. Oxidative protein damage and the proteasome. *Amino Acids* 2012, *42*, 23–38.

196 Daugherty, A., Dunn, J.L., Rateri, D.L., and Heinecke, J.W. Myeloperoxidase, a catalyst for lipoprotein oxidation, is expressed in human atherosclerotic lesions. *J. Clin. Invest.* 1994, *94*, 437–444.

197 Morgan, P.E., Pattison, D.I., Talib, J., Summers, F.A., Harmer, J.A., Celermajer, D.S., Hawkins, C.L., and Davies, M.J. High plasma thiocyanate levels in smokers are a key determinant of thiol oxidation induced by myeloperoxidase. *Free Radic. Biol. Med.* 2011, *51*, 1815–1822.

198 Pattison, D.I. and Davies, M.J. Reactions of myeloperoxidase-derived oxidants with biological substrates: gaining chemical insight into human inflammatory diseases. *Curr. Med. Chem.* 2006, *13*, 3271–3290.

199 Van Der Veen, B.S., De Winther, M.P.J., and Heeringa, P. Myeloperoxidase: molecular mechanisms of action and their relevance to human health and disease. *Antioxid. Redox Signal.* 2009, *11*, 2899–2937.

200 Davies, M.J., Hawkins, C.L., Pattison, D.I., and Rees, M.D. Mammalian heme peroxidases: from molecular mechanisms to health implications. *Antioxid. Redox Signal.* 2008, *10*, 1199–1234.

201 Ashfaq, S., Abramson, J.L., Jones, D.P., Rhodes, S.D., Weintraub, W.S., Hooper, W.C., Vaccarino, V., Harrison, D.G., and Quyyumi, A.A. The relationship between plasma levels of oxidized and reduced thiols and early atherosclerosis in healthy adults. *J. Am. Coll. Cardiol.* 2006, *47*, 1005–1011.

202 Sharma, V.K. and Sohn, M. Oxidation of amino acids, peptides, and proteins by chlorine dioxide. Implications for water treatment. In *Environmental Chemistry of Sustainable World*, Vol. 2, E. Lichtfouse, J. Schwarzbauer, and D. Robert, eds. Springer Science+Business Media, New York, 2012.

203 Umile, T.P. and Groves, J.T. Catalytic generation of chlorine dioxide from chlorite using a water-soluble manganese porphyrin. *Angew. Chem. Int. Ed.* 2011, *50*, 695–698.

204 Lippert, A.R., Van De Bittner, G.C., and Chang, C.J. Boronate oxidation as a bioorthogonal reaction approach for studying the chemistry of hydrogen peroxide in living systems. *Acc. Chem. Res.* 2011, *44*, 793–804.

205 Murphy, M.P. How mitochondria produce reactive oxygen species. *Biochem. J.* 2009, *417*, 1–13.

206 Cleeter, M.W.J., Cooper, J.M., and Schapira, A.H.V. Nitric oxide enhances MPP^+ inhibition of complex I. *FEBS Lett.* 2001, *504*, 50–52.

207 Al Ghouleh, I., Khoo, N.K.H., Knaus, U.G., Griendling, K.K., Touyz, R.M., Thannickal, V.J., Barchowsky, A., Nauseef, W.M., Kelley, E.E., Bauer, P.M., Darley-Usmar, V., Shiva, S., Cifuentes-Pagano, E., Freeman, B.A., Gladwin, M.T., and Pagano, P.J. Oxidases and peroxidases in cardiovascular and lung disease: new concepts in reactive oxygen species signaling. *Free Radic. Biol. Med.* 2011, *51*, 1271–1288.

208 Nesnow, S., Grindstaff, R.D., Lambert, G., Padgett, W.T., Bruno, M., Ge, Y., Chen, P.J., Wood, C.E., and Murphy, L. Propiconazole increases reactive oxygen species levels in mouse hepatic cells in culture and in mouse liver by a cytochrome P450 enzyme mediated process. *Chem. Biol. Interact.* 2011, *194*, 79–89.

209 Buettner, G.R. Superoxide dismutase in redox biology: the roles of superoxide and hydrogen peroxide. *Anticancer Agents Med. Chem.* 2011, *11*, 341–346.

210 Whittaker, J.W. Metal uptake by manganese superoxide dismutase. *Biochem. Biophys. Acta—Proteins Proteomics* 2010, *1804*, 298–307.

211 Yamakura, F. and Kawasaki, H. Post-translational modifications of superoxide dismutase. *Biochem. Biophys. Acta—Proteins Proteomics* 2010, *1804*, 318–325.

212 Zielonka, J., Sarna, T., Roberts, J.E., Wishart, J.F., and Kalyanaraman, B. Pulse radiolysis and steady-state analyses of the reaction between hydroethidine and superoxide and other oxidants. *Arch. Biochem. Biophys.* 2006, *456*, 39–47.

213 Tabares, L.C., Gätjens, J., and Un, S. Understanding the influence of the protein environment on the Mn(II) centers in superoxide dismutases using high-field electron paramagnetic resonance. *Biochem. Biophys. Acta—Proteins Proteomics* 2010, *1804*, 308–317.

214 Perry, J.J.P., Shin, D.S., Getzoff, E.D., and Tainer, J.A. The structural biochemistry of the superoxide dismutases. *Biochem. Biophys. Acta—Proteins Proteomics* 2010, *1804*, 245–262.

215 Miller, A.F., Yikilmaz, E., and Vathyam, S. ^{15}N-NMR characterization of His residues in and around the active site of FeSOD. *Biochim Biophys. Acta Proteins Proteomics* 2010, *1804*, 275–284.

216 Holley, A.K., Dhar, S.K., and St. Clair, D.K. Manganese superoxide dismutase vs. p53: regulation of mitochondrial ROS. *Mitochondrion* 2010, *10*, 649–661.

217 Krause, M.E., Glass, A.M., Jackson, T.A., and Laurence, J.S. MAPping the chiral inversion and structural transformation of a metal-tripeptide complex having Ni-superoxide dismutase activity. *Inorg. Chem.* 2011, *50*, 2479–2487.

218 Ragsdale, S.W. Nickel-based enzyme systems. *J. Biol. Chem.* 2009, *284*, 18571–18575.

219 Shearer, J., Neupane, K.P., and Callan, P.E. Metallopeptide based mimics with substituted histidines approximate a key hydrogen bonding network in the metalloenzyme nickel superoxide dismutase. *Inorg. Chem.* 2009, *48*, 10560–10571.

220 Lugo-Huitrón, R., Blanco-Ayala, T., Ugalde-Muñiz, P., Carrillo-Mora, P., Pedraza-Chaverrí, J., Silva-Adaya, D., Maldonado, P.D., Torres, I., Pinzón, E., Ortiz-Islas, E., López, T., García, E., Pineda, B., Torres-Ramos, M., Santamaría, A., and La Cruz, V.P.D. On the antioxidant properties of kynurenic acid: free radical scavenging activity and inhibition of oxidative stress. *Neurotoxicol. Teratol.* 2011, *33*, 538–547.

221 Chatgilialoglu, C., D'Angelantonio, M., Kciuk, G., and Bobrowski, K. New insights into the reaction paths hydroxyl radicals with 2′-deoxyguanosine. *Chem. Res. Toxicol.* 2011, *24*, 2200–2206.

222 Guan, Y., Ma, J., Li, X., Fang, J., and Chen, L. Influence of pH on the formation of sulfate and hydroxyl radicals in the UV/peroxymonosulfate system. *Environ. Sci. Technol.* 2011, *95*, 9308–9314.

223 Moncada, S., Palmer, R.M.J., and Higgs, E.A. Nitric oxide: physiology, pathophysiology, and pharmacology. *Pharmacol. Rev.* 1991, *43*, 109–142.

224 Dedon, P.C. and Tannenbaum, S.R. Reactive nitrogen species in the chemical biology of inflammation. *Arch. Biochem. Biophys.* 2004, *423*, 12–22.

225 D'Ischia, M., Napolitano, A., Manini, P., and Panzella, L. Secondary targets of nitrite-derived reactive nitrogen species: nitrosation/nitration pathways, antioxidant defense mechanisms and toxicological implications. *Chem. Res. Toxicol.* 2011, *24*, 2071–2092.

226 Rocha, B.S., Gago, B., Barbosa, R.M., Lundberg, J.O., Radi, R., and Laranjinha, J. Intragastric nitration by dietary nitrite: implications for modulation of protein and lipid signaling. *Free Radic. Biol. Med.* 2012, *52*, 693–698.

227 Radi, R. Peroxynitrite and reactive nitrogen species: the contribution of ABB in two decades of research. *Arch. Biochem. Biophys.* 2009, *484*, 111–113.

228 Ferrer-Sueta, G., Manta, B., Botti, H., Radi, R., Trujillo, M., and Denicola, A. Factors affecting protein thiol reactivity and specificity in peroxide reduction. *Chem. Res. Toxicol.* 2011, *24*, 434–450.

229 Nagy, P. and Winterbourn, C.C. Redox chemistry of biological thiols. *Adv. Mol. Toxicol.* 2010, *4*, 183–222.

230 Winterbourn, C.C. and Hampton, M.B. Thiol chemistry and specificity in redox signaling. *Free Radic. Biol. Med.* 2008, *45*, 549–561.

231 Ferrer-Sueta, G. and Radi, R. Chemical biology of peroxynitrite: kinetics, diffusion, and radicals. *ACS Chem. Biol.* 2009, *4*, 161–177.

232 Ignasiak, M., Scuderi, D., De Oliveira, P., Pedzinski, T., Rayah, Y., and Houée Levin, C. Characterization by mass spectrometry and IRMPD spectroscopy of the sulfoxide group in oxidized methionine and related compounds. *Chem. Phys. Lett.* 2011, *502*, 29–36.

233 Manta, B., Hugo, M., Ortiz, C., Ferrer-Sueta, G., Trujillo, M., and Denicola, A. The peroxidase and peroxynitrite reductase activity of human erythrocyte peroxiredoxin 2. *Arch. Biochem. Biophys.* 2009, *484*, 146–154.

234 Winterbourn, C.C. and Metodiewa, D. Reactivity of biologically important thiol compounds with superoxide and hydrogen peroxide. *Free Radic. Biol. Med.* 1999, *27*, 322–328.

235 Schöneich, C. Cysteine residues as catalysts for covalent peptide and protein modification: a role for thiyl radicals? *Biochem. Soc. Trans.* 2011, *39*, 1254–1259.

236 Madej, E. and Wardman, P. The oxidizing power of the glutathione thiyl radical as measured by its electrode potential at physiological pH. *Arch. Biochem. Biophys.* 2007, *462*, 94–102.

237 Wardman, P. Reduction potentials of one-electron couples involving free radicals in aqueous solutions. *J. Phys. Chem. Ref. Data* 1989, *18*, 1637–1755.

238 Ford, E., Hughes, M.N., and Wardman, P. Kinetics of the reactions of nitrogen dioxide with glutathione, cysteine, and uric acid at physiological pH. *Free Radic. Biol. Med.* 2002, *32*, 1314–1323.

239 Forni, L.G., Mönig, J., Mora-Arellano, V.O., and Willson, R.L. Thiyl free radicals: direct observations of electron transfer reactions with phenothiazines and ascorbate. *J. Chem. Soc. Perkin Trans. 2* 1983, 961–965.

240 Steenken, S., Jovanovic, S.V., Bietti, M., and Bernhard, K. The trap depth (in DNA) of 8-oxo-7,8-dihydro-2′deoxyguanosine as derived from electron-transfer equilibria in aqueous solution. *J. Am. Chem. Soc.* 2000, *122*, 2373–2374.

241 Stoyanovsky, D., Maeda, A., Atkins, J.L., and Kagan, V.E. Assessment of thiyl radicals in biosystems: difficulties and new applications. *Anal. Chem.* 2011, *83*, 6432–6438.

242 Nauser, T. and Schöneich, C. Thiyl radicals abstract hydrogen atoms from the αC-H bonds in model peptides: absolute rate constants and effect of amino acid structure. *J. Am. Chem. Soc.* 2003, *125*, 2042–2043.

243 Nauser, T., Pelling, J., and Schöneich, C. Thiyl radical reaction with amino acid side chains: rate constants for hydrogen transfer and relevance for posttranslational protein modification. *Chem. Res. Toxicol.* 2004, *17*, 1323–1328.

244 Nauser, T., Casi, G., Koppenol, W.H., and Schöneich, C. Reversible intramolecular hydrogen transfer between cysteine thiyl radicals and glycine and alanine in model peptides: absolute rate constants derived from pulse radiolysis and laser flash photolysis. *J. Phys. Chem. B* 2008, *112*, 15034–15044.

245 SchöNeich, C. Mechanisms of protein damage induced by cysteine thiyl radical formation. *Chem. Res. Toxicol.* 2008, *21*, 1175–1179.

246 Madej, E., Folkes, L.K., Wardman, P., Czapski, G., and Goldstein, S. Thiyl radicals react with nitric oxide to form S-nitrosothiols with rate constants near the diffusion-controlled limit. *Free Radic. Biol. Med.* 2008, *44*, 2013–2018.

247 Domazou, A.S., Koppenol, W.H., and Gebicki, J.M. Efficient repair of protein radicals by ascorbate. *Free Radic. Biol. Med.* 2009, *46*, 1049–1057.

248 Folkes, L.K., Trujillo, M., Bartesaghi, S., Radi, R., and Wardman, P. Kinetics of reduction of tyrosine phenoxyl radicals by glutathione. *Arch. Biochem. Biophys.* 2011, *506*, 242–249.

249 Carballal, S., Trujillo, M., Cuevasanta, E., Bartesaghi, S., Möller, M.N., Folkes, L.K., García-Bereguiaín, M.A., Gutiérrez-Merino, C., Wardman, P., Denicola, A., Radi, R., and Alvarez, B. Reactivity of hydrogen sulfide with peroxynitrite and other oxidants of biological interest. *Free Radic. Biol. Med.* 2011, *50*, 196–205.

250 Shatalin, K., Shatalina, E., Mironov, A., and Nudler, E. H_2S: a universal defense against antibiotics in bacteria. *Science* 2011, *334*, 986–990.

251 Lippert, A.R., New, E.J., and Chang, C.J. Reaction-based fluorescent probes for selective imaging of hydrogen sulfide in living cells. *J. Am. Chem. Soc.* 2011, *133*, 10078–10080.

252 Bauer, G., Chatgilialoglu, C., Gebicki, J.L., Gebicka, L., Gescheidt, G., Golding, B.T., Goldstein, S., Kaizer, J., Merenyi, G., Speien, G., and Wardman, P. Biologically relevant small radicals. *Chimia* 2008, *62*, 704–712.

253 Yin, G. Active transition metal oxo and hydroxo moieties in nature's redox, enzymes and their synthetic models: structure and reactivity relationships. *Coord. Chem. Rev.* 2010, *254*, 1826–1842.

254 Bollinger, J.M., Jr. and Matthews, M.L. Remote enzyme microsurgery. *Science* 2010, *327*, 1337–1338.

255 Friedle, S., Reisner, E., and Lipard, S.J. Current challenges of modeling diiron enzyme active sites for dioxygen activation by biometric synthetic complexes. *Chem. Soc. Rev.* 2010, *39*, 2768–2779.

256 Donald, W.A., McKenzie, C.J., and O'Hair, R.A.J. C-H bond activation of methanol and ethanol by a high-spin $Fe^{IV}O$ biomimetic complex. *Angew. Chem. Int. Ed.* 2011, *50*, 8379–8383.

257 Dzhabiev, T.S. and Shilov, A.E. Concerted reactions of polynuclear metalloenzymes and their functional chemical models. *Russ. J. Phys. Chem. A* 2011, *85*, 397–401.

258 Nam, W. High-valent iron(IV)-oxo complexes of heme and non-heme ligands in oxygenation reactions. *Acc. Chem. Res.* 2007, *40*, 522–531.

259 Fukuzumi, S., Kishi, T., Kotani, H., Lee, Y.M., and Nam, W. Highly efficient photocatalytic oxygenation reactions using water as an oxygen source. *Nat. Chem.* 2011, *3*, 38–41.

260 Groves, J.T. High-valent iron in chemical and biological oxidations. *J. Inorg. Biochem.* 2006, *100*, 434–447.

261 Morimoto, Y., Kotani, H., Park, J., Lee, Y.M., Nam, W., and Fukuzumi, S. Metal ion-coupled electron transfer of a nonheme oxoiron(IV) complex: remarkable enhancement of electron-transfer rates by Sc^{3+}. *J. Am. Chem. Soc.* 2011, *133*, 403–405.

262 Fu, R., Gupta, R., Geng, J., Dornevil, K., Wang, S., Zhang, Y., Hendrich, M.P., and Liu, A. Enzyme reactivation by hydrogen peroxide in heme-based tryptophan dioxygenase. *J. Biol. Chem.* 2011, *286*, 26541–26554.

263 Solomon, E.I., Brunold, T.C., Davis, M.I., Kemsley, J.N., Lee, S.K., Lehnert, N., Neese, F., Skulan, A.J., Yang, Y.S., and Zhou, J. Geometric and electronic structure/function correlations in non-heme iron enzymes. *Chem. Rev.* 2000, *100*, 235–349.

264 Krebs, C., Fujimori, D.G., Walsh, C.T., and Bollinger, J.M., Jr. Non-heme Fe(IV)-oxo intermediates. *Acc. Chem. Res.* 2007, *40*, 484–492.

265 Company, A., Prat, I., Frisch, J.R., Mas-Ballesté, D.R., Güell, M., Juhász, G., Ribas, X., Münck, D.E., Luis, J.M., Que, L., Jr., and Costas, M. Modeling the cis-oxo-labile binding site motif of non-heme iron oxygenases: water exchange and oxidation reactivity of a non-heme iron(IV)-oxo compound bearing a tripodal tetradentate ligand. *Chem. Eur. J.* 2011, *17*, 1622–1634.

266 Que, L., Jr. The road to non-heme oxoferryls and beyond. *Acc. Chem. Res.* 2007, *40*, 493–500.

267 Menton, J.D. and Bielski, B.H.J. Studies of the kinetics, spectral and chemical properties of Fe(IV) pyrophosphate by pulse radiolysis. *Radiat. Phys. Chem.* 1990, *36*, 725–733.

268 Berry, J.F., Bill, E., Bothe, E., George, S., Mienert, B., Neese, F., and Wieghardt, K. An octahedral coordination complex of iron(VI). *Science* 2006, *312*, 1937–1941.

269 Aliaga-Alcalde, N., DeBeer George, S., Mienert, B., Bill, E., Wieghardt, K., and Neese, F. The geometric and electronic structure of [(cyclam-acetato)Fe(N)]$^+$: a genuine iron(V) species with a ground-state spin S = 1/2. *Angew. Chem. Int. Ed.* 2005, *44*, 2908–2912.

270 Scepaniak, J.J., Vogel, C.S., Khusniyarov, M.M., Heinemann, F.W., Meyer, K., and Smith, J.M. Synthesis, structure, and reactivity of an iron(V) nitride. *Science* 2011, *331*, 1049–1052.

271 Francisco, J.S., Muckerman, J.T., and Yu, H.G. HOCO radical chemistry. *Acc. Chem. Res.* 2010, *43*, 1519–1526.

272 Bloss, C., Wagner, V., Bonzanini, A., Jenkin, M.E., Wirtz, K., Martin-Reviejo, M., and Pilling, M.J. Evaluation of detailed aromatic mechanisms (MCMv3 and MCMv3.1) against environmental chamber data. *Atmos. Chem. Phys.* 2005, *5*, 623–639.

273 Atkinson, R. and Arey, J. Atmospheric degradation of volatile organic compounds. *Chem. Rev.* 2003, *103*, 4605–4638.

274 Aschmann, S.M., Arey, J., and Atkinson, R. Kinetics and products of the reaction of OH radicals with 3-methoxy-3-methyl-1-butanol. *Environ. Sci. Technol.* 2011, *45*, 6896–6901.

275 Lin, J.J., Chen, A.F., and Lee, Y.T. UV photolysis of ClOOCl and the ozone hole. *Chem. Asian J.* 2011, *6*, 1664–1678.

276 Orkin, V.L., Poskrebyshev, G.A., and Kurylo, M.J. Rate constants for the reactions between OH and perfluorinated alkenes. *J. Phys. Chem. A* 2011, *115*, 6568–6574.

277 Shiraiwa, M., Sosedova, Y., Rouvière, A., Yang, H., Zhang, Y., Abbatt, J.P.D., Ammann, M., and Pöschl, U. The role of long-lived reactive oxygen intermediates in the reaction of ozone with aerosol particles. *Nat. Chem.* 2011, *3*, 291–295.

278 Aschmann, S.M. and Atkinson, R. Effect of structure on the rate constants for reaction of NO_3 radicals with a series of linear and branched C5-C7 1-alkenes at 296 ± 2 K. *J. Phys. Chem. A* 2011, *115*, 1358–1363.

279 Wayne, R.P., Barnes, I., Biggs, P., Burrows, J.P., Canosa-Mas, C.E., Hjorth, J., Le Bras, G., Moortgat, G.K., Perner, D., Poulet, G., Restelli, G., and Sidebottom, H. The nitrate radical: physics, chemistry, and the atmosphere. *Atmos. Environ. A* 1991, *25*, 1–203.

280 Goeschen, C., Wibowo, N., White, J.M., and Wille, U. Damage of aromatic amino acids by the atmospheric free radical oxidant NO_3 in the presence of NO_2, N_2O_4, O_3 and O_2. *Org. Biomol. Chem.* 2011, *9*, 3380–3385.

281 Sigmund, D.C.E. and Wille, U. Can the night-time atmospheric oxidant NO_3? damage aromatic amino acids? *Chem. Commun.* 2008, 2121–2123.

282 Fimmen, R.L., Trouts, T.D., Richter D.D., Jr., and Vasudevan, D. Improved speciation of dissolved organic nitrogen in natural waters: amide hydrolysis with fluorescence derivatization. *J. Environ. Sci.* 2008, *20*, 1273–1280.

283 Vitousek, P.M., Aber, J.D., Howarth, R.W., Likens, G.E., Matson, P.A., Schindler, D.W., Schlesinger, W.H., and Tilman, D.G. Human alternation of the global nitrogen cycle: sources and consequences. *Ecol. Appl.* 1997, *7*, 737–750.

284 Lee, W., Westerhoff, P., and Esparza-Soto, M. Occurrence and removal of dissolved organic nitrogen in US water treatment plants. *J. Am. Water Works Assoc.* 2006, *98*, 102–110+14.

285 Bond, T., Huang, J., Templeton, M.R., and Graham, N. Occurrence and control of nitrogenous disinfection by-products in drinking water—a review. *Water Res.* 2011, *45*, 4341–4354.

286 Bronk, D.A. Dynamics of DON. In *Biogeochemistry of Marine Dissolved Organic Matter*, D.A. Hansell and C.A. Carlson, eds. Elsevier, Inc., New York, 2002, pp. 153–247.

287 Trehy, M.L., Yost, R.A., and Miles, C.J. Chlorination by-products of amino acids in natural waters. *Environ. Sci. Technol.* 1986, *20*, 1117–1122.

288 Thurman, E.M. *Organic Geochemistry of Natural Waters*. Martinus Nijhoff Publishers, The Hague, The Netherlands, 1985, p. 512.

289 Dotson, A. and Westerhoff, P. Occurrence and removal of amino acids during drinking water treatment. *Am. Water Works Assoc. J.* 2009, *101*, 101–117.

290 Westgate, P.J. and Park, C. Evaluation of proteins and organic nitrogen in wastewater treatment effluents. *Environ. Sci. Technol.* 2010, *44*, 5352–5357.

291 Brillas, E., Sires, I., and Oturan, M.A. Electro-Fenton process and related electrochemical technologies based on Fenton's reaction chemistry. *Chem. Rev.* 2009, *109*, 6570–6631.

292 Sharma, V.K. Potassium ferrate(VI): environmental friendly oxidant. *Adv. Environ. Res.* 2002, *6*, 143–156.

293 Huie, R.E., Clifton, C.L., and Neta, P. Electron transfer reaction rates and equilibria of the carbonate and sulfate radical anions. *Radiat. Phys. Chem.* 1991, *38*, 477–481.

294 Stanbury, D.M. Reduction potentials involving inorganic free radicals in aqueous solution. *Adv. Inorg. Chem.* 1989, *33*, 69–138.

295 Richardson, S.D. and Ternes, T.A. Water analysis: emerging contaminants and current issues. *Anal. Chem.* 2011, *83*, 4616–4648.

296 Shah, A.D. and Mitch, W.A. Halonitroalkanes, halonitriles, haloamides, and N-nitrosamines: a critical review of nitrogenous disinfection by-product formation pathways. *Environ. Sci. Technol.* 2012, *46*, 119–131.

297 Sharma, V.K. Kinetics and mechanism of formation and destruction of N-nitrosodimethylamine in water—a review. *Sep. Purif. Technol.* 2012, *88*, 1–10.

298 Richardson, S.D. Environmental mass spectrometry: emerging contaminants and current issues. *Anal. Chem.* 2012, *84*, 747–778.

299 Laingam, S., Froscio, S.M., Bull, R.J., and Humpage, A.R. In vitro toxicity and genotoxicity assessment of disinfection by-products, organic N-chloramines. *Environ. Mol. Mutagen.* 2012, *53*, 83–93.

300 Sharma, V.K. Oxidation of inorganic compounds by ferrate (VI) and ferrate(V): one-electron and two-electron transfer steps. *Environ. Sci. Technol.* 2010, *44*, 5148–5152.

301 Brosillon, S., Lemasle, M., Renault, E., Tozza, D., Heim, V., and Laplanche, A. Analysis and occurrence of odorous disinfection by-products from chlorination of amino acids in three different drinking water treatment plants and corresponding distribution networks. *Chemosphere* 2009, *77*, 1035–1042.

302 Chu, W., Gao, N., Deng, Y., and Dong, B. Formation of chloroform during chlorination of alanine in drinking water. *Chemosphere* 2009, *77*, 1346–1351.

303 Chu, W.H., Gao, N.Y., Deng, Y., and Krasner, S.W. Precursors of dichloroacetamide, an emerging nitrogenous DBP formed during chlorination or chloramination. *Environ. Sci. Technol.* 2010, *44*, 3908–3912.

304 Cardador, M.J. and Gallego, M. Haloacetic acids in swimming pools: swimmer and worker exposure. *Environ. Sci. Technol.* 2011, *45*, 5783–5790.

305 Pals, J.A., Ang, J.K., Wagner, E.D., and Plewa, M.J. Biological mechanism for the toxicity of haloacetic acid drinking water disinfection by-products. *Environ. Sci. Technol.* 2011, *45*, 5791–5797.

306 Sharma, V.K. Disinfection performance of Fe(VI) in water and wastewater: a review. *Water Sci. Technol.* 2007, *55*, 225–232.

307 Hutchings, J.W., Ervens, B., Straub, D., and Herckes, P. N-nitrosodimethylamine occurrence, formation and cycling in clouds and fogs. *Environ. Sci. Technol.* 2010, *44*, 8128–8133.

308 Schäfer, A.I., Mitch, W., Walewijk, S., Munoz, A., Teuten, E., and Reinhard, M. Chapter 7 Micropollutants in water recycling: a case study of N-Nitrosodimethylamine (NDMA) exposure from water versus food. *Sustain. Sci. Eng.* 2010, *2*, 203–228.

309 Ripollés, C., Pitarch, E., Sancho, J.V., López, F.J., and Hernández, F. Determination of eight nitrosamines in water at the ng L^{-1} levels by liquid chromatography coupled to atmospheric pressure chemical ionization tandem mass spectrometry. *Anal. Chim. Acta* 2011, *702*, 62–71.

310 Le Roux, J., Gallard, H., and Croué, J.P. Chloramination of nitrogenous contaminants (pharmaceuticals and pesticides): NDMA and halogenated DBPs formation. *Water Res.* 2011, *45*, 3164–3174.

311 Mitch, W.A. and Sedlak, D.L. Formation of N-nitrosodimethylamine (NDMA) from dimethylamine during chlorination. *Environ. Sci. Technol.* 2002, *36*, 588–595.

312 Choi, J. and Valentine, R.L. Formation of N-nitrosodimethylamine (NDMA) from reaction of monochloramine: a new disinfection by-product. *Water Res.* 2002, *36*, 817–824.

313 Padhye, L., Luzinova, Y., Cho, M., Mizaikoff, B., Kim, J.H., and Huang, C.H. PolyDADMAC and dimethylamine as precursors of N -nitrosodimethylamine during ozonation: reaction kinetics and mechanisms. *Environ. Sci. Technol.* 2011, *45*, 4353–4359.

314 von Gunten, U., Salhi, E., Schmidt, C.K., and Arnold, W.A. Kinetics and mechanisms of N-nitrosodimethylamine formation upon ozonation of N,N-dimethylsulfamide-containing waters: bromide catalysis. *Environ. Sci. Technol.* 2010, *44*, 5762–5768.

315 Sharma, V.K. Kinetics and mechanism of formation and destruction of N-nitrosodimethylamine in water—a review. *Sep. Purif. Technol.* 2012, *88*, 1–10.

316 Bond, T., Goslan, E.H., Parsons, S.A., and Jefferson, B. Treatment of disinfection by-product precursors. *Environ. Technol.* 2011, *32*, 1–25.

317 Choi, H., Al-Abed, S.R., Dionysiou, D.D., Stathatos, E., and Lianos, P. Chapter 8 TiO_2-Based Advanced Oxidation Nanotechnologies for Water Purification and Reuse, In Anonymous, 2010, Vol. 2, pp. 229–254.

318 Sillanpää, M.E.T., Kurniawan, T.A., and Lo, W.H. Degradation of chelating agents in aqueous solution using advanced oxidation process (AOP). *Chemosphere* 2011, *83*, 1443–1460.

319 Sharma, V.K. Oxidative transformations of environmental pharmaceuticals by Cl_2, ClO_2, O_3, and Fe(VI): kinetics assessment. *Chemosphere* 2008, *73*, 1379–1386.

320 Lee, Y. and Gunten, U.V. Oxidative transformation of micropollutants during municipal wastewater treatment: comparison of kinetic aspects of selective (chlorine, chlorine dioxide, ferrate^{VI}, and ozone) and non-selective oxidants (hydroxyl radical). *Water Res.* 2010, *44*, 555–566.

321 Pan, J.H., Dou, H., Xiong, Z., Xu, C., Ma, J., and Zhao, X.S. Porous photocatalysts for advanced water purifications. *J. Mater. Chem.* 2010, *20*, 4512–4528.

322 Fujishima, A., Zhang, X., and Tryk, D.A. TiO_2 photocatalysis and related surface phenomena. *Surf. Sci. Rep.* 2008, *63*, 515–582.

323 Chen, X., Shen, S., Guo, L., and Mao, S.S. Semiconductor-based photocatalytic hydrogen generation. *Chem. Rev.* 2010, *110*, 6503–6570.

324 Han, F., Kambala, V.S.R., Srinivasan, M., Rajarathnam, D., and Naidu, R. Tailored titanium dioxide photocatalysts for the degradation of organic dyes in wastewater treatment: a review. *Appl. Catal. A* 2009, *359*, 25–40.

325 Nfodzo, P. and Choi, H. Sulfate radicals destroy pharmaceuticals and personal care products. *Environ. Eng. Sci.* 2011, *28*, 605–609.

326 Matta, R., Tlili, S., Chiron, S., and Barbati, S. Removal of carbamazepine from urban wastewater by sulfate radical oxidation. *Environ. Chem. Lett.* 2011, *9*, 347–353.

327 Deng, Y. and Ezyske, C.M. Sulfate radical-advanced oxidation process (SR-AOP) for simultaneous removal of refractory organic contaminants and ammonia in landfill leachate. *Water Res.* 2011, *45*, 6189–6194.

328 Sharma, V.K. Oxidation of inorganic contaminants by ferrates(Fe(VI), Fe(V), and Fe(IV))—kinetics and mechanisms: a review. *J. Environ. Manage.* 2011, *92*, 1051–1073.

329 Sharma, V.K., Luther, G.W., III, and Millero, F.J. Mechanisms of oxidation of organosulfur compounds by ferrate(VI). *Chemosphere* 2011, *82*, 1083–1089.

330 Sharma, V.K., Sohn, M., Anquandah, G., and Nesnas, N. Kinetics of the oxidation of sucralose and related carbohydrates by ferrate(VI). *Chemosphere* 2012, *87*, 644–648.

331 Delaude, L. and Laszlo, P.A. Novel oxidizing reagent based on potassium ferrate(VI). *J. Org. Chem.* 1996, *61*, 6360–6370.

332 Sharma, V.K. Oxidation of nitrogen-containing pollutants by novel ferrate(VI) technology: a review. *J. Environ. Sci. Health Part A Tox. Hazard. Subst. Environ. Eng.* 2010, *45*, 645–667.

333 Sharma, V.K., Yngard, R.A., Cabelli, D.E., and Clayton Baum, J. Ferrate(VI) and ferrate(V) oxidation of cyanide, thiocyanate, and copper(I) cyanide. *Radiat. Phys. Chem.* 2008, *77*, 761–767.

334 Shappell, N.W., Vrabel, M.A., Madsen, P.J., Harrington, G., Billey, L.O., Hakk, H., Larsen, G.L., Beach, E.S., Horwitz, C.P., Ro, K., Hunt, P.G., and Collins, T.J.

Destruction of estrogens using Fe-TAML/peroxide catalysis. *Environ. Sci. Technol.* 2008, *42*, 1296–1300.

335 Sharma, V.K., Graham, N.J.D., Li, X.Z., and Yuan, B.L. Ferrate(VI) enhanced photocatalytic oxidation of pollutants in aqueous TiO_2 suspensions. *Environ. Sci. Pollut. Res.* 2010, *17*(2), 453–461.

336 Collins, T.J., Khetan, S.K., and Ryabov, A.D. Chemistry and applications of iron-TAML catalysts in green oxidation processes based on hydrogen peroxide. *Handb. Green Chem. Green Catal.* 2009, *1*, 39–77.

337 Tanaka, N., Ikeda, C., Kanaori, K., Hiraga, K., Konno, T., and Kunugi, S. Fluctuation of apomyoglobin monitored from H/D exchange and proteolysis under high pressure. *Prog. Biotechnol.* 2002, *19*, 47–54.

338 Fraczkiewicz, R. and Braun, W. Exact and efficient analytical calculation of the accessible surface areas and their gradients for macromolecules. *J. Comput. Chem.* 1998, *19*, 319–333.

339 Cornilescu, G., Marquardt, J.L., Ottiger, M., and Bax, A. Validation of protein structure from anisotropic carbonyl chemical shifts in a dilute liquid crystalline phase. *J. Am. Chem. Soc.* 1998, *120*, 6836–6837.

2

ACID–BASE PROPERTIES

Acid–base chemistry of amino acids and reactive intermediates is prerequisite to understand oxidative mechanisms in proteins and enzymes. For example, metal binding to the biological ligand and its subsequent oxidation by reactive species is determined by whether functional groups of amino acids are protonated or unprotonated. This is the case in the redox activity of glutathione, which is exclusively associated with the SH groups of cysteine under physiological conditions [1, 2]. The sulfhydryl group of cysteine also plays a crucial role in both the structure and function of proteins [1, 2]. Moreover, the oxidative chemistry of proteins is influenced by many factors such as hydrogen bonds, noncovalent interactions, aromatic–aromatic interactions, and metal–protein binding, which vary with protonation of amino acid side-chain groups. Reaction rates of amino acid chains are thus affected by the pH of the reaction medium. This is described in Chapters 3–6. Besides protonation of amino acids, the acid–base equilibria of reactive intermediates (halogens, oxygen, nitrogen, sulfur, and phosphate species) also influence the rates of the reactions. For example, protonated halogen species (e.g., HOCl) is more reactive than unprotonated species (e.g., OCl^-). More on the role of protonated/unprotonated species of reactive intermediates is discussed in subsequent chapters.

Basically, the pH dependence of reaction rates can be understood by considering equilibrium species of both amino acids and reactive intermediates in the system (see Chapter 6). The distributions of different forms of amino

Oxidation of Amino Acids, Peptides, and Proteins: Kinetics and Mechanism, First Edition.
Virender K. Sharma.

acids/peptides (i.e., speciation) can be calculated by applying their protonation/dissociation constants. The next section in this chapter presents the procedure to determine speciation. Cysteine was chosen as an example because of its importance in the oxidative chemistry of proteins. Generally, cysteine was found the most reactive amino acid in proteins with reactive species (Chapter 4). Protonation of amino acids depends on temperature, ionic strength, and polarity of solvent (see the next section) and thus the speciation. The interactions between amino acids and solvent with different dielectric constants provide information on the chemistry of *in vivo* processes [3, 4].

In the biological environment, rates of the reactions depend on the conformational dynamics of proteins. The degree of exposure of proteins to solvent media (e.g., H_2O) may therefore affect the oxidation of amino acid side chains by reactive species. The dynamics of proteins has been studied extensively using the hydrogen/deuterium exchange approach [5, 6]. The use of the hydrogen exchange (HX) phenomenon in protein measurements has a long history, which was first described in 1950 to measure deuteration using density gradients [5, 7]. Hydrogen atoms of S–H, O–H, and N–H groups in proteins can exchange with the surrounding water. The degree of deuteration in continuous-labeling experiments is monitored as a function of time. The slow hydrogen/deuterium exchange (HDX) occurs if sites in the protein are involved in stable hydrogen bonds or exist inside the protein core. The details and mechanisms of the exchange of hydrogen between solvent and protein have been reviewed [8–11]. The exchange of deuterium for hydrogen can be followed by different spectroscopic techniques. Of the several spectroscopic tools, nuclear magnetic resonance (NMR) has been extensively used for the analysis of HX in proteins [9, 12]. Mass spectrometry (MS)-based HDX measurements are also appropriate because the mass of deuterium is almost twice the mass of hydrogen [10, 11, 13, 14]. In these measurements, only a miniscule amount of the protein is needed. As little as 500–1000 pmol is needed for an entire experiment, which includes the analysis of 10 points of exchange. Moreover, protein concentration can be as low as 0.1 μM, which is at least an order of magnitude smaller than those of many other techniques used in studying the conformation and dynamics of a protein. In the HX-MS approach, samples having multiple proteins and ligands can be analyzed [10].

The hydrogen at the peptide backbone amide linkages in proteins can be easily measured. Many hydrogen atoms on the side chains of proteins exchange so rapidly that the measurement of them by MS is difficult. Hydrogens bonded to carbons in proteins almost never exchange. The advantage of measuring amide hydrogens is that they hold secondary structure elements (β-sheets, α-helices) together. Moreover, every amino acid except proline has an amide hydrogen located on the backbone. The exchange rates of NHs in folded proteins are approximately eight orders of magnitude slower than those in unfolded proteins. Thus, hydrogen bonding and solvent exposure determine the exchange between the two interwoven proteins. Fast exchange rates are suggestive of solvent exposure and/or no hydrogen bonding, while the reverse is true for slow exchange [10]. HDX-MS spectrometry is applied to determine the conformation of membrane proteins in phospholipid bilayer nanodiscs [15].

In addition to the involvement of protonation in redox reactions, interactions of metals–amino acids and metals–peptides also play significant role in the oxidative chemistry of proteins. For example, the addition of glutathione to the Cu(II) ion immediately formed the Cu(I)–$(GSH)_2$ complex [16], which was able to generate superoxide by its reaction with oxygen [17, 18]. The formation of iron, cobalt, copper, and nickel complexes has been shown to influence the production of superoxide and hydroxyl radicals [19, 20]. Complexation can also enhance or retard the reactivity of reactive species with metal centers, which is described in Chapters 4–6. Complexes of metals may stabilize reactive intermediates such as the formation of high-valent iron (Fe(IV) and Fe(V)) complexes with oxo- and nitrogen donor ligands [21], which then react with amino acid side chains of proteins (see Chapter 6). The chemistry of metal complexation is discussed later in this chapter.

2.1 DISSOCIATION CONSTANTS

The acid–base properties of amino acids/peptides are generally characterized using macroscopic protonation (or dissociation) constants, log K_i (or pK_i), which are composites of the microscopic constants (log k_i or pk_i) for the individual groups [22]. The microscopic constants represent a particular molecular subunit in a defined protonation state for all other moieties in that ligand [22]. Microspecies remain in incessant interconversion due to the instantaneous nature of protonation reactions; therefore, a direct approach to determine log k_i is not feasible. Thus, an indirect approach is applied with the use of at least two experimental techniques and/or auxiliary compounds, followed by an appropriate evaluation technique. This approach has been described in detail in the microequilibrium analysis of tetrabasic acid [22, 23].

The dissociation of a triprotic acid (e.g., cysteine) is described by Equations (2.1)–(2.6):

$$AH_3^+ \rightleftharpoons H^+ + AH_2^\pm \tag{2.1}$$

$$AH_2^\pm \rightleftharpoons H^+ + AH^- \tag{2.2}$$

$$AH^- \rightleftharpoons H^+ + A^{2-} \tag{2.3}$$

$$K_{a1} = a_{H+}a_{AH2}/a_{AH3+} = (\gamma_H\gamma_{AH2}/\gamma_{AH3})([H^+][AH_2^\pm]/[AH_3^+]) \tag{2.4}$$

$$K_{a2} = a_{H+}a_{AH-}/a_{AH2\pm} = (\gamma_{H+}\gamma_{AH}/\gamma_{AH2})([H^+][AH^-]/[AH_2^\pm]) \tag{2.5}$$

$$K_{a3} = a_{H+}a_{A2-}/a_{AH-} = (\gamma_H\gamma_A/\gamma_{AH})([H^+][A^{2-}]/[AH^-]), \tag{2.6}$$

where K_i, a_i, γ_i, and terms in brackets are thermodynamic dissociation constants, activities, activity coefficients, and molarities of the respective species.

Numerous experimental methods have been applied to determine dissociation constants, which include liquid–liquid partitioning, high-pressure liquid chromatography, capillary zone electrophoresis, matrix-assisted laser desorption/ionization time-of-flight MS, and titration methods [24–31]. Potentiometric titrations are frequently applied to determine dissociation constants [32, 33]. However, when dissociation constants are particularly low or high, or in cases where the substance has very low solubility, spectrophotometric methods are used. Theoretical calculations using quantum theoretical techniques have also been applied to predict the values of dissociation constants [26, 34–36]. These calculations have shown to successfully predict the dissociation constants of amino acids and peptides in water [26].

The values of thermodynamic constants of amino acids and simple peptides are presented in Tables 2.1 and 2.2 [28, 32, 37–45]. Examination of simple peptides allowed understanding of reactions of various peptides with reactive species [46, 47] (Chapters 5 and 6). Most of the amino acids have an α-carboxylic acid group with pK_{a1} values of ~2.0 to 2.5 and an α-amino group with pK_{a2} values of ~9 to 10. The values are lower than those of monofunctional carboxylic acids and amines, respectively, due to the negative charge on the carboxylate group at high pH and the positive charge on the amino group at low pH, which stabilizes amino acids [48]. Both Asp and Glu have a carboxyl group in the side chain, which have pK_a values in the range of 3.5–4.5 when incorporated into a protein. Arg, Lys, and His have basic side chains, and their pK_a values in proteins are in the range of 9.5–10.5, 12.0–13.0, and 6.0–7.0, respectively. The resonance stabilization of Arg makes it a relatively strong base. Because His has a pK_a value close to the pH of a living cell, the pH determines the charge on a particular His in a protein. The side chains of Asn and Gln have a resonance-stabilized unprotonated amide group and are not basic. This is similar to the aromatic side chain of Trp. The pK_a value of Cys is ~9. Proteins usually have uncharged thiols; however, ionization may take place in some enzymes [48]. Cystine has a disulfide bridge and has four protonation sites (Table 2.1).

The formation of a peptide involves condensation of the carboxylic group of one amino acid with the amino group of the next amino acid to yield a peptide bond (amide linkage). The amide group itself does not demonstrate properties of either the amino group or the carboxylic group. Generally, a simple peptide (e.g., diglycine and triglycine) has two pK_a values, one from the C-terminus and the other from the N-terminus (Table 2.2). Glutathione is a tripeptide (γ-L-glutamyl-L-cysteinyl-glycine [GSH]) and has four pK_a values. The oxidized form of GSH is glutathione disulfide (GSSG) with four carboxylic and two amino groups, producing six pK_a values (Table 2.2). Metal binding to GSH and GSSG thus varies with pH significantly. Moreover, the variation of redox potential of the two GSH-GSSG pairs can be understood in depth by knowing accurately the pK_a values of GSH and GSSG [49].

As stated earlier, temperature influences dissociation of amino acids (i.e., speciation), which, in turn, affects the reaction rates of functional groups of

TABLE 2.1. Structure and Dissociation Constants of Amino Acids at 25°C at Zero Ionic Strength

Type of Side Chain	Name	Abbreviated Name (One Letter)	Structure of the Side Chain	pK_{a1}	pK_{a2}	pK_{a}
Aliphatic	Glycine	Gly (G)	—H	2.43[a]	9.60[a]	–
	Alanine	Ala (A)	—Me	2.55[b]	10.08[b]	–
	Valine	Val (V)	—CH(Me)Me	2.28[c]	9.54[c]	–
	Leucine	Leu (L)	$—CH_2CH(Me)Me$	2.58[b]	9.93[b]	–
	Isoleucine	Ile (I)	—CH(Me)Et	2.32[d]	9.76[d]	–
Alcohols	Serine	Ser (S)	$—CH_2OH$	2.09[e]	9.05[e]	–
	Threonine	Thr (T)	—CH(OH)Me	2.45[b]	9.28[b]	–
Acids	Aspartic acid	Asp (D)	$—CH_2CO_2H$	2.10[c]	3.86[c]	9.8
	Glutamic acid	Glu (E)	$—CH_2CH_2CO_2H$	2.19[f]	4.45[f]	10.1[f]
Amides	Asparagine	Asn (N)	$—CH_2CONH_2$	2.16[g]	8.73[g]	–
	Glutamine	Gln (Q)	$—CH_2CH_2CONH_2$	2.18[g]	9.00[g]	–
Bases	Lysine	Lys (K)	$—CH_2CH_2CH_2CH_2NH_2$	1.85[f]	9.42[f]	11.1
	Arginine	Arg (R)	$—CH_2CH_2CH_2NH—C(=NH)—NH_2$	1.97[f]	9.05[f]	11.9
	Histidine	His (H)	$—H_2C$– (imidazole ring: HN, N)	1.56[f]	6.00[f]	9.2
Thiol	Cysteine	Cys (C)	$—CH_2SH$	2.14[h]	8.18[h]	10.2
Sulfide	Methionine	Met (M)	$—CH_2CH_2S—Me$	2.14[i]	9.61[i]	–
Aromatic	Tryptophan	Trp (W)	$—H_2C$– (indole ring: N, H)	2.37[f]	9.36[f]	–
	Tyrosine	Tyr (Y)	$—CH_2—C_6H_4—OH$	2.04[f]	9.09[f]	10.6

TABLE 2.1. (*Continued*)

Type of Side Chain	Name	Abbreviated Name (One Letter)	Structure of the Side Chain	pK_{a1}	pK_{a2}	pK_{a3}
	Phenylalanine	Phe (F)	$-CH_2-$ (phenyl ring)	2.28[f]	9.34[f]	–
	Proline[j]	Pro (P)	H N C O OH (pyrrolidine ring)	[k]	10.6[l]	–

From Fiol et al. [43].
From Fiol et al. [44].
From Bastug et al. [74] at $I = 0.10\,M$ ($NaClO_4$).
From Martell and Smith [42].
From Demirelli and Köseoĝlu [45].
From Nagai et al. [32].
From Martell and Smith [42] at $I = 0.1\,M$.
From Sharma et al. [39].
From Sharma et al. [41].
This is drawn in full.
From Martell and Smith [42].
From Fazary et al. [3].

TABLE 2.2. Dissociation Constants of Cystine, Glutathione (Reduced and Oxidized) at Zero Ionic Strength at 25°C

	Cystine[a]	Glutathione (Reduced)[b]	Glutathione (Oxidized)[c]	Diglycine[d]	Triglycine[e]
pK_{a1}	1.36	2.14	1.22	3.06	3.26
pK_{a2}	1.61	3.71	2.12	8.1	7.9
pK_{a3}	8.64	9.08	3.28	–	–
pK_{a4}	9.41	10.13	4.21	–	–
pK_{a5}	–	–	9.45	–	–
pK_{a6}	–	–	10.49	–	–

[a] From Furia et al. [28].
[b] From Crea et al. [37].
[c] From Crea et al. [38].
[d] From Armesto et al. [124].
[e] From Simic et al. [125].

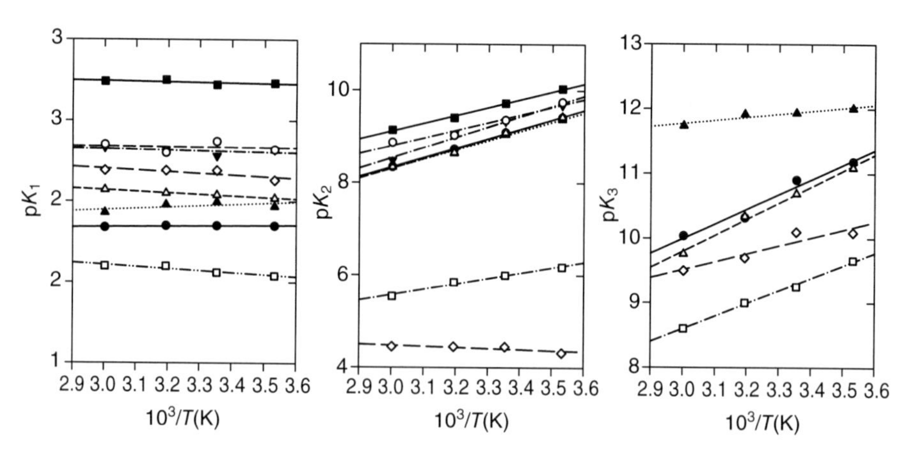

Figure 2.1. pK_1, pK_2, and pK_3 values of lysine, histidine, arginine, glutamic acid, and tyrosine extrapolated to zero ionic strength at 283.1, 298.1, 313.1, and 333.1 K. Symbols are ●, Lys; □, His; ▲, Arg; ◇, Glu; Δ, Tyr; ■, Thr; ▼, Phe; and ○, Trp (data were taken from Nagai et al. [32]).

amino acids with reactive species (Chapters 3–6). Most of available experimental values of pK_a are at 25°C in aqueous solutions and knowing values at 37°C would be closer to human *in vivo* conditions. The effect of temperature on dissociation constants can be determined by van't Hoff's equation:

$$K_a(T) = K_a(T_o)[-(\Delta h_{diss}/R)(1/T - 1/T_o), \tag{2.7}$$

where T_o = 298.15 K and Δh_{diss} is the dissociation enthalpy. Several studies have measured the effect of temperature on the pK_a of amino acids in aqueous solution [32, 50–53]. The temperature dependence of the thermodynamic dissociation constants of amino acids is shown in Figure 2.1. With the exception of Glu, the effect of temperature on the pK_a values was insignificant regarding deprotonation of the carboxylic acid group of amino acids (Fig. 2.1). Glu appeared to have temperature influence on the dissociation of the carboxylic group attached to the side chain. The values of $\Delta h_{diss}/R$ at zero ionic strength varied from −285 to 40 K [32]. Arg had a decrease in pK_{a1} with an increase in temperature. The pK_{a2} values varied strongly with temperature, and the $\Delta h_{diss}/R$ values at zero ionic strength ranged from 3380 to 5155 K (Fig. 2.1). The pK_{a3} values decreased with an increase in temperature (Fig. 2.1). The temperature greatly affected the pK_{a3} values of the aromatic hydroxyl group on the side chain of Tyr, and $\Delta h_{diss}/R$ at zero ionic strength was obtained as 5713 K. Comparatively, the values of $\Delta h_{diss}/R$ at zero ionic strength were 1604, 2789, 4510, and 5226 K for Arg, Glu, His, and Lys, respectively. Knowing the values of $\Delta h_{diss}/R$ at zero ionic strength can be used to calculate pK_i at zero ionic strength (Eq. 2.7), which are needed to determine accurately the dissociation of amino acids at ionic strength of the biological environment in which amino acid side chains are exposed (see below). The enthalpy of some amino acids in aqueous NaCl solution has also been directly determined [54].

The dissociation constants are also affected by the ionic strength (I), defined by Equation (2.8):

$$I = 1/2\sum m_i z_i^2, \tag{2.8}$$

where m_i and z_i are the molality (mole per kilogram of solvent) and charge of ion i. The dissociation constants at a given ionic strength of species i (pK_i^*) in NaCl solutions for Cys, as an example, are presented in Figure 2.2 [40]. Least-squares fitting of the pK_i^* results as a function of I is given in Equations (2.9)–(2.11):

$$pK_1^* = 1.378 + 228.4/T - 0.4044\, I^{0.5} + 0.2472\, I \tag{2.9}$$

$$pK_2^* = 2.031 + 1833.4/T - 0.1847\, I^{0.5} + 0.2190\, I \tag{2.10}$$

$$pK_3^* = 2.861 + 2191.3/T - 0.2170\, I^{0.5} + 0.2217\, I. \tag{2.11}$$

The pK_i^* values at different I have been interpreted using the effect of ionic strength on the activity coefficients [37–41, 55–58]. Various models have been used to estimate activity coefficients of amino acids and peptides in aqueous solutions [28, 32, 37, 38, 40, 41, 55, 57, 59–66]. The activity coefficients at different ionic strengths can be determined by using the suitable model. At low ionic strength, the Davies equation has been applied (Eq. 2.12) [67]:

$$\log \gamma_i = -z_i^2[AI^{0.5}/(1 + I^{0.5}) - bI], \tag{2.12}$$

where A is the Debye–Huckel constant and $b = 0.1$. A is a function of temperature and is given by [68]

$$A = 1.8252 \cdot 10^6 (\rho_W / \varepsilon^3 T^3)^{1/2}, \tag{2.13}$$

where ρ_W and ε are the density (gram per cubic centimeter) and the dielectric constants of water, respectively. The dielectric constant depends on temperature and is expressed by Equation (2.14) [69]:

$$\varepsilon = 5321/T + 233.76 - 0.9297\, T + 1.417 \cdot 10^{-3}\, T^2 - 8.292 \cdot 10^{-7}\, T^3. \tag{2.14}$$

In addition to ionic strength, cations such as Na^+, Rb^+, Mg^{2+}, and Ca^{2+} also influence pK_i values, which can also be interpreted using these models [37–39, 55, 56, 59, 60]. Overall, such information on the dissociation of amino acids can be used to enhance understanding of complicated biological systems. For example, quantitative information on the affinity of the sulfur side chain of

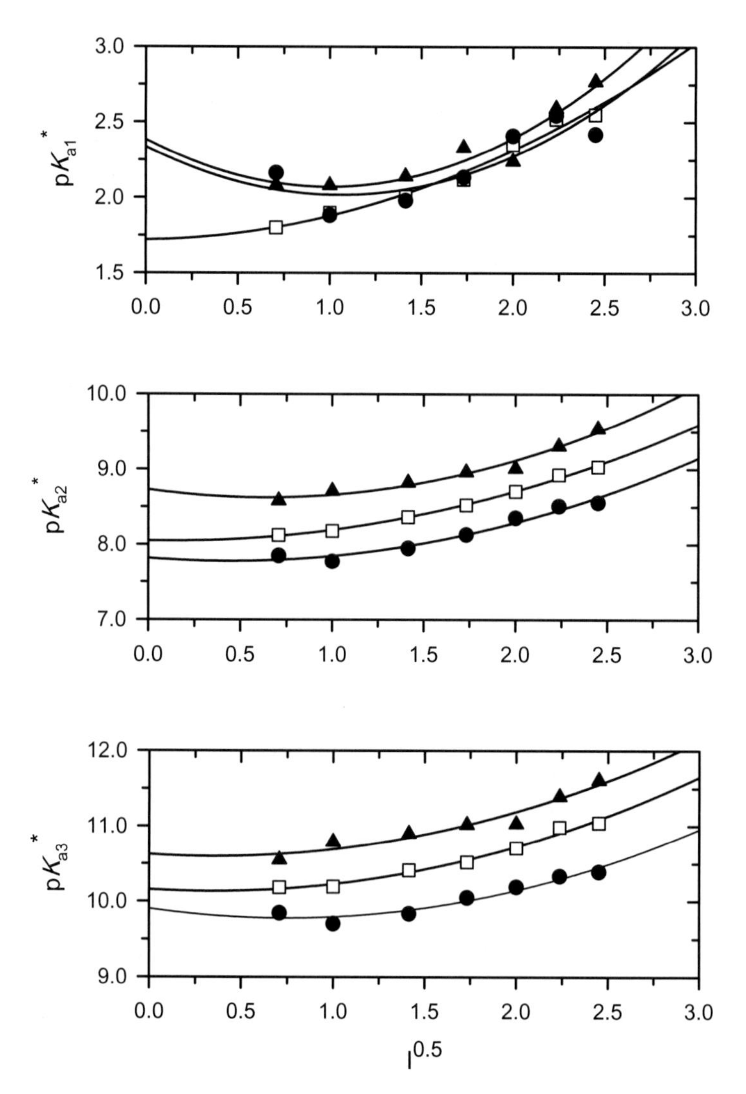

Figure 2.2. Values of dissociation constants of cysteine in NaCl solution at different temperatures (●, 5°C; □, 25°C; ▲, 45°C; solid lines are a second-order fit of the pK_i^* vs. $I^{0.5}$) (adapted from Sharma et al. [40] with the permission of Springer America).

proteins for alkali cations can evaluate competition from trace metals (e.g., Fe, Cd, Zn, and Co) to bind Cys at protein sites.

Studies on the effect of polarity of organic solvents on the pK_i of amino acids have been performed [3, 70]. Understanding of solvent effect is essential considering the influence of polarity on stabilization/destabilization on protein structure. Polarity of water is expected to be lower at the active sites of proteins in the biological systems. Dissociation of amino acids has therefore been

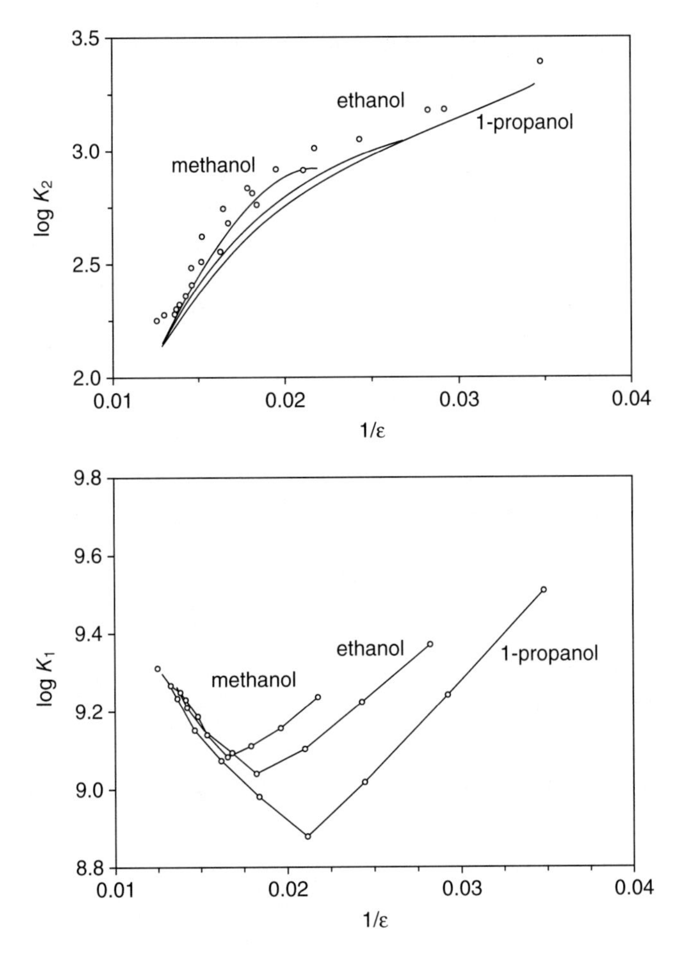

Figure 2.3. Plots of the experimental values of log K_2 (pK_{a1}) (upper) and log K_1 (pK_{a2}) (lower) versus the reciprocal of dielectric constant of different mixed solvents at 25°C and an ionic strength of 0.1 mol/dm^3 $NaClO_4$ (adapted from Gharib et al. [70] with the permission of the American Chemical Society).

examined in water containing organic solvents. An example of polarity effect on dissociation is demonstrated in Figure 2.3 for Trp in mixtures of water and alcohol [70]. The pK_{a1} of the carboxylic group of Trp increased, but the pK_{a2} of the amino group first decreased and then increased with an increasing percentage of alcohols in the mixed solvents. In the case of carboxyl group dissociation, no net change in charge on the molecule occurred and the change in polarity of the medium had a minor effect. Comparatively, dissociation of the amino group involving changes in charge and polarity greatly influenced the pK_{a2} of Trp in water–alcohol mixtures [70]. The microscopic protonation constants of Met, Gly, L-Ala, L-Leu, L-Ile, L-Ser, and L-Phe in water–ethanol and water–acetonitrile mixtures have also been determined to learn speciation of these amino acids in a water–organic solvent medium [71, 72]. The ratios of dipolar ionic form to neutral form vary with the polarity of the medium. As discussed in subsequent chapters, the rates of the reactions vary with the speciation of functional groups of amino acids of proteins, and therefore, kinetic studies under water–organic solvent conditions represent a more accurate biological system.

2.2 SPECIATION

2.2.1 Protonation

The total amount of each amino acid dissolved in solution ($[HA)]_{total}$) is the sum of all existing dissolved amino acid (e.g., Cys) species (Eq. 2.15):

$$[HA]_{total} = [H_3A^+] + [H_2A^\pm] + [HA^-] + [A^{2-}]. \tag{2.15}$$

The fraction of these species (α_i) is a function of pH and can be calculated using Equations (2.16)–(2.19):

$$\alpha_{H3A+} = [H^+]^3/([H^+]^3 + [H^+]^2 K_{a1} + [H^+]K_{a1}K_{a2} + K_{a1}K_{a2}K_{a3}) \tag{2.16}$$

$$\alpha_{H2A\pm} = [H^+]^2 K_{a1}/([H^+]^3 + [H^+]^2 K_{a1} + [H^+]K_{a1}K_{a2} + K_{a1}K_{a2}K_{a3}) \tag{2.17}$$

$$\alpha_{HA-} = [H^+]K_{a1}K_{a2}/([H^+]^3 + [H^+]^2 K_{a1} + [H^+]K_{a1}K_{a2} + K_{a1}K_{a2}K_{a3}) \tag{2.18}$$

$$\alpha_{A2-} = K_{a1}K_{a2}K_{a3}/([H^+]^3 + [H^+]^2 K_{a1} + [H^+]K_{a1}K_{a2} + K_{a1}K_{a2}K_{a3}). \tag{2.19}$$

The calculated fractions of species of Cys ($NH_3^+CH(CH_2SH)COOH$, H_3A^+; $NH_3^+CH(CH_2SH)COO^-$, H_2A; $NH_3^+CH(CH_2S^-)COO^-$, HA^-; $NH_2CH(CH_2S^-)COO^-$, A^{2-}) are depicted in Figure 2.4. The speciation of Cys at zero ionic strength suggests the neutral and negatively charged forms occur at pH 7.4. The neutral form predominates under this condition in which both the sulfur and amino groups of the Cys molecule would be protonated. The speciation of Cys would change slightly with an increase in ionic strength, but the neutral species of Cys would still remain the major species at pH 7.4 (Fig. 2.4).

The calculated fractions of other selected amino acids relevant to redox chemistry in the biological environment at pH 7.4 are given in Table 2.3. Most of the amino acids exist in their monoprotonated forms (HA) as the major species at nearly neutral pH. The major species of Tyr and Cys are in their diprotonated forms (H_2A). Deprotonated species (A) of Gly, His, Met, Trp, and Phe are in very minor fractions. Tyr and Cys had an insignificant presence of deprotonated forms (Table 2.3).

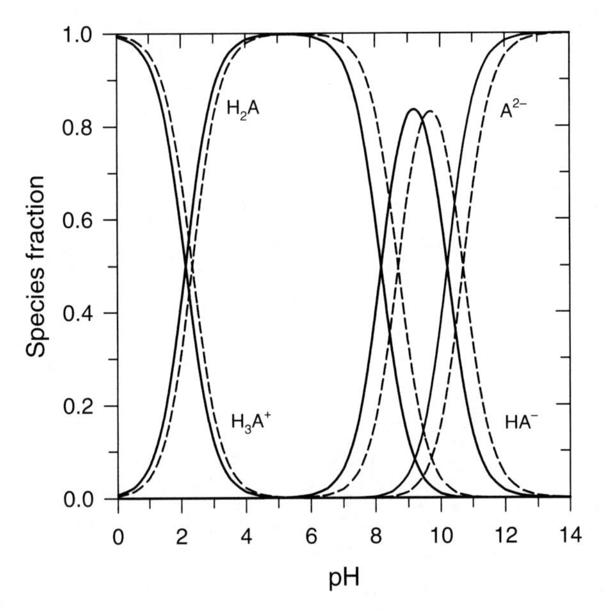

Figure 2.4. Calculated species distribution of cysteine at $I = 0.0$ (solid lines) and I = 4.0 (dashed lines).

TABLE 2.3. Speciation of Selected Amino Acids at pH 7.4

Amino Acid	α_{H3A}	α_{H2A}	α_{HA}	α_A
Gly	–	1.1×10^{-5}	0.9937	0.0063
Pro	–			
His	5.5×10^{-8}	0.0378	0.9488	0.0134
Cys	4.7×10^{-6}	0.8576	0.1423	2.2×10^{-4}
Met	–	5.5×10^{-6}	0.9939	0.0061
Trp	–	9.2×10^{-6}	0.9891	0.0108
Tyr	4.3×10^{-6}	0.9800	0.0200	1.0×10^{-5}
Phe	–	7.5×10^{-6}	0.9886	0.0114

2.2.2 Metal Complexes

Many proteins and enzymes contain metals ions in which peptide bonds (or amino acid groups) are coordinated with metal ions [73]. The nature of coordination depends on the protonation of amino acids (see discussion in the previous section). For example, two kinds of electron-pair donors in α-amino acids, nitrogen of amino groups (NH_3^+ and NH_2) and oxygen of carboxylate (COO^-), can form *mono*, *bis*, or *tris* complexes with Cu(II) [74]. Because of the role of coordination in the redox chemistry of metals in enzymes [75], the solution equilibria and structures of metal–amino acid and metal–peptide complexes have been studied for the last several decades [76–84]. Examples of such studies include the interaction of metals with sulfur-containing amino acids in the 1950s [76–79] and Cu(II) complexes of dipeptides and amino acids in the 1970s [80, 81]. In a recent study, the coordination dynamics of Zn in proteins has been reviewed [75]. In enzymes, the imidazole nitrogens from His, the carboxylate oxygen(s) from the side chain of Asp and Glu, and sulfur from the sulfhydryl groups of Cys were the donors of proteins to coordinate Zn (Fig. 2.5a). The sulfhydryl group provides the unique reactivity in the coordination environment [85]. The importance of amino acids in the coordination sphere is shown in Figure 2.5b. A role of hydrogen bonds in the interactions of zinc-bound hydroxide ion and His ligands with other amino acids was suggested.

His plays a significant role in the coordination of metals in biological environment such as superoxide dismutase, the prion protein, amyloid β-peptides, and histones [73, 86–90]. Metal-binding sites in proteins may also include the phenol ring of Tyr [91]. Peroxynitrometal complexes have also been evoked in hemoglobin and myoglobin [92]. Furthermore, the interaction of a metal ion with an amino acid ligand increases significantly with a decrease in the solvent polarity of the media [70]. The following sections represent selected examples of complex formation of metals with amino acids and proteins, which have shown roles in generating reactive species and in mimicking structures of metalloenzymes [93].

2.2.2.1 Iron. Complexes of iron have significant relevance in natural and biological environments [94–96]. Marine siderophores have a high affinity for the ferric iron, which are produced when a demand of iron arises [97]. Studies on iron–catecholate complexes and iron–peptide of sequence Ac-Ala-DOPA-Thr-Pro-$CONH_2$ (DOPA = 3,4-dihydroxyphenylalanine) have been performed [98]. DOPA-containing peptide mimics mussel adhesive proteins. The complexes with iron were $Fe(L)_n$ (n = 1–3), which could react with oxygen to yield organic radicals. Such complexes may also react with H_2O_2 to result in reactive species (see Chapter 6). The coordination chemistry of the Fe(III) complex with alterobactin A, a siderophore from the marine bacterium *Alteromonas luteoviolacea*, has also been evaluated [97]. High affinity of Fe^{3+} ion with alterobactin A was suggested based on the estimation of high stability constants for the formation of Fe(III)–alterobactin A complex. Complexation of Fe(III) with the biologically important ligand, cystine, has also been studied [99]. This study showed the existence of FeL^+ and FeL_2^- species in at least 30% of total Fe. Free iron and complex iron have shown different reactivities with reactive species (Chapter 4). Complexation of Fe(III) with poly(aminecarboxylate) ligands has been studied in detail to understand their rates with reactive species (e.g., Chapter 4) [94].

Formation of high-valent species (ferryl and perferryl species) in Fenton reaction and Fenton-type reactions is related to complexation of low-valent iron species (Chapter 4). Structural studies on high-valent iron complexes

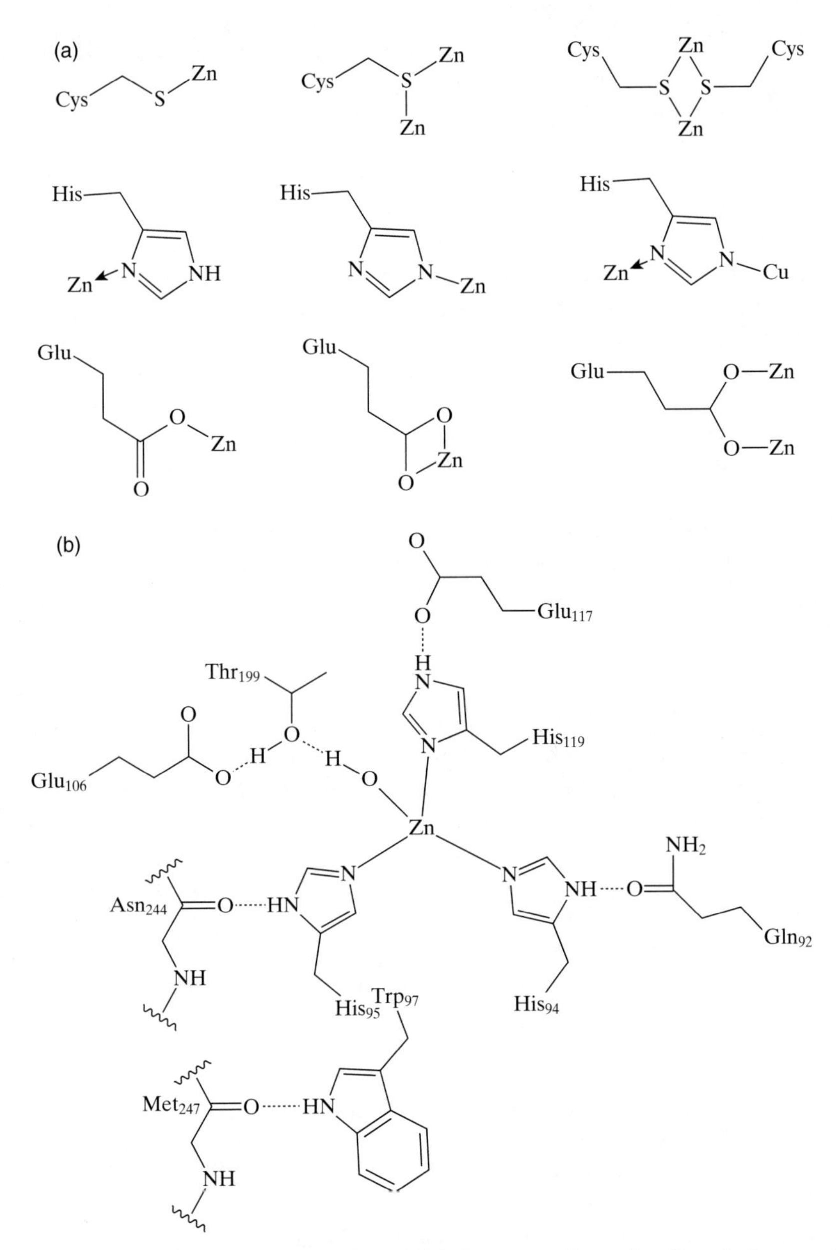

Figure 2.5. (a) Zinc–ligand interactions. With few exceptions, zinc ligands in proteins are cysteine (Cys) (S-donor), histidine (His) (N-donor), or glutamate (Glu)/aspartate (Asp) (O-donor). For cysteine, there is a single mode of interaction. For histidine, binding occurs with either one of the two nitrogen atoms of the imidazole ring. For glutamate/aspartate, binding modes include one oxygen, which can be syn or anti (not shown), or both oxygens of the carboxylate. All three ligands can bridge one or several zinc ions. For cysteine, there are one, two, or three bridges. For histidine, the only bridges known are those between zinc and copper in superoxide dismutase (SOD1) and between zinc and zinc in the zinc transporter CzrB (*Thermus thermophilus*). Aspartate/glutamate can form a bridge with the two oxygens of the carboxylate. (b) Zinc binding to amino acids (adapted from Maret and Li [75] with the permission of the American Chemical Society).

provide insight into active sites involved in reaction mechanisms of many heme and nonheme enzymes [100–103]. High-valent iron species as intermediates in the activation of oxygen by iron enzymes has also been invoked [101, 104]. Examples include aromatic dihydroxylation by arene-containing protein-bound iron complex [105]. Complexes of high-valent iron have therefore been synthesized, followed by their structural characterization using spectroscopic techniques such as Mössbauer spectroscopy, resonance Raman spectroscopy, and X-ray absorption spectroscopy (XAS) [100, 103, 106–108]. These techniques have identified species of Fe(IV), Fe(V), and Fe(VI) such as 6-coordinate Fe(IV)-oxo(N_4Py) and Fe(VI)-nitride complexes and 5-coordinate Fe(V)-oxo(TAML) (TAML = tetraamidomacrocyclic) species [109–111]. The reactivity of high-valent iron-oxo and -nitride species is presented in Chapter 6.

2.2.2.2 Copper, Zinc, and Nickel. The complexation of Cu(II) with bioligands is imperative in biological systems, and hence the formation constants of 1:1 binary complexes of Cu(II)–Gly, Cu(II)–Ala, Cu(II)–Val, Cu(II)–Leu, Cu(II)–Glu, and Cu(II)–Asp and 1:2 binary complexes of Cu–Gly_2 and Cu–Glu_2 have been determined using potentiometric and spectrometric techniques [74]. The enthalpy changes of complexes were negative, except Cu(II)–Glu and Cu(II)–Asp complexes, while the positive entropy changes for all of the complexes were obtained. This suggests that both enthalpy and entropy changes derived the complexation. The complexation of similar amino complexes with Zn(II) and Ni(II) has also been studied [45]. Knowledge on Cu(II), Zn(II), and Ni(II) complexation with studied amino acids may provide insight on structural differences in different superoxide dismutases in order to describe oxidation and reduction reactions of their catalytic metal ions [112]. Cu(II), Zn(II), and Ni(II) complexes of salicylaldehyde (Sal)-amino acid Schiff bases were also studied as nonenzymatic models for pyridoxal–amino acid systems [45]. The stability constants, $\log\beta_1$ and $\log\beta_2$, were in the order Sal-Gly > Sal-Ala > Sal-Ser > Sal-Tyr > Sal-Phe, except Sal-Gly for the Cu(II) complex. This indicates the steric effect and basicity of the Schiff base controls the Cu(II) complexation process [45]. Results may help in understanding complicated metal–protein interactions.

Interactions of Cu(II), Zn(II), and Ni(II) with terminally blocked (CH_3CONH- and –$CONH_2$) peptides (-TESHHK-, -TASHHK-, -TEAHHK-, -TESAHK-, and –TESHAK-) have been studied [113]. The selected His-containing peptides are models of histone H2A and the complexation study may elucidate toxicity of metals [113]. For example, Ni(II) binding to bioligands may lead to oxidative degradation of biomolecules via reactions with $^{\bullet}OH$ radicals. The studied peptides interacted strongly with metal ions, particularly Cu(II) and Ni(II) ions, which, on reactions with H_2O_2, generate reactive oxygen species, which can efficiently oxidize biomolecules such as 2′-deoxyguanosine [114].

Binary and ternary complexes of Cu(II), Zn(II), and Ni(II) (M) with dicarboxylic amino acids (Asp and Glu, B) and adenosine-5′-triphosphate (A) as ligands have been reported due to the importance of M–A–B interactions in biological reactions [115]. Stability of complexes, formed from these interactions, may influence reactions of metal centers and proteins with reactive species. Potentiometric titration curves suggested 1:1 and 1:2 binary complexes. Tertiary complexes were in a molar ratio of 1:1:1 (MAB), in which ATP was the primary ligand and Asp or Glu was the secondary ligand (e.g., M(II)ATP(Asp)). The stability of ternary complexes was of the order of Cu(II) > Ni(II) > Zn(II). Furthermore, ternary complexes containing Asp were more stable than those with Glu. The formation of binary and ternary complexes of Cu(II) and Ni(II) occurred with bicine (*N*,*N*′-bis(2-hydroxyethyl) glycine; zwitterionic buffer) and selected amino acids containing mono- and dicarboxylic acids (Gly, α-Ala, β-Ala, Val, Leu, Asp, Glu, Asn, and Phe) in aqueous solution because of function of ternary coordination in biological processes. Several ligands tend to compete for metals in biological fluids. The studied system may therefore mimic biological reactions (enzyme–metal ion–buffer interactions) [116].

Metal complexation with GSH (L) has been evaluated due to importance of metal–GSH interactions in living systems [16, 117, 118]. GSH is a major metal-binding ligand in cells, and the complexes may serve as carriers to metal-dependent proteins [17]. For example, GSH reduces first Cu(II) ion to Cu(I) and then forms a Cu(I)–$[GSH]_2$ complex, which has been characterized by ^{1}H-NMR and electron paramagnetic resonance (EPR) techniques [119]. This complex may react with O_2 to ultimately form superoxide radicals [119]. The generation of hydroxyl radicals from the reduction and release of iron from ferritin by the Cu(I)–$[GSH]_2$ complex has also been demonstrated [18]. The interaction of copper with GSSG (L), a primary oxidized product of GSH, has also been studied due to the importance of Cu(I)/Cu(II)–glutathione system in operation of enzymes (e.g., glutathione reductase and glutathione peroxidase) and in active transport of amino acids (γ-glutamyl cycle) [117, 120, 121]. In the structure elucidation of complexes, $CuLH_2$ and $CuLH^-$, the complexes contain two isomers and one of the copper atoms is bound solely to the carbonyl or carboxylate groups in the Cu_2L and $Cu_3L_2^{3-}$ complexes [121]. Structural information of the complexes may provide insight into their participation in the redox cycle of oxygen [18].

The Ni(II) complexes, NiHL, $Ni_2L_2^{2-}$, $NiHL_2^{3-}$, NiL_2^{4-}, and $NiH_{-1}L_2^{5-}$ have been obtained in the pH range of 6–12 when fourfold GSH in excess was added to Ni(II) [117]. The structures of the complexes were characterized spectroscopically. These complexes of Ni(II) were effective in causing damage to DNA in the presence of H_2O_2 [31]. A study on the complexation of Zn(II) and Ni(II) with an important intracellular NO carrier, nitrosoglutathione (GSNO, L′) [122], showed the formation of the ML′ and ML_2' complexes [123]. Zn(II) increased the stability of GSNO in the buffered solution at pH 7.4. However, Ni(II) ion destabilized GSNO, which depended on the concentration of NiL. This ability of Ni(II) to damage GSNO may ultimately result in loss of cellular redox signaling [123]. In contrast, stable complexes of GSNO with Zn(II) may provide some protection to GSNO from the reactive species.

Zn(II) is not redox-active in biological environment but becomes active in coordination environment when complexed with thiolate ligand. Furthermore, the coordination environment of the complexes critically control the availability of Zn [75]. The speciation study on Zn(II)–GSH (H_3G) and Zn(II)-*N*-acetylcysteineglycine (NaACCG, H_2L) suggested only mononuclear species, [ZnL], $[ZnL_2^{2-}]$, $[ZnL_3]^{4-}$, $[ZnL_2H_{-1}]^{3-}$, $[ZnL_2H_{-1}]^{3-}$, and $[ZnL_2H_{-2}]^{4-}$ [118]. However, nine species in the Zn(II)–GSH system were reported (Fig. 2.6a), which include both mononuclear and binuclear species under the

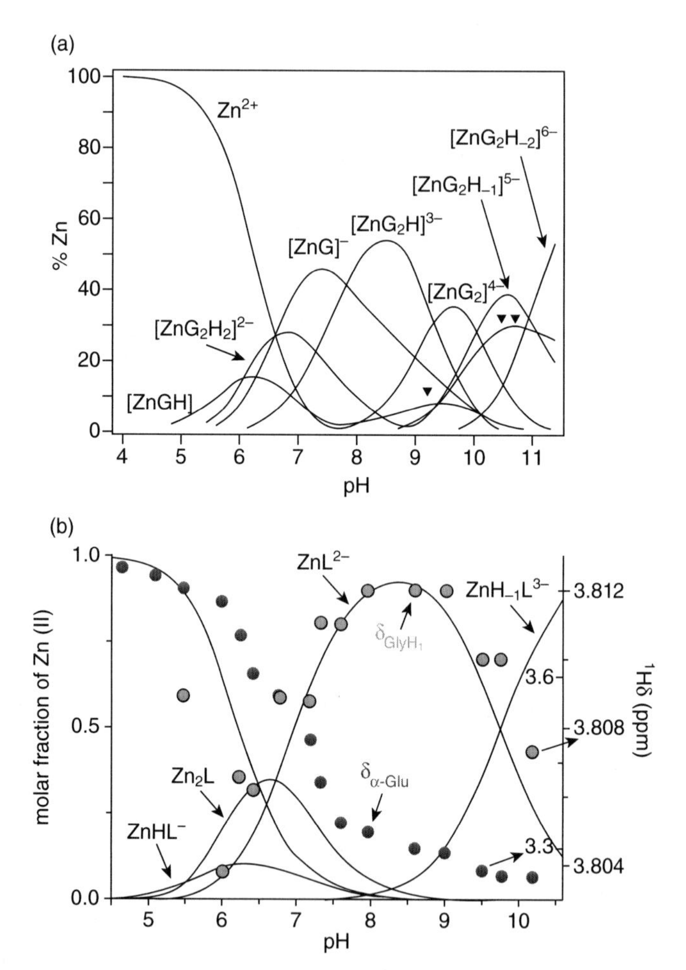

Figure 2.6. (a) Distribution diagram for the Zn^{2+}/GSH (H_3G) system (Zn:G = 1.0:1.81, $C_{Zn} = 1.6 \times 10^{-3}$ M): ▼ = $[Zn_2L_2H_{-1}]^{3-}$; ▼▼ = $[Zn_2L_2H_{-2}]^{4-}$ (adapted from Ferretti et al. [118] with the permission of the Elsevier Inc.). (b) The species distribution for Zn(II)–GSSG complexes, calculated for the conditions of NMR experiments (10 mM GSSG, 10 mM Zn(II), 25°C). Chemical shifts of α-Glu and $GlyH_1$ are overlaid for comparison (adapted from Krezel et al. [120] with the permission of the American Chemical Society). See color insert.

stoichiometric molar ratio of 1:2 ([Zn(II)]:[G]). The spectral measurements provided evidence of the formation of dinuclear complexes. The species, [ZnG], $[ZnG_2H_2]^{2-}$, and $[ZnG_2H]^{3-}$, contributed to the total Zn at pH 7.4. In the proposed structure of [ZnG], $[ZnG_2H_2]^{2-}$, Zn(II) coordinated to the thiolate of cysteine in the HG^{2-} ligand. Recently, Zn(II) complexation with GSSG and its nine analogues with C-terminal modifications has been studied using potentiometric and NMR spectroscopic techniques [120]. Four different complex species, three monomeric ($ZnHL^-$, ZnL^{2-}, and $ZnH_{-1}L^{3-}$) and a bimetallic (Zn_2L), were shown at different pH values (Fig. 2.6b). The major species present at the physiological pH is ZnL_2^- ($[Zn-GSSG]_2^-$) (Fig. 2.6b). Also, the $GlyH_1$ proton was sensitive to the formation of the ZnL^{2-} complex. The role of the Glu residue in the complex formation was observed when comparing the chemical shift profiles with the speciation of the complexes (Fig. 2.6b). The removal equivalence of β-Glu protons in the complex formation increased the signal separation of β-Cys protons. In summary, binding of Zn^{2+} with thiol ligands may be important in intracellular Zn(II) transport and storage.

In summary, protonation of amino acid side chains of proteins vary with pH, temperature, ionic strength, and polarity of the system. Understanding the influence of these variables on equilibrium provides distribution of protonated/unprotonated species under biological environment, which can then be used to evaluate species-specific reaction rates of redox reactions in order to learn the mechanism of damage to proteins caused by reactive intermediates (Chapters 3–6). Furthermore, information on metal binding to bioligands may give an insight on the role of metals in generating reactive species and in affecting rates of redox reactions.

REFERENCES

1 Huxtable, R.J. *Biochemistry of Sulfur*. In Anonymous. Plenum Press, New York, 1986.

2 Sies, H., Brigelius, R., and Akerboom, T.P.M. Intrahepatic glutathione status. In *Function of Glutathione: Biochemical, Physiological, Toxicological, and Clinical Aspects*. A. Larsson, S. Orrenius, A. Holmgren, and B. Mannervik, eds. Raven Press, New York, 1983, pp. 51–64.

3 Fazary, A.E., Mohamed, A.F., and Lebedeva, N.S. Protonation equilibria studies of the standard α-amino acids in $NaNO_3$ solutions in water and in mixtures of water and dioxane. *J. Chem. Thermodyn.* 2006, *38*, 1467–1473.

4 Blokzijl, W. and Engberts, J.B.F.N. Hydrophobic effects. Opinions and facts. *Angew. Chem. Int. Ed.* 1993, *32*, 1545–1579.

5 Englander, S.W., Mayne, L., Bai, Y., and Sosnick, T.R. Hydrogen exchange: the modem legacy of Linderstrom-Lang. *Protein Sci.* 1997, *6*, 1101–1109.

6 Percy, A.J., Rey, M., Burns, K.M., and Schriemer, D.C. Probing protein interactions with hydrogen/deuterium exchange and mass spectrometry—a review. *Anal. Chim. Acta* 2012, *721*, 7–21.

7 Hvidt, A. and Linderstrøm-Lang, K. Exchange of hydrogen atoms in insulin with deuterium atoms in aqueous solutions. *Biochim. Biophys. Acta* 1954, *14*, 574–575.

8 Hvidt, A. and Nielsen, S.O. Hydrogen exchange in proteins. *Adv. Protein Chem.* 1966, *21*, 287–386.

9 Englander, S.W. and Kallenbach, N.R. Hydrogen exchange and structural dynamics of proteins and nucleic acids. *Q. Rev. Biophys.* 1983, *16*, 521–655.

10 Engen, J.R. Analysis of protein conformation and dynamics by hydrogen/deuterium exchange MS. *Anal. Chem.* 2009, *81*, 7870–7875.

11 Konermann, L., Pan, J., and Liu, Y.H. Hydrogen exchange mass spectrometry for studying protein structure and dynamics. *Chem. Soc. Rev.* 2011, *40*, 1224–1234.

12 Englander, S.W., Downer, N.W., and Teitelbaum, H. Hydrogen exchange. *Annu. Rev. Biochem.* 1972, *41*, 903–924.

13 Konermann, L., Stocks, B.B., Pan, Y., and Tong, X. Mass spectrometry combined with oxidative labeling for exploring protein structure and folding. *Mass Spectrom. Rev.* 2010, *29*, 651–667.

14 Pan, Y., Brown, L., and Konermann, L. Hydrogen/deuterium exchange mass spectrometry and optical spectroscopy as complementary tools for studying the structure and dynamics of a membrane protein. *Int. J. Mass Spectrom.* 2011, *302*, 3–11.

15 Hebling, C.M., Morgan, C.R., Stafford, D.W., Jorgenson, J.W., Rand, K.D., and Engen, J.R. Conformational analysis of membrane proteins in phospholipid bilayer nanodiscs by hydrogen exchange mass spectrometry. *Anal. Chem.* 2010, *82*, 5415–5419.

16 Speisky, H., Gómez, M., Burgos-Bravo, F., López-Alarcón, C., Jullian, C., Olea-Azar, C., and Aliaga, M.E. Generation of superoxide radicals by copper-glutathione complexes: redox-consequences associated with their interaction with reduced glutathione. *Bioorg. Med. Chem* 2009, *17*, 1803–1810.

17 Aliaga, M.E., López-Alarcón, C., Barriga, G., Olea-Azar, C., and Speisky, H. Redox-active complexes formed during the interaction between glutathione and mercury and/or copper ions. *J. Inorg. Biochem.* 2010, *104*, 1084–1090.

18 Aliaga, M.E., Carrasco-Pozo, C., López-Alarcón, C., Olea-Azar, C., and Speisky, H. Superoxide-dependent reduction of free Fe^{3+} and release of Fe^{2+} from ferritin by the physiologically-occurring Cu(I)-glutathione complex. *Bioorg. Med. Chem.* 2011, *19*, 534–541.

19 Leed, M.G.D., Wolkow, N., Pham, D.M., Daniel, C.L., Dunaief, J.L., and Franz, K.J. Prochelators triggered by hydrogen peroxide provide hexadentate iron coordination to impede oxidative stress. *J. Inorg. Biochem.* 2011, *105*, 1161–1172.

20 Joyner, J.C., Reichfield, J., and Cowan, J.A. Factors influencing the DNA nuclease activity of iron, cobalt, nickel, and copper chelates. *J. Am. Chem. Soc.* 2011, *133*, 15613–15626.

21 Chandrasekaran, P., Chanta, S., Stieber, E., Collins, T.J., Que, J.L., Neese, F., and DeBeer, S. Prediction of high-valent iron K-edge absorption spectra by time-dependent density functional theory. *Dalton Trans.* 2011, *40*, 11070–11079.

22 Noszál, B. and Szakács, Z. Microscopic protonation equilibria of oxidized glutathione. *J. Phys. Chem. B* 2003, *107*, 5074–5080.

23 Szakács, Z., Kraszni, M., and Noszál, B. Determination of microscopic acid-base parameters from NMR-pH titrations. *Anal. Bioanal. Chem.* 2004, *378*, 1428–1448.

24 Canel, E., Gültepe, A., Dogan, A., and Kiliç, E. The determination of protonation constants of some amino acids and their esters by potentiometry in different media. *J. Solut. Chem.* 2006, *35*, 5–19.

25 Daldrup, J.-G., Held, C., Ruether, F., Schembecker, G., and Sadowski, G. Measurement and modeling solubility of aqueous multisolute amino-acid solutions. *Ind. Eng. Chem. Res.* 2010, *49*, 1395–1401.

26 Kiani, F., Rostami, A.A., Sharifi, S., Bahadori, A., and Chaichi, M.J. Determination of acidic dissociation constants of glycine, valine, phenylalanine, glycylvaline, and glycylphenylalanine in water using ab initio methods. *J. Chem. Eng. Data* 2010, *55*, 2732–2740.

27 Grosse Daldrup, J.-B., Held, C., Sadowski, G., and Schembecker, G. Modeling pH and solubilities in aqueous multisolute amino acid solutions. *Ind. Eng. Chem. Res.* 2011, *50*, 3503–3509.

28 Furia, E., Falvo, M., and Porto, R. Solubility and acidic constants of L-cystine in $NaClO_4$ aqueous solutions at 25°C. *J. Chem. Eng. Data* 2009, *54*, 3037–3042.

29 Nelson, K.J., Day, A.E., Zeng, B.-B., King, S.B., and Poole, L.B. Isotope-coded, iodoacetamide-based reagent to determine individual cysteine pK_a values by matrix-assisted laser desorption/ionization time-of-flight mass spectrometry. *Anal. Biochem.* 2008, *375*, 187–195.

30 Orgován, G. and Noszál, B. The complete microspeciation of arginine and citrulline. *J. Pharm. Biomed. Anal.* 2011, *54*, 965–971.

31 Krezel, A. and Bal, W. Structure-function relationships in glutathione and its analogues. *Org. Biomol. Chem.* 2003, *1*, 3885–3890.

32 Nagai, H., Kuwabara, K., and Carta, G. Temperature dependence of the dissociation constants of several amino acids. *J. Chem. Eng. Data* 2008, *53*, 619–627.

33 Hamborg, E.S., Niederer, J.P.M., and Versteeg, G.F. Dissociation constants and thermodynamic properties of amino acids used in CO_2 absorption from (293 to 353) K. *J. Chem. Eng. Data* 2007, *52*, 2491–2502.

34 De Abreu, H.A., De Almeida, W.B., and Duarte, H.A. pK_a calculation of poliprotic acid: histamine. *Chem. Phys. Lett.* 2004, *383*, 47–52.

35 Hajjar, E., Dejaegere, A., and Reuter, N. Challenges in pK_a predictions for proteins: the case of Asp213 in human proteinase 3. *J. Phys. Chem. A* 2009, *113*, 11783–11792.

36 Lee, A.C. and Crippen, G.M. Predicting pK_a. *J. Chem. Inf. Model.* 2009, *49*, 2013–2033.

37 Crea, P., De Robertis, A., De Stefano, C., Milea, D., and Sammartano, S. Modeling the dependence on medium and ionic strength of glutathione acid-base behavior in $LiCl_{aq}$, $NaCl_{aq}$, KCl_{aq}, $RbCl_{aq}$, $CsCl_{aq}$, $(CH_3)_4NCl_{aq}$, and $(C_2H_5)_4NI_{aq}$. *J. Chem. Eng. Data* 2007, *52*, 1028–1036.

38 Crea, P., De Stefano, C., Kambarami, M., Millero, F.J., and Sharma, V.K. Effect of ionic strength and temperature on the protonation of oxidized glutathione. *J. Solut. Chem.* 2008, *37*, 1245–1259.

39 Sharma, V.K., Moulin, A., Millero, F.J., and De Stefano, C. Dissociation constants of protonated cysteine species in seawater media. *Mar. Chem.* 2006, *99*, 52–61.

40 Sharma, V.K., Casteran, F., Millero, F.J., and De Stefano, C. Dissociation constants of protonated cysteine species in NaCl media. *J. Solut. Chem.* 2002, *31*, 783–792.

41 Sharma, V.K., Zinger, A., Millero, F.J., and De Stefano, C. Dissociation constants of protonated methionine species in NaCl media. *Biophys. Chem.* 2003, *105*, 79–87.

42 Martell, A.E. and Smith, R.M. *Critical Stability Constants*. Plenum Press, New York, 1974, Vol. 1.

43 Fiol, S., Brandariz, I., and de Vicente, M.S. The protonation constants of glycine in artificial seawater at 25°C. *Mar. Chem.* 1995, *49*, 215–219.

44 Fiol, S., Brandariz, I., Herrero, R.F., Vilariño, T., and Sastre De Vicente, M.E. Protonation constants of amino acids in artificial sea water at 25°C. *J. Chem. Eng. Data* 1995, *40*, 117–119.

45 Demirelli, H. and Köseoĝlu, F. Equilibrium studies of Schiff bases and their complexes with Ni(II), Cu(II) and Zn(II) derived from salicylaldehyde and some α-amino acids. *J. Solut. Chem.* 2005, *34*, 561–577.

46 Garrison, W.M. Reaction mechanisms in the radiolysis of peptides, polypeptides, and proteins. *Chem. Rev.* 1987, *87*, 381–398.

47 Xu, G. and Chance, M.R. Hydroxyl radical-mediated modification of proteins as probes for structural proteomics. *Chem. Rev.* 2007, *107*, 3514–3543.

48 Doonan, S. *Peptides and Proteins*. John Wiley & Sons, New York, 2002, pp. 7–9.

49 Schafer, F.Q. and Buettner, G.R. Redox environment of the cell as viewed through the redox state of the glutathione disulfide/glutathione couple. *Free Radic. Biol. Med.* 2001, *30*, 1191–1212.

50 Borst, C.L., Grzegorczyk, D.S., Strand, S.J., and Carta, G. Temperature effects on equilibrium and mass transfer of phenylalanine in cation exchangers. *React. Funct. Polym.* 1997, *32*, 25–41.

51 Gillespie, S.E., Oscarson, J.L., Izatt, R.M., Wang, P., Renuncio, J.A.R., and Pando, C. Thermodynamic quantities for the protonation of amino acid amino groups from 323.15 to 398.15 K. *J. Solut. Chem.* 1995, *24*, 1219–1247.

52 Izatt, R.M., Oscarson, J.L., Gillespie, S.E., Grimsrud, H., Renuncio, J.A.R., and Pando, C. Effect of temperature and pressure on the protonation of glycine. *Biophys. J.* 1992, *61*, 1394–1401.

53 Wang, P., Oscarson, J.L., Gillespie, S.E., Izatt, R.M., and Cao, H. Thermodynamics of protonation of amino acid carboxylate groups from 50 to 125°C. *J. Solut. Chem.* 1996, *25*, 243–266.

54 Pałecz, B., Dunal, J., and Waliszewski, D. Enthalpic interaction coefficients of several L-α-amino acids in aqueous sodium chloride solutions at 298.15 K. *J. Chem. Eng. Data* 2010, *55*, 5216–5218.

55 Crea, P., De Stefano, C., Millero, F.J., Sammartano, S., and Sharma, V.K. Dissociation constants of protonated oxidized glutathione in seawater media at different salinities. *Aquat. Geochem.* 2010, *16*, 447–466.

56 Sharma, V.K., Millero, F.J., De Stefano, C., and Crea, P. Dissociation constants of protonated methionine species in seawater media. *Mar. Chem.* 2007, *106*, 463–470.

57 Brandariz, I., Castro, P., Montes, M., Penedo, F., and Sastre de Vicente, M.E. Equilibrium constants of triethanolamine in major seawater salts. *Mar. Chem.* 2006, *102*, 291–299.

58 Xu, X., Pinho, S.P., and Macedo, E.A. Activity coefficient and solubility of amino acids in water by the modified Wilson model. *Ind. Eng. Chem. Res.* 2004, *43*, 3200–3204.

59 Chung, Y.-M. and Vera, J.H. Activity of the electrolyte and the amino acid in the systems water + DL,α-aminobutyric acid + NaCl, +NaBr, +KCl, and +KBr at 298.2 K. *Fluid Phase Equilib.* 2002, *203*, 99–110.

60 Chung, Y.-M. and Vera, J.H. Activity coefficients of the peptide and the electrolyte in ternary systems water + glycylglycine + NaCl, +NaBr, +KCl and +KBr at 298.2 K. *Biophys. Chem.* 2001, *92*, 77–88.

61 Mortazavi-Manesh, S., Ghotbi, C., and Taghikhani, V. A new model for predicting activity coefficients in aqueous solutions of amino acids and peptides. *J. Chem. Thermodyn.* 2003, *35*, 101–112.

62 Soto-Campos, A.M., Khoshkbarchi, M.K., and Vera, J.H. Activity coefficients of the electrolyte and the amino acid in water + $NaNO_3$ + glycine and water + NaCl + DL-methionine systems at 298.15 K. *Biophys. Chem.* 1997, *67*, 97–105.

63 Khoshkbarchi, M.K. and Vera, J.H. Measurement and modeling of activities of amino acids in aqueous salt systems. *AIChE J.* 1996, *42*, 2354–2363.

64 Khoshkbarchi, M.K. and Vera, J.H. A perturbed hard-sphere model with mean spherical approximation for the activity coefficients of amino acids in aqueous electrolyte solutions. *Ind. Eng. Chem. Res.* 1996, *35*, 4755–4766.

65 Khoshkbarchi, M.K., Soto-Campos, A.M., and Vera, J.H. Interactions of DL-serine and L-serine with NaCl and KCl in aqueous solutions. *J. Solut. Chem.* 1997, *26*, 941–955.

66 Held, C., Cameretti, L.F., and Sadowski, G. Measuring and modeling activity coefficients in aqueous amino-acid solutions. *Ind. Eng. Chem. Res.* 2011, *50*, 131–141.

67 Davies, C.W. *Ion Association*. Butterworths, London, 1962.

68 Debye, P. and Huckel, E. Theory of electrolytes. I. Lowering of freezing point and related phenomena. *Z. Phys.* 1923, *24*, 185–206.

69 Akerlof, G.C. and Oshry, H.I. The dielectric constant of water at high temperatures and in equilibrium with its vapor. *J. Am. Chem. Soc.* 1950, *72*, 2844–2847.

70 Gharib, F., Farajtabar, A., Farahani, A.M., and Bahmani, F. Solvent effects on protonation constants of tryptophan in some aqueous aliphatic alcohol solutions. *J. Chem. Eng. Data* 2010, *55*, 327–332.

71 Dogan, A. and Kiliç, E. Tautomeric and microscopic protonation equilibria of some α-amino acids. *Anal. Biochem.* 2007, *365*, 7–13.

72 Pillai, L., Boss, R.D., and Greenberg, M.S. On the role of solvent in complexation equilibiria. II. The acid-base chemistry of some sulfhydryl and ammonium-containing amino acids in water-acetonitrile mixed solvents. *J. Solut. Chem.* 1979, *8*, 635–646.

73 Timári, S., Cerea, R., and Várnagy, K. Characterization of CuZnSOD model complexes from a redox point of view: redox properties of copper(II) complexes of imidazole containing ligands. *J. Inorg. Biochem.* 2011, *105*, 1009–1017.

74 Bastug, A.S., Goz, S.E., Talman, Y., Gokturk, S., Asil, E., and Caliskan, E. Formation constants and coordination thermodynamics for binary complexes of Cu(II) and some α-amino acids in aqueous solution. *J. Coord. Chem.* 2011, *64*, 281–292.

75 Maret, W. and Li, Y. Coordination dynamics of zinc in proteins. *Chem. Rev.* 2009, *109*, 4682–4707.

76 Li, N.C., Doody, B.E., and White, J.M. Some metal complexes of glycine peptides, histidine and related substances. *J. Am. Chem. Soc.* 1957, *79*, 5859–5863.

77 Tanaka, N., Kolthoff, I.M., and Stricks, W. Iron-cysteinate complexes. *J. Am. Chem. Soc.* 1955, *77*, 1996–2004.

78 White, J.M., Manning, R.A., and Li, N.C. Metal interaction with sulfur-containing amino acids. II. Nickel and copper(II) complexes. *J. Am. Chem. Soc.* 1956, *78*, 2367–2370.

79 Li, N.C. and Manning, R.A. Some metal complexes of sulfur-containing amino acids. *J. Am. Chem. Soc.* 1955, *77*, 5225–5228.

80 Gergely, A. and Nagypál, I. Studies on transition-metal-peptide complexes. Part 1. Equilibrium and thermochemical study of the copper(II) complexes of glycylglycine, glycyl-DL-α-alanine, DL-α-alanylglycine, and DL-α-alanyl-DL-α-alanine. *J. Chem. Soc. Dalton Trans.* 1977, 1104–1108.

81 Nagypál, I. and Gergely, A. Studies on transition-metal-peptide complexes. Part 2. Equilibrium study of the mixed complexes of copper(II) with aliphatic dipeptides and amino-acids. *J. Chem. Soc. Dalton Trans.* 1977, 1109–1111.

82 Joseph, J., Nagashri, K., and Janaki, G.B. Novel metal based anti-tuberculosis agent: synthesis, characterization, catalytic and pharmacological activities of copper complexes. *Eur. J. Med. Chem.* 2012, *49*, 151–163.

83 Duncan, C. and White, A.R. Copper complexes as therapeutic agents. *Metallomics* 2012, *4*, 127–138.

84 Wood, B.A. and Feldmann, J. Quantification of phytochelatins and their metal(loid) complexes: critical assessment of current analytical methodology. *Anal. Bioanal. Chem.* 2012, *402*, 3299–3309.

85 Maret, W. Zinc and sulfur: a critical biological partnership. *Biochemistry* 2004, *43*, 3301–3309.

86 Kowalski, J.M. and Bennett, B. Spin Hamiltonian parameters for Cu(II)-prion peptide complexes from l-band electron paramagnetic resonance spectroscopy. *J. Am. Chem. Soc.* 2011, *133*, 1814–1823.

87 Karambelkar, V.V., Xiao, C., Zhang, Y., Narducci Sarjeant, A.A., and Goldberg, D.P. Geometric preferences in iron(II) and zinc(II) model complexes of peptide deformylase. *Inorg. Chem.* 2006, *45*, 1409–1411.

88 Tainer, J.A., Getzoff, E.D., Richardson, J.S., and Richardson, D.C. Structure and mechanism of copper, zinc superoxide dismutase. *Nature* 1983, *306*, 284–287.

89 Pappalardo, G., Impellizzeri, G., and Campagna, T. Copper(II) binding of prion protein's octarepeat model peptides. *Inorganica Chim. Acta* 2004, *357*, 185–194.

90 Mylonas, M., Plakatouras, J.C., and Hadjiliadis, N. Interactions of Ni(II) and Cu(II) ions with the hydrolysis products of the C-terminal -ESHH- Motif of histone H2A model peptides. Association of the stability of the complexes formed with the cleavage of the -E-S- bond. *Dalton Trans.* 2004, 4152–4160.

91 Yamauchi, O., Odani, A., and Takani, M. Metal-amino acid chemistry. Weak interactions and related functions of side chain groups. *J. Chem. Soc. Dalton Trans.* 2002, 3411–3421.

92 Herold, S. and Koppenol, W.H. Peroxynitritometal complexes. *Coord. Chem. Rev.* 2005, *249*, 499–506.

93 Feaga, H.A., Maduka, R.C., Foster, M.N., and Szalai, V.A. Affinity of Cu^+ for the copper-binding domain of the amyloid-β peptide of Alzheimer's disease. *Inorg. Chem.* 2011, *50*, 1614–1618.

94 Brausam, A., Maigut, J., Meier, R., Szilágyi, P.A., Buschmann, H.-J., Massa, W., Homonnay, Z., and Van Eldik, R. Detailed spectroscopic, thermodynamic, and kinetic studies on the protolytic equilibria of Fe^{III}cydta and the activation of hydrogen peroxide. *Inorg. Chem.* 2009, *48*, 7864–7884.

95 Katona, G., Carpentier, P., Nivière, V., Amara, P., Adam, V., Ohana, J., Tsanov, N., and Bourgeois, D. Raman-assisted crystallography reveals end-on peroxide intermediates in a nonheme iron enzyme. *Science* 2007, *316*, 449–453.

96 Sandy, M. and Butler, A. Microbial iron acquisition: marine and terrestrial siderophores. *Chem. Rev.* 2009, *109*, 4580–4595.

97 Holt, P.D., Reid, R.R., Lewis, B.L., Luther, G.W., III, and Butler, A. Iron(III) coordination chemistry of alterobactin A: a siderophore from the marine bacterium *Alteromonas luteoviolacea*. *Inorg. Chem.* 2005, *44*, 7671–7677.

98 Weisser, J.T., Nilges, M.J., Sever, M.J., and Wilker, J.J. EPR investigation and spectral simulations of iron-catecholate complexes and iron-peptide models of marine adhesive cross-links. *Inorg. Chem.* 2006, *45*, 7736–7747.

99 Furia, E. and Sindona, G. Complexation of L-cystine with metal cations. *J. Chem. Eng. Data* 2010, *55*, 2985–2989.

100 Rittle, J., Younker, J.M., and Green, M.T. Cytochrome P450: the active oxidant and its spectrum. *Inorg. Chem.* 2010, *49*, 3610–3617.

101 Costas, M. Selective C-H oxidation catalyzed by metalloporphyrins. *Coord. Chem. Rev.* 2011, *255*, 2912–2932.

102 Denisov, I.G., Makris, T.M., Sligar, S.G., and Schlichting, I. Structure and chemistry of cytochrome P450. *Chem. Rev.* 2005, *105*, 2253–2277.

103 Green, M.T., Dawson, J.H., and Gray, H.B. Oxoiron(IV) in chloroperoxidase compound II is basic: implications for P450 chemistry. *Science* 2004, *304*, 1653–1656.

104 Que, L., Jr. The road to non-heme oxoferryls and beyond. *Acc. Chem. Res.* 2007, *40*, 493–500.

105 Cavazza, C., Bochot, C., Rousselot-Pailley, P., Carpentier, P., Cherrier, M.V., Martin, L., Marchi-Delapierre, C., Fontecilla-Camps, J.C., and Ménage, S. Crystallographic snapshots of the reaction of aromatic C-H with O_2 catalysed by a protein-bound iron complex. *Nat. Chem.* 2010, *2*, 1069–1076.

106 Rittle, J. and Green, M.T. Cytochrome P450 compound I: capture, characterization, and C-H bond activation kinetics. *Science* 2010, *330*, 933–937.

107 Scepaniak, J.J., Vogel, C.S., Khusniyarov, M.M., Heinemann, F.W., Meyer, K., and Smith, J.M. Synthesis, structure, and reactivity of an iron(V) nitride. *Science* 2011, *331*, 1049–1052.

108 Krebs, C., Fujimori, D.G., Walsh, C.T., and Bollinger, J.M., Jr. Non-heme Fe(IV)-oxo intermediates. *Acc. Chem. Res.* 2007, *40*, 484–492.

109 Ryabov, A.D. and Collins, T.J. Mechanistic considerations on the reactivity of green Fe^{III}-TAML activators of peroxides. *Adv. Inorg. Chem.* 2009, *61*, 471–521.

110 Oliveria, F.T.D., Chanda, A., Benerjee, D., Shan, X., Mondal, S., Que, J.L., Bominaar, E.L., Munck, E., and Collins, T.J. Chemical and spectroscopic evidence for an Fe^{V}-Oxo complex. *Science* 2007, *315*, 835–839.

111 Kaizer, J., Klinker, E.J., Oh, N.Y., Rhode, J., Song, W.J., Stubna, A., Kim, J., Munck, E., Nam, W., and Que, J.L. Nonheme $Fe^{IV}O$ complexes that can oxidize the C-H bonds of cyclohexane at room temperature. *J. Am. Chem. Soc.* 2004, *126*, 472–473.

112 Perry, J.J.P., Shin, D.S., Getzoff, E.D., and Tainer, J.A. The structural biochemistry of the superoxide dismutases. *Biochem. Biophys. Acta—Proteins Proteomics* 2010, *1804*, 245–262.

113 Mylonas, M., Krezel, A., Plakatouras, J.C., Hadjiliadis, N., and Bal, W. Interactions of transition metal ions with His-containing peptide models of histone H2A. *J. Mol. Liq.* 2005, *118*, 119–129.

114 Bal, W., Liang, R., Lukszo, J., Lee, S.-H., Dizdaroglu, M., and Kasprzak, K.S. Ni(II) specifically cleaves the C-terminal tail of the major variant of histone H2A and forms an oxidative damage-mediating complex with the cleaved- off octapeptide. *Chem. Res. Toxicol.* 2000, *13*, 616–624.

115 Aydin, R. and Yirikogullari, A. Potentiometric study on complexation of divalent transition metal ions with amino acids and adenosine 5′-triphosphate. *J. Chem. Eng. Data* 2010, *55*, 4794–4800.

116 Taha, M. and Khalil, M.M. Mixed-ligand complex formation equilibria of cobalt(II), nickel(II), and copper(II) with N,N-bis(2-hydroxyethyl)glycine (bicine) and some amino acids. *J. Chem. Eng. Data* 2005, *50*, 157–163.

117 Krezel, A., Szczepanik, W., Sokołowska, M., Jezowska-Bojczuk, M., and Bal, W. Correlations between complexation modes and redox activities of Ni(II)-GSH complexes. *Chem. Res. Toxicol.* 2003, *16*, 855–864.

118 Ferretti, L., Elviri, L., Pellinghelli, M.A., Predieri, G., and Tegoni, M. Glutathione and N-acetylcysteinylglycine: protonation and Zn^{2+} complexation. *J. Inorg. Biochem.* 2007, *101*, 1442–1456.

119 Ciriolo, M.R., Desideri, A., Paci, M., and Rotilio, G. Reconstitution of Cu,Zn-superoxide dismutase by the Cu(I) glutathione complex. *J. Biol. Chem.* 1990, *265*, 11030–11034.

120 Krezel, A., Wójcik, J., Maciejczyk, M., and Bal, W. Zn(II) complexes of glutathione disulfide: structural basis of elevated stabilities. *Inorg. Chem.* 2011, *50*, 72–85.

121 Shtyrlin, V.G., Zyavkina, Y.I., Ilakin, V.S., Garipov, R.R., and Zakharov, A.V. Structure, stability, and ligand exchange of copper(II) complexes with oxidized glutathione. *J. Inorg. Biochem.* 2005, *99*, 1335–1346.

122 Mayer, B., Pfeiffer, S., Schrammel, A., Koesling, D., Schmidt, K., and Brunner, F. A new pathway of nitric oxide/cyclic GMP signaling involving S-nitrosoglutathione. *J. Biol. Chem.* 1998, *273*, 3264–3270.

123 Krezel, A. and Bal, W. Contrasting effects of metal ions on S-Nitrosoglutathione, related to coordination equilibria: GSNO decomposition assisted by Ni(II) vs stability increase in the presence of Zn(II)and Cd(II). *Chem. Res. Toxicol.* 2004, *17*, 392–403.

124 Armesto, X.L., Canle, M.L., Fernandez, M.I., Garcia, M.V., Rodriguez, S., and Santaballa, J.A. Intracellular oxidation of dipeptides. Very fast halogenation of the amino-terminal residue. *J. Chem. Soc. Perkin Trans.* 2001, *2*, 608–612.

125 Simic, M., Neta, P., and Hayon, E. Selectivity in the reactions of Normal e_{aq}^{-} and OH radicals with simple peptides in aqueous solution. Optical absorption spectra of intermediates. *J. Phys. Chem.* 1970, *92*, 4763–4768.

3

HALOGENATED SPECIES

Myeloperoxidase (MPO), a heme enzyme, when reacted with H_2O_2 and halide (Cl^-, Br^-) or pseudohalide (SCN^-) ions produces hypochlorous acid (HOCl), hypobromous acid (HOBr), and hypothiocyanous acid (HOSCN) (Fig. 3.1) [1]. Initially, the reaction of H_2O_2 with the resting Fe(III) form of the enzyme produces compound I, which undergoes either two successive one-electron reductions, via compound II, to result in a radical (the peroxidase cycle), or a two-electron reduction with Cl^-, Br^-, and SCN^- ions to yield oxidants. Under the physiological concentrations of Cl^- (100–150 mM), Br^- (10–100 μM), and SCN^- (10–100 μM), a combination of MPO and H_2O_2 can produce HOCl, HOBr, and HOSCN [1–6]. The enzyme may also undergo a one-electron reduction with $O_2^{\bullet -}$ to form compound III, which contributes to the superoxide dismutase (SOD) mimetic activity of MPO (Fig. 3.1). The generation of HOCl may also be accomplished by NADPH oxidase [7]. The hypohalous acid and HOSCN are among the species responsible for the antibactericidal activity of neutrophils [8]. However, the possibility of damaging the tissues by such oxidants also exists if they are produced in excessive levels at the inappropriate place or time [9, 10].

The oxidant species exist as acidic and anionic forms [11–13]:

$$HOCl \rightleftharpoons H^+ + OCl^- \quad pK_a = 7.6 \tag{3.1}$$

Oxidation of Amino Acids, Peptides, and Proteins: Kinetics and Mechanism, First Edition. Virender K. Sharma.

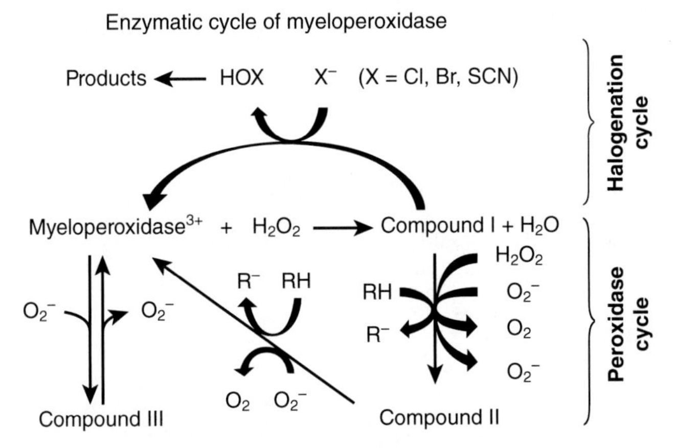

Figure 3.1. The enzymatic cycles of myeloperoxidase (adapted from Davies [1] with the permission of the Society of Free Radical Research Japan).

$$HOBr \rightleftharpoons H^+ + OBr^- \quad pK_a = 8.7 \tag{3.2}$$

$$HOSCN \rightleftharpoons H^+ + OSCN^- \quad pK_a = 5.3. \tag{3.3}$$

Thus, the speciation of the oxidants is pH dependent (Fig. 3.2). At physiological pH, the acidic forms of chlorine and bromine are present as the dominating species. In the case of HOSCN, the anionic form, $OSCN^-$, is the major species.

The reactivity of HOCl and HOBr with amino acids, peptides, and proteins is presented in this chapter. As a result of HOSCN as a significant product of MPO-mediated reactions [14], its reactivity is also presented. The HOCl- and HOBr-mediated oxidation of proteins generates moderately stable halamines [15–18], which further oxidizes compounds (e.g., thiols); therefore, the reactivity of halamines is provided. Finally, chlorine dioxide (ClO_2) has the oxidative ability to modify proteins as well as the ability to perform antibacterial activity against bacteria, viruses, and protozoa, and hence, the oxidations carried out by ClO_2 are summarized.

3.1 HYPOHALOGENS

3.1.1 Hypochlorite

3.1.1.1 Kinetics of HOCl. HOCl has been shown to be the major reactive Cl_2 species in chlorination of a number of compounds [19]. The rate law for the reactivity of Cl_2 with reported inorganic and organic compounds was first-order with respect to $[HOCl]_{Total}$ and first-order with respect to the total concentration

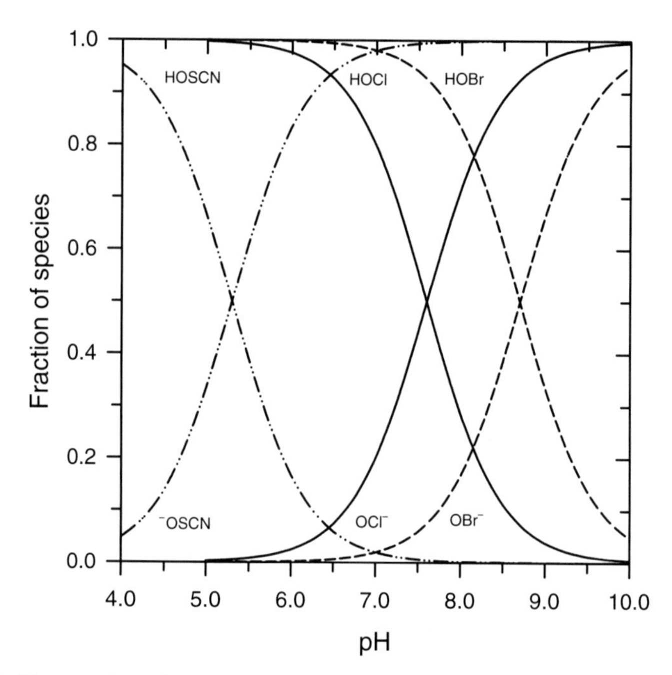

Figure 3.2. The species of aqueous Cl_2, Br_2, and HOSCN as a function of pH at 25°C.

of the compound, which contribute to an overall second-order reaction. The second-order rate constants (k) of HOCl and OCl^- for a particular compound varied significantly with pH in the chlorination reactions [19]. HOCl was generally the major reactive species. The variation of HOCl species as a function of pH is presented in Figure 3.2. The reactivity of Cl_2 with inorganic molecules generally derived from an initial electrophilic attack on HOCl and k for organic compounds varied from $<10^{-1}$ to 10^9/M/s. The possible pathways in the reaction mechanism include oxidation, addition, and electrophilic substitutions [19]. In a recent study, the roles of chlorine monoxide (Cl_2O) and Cl_2 have been demonstrated in the chlorination of molecules that are of environmental interest [20]. The proposed reactive species, Cl_2O and Cl_2, were generally present at low concentrations but were shown to play a greater role than HOCl in the chlorination process.

Table 3.1 reports the second-order rate constants for the chlorination of amino acids and peptides [21–32]. Met and Cys reacted most rapidly with chlorine ($k > 10^7$/M/s). Aromatic acids, with the exception of Tyr, showed reasonable reactivity ($k \sim 10^4$–10^5/M/s). Tyr had reactivity of the order of 10^1/M/s. Ser and Lys reacted similarly. Arg, Asn, and Gln showed the lowest reactivity ($k \sim 10^{-2}$–10^1/M/s). The rate constants for the chlorination of Gly, Ala, β-Ala, Val, and *iso*-Leu were on the order of 10^4–10^5/M/s, similar to the observed rate constants for basic aliphatic amines [19]. The order of

TABLE 3.1. Second-Order Rate Constants for the Chlorination of Amino Acids, Peptides, and Proteins

Compound	pH	T (°C)	k (/M/s)	Reference
Gly	7.0	25	6.4×10^{4a}	[21]
	7.0	25	9.9×10^{4a}	[32]
	6.8	22	7.1×10^{4}	[22]
	7.0	25	6.4×10^{4a}	[25]
	7.2–7.4	10	6.1×10^{4}	[28]
	7.2–7.4	22	1.0×10^{5}	[28]
(*N*,*N*)-di-Me-Gly	7.0	25	1.0×10^{1a}	[23]
Ala	7.0	25	5.6×10^{4a}	[21]
	7.0	25	5.6×10^{4a}	[25]
	7.2–7.4	10	3.3×10^{4}	[28]
	7.2–7.4	22	5.4×10^{4}	[28]
β-Ala	7.0	25	5.0×10^{4a}	[21]
Val	7.2–7.4	10	4.5×10^{4}	[28]
	7.2–7.4	22	7.4×10^{4}	[28]
iso-Leu	7.0	25	5.0×10^{5a}	[31]
Ser	7.2–7.4	10	1.0×10^{5}	[28]
	7.2–7.4	22	1.7×10^{5}	[28]
L-Arg	7.4	22	7.1×10^{5}	[30]
Lys	7.2–7.4	22	5.0×10^{3b}	[28]
Asn	7.2–7.4	22	3.0×10^{-2b}	[28]
Gln	7.2–7.4	22	3.0×10^{-2b}	[28]
Pro	7.0	25	1.6×10^{5a}	[31]
His	7.2–7.4	22	1.0×10^{5b}	[28]
Tyr	7.2–7.4	22	4.4×10^{1b}	[28]
Trp	7.2–7.4	22	1.1×10^{4b}	[28]
Met	7.0	25	6.8×10^{8a}	[26]
	7.2–7.4	22	3.8×10^{7b}	[28]
Cys	7.0	25	6.2×10^{7a}	[26]
	7.2–7.4	22	3.0×10^{7b}	[28]
Cystine	7.2–7.4	22	1.6×10^{5}	[28]
Gly-Gly	7.0	20	3.4×10^{5a}	[25]
	6.8	22	2.0×10^{5}	[22]
	7.0	25	2.2×10^{5a}	[21]
	7.0	25	2.2×10^{5a}	[27]
Gly-Sar	7.0	25	2.5×10^{5a}	[27]
Gly-Ala	7.0	25	3.2×10^{5a}	[27]
Gly-Val	7.0	25	3.5×10^{5a}	[27]
Gly-Ile	7.0	25	3.7×10^{5a}	[27]
Gly-Leu	7.0	25	3.4×10^{5a}	[27]
Gly-Pro	7.0	25	2.5×10^{5a}	[27]
Ala-Gly	7.0	25	3.4×10^{5a}	[27]
Val-Gly	7.0	25	4.8×10^{5a}	[27]
Leu-Ala	7.0	25	6.1×10^{5a}	[27]

(*Continued*)

TABLE 3.1. (*Continued*)

Compound	pH	T (°C)	k (/M/s)	Reference
Pro-Gly	7.0	25	2.7×10^{5a}	[27]
c-Ser-Tyr	7.8	25	4.9×10^{2a}	[29]
GSH	5.0, 7.4, and 9.0	25	$\geq 1 \times 10^{7}$	[24]
Fe^{III}cytc	7.6	25	$>1 \times 10^{5}$	[29]

[a] Calculated.
[b] Reported for modeling the reactivity of proteins.

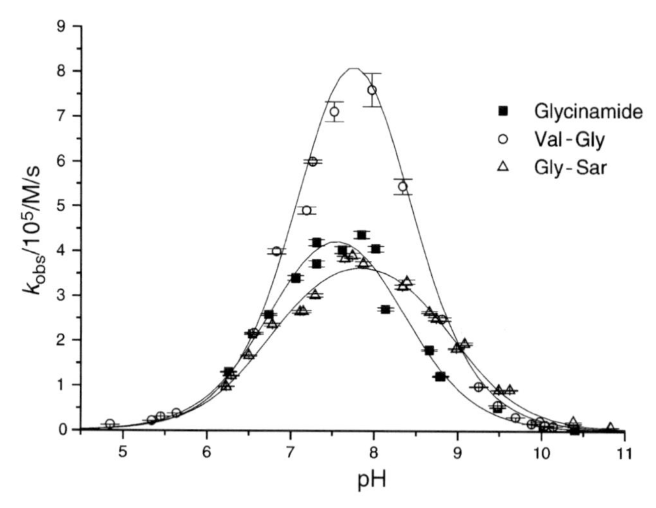

Figure 3.3. Influence of the pH on the rate of chlorination of dipeptides. Experimental conditions: [chlorinating agent] = [dipeptide] = 1.2 mM; I (NaCl) = 0.50 M; T = 298.15 K (adapted from Armesto et al. [27] with the permission of the Royal Society of Chemistry).

reactivity is Met > Cys >> Cystine ~ His ~ α-amino ~ Arg > Trp > Lys >> Tyr > Gln ~ Asn (Table 3.1).

The second-order rate constants for the chlorination of dipeptides have also been studied [27, 28, 33–37]. The rate constants for some of the dipeptides indicate pH dependence, shown in Figure 3.3 [27]. The maximum rate constants were observed at the arithmetic means of the pK_a values of HOCl and the terminal amino groups of the dipeptide [27]. The rate constants for the studied peptides at neutral pH are summarized in Table 3.1. The rate constants were on the order of 10^5/M/s (Table 3.1). The reactivity of HOCl with numerous model proteins showed the rates were highly dependent on the peptide environment [28]. The reaction of HOCl with cyclic peptides was rapid

(2.0×10^1/M/s per amide group). However, the incorporation of a charged side chain decreased the reactivity roughly by an order of magnitude. Similar charge dependence was observed for the rates of reaction of *N*-acetyl blocked amino acids, di- and tripeptides. The rates were generally much slower ($k = 10^{-3}$–10^{-1}/M/s) [28]. The chlorination rates of the amide bonds in the tripeptide models were greater than those of the blocked amino acids. Generally, chlorination primarily occurs at the side chain (and *N*-terminal α-amino group) rather than at the backbone sites of the protein. The second-order rate constants for the oxidations of glutathione (GSH) and Fe^{III}cytc by HOCl/OCl^- showed rapid reactivity in the acidic to basic pH range (Table 3.1).

3.1.1.2 Products of HOCl Oxidation. Products for the reactions of amines, α-amino groups, and Lys with HOCl have been studied in detail [33]. The reactions yielded unstable monochloramines. In the presence of excess HOCl, dichloroamine species were formed [38, 39]. Recently, a detailed product analysis for the chlorination reaction of Gly in water was studied by applying isotopically enriched (^{13}C and ^{15}N) samples of Gly and ^1H, ^{13}C, and ^{15}N NMR spectroscopy [40]. The results showed the C1 carboxylic acid carbon of Gly was quantitatively converted to CO_2, while the C2 methylene carbon of Gly was transformed to CO_2 and formaldehyde monohydrate. CO_2 was the predominant product at pH 6–9. However, the formaldehyde monohydrate was the likely product under acidic reaction conditions (pH < 6). The proposed mechanism for the chlorination of Gly is provided in Figure 3.4 [40]. Initially, *N*-chloroglycine (I) was obtained when one equivalent of aqueous chlorine reacted with Gly. *N*-Chloroglycine was converted to *N*,*N*-dichloroglycine (II) when more than 1 equiv of aqueous chlorine was used. The formation of *N*-chloromethanimine (III) as a transitory product occurred through the decarboxylation and elimination of HCl of labile *N*,*N*-dichloroglycine (II). Under acidic pH, *N*-chloromethanimine (III) may be hydrated to form *N*-chloroaminomethanol (IV), which is quantitatively transformed to formaldehyde monohydrate (VI). One other possibility to consider is the elimination of the second mole of HCl from *N*-chloromethanimine (III) to likely form cyanide (V), which, upon chlorination, yielded CNCl (VII). The hypochlorite-assisted catalytic hydrolysis of CNCl (VII) formed CO_2, also previously reported [41]. Further hydrolysis of CNCl (VII) and/or decomposition of chloroaminomethanol (IV) in the presence of excess chlorine yielded N_2 and nitrate.

The products of the chlorination of other amino acids have been studied mostly for His and Trp. Product analyses for the reactions of Gln and Asn with HOCl have not been carried out, possibly due to the slow rates of reactions (see Table 3.1). The reaction of L-Arg with HOCl was rapid, and the products, identified by MS, were monochlorinated and dichlorinated adducts of L-Arg [30]. The proposed structure of the chlorinated adducts (Fig. 3.5) showed the formation of chloramine at the guanidine group [30]. An important role of chloramines in endothelial cell dysfunction was suggested because such species

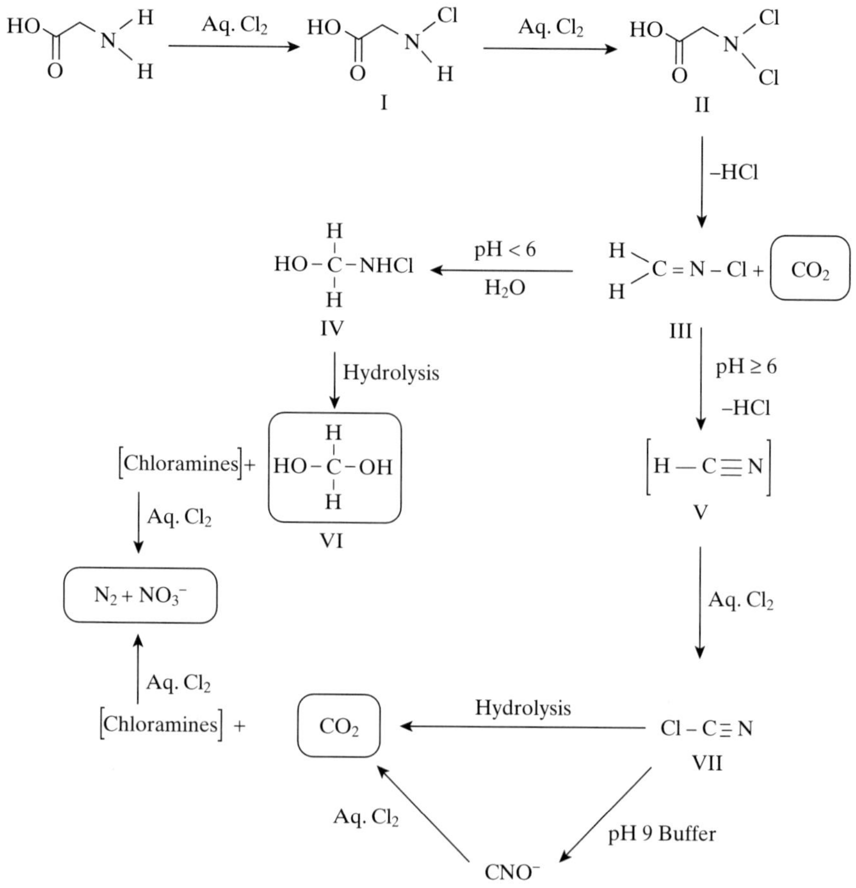

Figure 3.4. Proposed mechanism of glycine chlorination (compounds drawn in boxes are the terminal products; intermediates drawn in brackets were not directly detected by NMR and are only proposed as logical intermediates in the formation of cyanogen chloride (VII), N_2, and nitrate) (adapted from Mehrsheikh et al. [40] with the permission of Elsevier Inc.).

inhibit the activity of endothelial cell nitric oxide synthase and, hence, the formation of $NO^{\bullet}$ [30]. His also reacted rapidly with HOCl to form short-lived chloramine, which then transferred chlorine to other amine groups to produce more stable chloramines [33]. The resulting chloramines may further produce carbonyl products (e.g., 2-oxo-histidine) through either one- or two-electron processes [30].

Identification of products from the oxidation of Trp by HOCl remains unclear. The formation of kynurenine and *N*-formylkynurenine (NFK) has been determined by UV–visible (UV–vis) spectroscopy [42]. The proposed radical-mediated mechanism yielded these products of the reaction. Initially,

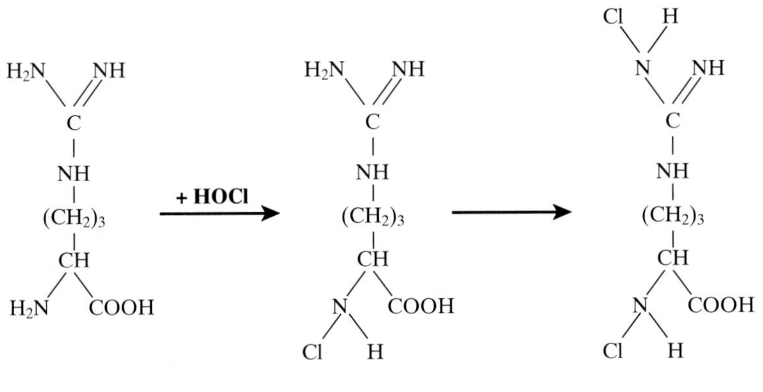

H_2N, NH, C, NH, $(CH_2)_3$, CH, H_2N, COOH

+ HOCl →

H_2N, NH, C, NH, $(CH_2)_3$, CH, N, COOH, Cl, H

→

Cl, H, N, NH, C, NH, $(CH_2)_3$, CH, N, COOH, Cl, H

L-arginine

***N*$^{\alpha}$-chloro-L-arginine**

***N*$^{\alpha}$, *N*$^{\omega}$-dichloro-L-arginine**

Figure 3.5. Proposed structures of chlorinated L-Arg products. (Adapted from [30] with the permission of American Society of Biochemistry and Molecular Biology).

chloramine was formed, which went through either a thermal or metal-catalyzed process to yield radicals. The other Trp side chain then reacted with radicals intermolecularly to produce kynurenine. Recently, the formation of chloramine in the enzymatic chlorination of Trp has also been proposed [43]. The direct oxidation of the side chain of Trp to yield 2-hydroxyindole derivatives of Trp has also been suggested [44–46]. The chlorination products of Tyr were 3-chloro-Tyr and 3.5-dichlorotyrosine [47–50]. The direct attack of HOCl and a chlorine transfer from the initially produced chloramine are the two proposed mechanisms for chlorination of the phenolic side chain of Tyr [33].

The chlorination of thiols, including sulfur-containing amino acids, has been studied extensively [26, 51–54]. Methionine sulfoxide was the reaction product in the oxidation of Met by HOCl. Evidence for the formation of an intermediate, a chlorosulfonium cation, was provided (Eq. 3.4) [26]. The hydrolysis of the resulting cation forms sulfoxide. Oxidation of Met and Tyr residues in myoglobin by HOCl has been observed [55]:

$$\begin{aligned} &CH_3S(CH_2)_2CH(NH_2)COO^- + HOCl \\ &\rightarrow CH_3S^+(Cl)(CH_2)_2CH(NH_2)COO^- + OH^-. \end{aligned} \tag{3.4}$$

The mechanism for the oxidation of Cys and cystine by HOCl is summarized in Figure 3.6 [33]. Similar to Met, the initial formation of sulfenyl chloride was proposed, which was ultimately converted to cysteic acid (RSO_3H) in aqueous solution through the intermediates, sulfenic (RSOH) and sulfinic (RSO_2H) acids [45, 51, 56]. Sulfenyl chlorides played a major role in the production of thiyl radicals in the reaction system, which were detected in the

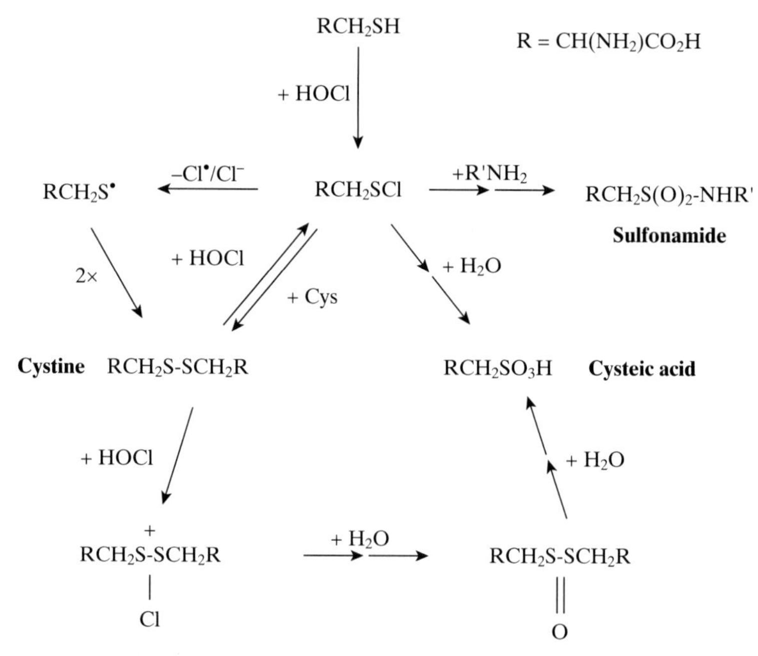

Figure 3.6. Mechanisms and products of chlorination of Cys and cystine (adapted from Hawkins et al. [33] with the permission of Springer America).

HOCl-induced oxidation of thiols [51]. Formation of sulfenyl chloride has also been proposed in the reactions of aryl sulfinates with HOCl ($k \sim 10^9$/M/s) [53]. The sulfenyl chloride intermediates may also react with other Cys molecules to yield cystine (Fig. 3.6). A recent study showed the formation of CySSCy, $CySO_2H$, and $CySO_3H$ in the reaction of a 1:1 molar ratio of CySH with HOCl at pH 11.4 [52]. Production of the cysteine thiosulfonate ester has also been suggested through the reaction of Cys with HOCl under alkaline media [52]. Further reaction of the cysteine thiosulfinate ester with Cys produces cysteine sulfenic acid [52]. *N,N′*-Dichlorocystine has been shown to form in the reaction of CySSCy with HOCl [57]. The oxidation of cystine by HOCl may also take place via a sulfenyl chloride intermediate $RS^+(Cl)SR$, which, upon hydrolysis, forms cysteic acid. Significantly, these intermediates can react with nitrogen centers to form additional products [54, 58–61]. For example, sulfenamide (RSN-) cross-links were formed in the reaction of sulfenyl chlorides with the side-chain amines of Lys. The reactions of sulfenamides with oxygen resulted in sulfinamides (RS(O)N-) and sulfonamides ($RS(O)_2N$-) [61].

Studies on the oxidation of GSH by HOCl showed the formation of three products, glutathione disulfide (GSSG), glutathione thiosulfonate, and glutathione sulfonamide [54, 59]. The treatment of human umbilical vein endothelial cells (HUVEC) with HOCl formed glutathione sulfonamide as the major

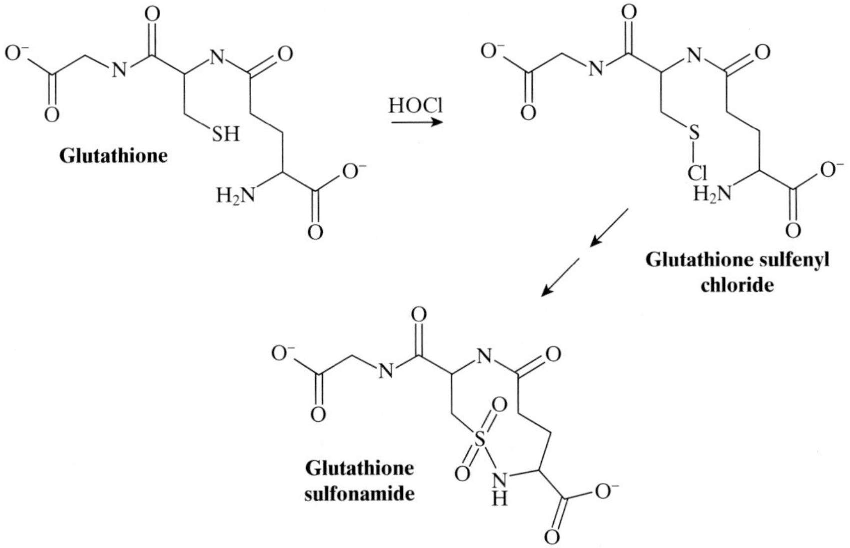

Figure 3.7. Chlorination of dipeptide (adapted from Armesto et al. [27] with the permission of the Royal Society of Chemistry).

product while GSSG and protein-mixed disulfides were determined only in minimal quantities. In the proposed mechanism, the formation of glutathione sulfonamide occurs through a sulfenyl chloride intermediate (Fig. 3.7) [54, 59]. The intramolecular condensation between an intermediate of the Cys side chain and the amino group of the glutamyl side chain was proposed to result in glutathione sulfonamide. This sulfonamide may be a specific marker of HOCl treatment in the biological system.

The oxidation of dipeptides by HOCl occurs through a chlorine atom transfer from the oxygen of HOCl to the terminal amino residue of the peptide (Fig. 3.8). Water molecules directly participate at the transition state. Participation of three water molecules in the oxidation of ammonia by HOCl has been theoretically suggested [62]. The cyclic structure of the transition state permits the synchronous transfer of Cl and H atoms to proceed without the formation of charged species and the reducing charge development on the reaction centers [27]. The determined activation parameters are consistent with the scheme given in Figure 3.8.

The second-order rate constants provided in Table 3.1 have been utilized in computational modeling of the reactions of free amino acids in plasma, apolipoprotein-AI, and human serum albumin (HSA) [28]. The model was successful in predicting which sites the majority of HOCl reacts with at different molar ratios of protein to HOCl. For example, at a low molar ratio of $<4:1$ of HOCl to HSA, the majority of HOCl is consumed by Cys and Met residues.

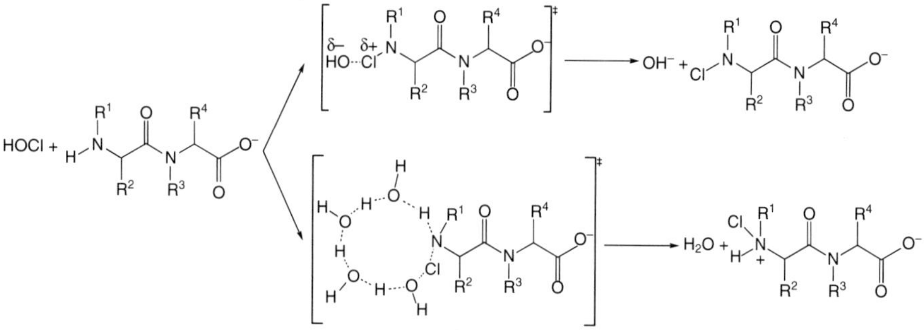

Figure 3.8. Mechanism of formation of glutathione sulfonamide following treatment of GSH with HOCl (adapted from Hawkins et al. [33] with the permission of Springer America).

However, at a larger molar ratio of 8:1 and 17:1, consumption of HOCl by Cys and Met is reduced significantly due to consumption also occurring by cystine, His, and Lys residues. At very high molar ratios of 67:1 and 170:1, consumptions of HOCl would be ~37%, 25%, 25%, and 9% by Lys, cystine, His, and Met residues, respectively. The consumption behavior of HOCl in reaction with Apo-A having no Cys residue would be different [28]. A mass analysis of HSA after treatment with HOCl identified five different Lys modification sites (Lys-130, Lys-257, Lys-438, Lys-499, and Lys-598) [63].

3.1.2 Hypobromite

3.1.2.1 Kinetics of HOBr. The second-order rate constants for reactions of amino acids and amides with HOBr are provided in Table 3.2 [64]. The rate constants for Gly, Ala, and Val were on the order of 10^6/M/s, while their cyclic compounds had much lower rate constants ($k \sim 10^2$–10^3/M/s). The variation in rate constants for various peptide amines was four orders of magnitude, similar to that of HOCl [28]. Such variations in rate constants could be attributed to a combination of charge (neutral vs. charged) and conformational (cyclic vs. linear) effect [28]. The rate constants of the side-chain sulfur-containing amino acids with HOBr are limited. A few studies reported the rate constants in the range of 10^5–10^7/M/s (Table 3.2). Overall, the rate constants varied by eight orders of magnitude and an increase in the order of Gln/Asn < backbone amides < Arg << Tyr ~ Lys < disulfide < α-amino ~ His ~ Met ~ Trp < Cys.

The rate constants for the reactions of amines and amides in amino acids with HOBr were 5–100 times higher than those for HOCl [9]. However, the rates for Cys and Met residues with HOBr were slightly slower than with HOCl. Disulfide bonds reacted similarly with both HOBr and HOCl. The rate constants for the oxidation of Tyr and Trp side chains with HOBr were 5000

TABLE 3.2. Second-Order Rate Constants (with 95% Confidence Limits) at 22°C and pH 7.2–7.5 for Reactions of Hypobromous Acid with Amino Acids, Free α-Amino Groups, and Amides

Substrate	k (/M/s)	Substrate	k (/M/s)
Gly	2.6×10^6	*N*-Acetyl-Cys	1.2×10^7
Cyclo(Gly)$_2$	9.0×10^2	(*N*-Acetyl-Cys)$_2$	3.4×10^5
Ala	1.6×10^6	*N*-Acetyl-Met-OMe	3.6×10^6
Cyclo(Ala)$_2$	2.5×10^2	*N*-Acetyl-Tyr	2.6×10^5
N-Acetyl-Ala	7.0×10^{-2}	*N*-Acetyl-Trp	3.7×10^6
N-Acetyl-Ala-OMe	2.1×10^0	Cyclo(Ser)$_2$	5.5×10^2
Val	1.7×10^6	Cyclo(Asp)$_2$	5.0×10^1
N-α-Acetyl-Lys	3.6×10^5	Propionamide	3.3×10^0
N-Acetyl-Arg-OMe	2.2×10^3	2-Methyl propionamide	1.5×10^0
		Trimethylacetamide	1.5×10^0

Data are taken from Pattison and Davies [64].

and ~450 times higher than with HOCl, respectively. Rates were also faster for HOBr than HOCl for the double bonds of unsaturated fatty acid chain model compounds such as pentenoic acid and sorbate [65].

3.1.2.2 Products of HOBr Oxidation. The reactions of HOBr are usually analogous to those of HOCl. HOBr is reactive with thioether, thiol, disulfide, and amine functions [1]. The oxidation of amino acids and dipeptides by HOBr in alkaline bromine resulted in nitriles with one less carbon atom [66]. The oxidative deamination of nitriles at pH 9.4 of di- and higher peptides yielded *N*-(α-ketoacetyl) peptides. Several studies showed the formation of bromoamines (R-NHBr) and bromamides (R-C(O)-NBr-R′) as major products in the bromination of amino acids, peptides, and proteins [64, 67–69]. Bromoamines are usually less stable ($t_{1/2} < 15$ minutes) than the bromamides ($t_{1/2} > 15$ minutes) [68, 70]. However, the amides of bromamides of Gly and Ala had half-lives of 12 and 35 minutes, respectively. Generally, *N*-bromo species were less stable than corresponding chloro-species [71, 72].

The decay products of *N*-bromo species of the soybean trypsin inhibitor (STI), lysozyme, and bovine serum albumin (BSA) were followed using different analytical techniques such as UV–vis spectroscopy, fluorescence, electron paramagnetic resonance (EPR) spectroscopy with spin trapping, and high-performance liquid chromatography (HPLC) [67, 68]. Radical intermediates in the oxidation of BSA by HOBr were observed. Initially, nitrogen-centered radicals on side-chain and backbone amide groups formed, which readily converted into C-centered radicals through rearrangement reactions. These radicals play an imperative role in the fragmentation of proteins. The final stable products include protein carbonyls, brominated Tyr residues,

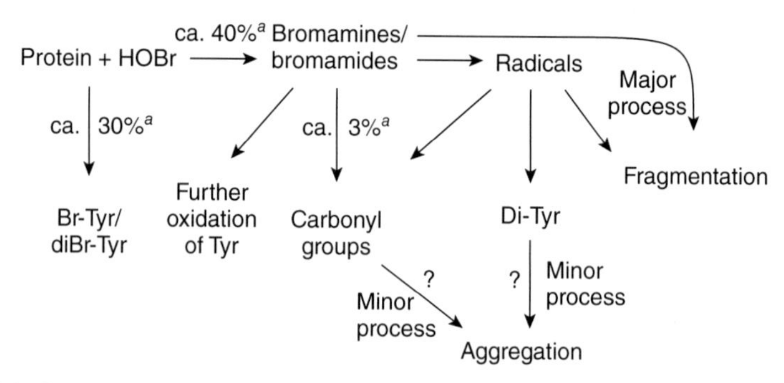

Figure 3.9. Summary of proposed reactions of HOBr with proteins such as BSA. [a] With a 50-fold molar excess HOBr (adapted from Hawkins and Davies [68] with the permission of Elsevier Inc.).

3,4,-dihydroxyphenylalanine (DOPA), and di-Tyr. A scheme for the formation of products is presented in Figure 3.9 [68]. Major products were Br-Tyr and diBr-Tyr, while other products were minor.

Several studies on the kinetics and mechanisms of the oxidation of organosulfur compounds by aqueous bromine and bromate in acidic solutions have also been performed [73–76]. Elementary steps within reactions of the mechanisms were modeled to explain the kinetic traces of oxidation. Oxidations occurred through S-oxygenation pathways. Final products were formed with and without cleavage of the C–S bond of the compound, and hence, products were sulfenic, sulfinic, and cysteic acids as well as the sulfate ion.

3.1.3 Hypothiocyanous

$HOSCN/OSCN^-$ decomposes rapidly at physiological pH to form oxyacids such as cyanosulfurous acid (HO_2SCN) and cyanosulfuric acid (HO_3SCN) (Eqs. 3.5 and 3.6) [77, 78]:

$$2HOSCN \rightarrow HO_2SCN + H^+ + SCN^- \tag{3.5}$$

$$HOSCN + HO_2SCN \rightarrow HO_3SCN + SCN^- + H^+. \tag{3.6}$$

Oxyacid may also be formed from the reaction of H_2O_2 with HOSCN (Eq. 3.7):

$$HOSCN + H_2O_2 \rightarrow HO_2SCN + H_2O. \tag{3.7}$$

The formation of OCN^-, CN^-, and thiocaramate-S-oxide has also been suggested in biological systems (Eqs. 3.8–3.10) [52, 79–83]:

$$HO_2SCN + H_2O_2 \rightarrow OCN^- + H_2SO_3 + H^+ \tag{3.8}$$

$$HO_3SCN + H_2O \rightarrow CN^- + H_2SO_4 + H^+ \tag{3.9}$$

$$OSCN^- + H_2O \rightarrow H_2NC(=O)SO^-. \quad (3.10)$$

The concentrations of HOSCN and $OSCN^-$ were determined using molar absorptivities of 95 and 3870/M/cm at 240 and 220 nm, respectively [77]. An additional spectrum of $OSCN^-$ at 376 nm has also been reported (ε = 26.5/M/cm) [82].

3.1.3.1 Kinetics of HOSCN. The reactions of HOSCN with Cys, Met, cystine, GSSG, Lys, His, and Trp were not detected ($k < 10^3$/M/s) [84]. However, thiol side chains and proteins reacted rapidly [84]. Table 3.3 provides the second-order rate constants for the reaction of HOSCN with thiol residues [84, 85]. Rate constants varied with the structure of the molecule. The rate constants for low-molecular-mass thiols varied from 7.3×10^3/M/s for *N*-acetyl-cysteine at pH 7.4 to 7.7×10^6/M/s for 2-nitro-5-thiobenzoic acid (TNB) at pH 6.0. The rate constants decreased with an increase in pH, which was more noticeable in the reaction of TNB than the reaction of GSH. The results suggest the HOSCN species is much more reactive than the $^-$OSCN species. The rate constants, with the exception of penicillamine, could be correlated with the pK_a of thiols, which had an inverse relationship (Fig. 3.10) [84]. Under the same conditions, anionic thiols are stronger nucleophiles than neutral thiols, and therefore, thiols with a low pK_a were expected to have higher reactivity. Pencillamine is a sterically hindered thiol, which may be the reason for not following the trend.

TABLE 3.3. Second-Order Rate Constants for the Reaction of Thiols with HOSCN in 0.1 M Phosphate at 22°C

Substrate	pH	*k* (/M/s)
5-Thio-2-nitrobenzoic acid	7.4	3.8×10^5
(TNB)	6.7	1.6×10^6
	6.0	7.7×10^6
L-Cys	7.4	7.8×10^4
Cysteamine	7.4	7.1×10^4
N-Acetyl-Cys	7.4	7.3×10^3
L-Cys-methyl ester	7.4	1.6×10^5
Penicillamine	7.4	3.4×10^5
GSH	7.4	2.5×10^4
	6.7	3.1×10^4
	6.0	4.0×10^4
BSA	7.4	7.6×10^4
	7.4	3.1×10^4
	7.4	1.0×10^4
	7.4	1.4×10^4

Data are taken from References 84 and 85.

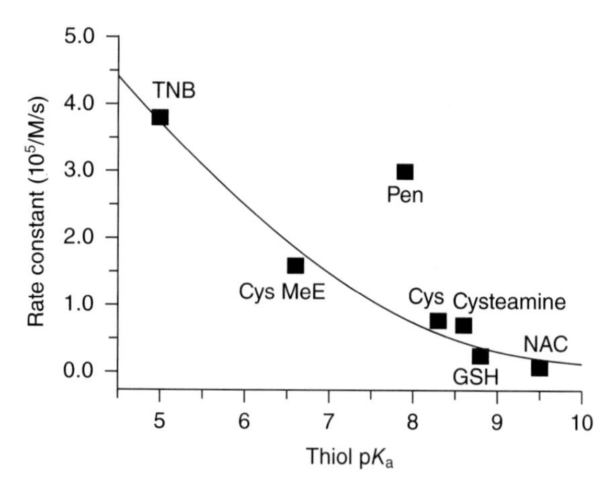

Figure 3.10. The relationship between the rate constants (k, /M/s) and pK_a for the reactions of thiols with HOSCN at pH 7.4 and 22°C. TNB, 5-thio-2-nitrobenzoic acid; Pen, penicillamine; Cys Me E, L-cysteine methyl ester; NAC, *N*-acetylcysteine (adapted from Skaff et al. [84] with the permission of the Biochemical Society).

The seleno-containing amino acids react rapidly with HOSCN, and the determined rate constants, k, were in the range from 2.8×10^3/M/s to 5.8×10^6/M/s for selenomethionine and selenocystamine, respectively [86]. These rate constants are much higher than the reactions of thiols with HOSCN. The selenium atom has more nucleophilicity than the sulfur atom and seleno-compounds would be more easily oxidized by HOSCN. Significantly, the rate constants are ~6000 times higher than the corresponding reactions of H_2O_2 with selenium-containing amino acids. These results have significance in the inhibition of TrxR and GPx enzyme activities [86].

3.1.3.2 Products of HOSCN Oxidation. HOSCN reacted with low-molecular-mass and protein thiols in different cell types, isolated proteins, and biological fluids, such as plasma, to yield unstable sulfenyl thiocyanate (RS-SCN) (Eq. 3.11) [87–90]. The adducts of RS-SCN showed pH dependence stability, and sulfenyl thiocyanates of Cys, GSH, and pencillamine were isolated [91–93]. However, the adducts showed reactivity with other thiols at the physiological pH resulted in the formation of disulfide via possible intermediates such as thiosulfinate esters and sulfenic acids (Eqs. 3.12 and 3.13):

$$\text{HOSCN} + \text{R-SH} \rightarrow \text{RS-SCN} + \text{H}_2\text{O} \quad (3.11)$$

$$\text{RS-SCN} + \text{H}_2\text{O} \rightarrow \text{RS-OH} + \text{SCN}^- + \text{H}^+ \quad (3.12)$$

$$\text{RS-SCN} + \text{R}'\text{-SH} \rightarrow \text{RS-SR} + \text{SCN}^- + \text{H}^+. \quad (3.13)$$

The formation of Cys dimers has also been suggested under acidic conditions [93].

A recent study showed HOSCN attacked the Cys residues present in protein tyrosine phosphatases (PTPs), causing a loss of PTP activity for the isolated enzyme, in cell lysates and intact J774A.1 macrophage-like cells [5]. This ultimately resulted in altered mitogen-activated protein kinase (MAPK) signaling. The HOSCN may also induce the inactivation of thiol-dependent enzymes in which sulfenyl thiocynate and sufenic acid were the intermediates [94]. Overall, HOSCN-mediated oxidation of thiol may play a role as an important mediator of inflammation-induced modulation of cellular signaling and oxidative damage. The role of HOSCN would depend on both the cellular environment and the specific cell type [94, 95]. Further studies are needed to fully understand how HOSCN interacts with cellular membranes and the specific cellular targets.

3.2 HALAMINES

Halamines (RNHCl and RHBr) are key intermediates, formed from the reactions of HOCl and HOBr, respectively, which may also have oxidizing power to induce further reactions in biological systems [16, 17, 71]. The reactions of halamines with nitrogenous compounds are also of interest as a disinfection process in water [32, 96–98]. Halamines undergo different pathways in their reactions such as hydrolysis, halogen transfer, and one-electron reduction. These pathways are described below.

3.2.1 Hydrolysis of Halamines

Chloramines, located at the α-amino site, decompose to unstable intermediates imines, which, upon hydrolysis, form aldehydes and ammonia (Eqs. 3.14 and 3.15) [49, 99, 100]:

$$\text{R-CH}_2\text{-NHCl} \rightarrow \text{R-CH} = \text{NH} + \text{HCl} \tag{3.14}$$

$$\text{R-CH} = \text{NH} + \text{H}_2\text{O} \rightarrow \text{R-CH} = \text{O} + \text{NH}_3. \tag{3.15}$$

Chloramines of the N-terminus of small peptides have shown to form aldehydes [101]. Carbonyls have been detected in the treatment of various proteins with HOCl [102–106]. The resulting ammonia may react with excess HOCl to yield NH_2Cl. Aldehydes may also undergo further reactions with free amino groups of proteins (e.g., Lys side chains) to produce a Schiff base [100, 107].

Reactions of *N*-bromosuccinamide with amino acids and their peptides have been studied in the acidic medium [108–110]. The rates of peptides are lower than those of free amino acids. The roles of the function group and hydrophobicity were suggested. For example, *N*-bromosuccinamide oxidized Val-Pro at a faster rate than the less hydrophobic dipeptides, Ala-Pro and

Val-Gly. Aldehydes as products of the reactions were observed. The suggested mechanism involves oxidative deamination and decarboxylation to form corresponding aldehydes [108–110].

3.2.2 Halogen Transfer by Halamines

The second-order rate constant determined for chlorine transfer reactions of NH_2Cl with amino acids are summarized in Table 3.4 [29, 60, 111–113]. The rate constants were about five orders of magnitude smaller than the corresponding constants for HOCl (see Table 3.1). Only a limited number of rate constants for NH_2Br are available (Table 3.4). The rate constants for the reactions with Met and GSH with NH_2Br were much faster than those with NH_2Cl. Based on the reaction of halamines with GSH, NH_2Br was suggested to be more selective than NH_2Cl [114]. This selectivity was also observed in the reaction of NH_2Br with ascorbic acid, which was 60 times faster than the corresponding reaction of NH_2Cl [28, 64].

Reactions of chloramines, formed on the biological molecules, with possible targets of proteins have been studied [60, 111, 114–118]. The chloramines formed on the imidazole groups of His and nucleobases (thymidine monophosphate [TMP], guanosine monophosphate [GMP], and inosine) were usually more reactive than the chloramines formed on primary amines [119]. The second-order rate constants of chloramines on amino acids and small peptides with amines and thiols are summarized in Table 3.4. Rate constants for Gly, His, Gly, and *N*-α-acetyl-Lys chloramines with Met, Cys, and GSH were more than five orders of magnitude lower than for the corresponding reactions of HOCl (see Table 3.1). The rate constants of taurine chloramine with thiols varied from 3.9×10^1/M/s to 5.6×10^2/M/s. Significantly, these rate constants had a strong inverse relationship with the pK_a values of thiols [60]. Chloramines of Gly and *N*-α-acetyl-Lys also had a similar relationship, but their rate constants were higher than for chloramines of Tau. The reactivity of His chloramines with Cys and *N*-Ac-Cys was higher than those of Gly, taurine, or *N*-α-acetyl-Lys chloramines (Table 3.4). The general trend of the reactivity of these chloramines species was histamine > *N*-acetyl-Lys > Gly > taurine. An increase in negative charge in the vicinity of chloramines decreased the rate. As expected, neutral Met thioether did not show such dependence on charge; therefore, the order of decreasing reactivity was Gly > histamine > *N*-α-acetyl-Lys > taurine (Table 3.4). The kinetics results summarized in Table 3.4 suggest the possible oxidation of thiols by chloramines *in vivo* may cause selective inactivation of the enzyme. For example, the chloramine of Tau was more effective than HOCl at inhibiting the thiol-dependent enzymes, creatine kinase and GAPDH [120].

The rate constants for the oxidation of GSH and peroxiredoxin 2 (Prx2) by HOCl, NH_2Cl, and Gly-Cl have been determined (Tables 3.1 and 3.4) [29, 60, 111–113]. GSH was about five times more reactive with NH_2Cl than Gly-Cl, but the reactivity of Prx2 was at least three orders of magnitude greater for

TABLE 3.4. Second-Order Rate Constants for the Reactions of Chloramines with Amino Acids and Related Compounds

Substrate	pH	*T* (°C)	*k* (/M/s)				
			NH_2Cl	NH_2Br			
Gly	7.1	20–25	1.5×10^0	–			
	7.0–7.5	20–25	1.0×10^0	–			
Ala	7.0–7.5	20–25	8.5×10^{-1}	–			
β-Ala	7.1	20–25	5.0×10^{-1}	–			
Gly-Gly	6.6	20–25	6.0×10^0	–			
	7.0–7.5	20–25	4.8×10^0	–			
Met	7.3	20–25	4.3×10^2	1.8×10^3			
GSH	7.2	20–25	1.0×10^3	$>1 \times 10^5$			
Prx2	7.4	24	1.5×10^4	–			
			Gly-Cl	His-Cl	Taurine-Cl	c-$(Gly)_2$-Cl	*N*-Ac-Lys-Cl
Gly	7.1	24	–	–	1.3×10^{-1}	–	–
His	7.4	24	4.0×10^{-1}	–	–	–	–
Met	7.4	24	2.0×10^2	9.1×10^1	3.9×10^1	2.7×10^{2}[a]	5.2×10^1
Cys	7.4	24	3.5×10^2	9.3×10^2	2.0×10^2	–	4.8×10^2
Cysteamine	7.4	24	–	–	5.6×10^2	–	–
N-Ac-Cys	7.4	24	–	1.1×10^2	4.6×10^1	–	–
Cys-OMe	7.4	24	–	–	4.2×10^2	–	–
Tau	6.8	24	3.2×10^{-1}	–	–	–	–
GSH	7.4	24	2.3×10^2	7.2×10^2	1.1×10^2	$>5.0 \times 10^4$	2.6×10^2
GSSG	7.2	20–25	–	–	–	$<5.0 \times 10^{-1}$	–
TNB	7.4	24	2.3×10^3	–	9.7×10^2	–	5.7×10^3
3,3′-Dithio-propionic acid	7.4	22	6.8×10^{-1}	–	–	–	1.4×10^{-1}
Pencillamine	7.4	24	–	–	3.1×10^2	–	–
Peroxiredoxin 2	7.4	24	8.0×10^0	–	3.0×10^0	–	–

[a] pH 7.2.

Data were taken from References 29, 60, and 111–113.

HOCl than for Gly-Cl. Chloramines oxidized Prx2 to a disulfide-linked dimer. An implication of these results in relation to the oxidation of Prx2 by H_2O_2 has been studied to understand the erythrocyte's ability to resist oxidative damage [112].

Bromamines are likely the major products of HOBr-mediated oxidation of biological molecules. A few studies showed NH_2Br had a much higher reactivity than the corresponding NH_2Cl (Table 3.4) [29, 121]. Melatonin was oxidized efficiently by Tau-Br, while limited oxidation activity was observed using Tau-Cl [122]. A study was also conducted to oxidize Trp as free or as a residue in albumin by Tau-Br [123]. Fluorescence analysis on the oxidation of free Trp by Tau-Br and other oxidants are shown in Figure 3.11 [123]. Similar reactivity of HOCl and HOBr with Trp was observed. Significantly, Tau-Br was a much more efficient oxidant than Tau-Cl (Fig. 3.11a). This was also supported by results of dose-dependent effects of Tau-Br on the consumption of Trp in comparison with Tau-Cl (Fig. 3.11b). The increased oxidation efficiency of Trp residues in albumin by Tau-Br in comparison to any other oxidants was also obtained [123]. The formation of formylkynurenine, an oxidized product of Trp, was more pronounced in the oxidation of albumin by Tau-Br than other oxidants. Depletion of Cys and the formation of a carbonyl group were not significantly different among the tested oxidants. These results clearly demonstrated that Tau-Br had a higher reactivity for Trp residues in proteins.

3.2.3 Reduction of Halamines

N-Halogenated species can be rapidly reduced by a one-electron process. The rates of reduction of a series of halamines and haloamides by e^-_{aq} and $O_2^{\bullet-}$ have been determined (Table 3.5) [124, 125]. The halogenated species were reduced by e^-_{aq} with $k > 10^9$/M/s, while only *N*-halogenated imides were reduced by $O_2^{\bullet-}$ ($k > 10^6$/M/s). The reaction scheme is presented in Figure 3.12 [124, 125]. Deleterious *N*-centered radicals and Cl^-/Br^- were produced in the reactions of chloramines/bromamines with e^-_{aq} and $O_2^{\bullet-}$ (Fig. 3.12). Comparatively, in the case of *N*-bromoimides, the heterolytic cleavage of the N–Br bond produced bromine atoms ($Br^\bullet$), which may initiate chain reactions. A chain length of 10–20 was suggested. These results indicate that at the elevated $O_2^{\bullet-}$ concentration, there may be tissue damage *in vivo* due to the production of destructive $Br^\bullet$ and organic radicals. This would be pronounced due to an efficient chain reaction [125].

3.3 CHLORINE DIOXIDE

Chlorine dioxide (ClO_2) is produced in the chlorite-mediated peroxidase cycle. This is demonstrated in Figure 3.13a [126]. The two-electron oxidation of ferric horseradish peroxidase (HRP) by chlorite produces compound I at a rate constant k_1. Compounds I and II are reduced by chlorite via a one-electron

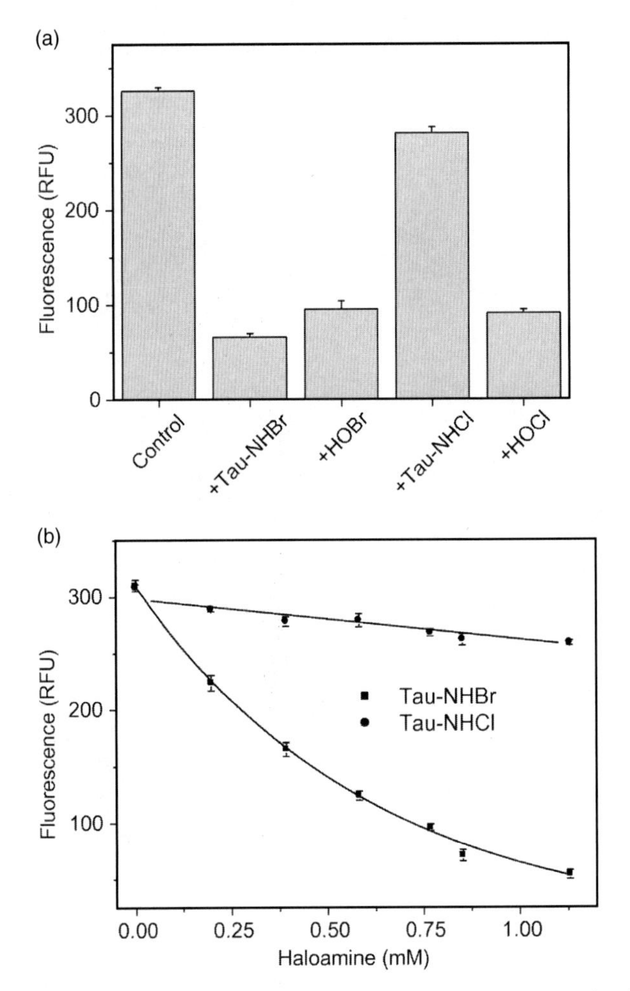

Figure 3.11. Comparison of tryptophan oxidation by measurement of intrinsic fluorescence. (a) The reaction mixtures consisted of 1.0 mM tryptophan (control) and 1.0 mM Tau-NHBr, Tau-NHCl, HOCl, or HOBr in phosphate buffer saline (PBS), pH 7.4 at 25°C. The intrinsic fluorescence was measured 2 minutes after the addition of oxidants (λ_{ex} 295 nm, λ_{em} 360 nm). (b) Dose-dependent effect of halamines on tryptophan consumption. The reaction mixtures consisted of 1.0 mM tryptophan and the indicated concentrations of halamines in PBS, pH 7.4 at 25°C. The results shown are representative of three experiments (adapted from Ximenes et al. [123] with the permission of Elsevier Inc.).

transfer with rate constants k_2 and k_3, respectively, to generate ClO_2. All three rate constants decrease in rate constants with an increase in pH (Fig. 3.13b). The values of k_1, k_2, and k_3 at pH 7.0 were determined as 7.5×10^3/M/s, 1.9×10^4/M/s, and 1.5×10^3/M/s, respectively. The production of ClO_2 as well as of HOCl may cause heme bleaching and the modification of amino acids.

TABLE 3.5. Second-Order Rate Constants for the Reduction of Chloramines and Amides by Hydrate Electron and Superoxide

Substrate	k (/M/s) e^-_{aq}	k (/M/s) $O_2^{\bullet -}$
GlyCl	6.5×10^9	
β-AlaCl	6.1×10^9	
$(Gly)_2Cl$	1.5×10^{10}	No reaction[a]
$(Gly)_2Br$	$>1 \times 10^{10}$	
$(Ala)_2Cl$	1.4×10^{10}	
N-Chlorosuccinimide	1.6×10^{10}	8.0×10^5
6-Aminohexanoic acid chloramines, CANCl	9.3×10^9	No reaction[b]
6-Aminohexanoic acid bromoamines, CANBr	1.2×10^{10}	No reaction[c]
N-Bromoglutarimide	$>1 \times 10^{10}$	$>3.0 \times 10^6$
N-Bromo-4-hydroxy-2-pyrrolidinone, NBr-HOP	6.4×10^9	No reaction[d]

[a] $[(Gly)_2Cl] = 170\,\mu M$.
[b] $[CANCl] = 265\,\mu M$.
[c] $[CANBr] = 90\,\mu M$.
[d] $[NBr\text{-}HOP] = 60\,\mu M$.

Data were taken from References 124 and 125.

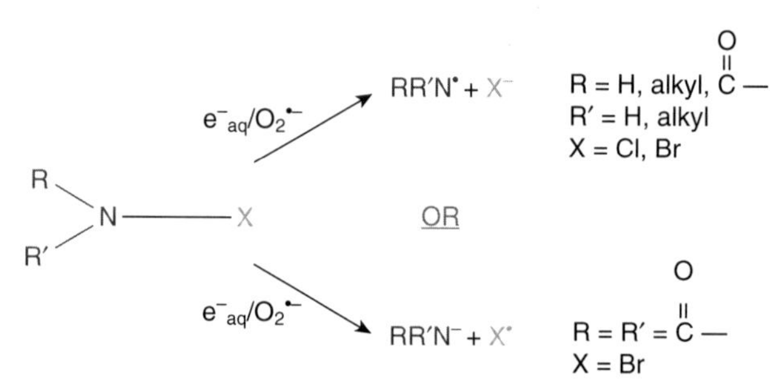

Figure 3.12. One-electron reduction of *N*-chlorinated and *N*-brominated species (adapted from Pattison et al. [125] with the permission of the American Chemical Society). See color insert.

The rate constants for the oxidation of ferric HRP by ClO_2 and HOCl were 2.7×10^4/M/s and 2.4×10^4/M/s, respectively.

ClO_2 is employed as an alternative to chlorine in the purification and disinfection of drinking water, bleaching of paper, sterilization of medical devices, and sanitization of food products [127–133]. The main advantage of using ClO_2 over chlorine is it controls the formation of harmful organochloro compounds. Table 3.6 shows the susceptibility of ClO_2 gas for several microorganisms [133]. The mean reduction of gram-negative bacteria and gram-positive bacteria

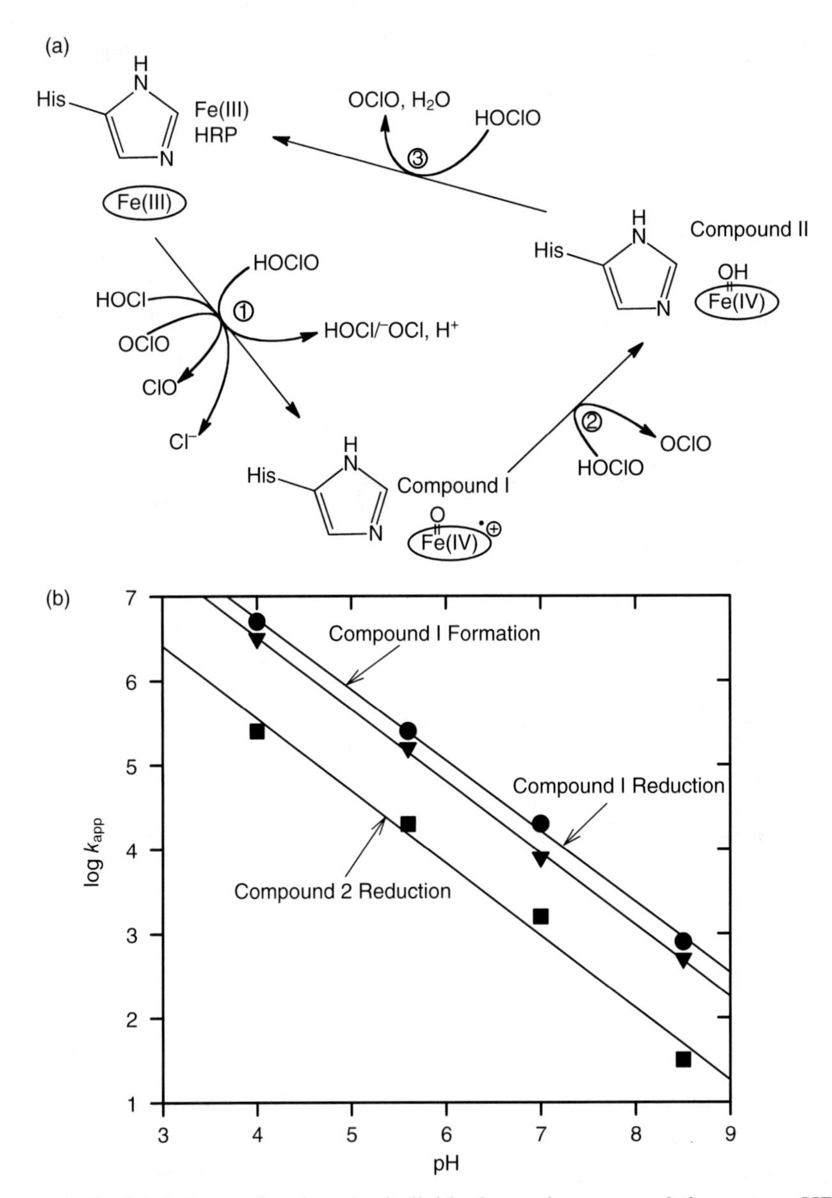

Figure 3.13. (a) Scheme showing the individual reaction steps of the system HRP/chlorite including the reactions of the reaction products chlorine dioxide and hypochlorite with ferric HRP. Note that the two-electron reduction product of chlorine dioxide (i.e., ClO) is only speculative. (b) pH dependence of apparent bimolecular rate constants of reactions involving chlorite, that is, compound I formation and reduction as well as compound II reduction (adapted from Jakopitsch et al. [126] with the permission of Elsevier Inc.).

TABLE 3.6. Log Reductions Obtained by a Treatment with 0.08 ± 0.02 mg ClO_2/L for 1 Minute at RH 90 ± 0.5% and Room Temperature

Microorganism	log Reduction[a] (log cfu)
Bacteria, gram-negative	
Aeromonas hydrophila	2.6 ± 0.9
Enterobacter aeruginosa	3.4 ± 0.8
Escherichia coli	2.7 ± 0.6
Klebsiella oxytoca	3.1 ± 0.7
Pseudomonas fluorescens	0.5 ± 0.5
Salmonella typhimurium	3.2 ± 0.3
Shigella flexnii	3.4 ± 0.5
Yersinia entercolitica	3.5 ± 0.6
Bacteria, gram-positive	
Brochotrix thermosphacta	1.4 ± 0.7
Lactobacillus sakei	2.6 ± 0.6
Leuconostoc mesenteroïdes	3.2 ± 0.4
Listeria monocytogenes	2.8 ± 0.6
Staphylococcus aureus	3.1 ± 0.5
Yeasts	
Candida lambica	1.1 ± 0.2
Pichia burtonii	2.0 ± 0.5
Saccharomyces cerevisiae	0.6 ± 0.1
Mold spores	
Aspergillus niger	0.1 ± 0.1
Penicillium roqueforti	0.4 ± 0.5
Botrytis cinerea	0.9 ± 0.4
Bacterial spores	
Bacillus cereus	0.2 ± 0.2

[a] Mean ± standard deviation, $n = 6$.

Adapted from Vandekinderen et al. [133] with the permission of Elsevier Inc.

were 3.5 and 2.6 log cfu/cm^2 (Table 3.6). Yeasts were more resistant to ClO_2 than gram-negative bacteria and gram-positive bacteria. Significantly, ClO_2 has been shown to be effective in deactivating *Bacillus anthracis* spores in governmental and commercial buildings [134–137]; however, *Bacillus cereus* was little affected by ClO_2 (Table 3.6).

The bactericidal capacity of ClO_2 was shown to be higher than that of liquid chlorine [138]. The bactericidal rate of ClO_2 was found to be faster than that of liquid chlorine. Significantly, *Bacillus* could be killed by ClO_2 in the pH range from 3.0 to 8.0. However, the pH range in using liquid chlorine to achieve a similar effect for killing was much narrower (6.8–8.5). Values of Ct are used to express the level of disinfection, which is the concentration of disinfectant (in milligram per liter) multiplied by the contact time (in minutes). Ct values for inactivation of murine norovirus (MNV) and coliphage MS2 by ClO_2 were lower than chlorine [139]. At 5°C, Ct values for 4-log reduction in MNV were determined as 0.314 and 0.247 mg/L min for chlorine and ClO_2, respectively. Another example is the inactivation of *Cryptosporidium parvum* [140]. Ninety percentage inactivation of *C.parvum* required 1.3 mg/L ClO_2 for 1 hour, but 80 mg/L of chlorine and monochloramine for 1 hour was needed to obtain same percentage of inactivation [140]. ClO_2 was also shown to be equal or superior to free available chlorine to inactivate bacterial threat agents [141].

An overview of the generation and the stability of ClO_2 in aqueous solution, followed by its reactivity with amino acids, peptides, and proteins are discussed below.

3.3.1 Generation of ClO_2

Reduction of the chlorate ion in acidic media with and without H_2O_2 has been used to produce excessive quantities of ClO_2. Generally, ClO_2 is generated by the reaction of the chlorite ion with an acid and/or chlorine [132]. These methods involve the use of concentrated acids and/or externally added oxidants such as Cl_2, OCl^-, and H_2O_2. As an alternative, the electrochemical generation by a one-electron transfer from ClO_2^- to ClO_2 requires a considerable input of electrical energy. Recently, a catalytic process using the water-soluble manganese porphyrin, tetrakis-5,10,15,20-(*N,N*-dimethylimidazolium) porphyrinatomanganese(III) ([Mn-(TDMImP)]), demonstrated the generation of ClO_2 efficiently from chlorite ion under mild ambient conditions [142]. These results may have implications in understanding the reactions of the chlorite ion with heme proteins and metalloporphyrins. Water-soluble synthetic iron porphyrin complexes have been shown to generate O_2 from the chlorite ion [143, 144]. The heme-thiolate enzyme chloroperoxidase transferred ClO_2^- to a mixture of ClO_3^-, Cl^-, and O_2 through an intermediate ClO_2 [145]. The chemistry of the O_2-evolving enzyme chlorite dismutase (Cld) has also been studied [142–144, 146–149]. For example, the decomposition of chlorite ion by *Dechloromonas aromatica* chlorite dismutase (DA-Cld) initially produce O_2DA-Cld, which ultimately yields 1 mol of O_2 and 1 mol of Cl^- for each mole of chlorite consumption [148, 150]. A proposed mechanism is presented in Figure 3.14 [148]. Initially, a Cld–chlorite Michaelis complex formed, which subsequently cleaved via heterolytic and hemolytic cleavages, respectively. Heterolytic cleavage yielded compound I and hypochlorite as the leaving group, whereas compound II and the hypochloryl radical were formed in hemolytic cleavage. The peroxychlorite ion was produced before the formation of products of the reaction.

3.3.2 Decomposition of ClO_2

The decomposition rate of ClO_2 in neutral solution is slow. However, in alkaline solution, the rate is enhanced with the formation of ClO_2^- and ClO_3^- (Eq. 3.16) [151]:

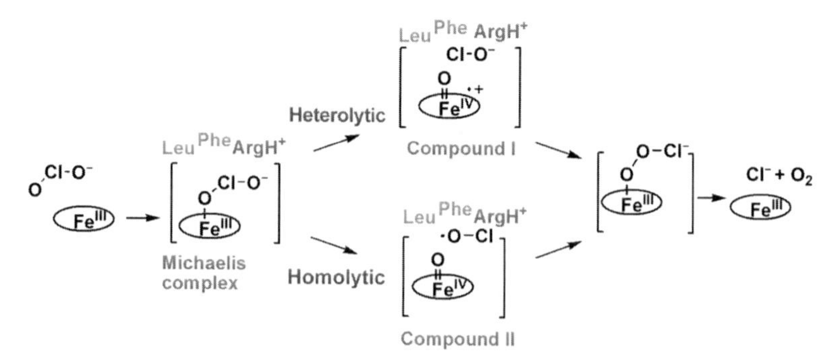

Figure 3.14. Proposed mechanisms for chlorite decomposition and O_2 evolution catalyzed by *Dechloromonas aromatica* (DA-Cld) (adapted from Goblirsch et al. [148] with the permission of Elsevier Inc.). See color insert.

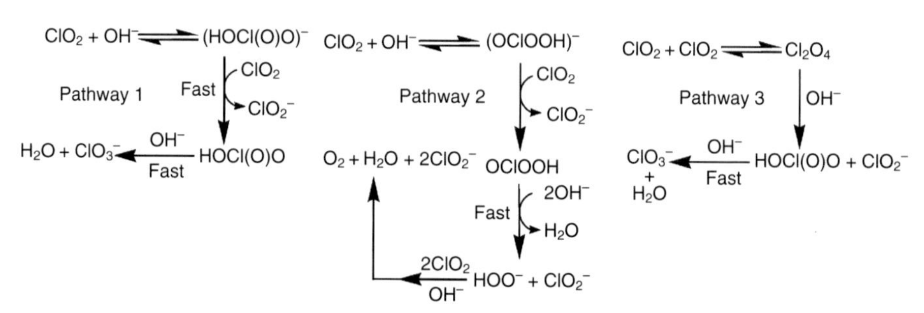

Figure 3.15. Decomposition pathways of ClO_2 in alkaline solution (adapted from Odeh et al. [151] with the permission of the American Chemical Society).

$$2ClO_2 + 2OH^- \rightarrow ClO_2^- + ClO_3^- + H_2O. \tag{3.16}$$

Significantly, equal amounts of ClO_2^- and ClO_3^- were formed at millimolar or higher concentrations of ClO_2, but the yield of ClO_2^- was greater than the yield of ClO_3^- at micromolar levels of ClO_2. The additional reaction (Eq. 3.17) could account for the change in the molar stoichiometry of ClO_2^- to ClO_3^- (Eq. 3.17):

$$4ClO_2 + 4OH^- \rightarrow ClO_2^- + O_2 + 2H_2O. \tag{3.17}$$

Both first-order and second-order dependence on the concentration of ClO_2 have been suggested for the disproportionation of ClO_2 [151]. The rate expression (Eq. 3.18) for the second-order dependence is given as

$$-d[ClO_2]/dt = k[ClO_2]^2[OH^-]. \tag{3.18}$$

Three mechanisms, shown in Figure 3.15, were given to explain the kinetics and stoichiometry of the decomposition of ClO_2 in alkaline solution [151]. All three concurrent pathways involved base-assisted electron transfer steps. Pathway 1 involved the formation of an intermediate adduct, $(HOCl)(O)O^-$, which reacted rapidly to yield ClO_2^- and $HOClO_2$. The formation of ClO_3^- occurred from the rapid reaction between $HOClO_2$ and OH^-. This pathway displayed first-order kinetics with respect to the concentrations of ClO_2 and OH^-. In pathway 2, a different intermediate, $(OClOOH)^-$, reacted with ClO_2 to form ClO_2^- and an additional intermediate, OClOOH. The reaction of OClOOH with OH^- produced HOClO and HOO^-. The reaction of the latter species with ClO_2 could give ClO_2^- and O_2 [152]. An alternative to these steps was also proposed in which OClOOH reacted with OH^- to produce $OClOH^-$, which reacted with a second molecule of ClO_2 to generate OClOH and ClO_2^-. Pathway 2 also gave a first-order rate expression with respect to the concentration of ClO_2 and was important at low levels of ClO_2. Finally, pathway 3 involved the formation of a dimer intermediate, Cl_2O_4, which reacted with OH^- (an electron transfer step). This pathway displayed a second-order rate expression with respect to the concentration of ClO_2 and was important at high concentration of ClO_2.

The catalytic effect of the hypohalite ion, OX^- (X = Cl, Br), on the disproportionation of ClO_2 has also been studied [153]. A first-order dependence in both $[ClO_2]$ and $[OX^-]$ was observed at low concentrations of ClO_2^-. Reactions became second-order in $[ClO_2]$ at excess $[ClO_2^-]$, and observed rates were inversely proportional to the concentration of ClO_2^-. In the first step of the proposed steps of catalysis of the hypohalite, the reaction between ClO_2 and OX^- involved an electron transfer to form ClO_2^- and OX (Eqs. 3.19 and 3.20):

$$ClO_2 + OCl^- \rightleftharpoons ClO_2^- + ClO \quad k_{19} = 9.1 \times 10^{-1}/M/s; K_{19} = 5.1 \times 10^{-10} \tag{3.19}$$

$$ClO_2 + OBr^- \rightleftharpoons ClO_2^- + BrO \quad k_{20} = 2.0/M/s; K_{20} = 1.3 \times 10^{-7}. \tag{3.20}$$

Values of K_{19} and K_{20} explain the observed suppression by ClO_2^-. The activation parameters, $\Delta H_1^{\ddagger}$ and $\Delta S^{\ddagger}$, of the first step were 61 kJ/mol and −43 J/mol/K, respectively, for OCl^-/ClO_2 and 55 kJ/mol and −49 J/mol/K, respectively, for OBr^-/ClO_2. The positive $\Delta H_1^{\ddagger}$ and negative $\Delta S^{\ddagger}$ values indicate that both ClO_2 and OX^- come together before electron transfer to form ClO_2^- and OX. In the second step, reactions between ClO_2 and XO were fast to form $XOClO_2$ (Eqs. 3.21 and 3.22):

$$ClO_2 + ClO \rightarrow ClOClO_2 \quad k_{21} = 7 \times 10^9 /M/s \tag{3.21}$$

$$ClO_2 + BrO \rightarrow BrOClO_2 \quad k_{22} = 1.0 \times 10^8 /M/s. \tag{3.22}$$

The hydrolysis of $XOClO_2$ was also fast to yield ClO_3^- and OX^-. The catalysis of ClO_2 disproportionation by BrO_2^- has also shown to be effective [153].

3.3.3 Reactivity of ClO_2

The reactivity of ClO_2 with a number of inorganic and organic compounds has been studied [152, 154–162]. The rate laws were first-order with respect to the concentrations of ClO_2, and these compounds and their resulting second-order rate constants varied from 10^{-5} to 10^5/M/s. The rate constants were high for I^-, NO_2^-, O_3, H_2O_2, and Fe(II). The rate constants for tertiary amines and phenols were also high at pH≥6. Ammonia, Br^-, primary and secondary amines, carbohydrates, aromatic hydrocarbons, and compounds containing olefic C=C double bonds were unreactive with ClO_2 at near neutral pH conditions.

In the oxidation of Fe(II) by ClO_2, a five-electron transfer has been proposed (Eq. 3.23) [157–159]:

$$ClO_2 + 5Fe^{2+} + 4H^+ \rightarrow Cl^- + 5Fe^{3+} + 2H_2O. \tag{3.23}$$

Both inner-sphere and outer-sphere electron transfer pathways were suggested for the reaction [159]. Among the organic compounds, a detailed analysis of the kinetics has been conducted for the phenolic compounds [156]. The rate constants for the reactivity of ClO_2 with phenoxide anions were approximately six orders of magnitude higher than their protonated analogues. The rate constants for phenoxide anions (k_{ArO^-}) were analyzed using the Hammett equation (Eq. 3.24):

$$\log k_{ArO^-} = 8.2(\pm 0.2) - 3.2(\pm 0.4)\sum \sigma^-_{o,m,p};\ n = 23,\ s = 0.39;\ r = 0.97. \tag{3.24}$$

where $\sigma^-_{o,m,p}$ are the constants for ortho-, meta-, and para-substituents. The negative sign of the slope of Equation (3.24) indicates that the oxidation by ClO_2 was through a free radical mechanism in which increase in electron density at the reaction center increased the reaction rate [156]. Furthermore, the rate-determining step involving primarily an outer-sphere single-electron transfer step from the phenoxide anion to ClO_2 was supported by the Marcus correlations [156].

3.3.3.1 Amino Acids, Peptides, and Proteins. The reactivity of biologically important molecules including amino acids and some peptides with ClO_2 has been performed [154, 155, 163–171]. A pseudo-first-order decay of chlorine dioxide was detected in these reactions [155]. Of the 21 amino acids and three peptides (L-aspartyl-L-phenylalanine methyl ester (aspartame), L-glycyl-L-tryptophan and L-tryptophylglycine) reacted, only a few showed reactivity with ClO_2 at pH6.0 [155]. Among the amino acids and peptides, Cys, Trp, Tyr, L-Gly-L-Trp, and L-Trp-Gly reacted rapidly, while His, Pro, and hydroxyproline had slower rates. The second-order rate constants for their reactions are presented in Table 3.7 [152, 155, 163, 168, 170, 171]. The order of the rate constants for amino acids and peptides was Cys > Tyr > Trp ~ *N*-acetyl-Tyr > GSH > OH-Pro > His > Pro.

TABLE 3.7. Reaction Rate Constants for Consumption of Chlorine Dioxide by Amino Acids and Peptides

Compound	pH	Temp (°C)	k (/M/s)	Reference
Gly	8.0	22–24	$<10^{-5}$	[152]
Pro	6.0	25.0	3.4×10^{-2}	[155]
OH-Pro	6.0	25.0	6.9×10^{-2}	[155]
His	6.0	25.0	5.4×10^{-2}	[155]
Cys	7.0	25.0	1.0×10^{7}	[163]
Tyr	7.0	25.0	1.8×10^{5}	[163]
N-Acetyl-Tyr	6.2	25.0	3.2×10^{4}	[168]
Trp	7.0	25.0	3.4×10^{4}	[171]
GSH	5.9	25.0	5.8×10^{2}	[163]
NADH	6–8	24.6	3.9×10^{6}	[170]

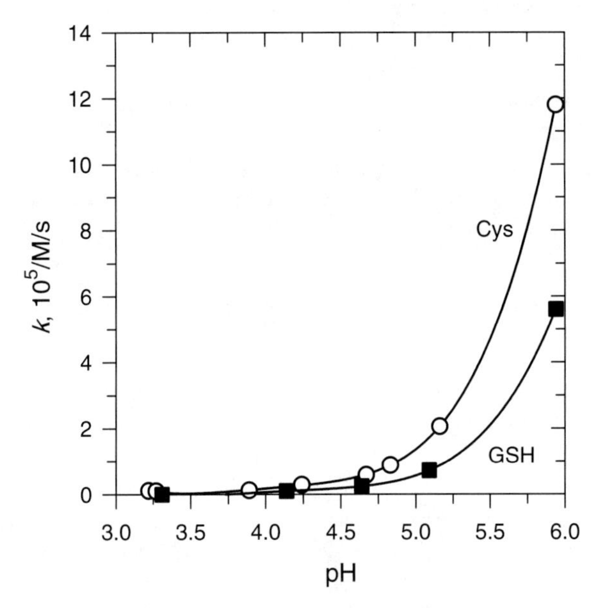

Figure 3.16. Effect of pH on the second-order rate constants for the oxidation of Cys and GSH by ClO_2 at 25°C (data were taken from Ison et al. [163]).

The oxidation of thiols (Cys and GSH) by ClO_2 as a function of pH has been performed [163, 164]. The rate constant for Cys and GSH increased with an increase in pH from 3.2 to 5.9 (Fig. 3.16) [163]. The pH-dependence behavior suggests the deprotonated species of thiols were the reactive species. The solid fitted lines of the experimental results gave rate constants for the reactions of ClO_2 with the cysteinyl anion (CS^-) and the glutathione anion (GS^-)

as 1.0×10^8/M/s and 1.4×10^8/M/s, respectively. Similar rate constants suggest common oxidation mechanisms for both thiols by ClO_2.

The stoichiometry of the reaction between ClO_2 and Cys varied from 6:5 at low pH 3.6 to 2:10 at high pH 9.5 [163]. Cysteic acid was the major product at low pH, while the disulfide was the major product at high pH [164, 172]. At pH 6.7, the rate constant for the oxidation of CS^- with ClO_2 was seven orders of magnitude faster than the corresponding reaction with cystine (CSSC) at pH 6.7. Therefore, no further oxidation of CSSC is expected. Based on the substrate consumed and the concentrations of the products formed, other products of the reaction may be thiosulfinate and thiosulfonate. A scheme shown in Figure 3.17 was proposed [163]. Initially, a cysteine radical was formed through a one-electron transfer process. The cysteine radicals further reacted with another molecule of Cys to produce a ClO_2-cysteinyl adduct. The adduct was proposed to disproportionate by two pH-dependent pathways to yield different products. The formation of cysteic acid from the oxidation of disulfide (CSSC) was ruled out based on its slow reactivity with ClO_2. In acidic solution, sulfinic acid hydrolyzed to form sulfinic acid and HOCl. The rapid reaction between these hydrolyzed products formed cysteine acid and Cl^-. The second pathway at high pH involved the reaction between the adduct and CS^- to give cystine and ClO_2^-.

The stoichiometry and products of the reactions of Trp, Tyr, and His by ClO_2 have been performed [166, 171]. The stoichiometry of the reaction between ClO_2 and Trp is presented in Equation (3.25):

$$2ClO_2 + Trp + H_2O \rightarrow ClO_2^- + NFK + HOCl + H^+. \tag{3.25}$$

In the proposed mechanism (Fig. 3.18), a one-electron transfer initially yields a tryptophan radical cation and chlorite ion. The radical cation deprotonates to produce a neutral tryptophan radical, which is readily combined with a second ClO_2 molecule to form a short-lived adduct (Fig. 3.18). This radical decomposes to give the observed product, NFK (Fig. 3.18). The adduct reacts with water to give an intermediate product A. The rearrangement of product A forms the final product, NFK. Several other unidentified products were also formed in the reaction of ClO_2 with Trp. More recently, product identification of this reaction was conducted using mass analysis [166]. Molar ratios of ClO_2 to Trp were changed from 0.25 to 4.0 in the presence and absence of oxygen in the reaction system. In excess ClO_2, fumaric and oxalic acids were the abundant products. Minor products were 2-aminobenzoic acid, *N*-formylanthranilic acid, and 2-(2-oxoindolin-3-ylidene) acetic acid. Such products suggest the carbon–carbon bond breaking of molecules during the reaction. Different kinds of products were also observed in the reactions of His and Tyr with ClO_2 [166].

The reactions of Tyr, *N*-acetyltyrosine (NAT), and DOPA have been examined in the pH range from 4 to 7 [168]. Similar to the reaction of Trp with ClO_2, the oxidation of 1 mol of Tyr and NAT consumed 2 mol of ClO_2 each and

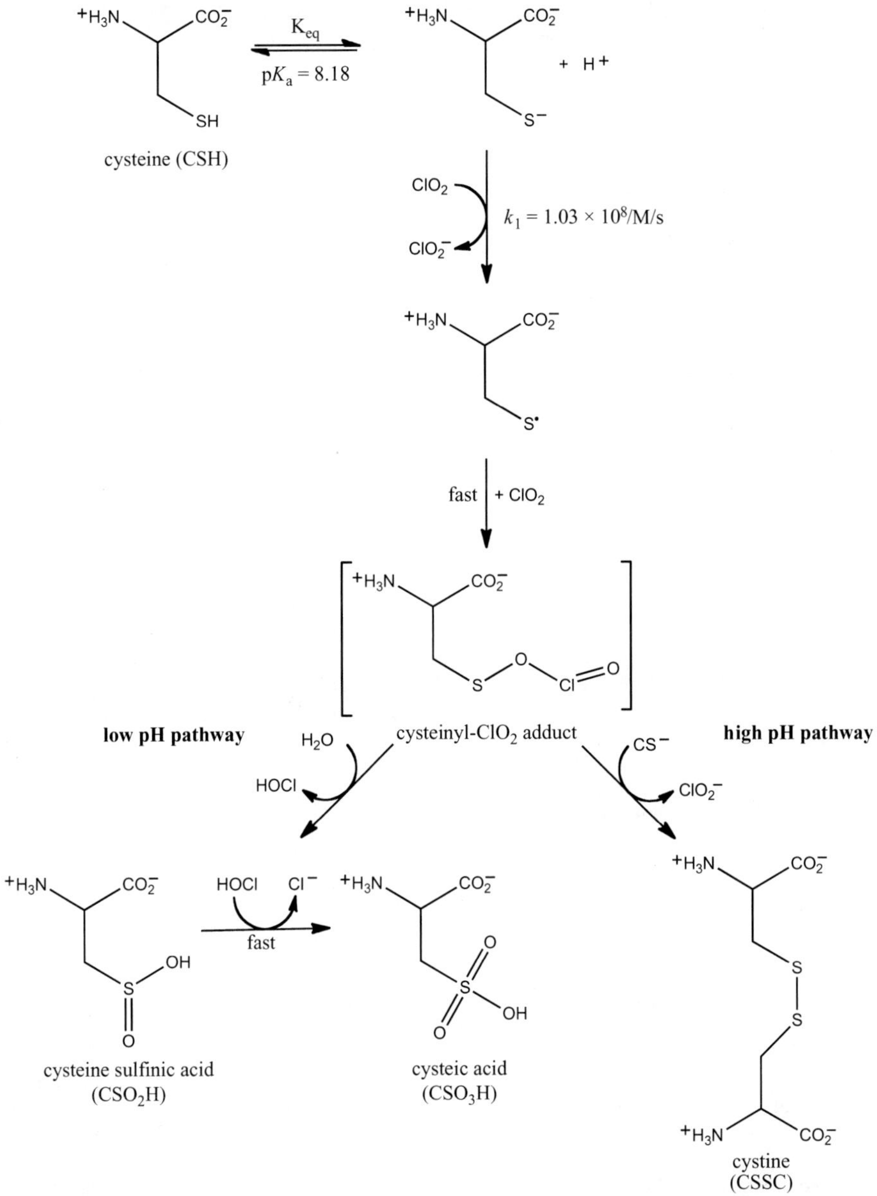

Figure 3.17. Proposed mechanism for the initial reactions between ClO_2 and CSH and the subsequent decay of the cysteinyl–ClO_2 adduct (adapted from Ison et al. [163] with the permission of the American Chemical Society).

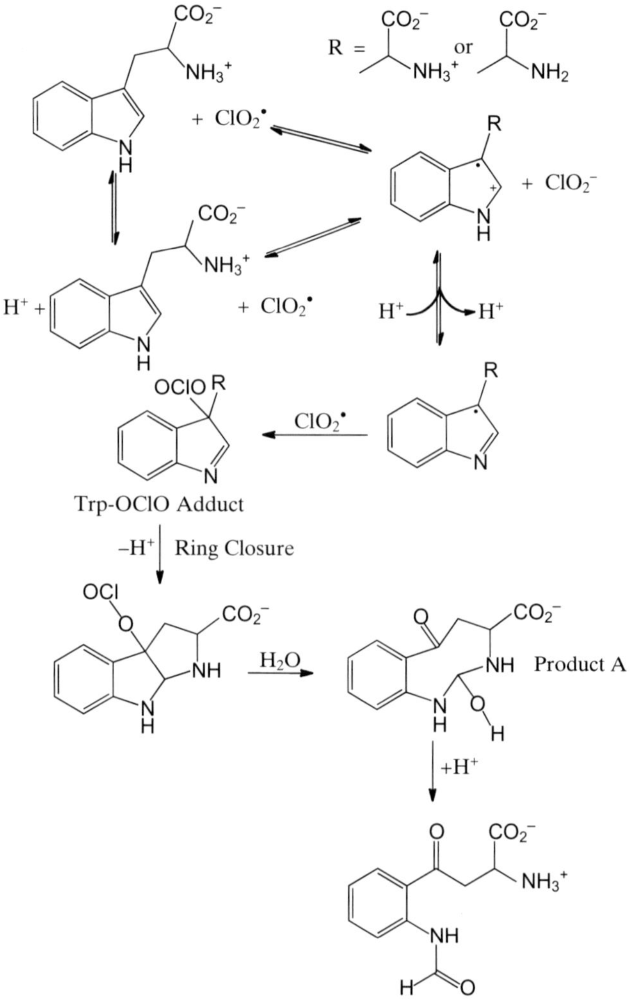

Figure 3.18. The proposed mechanism for the reaction of $ClO_2\cdot$ with Trp (adapted from Stewart et al. [171] with the permission of the American Chemical Society).

generated short-lived adducts, which decayed rapidly to yield dopaquinone (from Tyr) and *N*-acetyldopaquinone (from NAT). DOPA also consumed 2 mol of ClO_2 to yield dopaquinone and 2 mol of ClO_2^-. The cyclization of dopaquinone occurred at pH > 4 to form cyclodopa, which subsequently oxidized to dopachrome.

There are few studies on the oxidation of peptides and proteins by ClO_2 [167, 173, 174]. Recent work focused on BSA and glucose-6-phosphate dehydrogenase (G6PD) of baker's yeast (*Saccharomyces cerevisiae*) as model proteins [173]. The denaturation of proteins by ClO_2 decreased the α-helical content as well as in the transition temperature and endothermic transition enthalpy of heat-induced unfolding [173]. The decrease in the concentrations of ClO_2 with time in reactions with BSA and G6PD was exponential. The denaturing of proteins by ClO_2 was also confirmed by the absorbance and fluorescence spectra of proteins with and without ClO_2 [173]. Residues of Trp and Tyr were destroyed in the oxidation process. Product analysis showed the Trp residue was converted into NFK, DOPA, and 2,4,5-trihydroxyphenylalanine (TOPA) in the ClO_2-treated proteins [173]. Experiments with ^{18}O labeling of peptides in the reaction with ClO_2 provided evidence of the incorporation of oxygen from ClO_2 into the oxidation products.

3.4 CONCLUSIONS

Significant information on the HOCl reactivity with amino acids and proteins is known to assess targets for the reactions of proteins with HOCl. However, similar knowledge on other oxidants, HOBr and HOSCN, is lacking. The chemistry of HOBr appears to be similar to HOCl, but much higher reactivity of HOBr than HOCl has been observed, particularly for the aromatic amino acids. These results have important implications for the biological environment. Chlorination results in chloramines, which can either transform into products or react with side chains of proteins. The products include oxyacids and carbonyls. Side chains of thiols are highly reactive with HOCl, which first form sulfenyl chloride before converting into sulfoxides and sulfonamides. Halamines such as NH_2Cl, NH_2Br, Gly-Cl, taurine-Cl, and taurine-Br, formed from the reactions of HOCl and HOBr, may also play a role in the reactions of biological systems. Reduction of halamines may produce damaging halogen- and nitrogen-centered radicals in the process. However, there are limited rate studies of the reactions of these oxidants with proteins that have been conducted. HOSCN has not shown much reactivity with free amino acids, but the side chains in proteins have significant reactive rates with HOSCN. The reactions of HOSCN with protein thiols form a disulfide and a dimer. This suggests the need to examine reactions of HOSCN with different proteins in order to understand the role of HOSCN in protein oxidation *in vivo* and in cellular signaling.

ClO_2 is generated from the chlorite ion by using chemical, electrochemical, and biocatalysis methods. The use of Cld to catalyze the chlorite ion may be imperative not only in generating ClO_2 but also in gaining insight on the interaction of the chlorite ion with heme proteins and metalloporphrins and, thus, warrants further studies. ClO_2 showed reactivity with Cys, Trp, and Tyr, and the formation of oxidized products occurred through the incorporation of oxygen from ClO_2.

ClO_2 as a disinfectant has shown equal or better effectiveness than commonly used chlorine. The concern of halogenated disinfection by-products (e.g., trihalomethane and halo acids) can be minimized using ClO_2. Aside from temperature, pH, and strains of the organism, Ct (cycle threshold) values are also affected by turbidity and nutrient availability; thus, future research may include the evaluation of ClO_2 disinfection under different water conditions. The inactivation of foodborne microorganisms using ClO_2 may be affected by proteins and, therefore, mechanistic studies on the disinfection of food products by ClO_2 need to be performed.

REFERENCES

1 Davies, M.J. Myeloperoxidase-derived oxidation: mechanisms of biological damage and its prevention. *J. Clin. Biochem. Nutr.* 2011, *48*, 8–19.

2 Hampton, M.B., Kettle, A.J., and Winterbourn, C.C. Inside the neutrophil phagosome: oxidants, myeloperoxidase, and bacterial killing. *Blood* 1998, *92*, 3007–3017.

3 Hawkins, C.L., Morgan, P.E., and Davies, M.J. Quantification of protein modification by oxidants. *Free Radic. Biol. Med.* 2009, *46*, 965–988.

4 Hawkins, C.L. The role of hypothiocyanous acid (HOSCN) in biological systems. *Free Radic. Res.* 2009, *43*, 1147–1158.

5 Lane, A.E., Tan, J.T.M., Hawkins, C.L., Heather, A.K., and Davies, M.J. The myeloperoxidase-derived oxidant HOSCN inhibits protein tyrosine phosphatases and modulates cell signalling via the mitogen-activated protein kinase (MAPK) pathway in macrophages. *Biochem. J.* 2010, *430*, 161–169.

6 Ramos, D.R., Garcia, M.V., Canle, L.M., Santaballa, J.A., Furtmuller, P.G., and Obinger, C. Myeloperoxidase-catalyzed chlorination: the quest for the active species. *J. Inorg. Biochem.* 2008, *102*, 1300–1311.

7 Babior, B.M. Phagocytes and oxidative stress. *Am. J. Med.* 2000, *109*, 33–44.

8 Gebendorfer, K.M., Drazic, A., Le, Y., Gundlach, J., Bepperling, A., Kastenmüller, A., Ganzinger, K.A., Braun, N., Franzmann, T.M., and Winter, J. Identification of a hypochlorite-specific transcription factor from *Escherichia coli*. *J. Biol. Chem.* 2012, *287*, 6892–6903.

9 Davies, M.J., Hawkins, C.L., Pattison, D.I., and Rees, M.D. Mammalian heme peroxidases: from molecular mechanisms to health implications. *Antioxid. Redox Signal.* 2008, *10*, 1199–1234.

10 Van Der Veen, B.S., De Winther, M.P.J., and Heeringa, P. Myeloperoxidase: molecular mechanisms of action and their relevance to human health and disease. *Antioxid. Redox Signal.* 2009, *11*, 2899–2937.

11 Thomas, E.L. Lactoperoxidase-catalyzed oxidation of thiocyanate: equilibria between oxidized forms of thiocyanate. *Biochemistry* 1981, *20*, 3273–3280.

12 Prütz, W.A., Kissner, R., Koppenol, W.H., and Rüegger, H. On the irreversible destruction of reduced nicotinamide nucleotides by hypohalous acids. *Arch. Biochem. Biophys.* 2000, *380*, 181–191.

13 Carrell Morris, J. The acid ionization constant of HOCl from 5 to 35°C. *J. Phys. Chem.* 1966, *70*, 3798–3805.

14 Rees, M.D., Hawkins, C.L., and Davies, M.J. Hypochlorite and superoxide radicals can act synergistically to induce fragmentation of hyaluronan and chondroitin sulphates. *Biochem. J.* 2004, *381*, 175–184.

15 Hawkins, C.L. and Davies, M.J. Reaction of HOCl with amino acids and peptides: EPR evidence for rapid rearrangement and fragmentation reactions of nitrogen-centred radicals. *J. Chem. Soc. Perkin Trans. 2* 1998, 1937–1945.

16 Thomas, E.L., Grisham, M.B., and Jefferson, M.M. Cytotoxicity and chloramines. *Method Enzymol.* 1986, *132*, 585–593.

17 Thomas, E.L., Grisham, M.B., and Jefferson, M.M. Preparation and characterization of chloramines. *Method Enzymol.* 1986, *132*, 569–585.

18 Thomas, E.L. Myeloperoxidase-hydrogen peroxide-chloride antimicrobial system: effect of exogenous amines on antibacterial action against *Escherichia coli*. *Infect. Immun.* 1979, *25*, 110–116.

19 Deborde, M. and Gunten, U.V. Reactions of chlorine with inorganic and organic compounds during water treatment—kinetics and mechanism: a critical review. *Water Res.* 2008, *42*, 13–51.

20 Sivey, J.D., McCullough, C.E., and Roberts, A.L. Chlorine monoxide (Cl_2O) and molecular chlorine (Cl_2) as active chlorinating agents in reaction of dimethenamid with aqueous free chlorine. *Environ. Sci. Technol.* 2010, *44*, 3357–3362.

21 Margerum, D.W., Gray, E.T., Jr., and Huffman, R.P. Chlorination and the formation of *N*-chloro compounds in water treatment. *ACS Symp. Ser.* 1978, *82*, 278–291.

22 Yoon, J. and Jensen, J.N. Distribution of aqueous chlorine with nitrogenous compounds: chlorine transfer from organic chloramines to ammonia. *Environ. Sci. Technol.* 1993, *27*, 403–409.

23 Armesto, X.L., Canle, L.M., Gamper, A.M., Losada, M., and Santaballa, J.A. Acid-base equilibria and decomposition of secondary (N-Cl)-alpha-amino acids. *Tetrahedron* 1994, *50*, 10509–10520.

24 Folkes, L.K., Candeias, L.P., and Wardman, P. Kinetics and mechanisms of hypochlorous acid reactions. *Arch. Biochem. Biophys.* 1995, *323*, 120–126.

25 Armesto, X.L., Canle, M., Garcia, M.V., Losada, M., and Santaballa, J.A. Chlorination of dipetides by hypochlorous acid in aqueous solution. *Gazz. Chim. Ital.* 1994, *124*, 519–523.

26 Armesto, X.L., Canle, L.M., Fernandez, M.I., Garcia, M.V., and Santaballa, J.A. First steps in the oxidation of sulfur-containing amino acids by hypohalogenation: very fast generation of intermediate sulfenyl halides and halosulfonium cations. *Tetrahedron* 2000, *56*, 1103–1109.

27 Armesto, X.L., Canle, M.L., Fernandez, M.I., Garcia, M.V., Rodriguez, S., and Santaballa, J.A. Intracellular oxidation of dipeptides. Very fast halogenation of the amino-terminal residue. *J. Chem. Soc. Perkin Trans. 2* 2001, 608–612.

28 Pattison, D.I. and Davies, M.J. Absolute rate constants for the reaction of hypochlorous acid with protein side chains and peptide bonds. *Chem. Res. Toxicol.* 2001, *14*, 1453–1464.

29 Prütz, W.A., Kissner, R., Nauser, T., and Koppenol, W.H. On the oxidation of cytochrome *c* by hypohalous acids. *Arch. Biochem. Biophys.* 2001, *389*, 110–122.

30 Zhang, C., Reiter, C., Eiserich, J.P., Boersma, B., Parks, D.A., Beckman, J.S., Barnes, S., Kirk, M., Baldus, S., Darley-Usmar, V.M., and White, C.R. L-Arginine chlorination products inhibit endothelial nitric oxide production. *J. Biol. Chem.* 2001, *276*, 27159–27165.

31 Na, C. and Olson, T.M. Relative reactivity of amino acids with chlorine in mixtures. *Environ. Sci. Technol.* 2007, *41*, 3220–3225.

32 Isaac, R.A. and Morris, J.C. Transfer of active chlorine from chloramine to nitrogenous organic compounds. 2. Mechanism. *Environ. Sci. Technol.* 1985, *19*, 810–814.

33 Hawkins, C.L., Pattison, D.I., and Davies, M.J. Hypochlorite-induced oxidation of amino acids, peptides and proteins. *Amino Acids* 2003, *25*, 259–274.

34 Armesto, X.L., Canle, L.M., Garcia, V., and Santaballa, J.A. Solvent isotope effects in the oxidation of dipeptides by aqueous chlorine. *Can. J. Chem.* 1999, *77*, 997–1004.

35 Antelo, J.M., Arce, F., and Perez-Moure, J.C. Kinetics of the N-chlorination of 2-aminobutyric, 3-aminobutyric, 3-aminoisobutyric and 4-aminobutyric acids in aqueous solution. *Int. J. Chem. Kinet.* 1992, *24*, 1093–1101.

36 Jensen, J.S., Lam, Y.-F., and Helz, G.R. Role of amide nitrogen in water chlorination: proton NMR evidence. *Environ. Sci. Technol.* 1999, *33*, 3568–3573.

37 Jensen, J.S. and Hetz, G.R. Rates of reduction on N-chlorinated peptides by sulfite: relevance to incomplete dechlorination of wastewaters. *Environ. Sci. Technol.* 1998, *32*, 516–522.

38 Armesto, X.L., Canle, L.M., García, M.V., and Santaballa, J.A. Aqueous chemistry of N-halo-compounds. *Chem. Soc. Rev.* 1998, *27*, 453–460.

39 Hawkins, C.L. and Davies, M.J. Hygochlorite-induced oxidation of proteins in plasma: formation of chloramines and nitrogen-centred radicals and their role in protein fragmentation. *Biochem. J.* 1999, *340*, 539–548.

40 Mehrsheikh, A., Bleeke, M., Brosillon, S., Laplanche, A., and Roche, P. Investigation of the mechanism of chlorination of glyphosate and glycine in water. *Water Res.* 2006, *40*, 3003–3014.

41 Na, C. and Olson, T.M. Stability of cyanogen chloride in the presence of free chlorine and monochloramine. *Environ. Sci. Technol.* 2004, *38*, 6037–6043.

42 Aspée, A. and Lissi, E.A. Chemiluminescence associated with amino acid oxidation mediated by hypochlorous acid. *Luminescence* 2002, *17*, 158–164.

43 Flecks, S., Patallo, E.P., Zhu, X., Ernyei, A.J., Seifert, G., Schneider, A., Dong, C., Naismith, J.H., and Van Pée, K.-H. New insights into the mechanism of enzymatic chlorination of tryptophan. *Angew. Chem. Int. Ed.* 2008, *47*, 9533–9536.

44 Naskalski, J.W. Oxidative modification of protein structures under the action of myeloperoxidation and the hydrogen peroxide and chloride system. *Ann. Biol. Clin. (Paris)* 1994, *52*, 451–456.

45 Drozdz, R., Naskalski, J.W., and Sznajd, J. Oxidation of amino acids and peptides in reaction with myeloperoxidase, chloride and hydrogen peroxide. *Biochem. Biophys. Acta—Proteins Struct. Mol. Enzymol.* 1988, *957*, 47–52.

46 Dellegar, S.M., Murphy, S.A., Bourne, A.E., Dicesare, J.C., and Purser, G.H. Identification of the factors affecting the rate of deactivation of hypochlorous acid by melatonin. *Biochem. Biophys. Res. Commun.* 1999, *257*, 431–439.

47 Domigan, N.M., Charlton, T.S., Duncan, M.W., Winterbourn, C.C., and Kettle, A.J. Chlorination of tyrosyl residues in peptides by myeloperoxidase and human neutrophils. *J. Biol. Chem.* 1995, *270*, 16542–16548.

48 Hazen, S.L., Hsu, F.F., Mueller, D.M., Crowley, J.R., and Heinecke, J.W. Human neutrophils employ chlorine gas as an oxidant during phagocytosis. *J. Clin. Invest.* 1996, *98*, 1283–1289.

49 Fu, S., Wang, H., Davies, M., and Dean, R. Reactions of hypochlorous acid with tyrosine and peptidyl-tyrosyl residues give dichlorinated and aldehydic products in addition to 3-chlorotyrosine. *J. Biol. Chem.* 2000, *275*, 10851–10858.

50 Kang, J.I., Jr. and Neidigh, J.W. Hypochlorous acid damages histone proteins forming 3-chlorotyrosine and 3,5-dichlorotyrosine. *Chem. Res. Toxicol.* 2008, *21*, 1028–1038.

51 Davies, M.J. and Hawkins, C.L. Hypochlorite-induced oxidation of thiols: formation of thiyl radicals and the role of sulfenyl chlorides as intermediates. *Free Radic. Res.* 2000, *33*, 719–729.

52 Nagy, P. and Ashby, M.T. Reactive sulfur species: kinetics and mechanisms of the oxidation of cysteine by hypohalous acid to give cysteine sulfenic acid. *J. Am. Chem. Soc.* 2007, *129*, 14082–14091.

53 Ueki, H., Chapman, G., and Ashby, M.T. Reactive sulfur species: kinetics and mechanism of the oxidation of aryl sulfinates with hypochlorous acid. *J. Phys. Chem. A* 2010, *114*, 1670–1676.

54 Pullar, J.M., Vissers, M.C.M., and Winterbourn, C.C. Glutathione oxidation by hypochlorous acid in endothelial cells produces glutathione sulfonamide as a major product but not glutathione disulfide. *J. Biol. Chem.* 2001, *276*, 22120–22125.

55 Szuchman-Sapir, A.J., Pattison, D.I., Ellis, N.A., Hawkins, C.L., Davies, M.J., and Witting, P.K. Hypochlorous acid oxidizes methionine and tryptophan residues in myoglobin. *Free Radic. Biol. Med.* 2008, *45*, 789–798.

56 Pereira, W.E., Hoyano, Y., and Summons, R.E. Chlorination studies. II. The reaction of aqueous hypochlorous acid with α amino acids and dipeptides. *Biochim. Biophys. Acta* 1973, *313*, 170–180.

57 Nagy, P. and Ashby, M.T. Reactive sulfur species: kinetics and mechanism of the oxidation of cystine by hypochlorous acid to give *N,N′*-dichlorocystine. *Chem. Res. Toxicol.* 2005, *18*, 919–923.

58 Raftery, M.J., Yang, Z., Valenzuela, S.M., and Geczy, C.L. Novel intra- and intermolecular sulfinamide bonds in S100A8 produced by hypochlorite oxidation. *J. Biol. Chem.* 2001, *276*, 33393–33401.

59 Winterbourn, C.C. and Brennan, S.O. Characterization of the oxidation products of the reaction between reduced glutathione and hypochlorous acid. *Biochem. J.* 1997, *326*, 87–92.

60 Peskin, A.V. and Winterbourn, C.C. Kinetics of the reactions of hypochlorous acid and amino acid chloramines with thiols, methionine, and ascorbate. *Free Radic. Biol. Med.* 2001, *30*, 572–579.

61 Fu, X., Mueller, D.M., and Heinecke, J.W. Generation of intramolecular and intermolecular sulfenamides, sulfinamides, and sulfonamides by hypochlorous acid: a potential pathway for oxidative cross-linking of low-density lipoprotein by myeloperoxidase. *Biochemistry* 2002, *41*, 1293–1301.

62 Andrés, J., Canle, L.M., García, M.V., Rodríguez Vázquez, L.F., and Santaballa, J.A. A B3LYP/6- 31G** study on the chlorination of ammonia by hypochlorous acid. *Chem. Phys. Lett.* 2001, *342*, 405–410.

63 Temple, A., Yen, T.-Y., and Gronert, S. Identification of specific protein carbonylation sites in model oxidations of human serum albumin. *J. Am. Soc. Mass Spectrom.* 2006, *17*, 1172–1180.

64 Pattison, D.I. and Davies, M.J. Kinetic analysis of the reactions of hypobromous acid with protein components: implications for cellular damage and use of 3-bromotyrosine as a marker of oxidative stress. *Biochemistry* 2004, *43*, 4799–4809.

65 Skaff, O., Pattison, D.I., and Davies, M.J. Kinetics of hypobromous acid-mediated oxidation of lipid components and antioxidants. *Chem. Res. Toxicol.* 2007, *20*, 1980–1988.

66 McGregor, W.H. and Carpenter, F.H. Alkaline bromine oxidation of amino acids and peptides: formation of α-ketoacyl peptides and their cleavage by hydrogen peroxide. *Biochemistry* 1962, *1*, 53–60.

67 Hawkins, C.L. and Davies, M.J. The role of aromatic amino acid oxidation, protein unfolding, and aggregation in the hypobromous acid-induced inactivation of trypsin inhibitor and lysozyme. *Chem. Res. Toxicol.* 2005, *18*, 1669–1677.

68 Hawkins, C.L. and Davies, M.J. The role of reactive N-bromo species and radical intermediates in hypobromous acid-induced protein oxidation. *Free Radic. Biol. Med.* 2005, *39*, 900–912.

69 Simoyi, R.H., Streete, K., Mundoma, C., and Olojo, R. Complex kinetics in the reaction of taurine with aqueous bromine and acidic bromate: a possible cytoprotective role against hypobromous acid. *S. Afr. J. Chem.* 2002, *55*, 136–143.

70 Wajon, J.E. and Morris, J.C. Rates of formation of N-bromo amines in aqueous solution. *Inorg. Chem.* 1982, *21*, 4258–4263.

71 Thomas, E.L., Bozeman, P.M., Jefferson, M.M., and King, C.C. Oxidation of bromide by the human leukocyte enzymes myeloperoxidase and eosinophil peroxidase. Formation of bromamines. *J. Biol. Chem.* 1995, *270*, 2906–2913.

72 Carr, A.C., Winterbourn, C.C., and Van Den Berg, J.J.M. Peroxidase-mediated bromination of unsaturated fatty acids to form bromohydrins. *Arch. Biochem. Biophys.* 1996, *327*, 227–233.

73 Ajibola, R.O. and Simoyi, R.H. S-Oxygenation of thiocarbamides IV: kinetics of oxidation of tetramethylthiourea by aqueous bromine and acidic bromate. *J. Phys. Chem. A* 2011, *115*, 2735–2744.

74 Darkwa, J., Mundoma, C., and Simoyi, R.H. Antioxidant chemistry: reactivity and oxidation of DL-cysteine by some common oxidants. *J. Chem. Soc. Faraday Trans.* 1998, *94*, 1971–1978.

75 Darkwa, J., Olojo, R., Olagunju, O., Otoikhian, A., and Simoyi, R. Oxyhalogen-sulfur chemistry: oxidation of N-acetylcysteine by chlorite and acidic bromate. *J. Phys. Chem. A* 2003, *107*, 9834–9845.

76 Morakinyo, M.K., Chikwana, E., and Simoyi, R.H. Oxyhalogen-sulfur chemistry—kinetics and mechanism of the bromate oxidation of cysteamine. *Can. J. Chem.* 2008, *86*, 416–425.

77 Tenovuo, J., Pruitt, K.M., Mansson-Rahemtulla, B., Harrington, P., and Baldone, D.C. Products of thiocyanate peroxidation: properties and reaction mechanisms. *Biochem. Biophys. Acta—Protein Struct. Mol.* 1986, *870*, 377–384.

78 Aune, T.M. and Thomas, E.L. Accumulation of hypothiocyanite ion during peroxidase-catalyzed oxidation of thiocyanate ion. *Eur. J. Biochem.* 1977, *80*, 209–214.

79 Furtmuller, P.G., Burner, U., and Obinger, C. Reaction of myeloperoxidase compound I with chloride, bromide, iodide, and thiocyanate. *Biochemistry* 1998, *37*, 17923–17930.

80 Wang, Z., Nicholls, S.J., Rodriguez, E.R., Kummu, O., Hörkkö, S., Barnard, J., Reynolds, W.F., Topol, E.J., DiDonato, J.A., and Hazen, S.L. Protein carbamylation links inflammation, smoking, uremia and atherogenesis. *Nat. Med.* 2007, *13*, 1176–1184.

81 Wang, X. and Ashby, M.T. Reactive sulfur species: kinetics and mechanism of the reaction of thiocarbamate-S-oxide with cysteine. *Chem. Res. Toxicol.* 2008, *21*, 2120–2126.

82 Nagy, P., Alguindigue, S.S., and Ashby, M.T. Lactoperoxidase-catalyzed oxidation of thiocyanate by hydrogen peroxide: a reinvestigation of hypothiocyanite by nuclear magnetic resonance and optical spectroscopy. *Biochemistry* 2006, *45*, 12610–12616.

83 Lovaas, E. Free radical generation and coupled thiol oxidation by lactoperoxidase/ SCN^-/H_2O_2. *Free Radic. Biol. Med.* 1992, *13*, 187–195.

84 Skaff, O., Pattison, D.I., and Davies, M.J. Hypothiocyanous acid reactivity with low-molecular-mass and protein thiols: absolute rate constants and assessment of biological relevance. *Biochem. J.* 2009, *422*, 111–117.

85 Nagy, P., Jameson, G.N.L., and Winterbourn, C.C. Kinetics and mechanisms of the reaction of hypothiocyanous acid with 5-thio-2-nitrobenzoic acid and reduced glutathione. *Chem. Res. Toxicol.* 2009, *22*, 1833–1840.

86 Skaff, O., Pattison, D.I., Morgan, P.E., Bachana, R., Jain, V.K., Priyadarsini, K.I., and Davies, M.J. Selenium-containing amino acids are targets for myeloperoxidase-derived hypothiocyanous acid: determination of absolute rate constants and implications for biological damage. *Biochem. J.* 2012, *441*, 305–316.

87 Lloyd, M.M., van Reyk, D.M., Davies, M.J., and Hawkins, C.L. Hypothiocyanous acid is a more potent inducer of apoptosis and protein thiol depletion in murine macrophage cells than hypochlorous acid or hypobromous acid. *Biochem. J.* 2008, *414*, 271–280.

88 Arlandson, M., Decker, T., Roongta, V.A., Bonilla, L., Mayo, K.H., MacPherson, J.C., Hazen, S.L., and Slungaard, A. Eosinophil peroxidase oxidation of thiocyanate. Characterization of major reaction products and a potential sulfhydryl-targeted cytotoxicity system. *J. Biol. Chem.* 2001, *276*, 215–224.

89 Grisham, M.B. and Ryan, E.M. Cytotoxic properties of salivary oxidants. *Am. J. Physiol. -Cell Physiol.* 1990, *258*, C115–C121.

90 Hawkins, C.L., Pattison, D.I., Stanley, N.R., and Davies, M.J. Tryptophan residues are targets in hypothiocyanous acid-mediated protein oxidation. *Biochem. J.* 2008, *416*, 441–452.

91 Ashby, M.T., Aneetha, H., Carlson, A.C., Scott, M.J., and Beal, J.L. Bioorganic chemistry of hypothiocyanite. *Phosphorus Sulfur Silicon Relat. Elem.* 2005, *180*, 1369–1374.

92 Ashley, D.C., Brinkley, D.W., and Roth, J.P. Oxygen isotope effects as structural and mechanistic probes in inorganic oxidation chemistry. *Inorg. Chem.* 2010, *49*, 3661–3675.

93 Alguindigue Nimmo, S.L., Lemma, K., and Ashby, M.T. Reactions of cysteine sulfenyl thiocyanate with thiols to give unsymmetrical disulfides. *Heteroat. Chem.* 2007, *18*, 467–471.

94 Barrett, T.J., Pattison, D.I., Leonard, S.E., Carroll, K.S., Davies, M.J., and Hawkins, C.L. Inactivation of thiol-dependent enzymes by hypothiocyanous acid: role of sulfenyl thiocyanate and sulfenic acid intermediates. *Free Radic. Biol. Med.* 2012, *52*, 1075–1085.

95 Barrett, T.J. and Hawkins, C.L. Hypothiocyanous acid: benign or deadly? *Chem. Res. Toxicol.* 2012, *25*, 263–273.

96 Margerum, D.W., Schurter, L.M., Hobson, J., and Moore, E.E. Water chlorination chemistry: nonmetal redox kinetics of chloramine and nitrite ion. *Environ. Sci. Technol.* 1994, *28*, 331–337.

97 Sathasivan, A., Bal Krishna, K.C., and Fisher, I. Development and application of a method for quantifying factors affecting chloramine decay in service reservoirs. *Water Res.* 2010, *44*, 4463–4472.

98 Sathasivan, A., Fisher, J., and Kastl, G. Simple method for quantifying microbiologically assisted chloramine decay in drinking water. *Environ. Sci. Technol.* 2005, *39*, 5407–5413.

99 Hazen, S.L., D'Avignon, A., Anderson, M.M., Hsu, F.F., and Heinecke, J.W. Human neutrophils employ the myeloperoxidase-hydrogen peroxide-chloride system to oxidize α-amino acids to a family of reactive aldehydes: mechanistic studies identifying labile intermediates along the reaction pathway. *J. Biol. Chem.* 1998, *273*, 4997–5005.

100 Anderson, M.M., Hazen, S.L., Hsu, F.F., and Heinecke, J.W. Human neutrophils employ the myeloperoxidase-hydrogen peroxide-chloride system to convert hydroxy-amino acids into glycolaldehyde, 2-hydroxypropanal, and acrolein: a mechanism for the generation of highly reactive α-hydroxy and α,β-unsaturated aldehydes by phagocytes at sites of inflammation. *J. Clin. Invest.* 1997, *99*, 424–432.

101 Stelmaszynska, T. and Zgliczynski, J.M. N-(2-Oxoacyl)amino acids and nitriles of dipeptide chlorination mediated by myeloperoxidase/H_zO_z/Cl^- System. *Eur. J. Biochem.* 1978, *92*, 301–308.

102 Vissers, M.C.M. and Winterbourn, C.C. Oxidative damage to fibronectin. I. The effects of the neutrophil myeloperoxidase system and HOCl. *Arch. Biochem. Biophys.* 1991, *285*, 53–59.

103 Chapman, A.L.P., Senthilmohan, R., Winterbourn, C.C., and Kettle, A.J. Comparison of mono- and dichlorinated tyrosines with carbonyls for detection of hypochlorous acid modified proteins. *Arch. Biochem. Biophys.* 2000, *377*, 95–100.

104 Hawkins, C.L. and Davies, M.J. Hypochlorite-induced damage to proteins: formation of nitrogen-centred radicals from lysine residues and their role in protein fragmentation. *Biochem. J.* 1998, *332*, 617–625.

105 Yang, C., Gu, Z., Yang, H., Yang, M., Gotto, A.M., and Smith, C.V. Oxidative modification of apoB-100 by exposure of low density lipoproteins to HOCl in vivo. *Free Radic. Biol. Med.* 1997, *23*, 82–89.

106 Hazell, L.J., van den Berg, J.J.M., and Stocker, R. Oxidation of low-density lipoprotein by hypochlorite causes aggregation that is mediated by modification by lysine residues rather than lipid oxidation. *Biochem. J.* 1994, *302*, 297–304.

107 Hazen, S.L., Gaut, J.P., Hsu, F.F., Crowley, J.R., D'Avignon, A., and Heinecke, J.W. p-Hydroxyphenylacetaldehyde, the major product of L-tyrosine oxidation by the myeloperoxidase-H_2O_2-chloride system of phagocytes, covalently modifies ε-amino groups of protein lysine residues. *J. Biol. Chem.* 1997, *272*, 16990–16998.

108 Kumara, M.N., Linge Gowda, N.S., Mantelingu, K., and Rangappa, K.K.S. N-Bromosuccinimide assisted oxidation of tripeptides and their amino acid analogs: synthesis, kinetics, and product studies. *J. Mol. Catal. A Chem.* 2009, *309*, 172–177.

109 Linge Gowda, N.S., Kumara, M.N., Channe Gowda, D., Rangappa, K.K.S., and Made Gowda, N.M. *N*-Bromosuccinimide assisted oxidation of hydrophobic tetrapeptide sequences of elastin: a mechanistic study. *J. Mol. Catal. A Chem.* 2007, *269*, 225–233.

110 Linge Gowda, N.S., Kumara, M.N., Channe Gowda, D., and Rangappa, K.S. N-bromosuccinimide oxidation of dipeptides and their amino acids: synthesis, kinetics and mechanistic studies. *Int. J. Chem. Kinet.* 2006, *38*, 376–385.

111 Pattison, D.I., Hawkins, C.L., and Davies, M.J. Hypochlorous acid-mediated protein oxidation: how important are chloramine transfer reactions and protein tertiary structure? *Biochemistry* 2007, *46*, 9853–9864.

112 Stacey, M.M., Peskin, A.V., Vissers, M.C., and Winterbourn, C.C. Chloramines and hypochlorous acid oxidize erythrocyte peroxiredoxin 2. *Free Radic. Biol. Med.* 2009, *47*, 1468–1476.

113 Peskin, A.V., Midwinter, R.G., Harwood, D.T., and Winterbourn, C.C. Chlorine transfer between glycine, taurine, and histamine: reaction rates and impact on cellular reactivity. *Free Radic. Biol. Med.* 2004, *37*, 1622–1630.

114 Prütz, W.A., Kissner, R., and Koppenol, W.H. Oxidation of NADH by chloramines and chloramides and its activation by iodide and by tertiary amines. *Arch. Biochem. Biophys.* 2001, *393*, 297–307.

115 Pattison, D.I. and Davies, M.J. Evidence for rapid inter- and intramolecular chlorine transfer reactions of histamine and carnosine chloramines: implications for the prevention of hypochlorous-acid-mediated damage. *Biochemistry* 2006, *45*, 8152–8162.

116 Peskin, A.V. and Winterbourn, C.C. Histamine chloramine reactivity with thiol compounds, ascorbate, and methionine and with intracellular glutathione. *Free Radic. Biol. Med.* 2003, *35*, 1252–1260.

117 Prütz, W.A. Interactions of hypochlorous acid with pyrimidine nucleotides, and secondary reactions of chlorinated pyrimidines with GSH, NADH, and other substrates. *Arch. Biochem. Biophys.* 1998, *349*, 183–191.

118 Prütz, W.A. Consecutive halogen transfer between various functional groups induced by reaction of hypohalous acids: NADH oxidation by halogenated amide groups. *Arch. Biochem. Biophys.* 1999, *371*, 107–114.

119 Pattison, D.I. and Davies, M.J. Reactions of myeloperoxidase-derived oxidants with biological substrates: gaining chemical insight into human inflammatory diseases. *Curr. Med. Chem.* 2006, *13*, 3271–3290.

120 Peskin, A.V. and Winterbourn, C.C. Taurine chloramine is more selective than hypochlorous acid at targeting critical cysteines and inactivating creatine kinase and glyceraldehyde-3-phosphate dehydrogenase. *Free Radic. Biol. Med.* 2006, *40*, 45–53.

121 Weiss, S.J., Test, S.T., Eckmann, C.M., Roos, D., and Regiani, S. Brominating oxidants generated by human eosinophils. *Science* 1986, *234*, 197–200.

122 Ximenes, V.F., Padovan, C.Z., Carvalho, D.A., and Fernandes, J.R. Oxidation of melatonin by taurine chloramine. *J. Pineal Res.* 2010, *49*, 115–122.

123 Ximenes, V.F., Da Fonseca, L.M., and De Almeida, A.C. Taurine bromamine: a potent oxidant of tryptophan residues in albumin. *Arch. Biochem. Biophys.* 2011, *507*, 315–322.

124 Pattison, D.I., Davies, M.J., and Asmus, K. Absolute rate constants for the formation of nitrogen-centered radical from chloramins/amides and their reactions with antioxidants. *J. Chem. Soc. Perkin Trans. 2* 2002, 1461–1467.

125 Pattison, D.I., O'Reilly, R.J., Skaff, O., Radom, L., Anderson, R.F., and Davies, M.J. One-electron reduction of *N*-chlorinated and *N*-brominated species is a source of radicals and bromine atom formation. *Chem. Res. Toxicol.* 2011, *24*, 371–382.

126 Jakopitsch, C., Spalteholz, H., Furtmüller, P.G., Arnhold, J., and Obinger, C. Mechanism of reaction of horseradish peroxidase with chlorite and chlorine dioxide. *J. Inorg. Biochem.* 2008, *102*, 293–302.

127 Tsai, L., Higby, R., and Schade, J. Disinfection of poultry chiller water with chlorine dioxide: consumption and by-product formation. *J. Agric. Food Chem.* 1995, *43*, 2768–2773.

128 Kim, J.M., Huang, T.S., Marshall, M.R., and Wei, C.I. Chlorine dioxide treatment of seafoods to reduce bacterial loads. *Food Sci.* 1999, *64*, 1089–1093.

129 Kim, H., Kang, Y., Beuchat, L.R., and Ryu, J.-H. Production and stability of chlorine dioxide in organic acid solutions as affected by pH, type of acid, and concentration of sodium chlorite, and its effectiveness in inactivating *Bacillus cereus* spores. *Food Microbiol.* 2008, *25*, 964–969.

130 Kull, T.P.J., Backlund, P.H., Karlsson, K.M., and Meriluoto, J.A.O. Oxidation of the cyanobacterial heptotoxin microcystin-LR by chlorine dioxide: reaction kinetics, characterization, and toxicity of reaction products. *Environ. Sci. Technol.* 2004, *38*, 6025–6031.

131 Kull, T.P.J., Sjovall, O.T., Tammenkoski, M.K., Backlund, P.H., and Meriluoto, J.A.O. Oxidation of the cyanobacterial hepatoxin microcystin-LR by chlorine dioxide: influence of natural organic matter. *Environ. Sci. Technol.* 2006, *40*, 1504–1510.

132 Gordon, G. and Rosenblatt, A. Chlorine dioxide: the current state of the art. *Ozone Sci. Eng.* 2005, *27*, 203–207.

133 Vandekinderen, I., Devlieghere, F., Van Camp, J., Kerkaert, B., Cucu, T., Ragaert, P., De Bruyne, J., and De Meulenaer, B. Effects of food composition on the inactivation of foodborne microorganisms by chlorine dioxide. *Int. J. Food Microbiol.* 2009, *131*, 138–144.

134 Wilson, S.C., Wu, C., Andriychuck, L.A., Martin, J.M., Brasel, T.L., Jumper, C.A., and Straus, D.C. Effect of chlorine dioxide gas on fungi and mycotoxins associated with sick building syndrome. *Appl. Environ. Microbiol.* 2005, *71*, 5399–5402.

135 Southwell, K.L. Chlorine dioxide dry fumigation in special collection libraries—a case study. *Libr. Arch. Secur.* 2003, *18*, 39–49.

136 Buttner, M.P., Cruz, P., Stetzenback, L.D., Klima-Comba, A.K., Stevens, V.L., and Cronin, T.D. Determination of the efficacy of two building decontamination strategies by surface sampling with culture and quantitative PCR analysis. *Appl. Environ. Microbiol.* 2004, *70*, 4740–4747.

137 Hubbard, H., Poppendieck, D., and Corsi, R.L. Chlorine dioxide reactions with indoor materials during building disinfection: surface uptake. *Environ. Sci. Technol.* 2009, *43*, 1329–1335.

138 Junli, H., Li, W., Nenqi, R., Li, L.X., Fun, S.R., and Guanle, Y. Disinfection effect of chlorine dioxide on viruses, algae and animal planktons in water. *Water Res.* 1997, *31*, 455–460.

139 Lim, M.Y., Kim, J.-M., and Ko, G. Disinfection kinetics of murine norovirus using chlorine and chlorine dioxide. *Water Res.* 2010, *44*, 3243–3251.

140 Korich, D.G., Mead, J.R., Madore, M.S., Sinclair, N.A., and Sterling, C.R. Effects of ozone, chlorine dioxide, chlorine, and monochloramine on *Cryptosporidium parvum* oocyst viability. *Appl. Environ. Microbiol.* 1990, *56*, 1423–1428.

141 Shams, A.M., O'Connell, H., Arduino, M.J., and Rose, L.J. Chlorine dioxide inactivation of bacterial threat agents. *Lett. Appl. Microbiol.* 2011, *53*, 225–230.

142 Umile, T.P. and Groves, J.T. Catalytic generation of chlorine dioxide from chlorite using a water-soluble manganese porphyrin. *Angew. Chem. Int. Ed.* 2011, *50*, 695–698.

143 Zdilla, M.J., Lee, A.Q., and Abu-Omar, M.M. Bioinspired dismutation of chlorite to dioxygen and chloride catalyzed by a water-soluble iron porphyrin. *Angew. Chem. Int. Ed.* 2008, *47*, 7697–7700.

144 Zdilla, M.J., Lee, A.Q., and Abu-Omar, M.M. Concerted dismutation of chlorite ion: water-soluble iron-porphyrins as first generation model complexes for chlorite dismutase. *Inorg. Chem.* 2009, *48*, 2260–2268.

145 Shahangian, S. and Hager, L.P. The reaction of chloroperoxidase with chlorite and chlorine dioxide. *J. Biol. Chem.* 1981, *256*, 6034–6040.

146 Ingram, P.R., Homer, N.Z.M., Smith, R.A., Pitt, A.R., Wilson, C.G., Olejnik, O., and Spickett, C.M. The interaction of sodium chlorite with phospholipids and glutathione: a comparison of effects in vitro, in mammalian and in microbial cells. *Arch. Biochem. Biophys.* 2003, *410*, 121–133.

147 Lee, A.Q., Streit, B.R., Zdilla, M.J., Abu-Omar, M.M., and DuBois, J.L. Mechanism of and exquisite selectivity for O-O bond formation by the heme-dependent chlorite dismutase. *Proc. Natl Acad. Sci. U.S.A.* 2008, *105*, 15654–15659.

148 Goblirsch, B., Kurker, R.C., Streit, B.R., Wilmot, C.M., and DuBois, J.L. Chlorite dismutases, DyPs, and EfeB: 3 microbial heme enzyme families comprise the CDE structural superfamily. *J. Mol. Biol.* 2011, *408*, 379–398.

149 Streit, B.R., Blanc, B., Lukat-Rodgers, G.S., Rodgers, K.R., and Dubois, J.L. How active-site protonation state influences the reactivity and ligation of the heme in chlorite dismutase. *J. Am. Chem. Soc.* 2010, *132*, 5711–5724.

150 Goblirsch, B.R., Streit, B.R., DuBois, J.L., and Wilmot, C.M. Structural features promoting dioxygen production by *Dechloromonas aromatica* chlorite dismutase. *J. Biol. Inorg. Chem.* 2010, *15*, 879–888.

151 Odeh, I.N., Francisco, J.S., and Margerum, D.W. New pathways for chlorine dioxide decomposition in basic solution. *Inorg. Chem.* 2002, *41*, 6500–6506.

152 Hoigne, J. and Bader, H. Kinetics of reactions of chlorine dioxide (OClO) in water—I. Rate constants for inorganic and organic compounds. *Water Res.* 1994, *28*, 45–55.

153 Wang, L., Nicoson, J.S., Huff Hartz, K.E., Francisco, J.S., and Margerum, D.W. Bromite ion catalysis of the disproportionation of chlorine dioxide with nucleophile assistance of electron-transfer reactions between ClO_2 and BrO_2 in basic solution. *Inorg. Chem.* 2002, *41*, 108–113.

154 Yakupov, M.Z., Shereshovets, V.V., Imashev, U.B., and Ismagilov, F.R. Liquid-phase oxidation of thiols with chlorine dioxide. *Russ. Chem. Rev.* 2001, *50*, 2352–2355.

155 Tan, H.K., Wheeler, W.B., and Wei, C.I. Reaction of chlorine dioxide with amino acids and peptides: kinetics and mutagenicity studies. *Mutat. Res.* 1987, *188*, 259–266.

156 Tratnyek, P.G. and Hoigne, J. Kinetics of reactions of chlorine dioxide (OClO) in water—II. Quantitative structure-activity relationships for phenolic compounds. *Water Res.* 1994, *28*, 57–66.

157 Knocke, W.R., Van Benschoten, J.E., Kearney, M.J., Soborski, A.W., and Reckhow, D.A. Kinetics of manganese and iron oxidation by potassium permanganate and chlorine dioxide. *J. Am. Water Works Assoc.* 1991, *83*, 80–87.

158 Moore, E.R., Bourne, A.E., Hoppe, T.J., Abode, P.J., Boone, S.R., and Purser, G.H. Kinetics and mechanism of the oxidation of iron(II) ion by chlorine dioxide in aqueous solution. *Int. J. Chem. Kinet.* 2004, *36*, 554–565.

159 Wang, L., Odeh, I.N., and Margerum, D.W. Chlorine dioxide reduction by aqueous iron(II) through outer-sphere and inner-sphere electron-transfer pathways. *Inorg. Chem.* 2004, *43*, 7545–7551.

160 Rosenblatt, D.H. Oxidation of phenol and hydroquinone by chlorine dioxide. *Environ. Sci. Technol.* 1982, *16*, 396–402.

161 Burrows, E.P. and Rosenblatt, D.H. Conversion of acyclic amines to amides by chlorine dioxide. *J. Org. Chem.* 1982, *47*, 892–893.

162 Csekő, G. and Horváth, A.K. Kinetics and mechanism of the chlorine dioxide-trithionate reaction. *J. Phys. Chem. A* 2012, *116*, 2911–2919.

163 Ison, A., Odeh, I.N., and Margerum, D.W. Kinetics and mechanisms of chlorine dioxide and chlorite oxidations of cysteine and glutathione. *Inorg. Chem.* 2006, *45*, 8768–8775.

164 Darkwa, J., Olojo, R., Chikwana, E., and Simoyi, R.H. Antioxidant chemistry: oxidation of L-cysteine and its metabolites by chlorite and chlorine dioxide. *J. Phys. Chem. A* 2004, *108*, 5576–5587.

165 Taymaz, K., Williams, D.T., and Benoit, F.F. Reactions of aqueous chlorine dioxide with amino acids found in water. *Bull. Environ. Contam. Toxicol.* 1979, *23*, 456–463.

166 Navalon, S., Alvaro, M., and Garcia, H. Chlorine dioxide reaction with selected amino acids in water. *J. Hazard. Mater.* 2009, *164*, 1089–1097.

167 Noss, C.I., Dennis, W.H., and Olivieri, V.P. Reactivity of chlorine dioxide with nucleic acids and proteins. *Water Chlorination Environ. Impact Health Eff.* 1983, *4*, 1077–1086.

168 Napolitano, M.J., Green, B.J., Nicoson, J.S., and Margerum, D.W. Chlorine dioxide oxidations of tyrosine, *N*-acetyltyrosine, and dopa. *Chem. Res. Toxicol.* 2005, *18*, 501–508.

169 Masschelein, W.J. and Rice, R.G. *Chlorine Dioxide: Chemistry and Environmental Impact of Oxychlorine Compounds*. Ann Harbor Science Publishers, Ann Harbor, MI, 1979, p. 190.

170 Bakhmutova-Albert, E.V., Margerum, D.W., Auer, J.G., and Appelgate, B.M. Chlorine dioxide oxidaiton of dihydronicotinamide adenine dinucleotide (NADH). *Inorg. Chem.* 2008, *47*, 2205–2211.

171 Stewart, D.J., Napolitano, M.J., Bakhmutova-Albert, E.V., and Margerum, D.W. Kinetics and mechanisms of chlorine dioxide oxidation of tryptophan. *Inorg. Chem.* 2008, *47*, 1639–1647.

172 Lynch, E., Sheerin, A., Claxson, A.W.D., Atherton, M.D., Rhodes, C.J., Silwood, C.J.L., Naughton, D.P., and Grootveld, M. Multicomponent spectroscopic investigations of salivary antioxidant consumption by an oral rinse preparation containing the stable free radical species chlorine dioxide (ClO_2). *Free Radic. Res.* 1997, *26*, 209–234.

173 Ogata, N. Denaturation of protein by chlorine dioxide: oxidative modification of tryptophan and tyrosine residues. *Biochemistry* 2007, *46*, 4898–4911.

174 Noss, C.I., Hauchman, F.S., and Olivieri, V.P. Chlorine dioxide reactivity with proteins. *Water Res.* 1986, *20*, 351–356.

4

REACTIVE OXYGEN SPECIES

A number of enzymatic and nonenzymatic processes can produce reactive oxygen species (ROS) in mammalian cells. ROS include $O_2^{-\bullet}$, $^1O_2$, O_3, and $^{\bullet}OH$, which are involved in cardiovascular diseases, asthma, diabetes, and asthma [1–3]. Some of the important inputs of ROS are lipoxygenase, mitochondrial respiratory chain, xanthine oxidase, nicotinamide adenine dinucleotide phosphate (NADPH) oxidases, uncoupled nitric oxide synthase (NOS), and myeloperoxidase (MPO) [4]. Impairment of cellular functions, cellular apoptosis, and killing of pathogens may be associated with high levels of ROS. ROS can modify proteins by oxidizing amino acid residues, which can lead to fragmentation of the polypeptide chain, oxidation of amino acid chains, and generation of protein–protein cross-linking [5–7]. This chapter begins with the discussion of different approaches to generate ROS in order to study their reactivities with protein residues, followed by reaction pathways of the mechanisms involved in the oxidation of proteins by ROS [5, 8–15].

4.1 SUPEROXIDE

4.1.1 Generation

Several techniques have been used to produce superoxide ($O_2^{\bullet-}/HO_2^{\bullet}$) to study its reactions. These techniques include pulse radiolysis, UV photolysis, and

Oxidation of Amino Acids, Peptides, and Proteins: Kinetics and Mechanism, First Edition.
Virender K. Sharma.

electrolysis. The pulse radiolysis technique has been used widely to study enzymatic reactions in which the radiolysis of water is performed under high-energy electrons (Eq. 4.1):

$$H_2O \rightsquigarrow {}^{\bullet}OH, e^-_{aq}, H^{\bullet}, H_2, H_2O_2. \quad (4.1)$$

If water contains molecular oxygen and sodium formate, the primary radicals are converted to $O_2^{\bullet-}/HO_2^{\bullet}$ (reactions 4.2–4.5) [16]:

$$e^-_{aq} + O_2 \rightarrow O_2^{\bullet-} \quad k_2 = 2\times10^{10}\,/M/s \quad (4.2)$$

$$H^{\bullet} + O_2 \rightarrow HO_2^{\bullet} \quad k_3 = 5\times10^{9}\,/M/s \quad (4.3)$$

$${}^{\bullet}OH + HCOO^- \rightarrow COO^{\bullet-} + H_2O \quad k_4 = 5\times10^{9}\,/M/s \quad (4.4)$$

$$COO^{\bullet-} + O_2 \rightarrow CO_2 + O_2^{\bullet-} \quad k_5 = 5\times10^{9}\,/M/s. \quad (4.5)$$

Vacuum ultraviolet (VUV) photolysis (165–185 nm) generates the primary radicals, ${}^{\bullet}OH$ and $H^{\bullet}$, which undergo reactions (4.3)–(4.5) to produce $O_2^{\bullet-}$ [16]. UV photolysis of H_2O_2 forms $O_2^{\bullet-}/HO_2^{\bullet}$ by the following reactions (Eqs. 4.6–4.8):

$$H_2O_2 + h\nu \rightarrow 2{}^{\bullet}OH \quad (4.6)$$

$$H_2O_2 + {}^{\bullet}OH \rightarrow HO_2^{\bullet} + H_2O \quad (4.7)$$

$$HO_2^- + {}^{\bullet}OH \rightarrow O_2^{\bullet-} + H_2O. \quad (4.8)$$

Reactions of semiquinone radicals with O_2 resulted in superoxide ions [17]. A study on the electrochemical production of $O_2^{\bullet-}$ from O_2 in concentrated NaOH solutions has also been performed [18]. The electroreduction of O_2 in nonaqueous aprotic solvents, such as dimethyl formamide, dimethyl sulfoxide, acetonitrile, propylene, and acetone, also generates $O_2^{\bullet-}$ [19, 20]. An aprotic solution of $O_2^{\bullet-}$ can also be prepared by adding KO_2 into a "crown" ether. Recently, room-temperature ionic liquid as a solvent has been applied to produce $O_2^{\bullet-}$ [21, 22].

4.1.2 Properties

Superoxide is produced in several oxidation reactions occurring in chemical and biological systems. For example, superoxide is the first species produced in the respiratory chain by an electron transfer reduction of oxygen [23]. The thermodynamic parameters of formation of superoxide have been determined, which may help in evaluating processes that lead to its generation and disappearance. ΔH_f^o were determined as −36.0 and −24.7 kJ/mol at 298 K for $HO_2^{\bullet}$ and $O_2^{\bullet-}$, respectively [24]. S^o values were calculated as 138 and 79.5 J/mol [24]. Enthalpy, entropy, and free energy of hydration for the $O_2^{\bullet-}$ species have also been calculated as −398 kJ/mol, −124 J/K/mol, and −356 kJ/mol, respectively [25]. The diffusion coefficient for the $O_2^{\bullet-}$ species in 0.1 M NaOH and 10% (v)

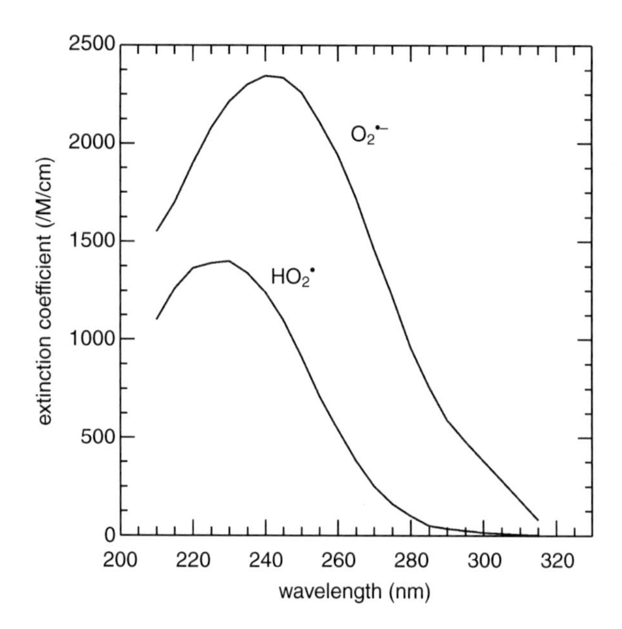

Figure 4.1. The spectra of O_2^-/HO_2 (adapted from Abreu and Cabelli [31] with the permission of Elsevier Inc.).

PrOH was estimated to be $1.5 \times 10^{-5}\,cm^2/s$ [26]. The reduction potentials of superoxide in water at 1 atm (vs. normal hydrogen electrode [NHE]) vary from –0.05 V (O_2, $H^+/HO_2^{\bullet}$) to –0.33 V ($O_2/O_2^{\bullet-}$) [25]. This suggests the dependence of reaction kinetics of superoxide on pH.

Spectroscopic methods have been applied to learn fundamental properties and the kinetics of the reactions of superoxide. Electron spin resonance (ESR) at ambient temperatures can detect $HO_2^{\bullet}$ in acidic solution, while the detection of $O_2^{\bullet-}$ is only possible at very low temperatures [27, 28]. The spin-trap method has also been used to detect $HO_2^{\bullet}/O_2^{\bullet-}$ [29, 30]. The most common spectroscopic method uses characteristic spectra of $HO_2^{\bullet}$ and $O_2^{\bullet-}$ (Fig. 4.1) [31]. In the low UV range, both species have maxima (ε_{225nm} $(HO_2^{\bullet}) = (1.40 \pm 0.08) \times 10^3$/M/s and ($\varepsilon_{245nm}$ $(O_2^{\bullet-}) = (2.35 \pm 0.12) \times 10^3$/M/s) [16]. The acid–base equilibrium of superoxide is represented by reaction (4.9). Thus, pH controls the distribution between $HO_2^{\bullet}$ and $O_2^{\bullet-}$:

$$HO_2^{\bullet} \rightleftharpoons H^+ + O_2^{\bullet-} \quad pK_a = 4.8. \tag{4.9}$$

In recent years, efforts have been made to detect low concentrations of superoxide. Reactions of $O_2^{\bullet-}$ with chemical indicators such as tetranitromethane, cytochrome *c*, and nitro blue tetrazolium, which form reduced products with intense optical absorbance, have also been utilized to indirectly quantify superoxide in aqueous solution [32–36]. Progress is also being made

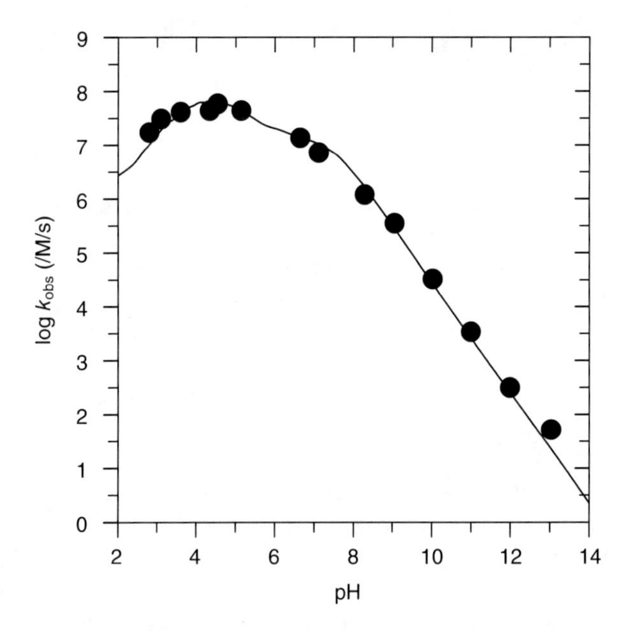

Figure 4.2. The pH dependence of the rate constants for disproportionation of O_2^-/HO_2 (adapted from Abreu and Cabelli [31] with the permission of Elsevier Inc.).

on detecting superoxide in biological processes. Examples include a fluorescent sensor using acridinium ion-linked porphyrin triad and single-cell imaging techniques in skeletal muscle fibers [37, 38]. In natural waters, the chemical reagent 2-methyl-6-(4-methoxyphenyl)-3,7-dihydroimidazo[1,2-a] pyrazin-3(7H)-one (MCLA) has been shown to detect superoxide in picomolar concentrations [39].

The spontaneous decay of superoxide in aqueous solution is pH dependent (Fig. 4.2). The plot in Figure 4.2 has a slope of –1 at pH > 6.0 and a maximum at pH ~ 4.8. The second-order disproportionation process was explained by reactions (4.10)–(4.12) [40]. The activation energies of reactions (4.10) and (4.11) were determined as 20.5–24.7 and 8.8 kJ/mol, respectively [25]:

$$HO_2^\bullet + HO_2^\bullet \rightarrow O_2 + H_2O_2 \quad k_{10} = 1\times 10^6\,/M/s \tag{4.10}$$

$$HO_2^\bullet + O_2^{\bullet-}(+H^+) \rightarrow O_2 + H_2O_2 \quad k_{11} = 1\times 10^8\,/M/s \tag{4.11}$$

$$O_2^{\bullet-} + O_2^{\bullet-}(+2H^+) \rightarrow O_2 + H_2O_2 \quad k_{12} = \text{no reaction.} \tag{4.12}$$

4.1.3 Reactivity

Superoxide goes through several mechanisms depending on the nature of the substrate [41]. $O_2^{\bullet-}$ is known to react with alkyl sulfide through nucleophilic substitution or as a moderate one-electron reducing agent with numerous transition metal complexes and organic compounds [19, 42–45]. $O_2^{\bullet-}$ also acts as a weak base with acidic compounds, and hence, a proton transfer is involved [42, 43]. Both proton transfer and radical transfer pathways occur when phenolic compounds react with $O_2^{\bullet-}$ [46, 47].

The rates of reactions of superoxide with amino acids were performed because amino acids are building blocks of large molecules (e.g., protein). The second-order rate constants were found to be relatively low [48]. The second-order rate constants for the reactions of $HO_2^{\bullet}$ were in the range of 1×10^1/M/s for aliphatic amino acids to ~6×10^2/M/s for Cys. Comparatively, the rate constants for the reactions of $O_2^{\bullet-}$ with amino acids were significantly smaller and ranged from 1.0×10^{-1}/M/s to ~2×10^1/M/s. Such difference in reactivity of two superoxide species may be that $HO_2^{\bullet}$ acts like a weak oxidizing agent, while $O_2^{\bullet-}$ behaves as both a weak oxidizing and reducing agent. Furthermore, a rate constant represents composite of many reactions depending on the fractions of $HO_2^{\bullet}/O_2^{\bullet-}$ and protonated/unprotonated forms of amino acids. Other sulfur-containing amino acids, such as cystine and Met, were unreactive with superoxide. However, thiols such as dithiothreitol, *N*-acetylcysteine, and glutathione (GSH) showed significant reactivity with rate constants of 3.5×10^1/M/s, 6.8×10^1/M/s, and 2.4×10^1/M/s, respectively [49–51]. The relative reactivity of thiols with superoxide inversely correlated with pK_a of thiols. The *ab initio* molecular calculations suggested the main pathway of formation of a three-electron-bonded adduct followed by elimination of a hydroxide ion to yield a sulfinyl radical as the reaction product [52, 53]. The formation of a sulfinyl radical in the oxidation of thiols by superoxide has also been supported by *ab initio* molecular orbital theory [52]. The disulfide was the only product in the reaction of *N*-acetylcystein with $O_2^{\bullet-}$, studied by steady-state radiolysis [49].

Phe was the least reactive compound among aromatic amino acids at alkaline pH ($k = (3.6 \pm 0.5) \times 10^{-1}$/M/s). His had a rate constant of 1.0 ± 0.2/M/s in the alkaline pH range. The rate constants of Tyr and Trp were $(1.0 \pm 0.2) \times 10^1$/M/s and $(2.4 \pm 0.3) \times 10^1$/M/s, respectively, in the alkaline pH range. These results suggest that the aromatic ring(s) in Tyr and Trp influenced reactivity with $O_2^{\bullet-}$.

A number of studies on the kinetics of superoxide with biological molecules such as ascorbic acids, vitamin E, Trolox, catalase, MPO, and horseradish peroxidase have been summarized and reviewed [19, 25]. Studies involving metal complexes and $O_2^{\bullet-}$ have been performed to understand the role of metals in the deleterious effects of superoxide in biological systems [19, 54–56]. The reactions between Cu^{2+}–amino acid complexes with $HO_2^{\bullet}/O_2^{\bullet-}$ are of interest because many of such copper complexes possess significant catalytic efficiency to disproportionate superoxide radicals [19, 57]. For example, a study on the reactivity of Cu^{2+}–histidine complexes in the pH range from 1 to 10 showed that one of the complexes, $(CuHist_2H)^{3+}$, out of possible six complexes was catalytic active [58]. The reaction of Cu^{2+}–arginine complexes with superoxide in the pH range of 2.5–11.0 also supported the catalytic effect of copper complexes [57]. Reactions of copper complexes with superoxide were postulated to proceed via an inner-sphere mechanism with the formation of a CuO_2^+ transient [57].

Reactions of various iron compounds and its complexes with $HO_2^\bullet/O_2^{\bullet-}$ have been carried out for the last several decades in order to evaluate the adverse effects of superoxide in the absence and presence of iron catalysts. The iron complexes allowed to carry out studies in neutral and basic solutions because corresponding free (or uncomplexed) iron ions form insoluble Fe(III) hydroxides at pH > 2. The reactivity between Fe(II)/Fe(III) complexes and $HO_2^\bullet/O_2^{\bullet-}$, where iron is complexed to polyaminocarboxylate ligands, is presented in Table 4.1 [54, 59–62]. The most studied complex is iron-ethylenediaminetetracetate (Fe-EDTA) complex. The oxidation–reduction potential also changes when iron is chelated, hence the rates. The reaction between the Fe^{2+}-aquo complex and hydrogen peroxide was found be about two orders of magnitude times slower than the Fe^{2+}EDTA complex [63, 64]. The type of ligand determined the rate constants of iron(III) complexes with $O_2^{\bullet-}$ (Table 4.1). Reactions (4.13) and (4.14) may explain the catalytic dismutation of $O_2^{\bullet-}$ [59]:

$$Fe^{3+}EDTA + O_2^{\bullet-} \rightarrow Fe^{2+}EDTA + O_2 \tag{4.13}$$

$$Fe^{2+}EDTA + O_2^{\bullet-} \rightarrow Fe^{3+}EDTA\text{-}O_2^{2-}. \tag{4.14}$$

A recent study using nuclear magnetic resonance (NMR) spectroscopy and pulse radiolysis technique suggested the reactions of Fe^{III} complexes with $O_2^{\bullet-}$ occur via inner-sphere mechanisms where the metal/$O_2^{\bullet-}$ association is the rate-limiting step [65]. The formation of a purple peroxo complex, $[Fe^{III}(EDTA)(\eta^2\text{-}O_2)]^{3-}$, in the reaction of Fe^{3+}EDTA and related complexes with H_2O_2 has

TABLE 4.1. Rate Constants (/M/s) of Ferrates with Hydrogen Peroxide and Superoxide

Species	$HO_2^\bullet$	$O_2^{\bullet-}$	Species	$O_2^{\bullet-}$
Fe(II)			Fe(V)	
Fe^{2+}	1.2×10^6	1.0×10^7	$HFeO_4^{2-}$	$1.0 \pm 0.3 \times 10^{7a}$
Fe^{2+}EDTA	–	2.0×10^6		
Fe(III)			Fe(VI)	
Fe^{3+}	$<10^3$	1.5×10^8	H_2FeO_4	7.0×10^8
$Fe^{3+}EDTA(H_2O)^-$	–	1.9×10^6	$HFeO_4^-$	1.7×10^7
$Fe^{3+}DTPA^-$	–	4.4×10^5	FeO_4^{2-}	$\leq 1.0 \times 10^2$
$Fe^{3+}DTPA^{2-}$	–	6.0×10^3		
$Fe_2^{3+}O(TTHA)^{2-}$	–	1.2×10^4		

[a] pH 8.2.

Data were taken from References 54, 59, 65, and 78–80.

been investigated by kinetic and spectroscopic methods [66–70]. The Fe^{III}-peroxo complex as an intermediate was also involved in a selective reduction of $O_2^{\bullet-}$ to H_2O_2 by superoxide reductase (SOR) [71, 72]. A recent kinetic study proposed a thiolate-ligated hydroperoxo intermediate $[Fe^{III}(S^{Me2}N_4(tren)(OOH)]^+$ in the proton-dependent reduction of $O_2^{\bullet-}$ by $[Fe^{III}(S^{Me2}N_4(tren)]^+$ [73].

The measured reactivity of $HO_2^{\bullet}/O_2^{\bullet-}$ with Fe^{2+} and Fe^{3+} as a function of pH was used to understand the reaction mechanisms of Fenton and Fenton-like reactions [74]. The Fenton-type reactions are presented as

$$L_mFe^{2+} + H_2O_2 \rightarrow L_mFe^{3+} + {}^{\bullet}OH + OH^- \quad (4.15)$$

$$L_mFe^{2+} + H_2O_2 \rightarrow [L_mFe{=}O]^{2+} + H_2O. \quad (4.16)$$

Oxidations of substrates carried out by $^{\bullet}OH$ and ferryl, $[L_mFe{=}O]^{2+}$, produced by reactions (4.15) and (4.16), respectively, suggested the formation of OH radical in acidic solutions, while both the hydroxyl radical and ferryl were responsible oxidants in neutral and basic solutions [75, 76]. Stopped-flow and pulse radiolysis methods have thus been applied to study the chemistry of ferryl species, which are described in detail in Chapter 6. Additionally, different pathways to high-valent iron species were also pursued by studying the reduction of stable ferrate(VI) ion (FeO_4^{2-}) to perferryl ion (FeO_4^{3-}). The perferryl as a model was used to learn the chemistry of ferryl species with different organic substrates [77]. The rate constants of reactions of high-valent iron species with $HO_2^{\bullet}/O_2^{\bullet-}$ are provided in Table 4.1 [54, 59, 65, 78–80].

The reactivity of ferrate(VI) species with superoxide as a function of pH has also been performed (Fig. 4.3) [78]. The values of k_{obs} for the reaction decreased with pH resulting in a slope of $\simeq 1$ from pH 8–13. A plateau region was observed between the pH range of ≈4.5–7.5, followed by a decrease in rates below pH 4.0 (Fig. 4.3). The rate constants for the individual reactions of ferrate(VI) species with superoxide were calculated and are provided in Table 4.1. The stoichiometries of the reactions ($[Fe(VI)]:[O_2^{\bullet-}]$) were found to be 1:1 and 1:2 at pH 8.2 and 10.0, respectively and reactions (4.17) and (4.18) explain the observed stoichiometries [78]:

$$Fe(VI) + O_2^{\bullet-} \rightarrow Fe(V) + O_2 \quad (4.17)$$

$$Fe(V) + O_2^{\bullet-} \rightarrow Fe(IV) + O_2. \quad (4.18)$$

At pH 8.2, only reaction (4.17) occurs, while both reactions (4.17) and (4.18) participate at pH 10.0. The formed Fe(V) at pH 8.2 self-decomposes rather than reacting further with $O_2^{\bullet-}$. Reactions of Fe(VI) with superoxide and ascorbate had similar rates at neutral pH (Fig. 4.3) [78, 81].

The reactivity of superoxide with various metal-centered porphyrins has been performed to understand the formation of metal-superoxo intermediates in the metalloenzyme-catalyzed activation of oxygen and hydrogen peroxide

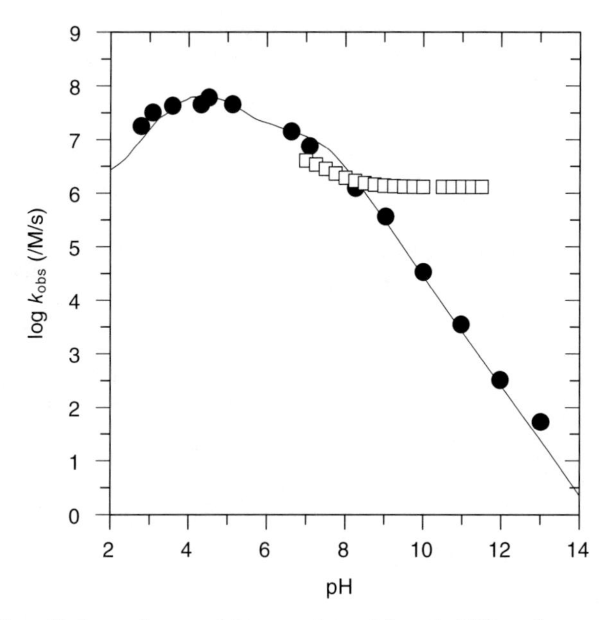

Figure 4.3. The pH dependence of the reaction of ferrate(VI) and superoxide ion and ascorbate (●, superoxide; □, ascorbate).

[19, 82, 83]. The centers of Cu(II), Zn(II), and Ni(II) were unreactive toward superoxide [84, 85]. The rates of the reactions of superoxide with Fe(II,III)-porphyrins with $O_2^{\bullet-}$ ranged from 3×10^5/M/s to 2.3×10^9/M/s and was dependent upon the type of ligand(s) in the axial positions of the porphyrins [19]. Recently, the reaction between superoxide and the Fe(II)–porphyrin complex has been studied in detail [82, 83]. The results demonstrated the reaction products consisted of Fe^{III}-peroxo and Fe^{II}-superoxo species, which were in equilibrium. The reactivity of $O_2^{\bullet-}$ with Mn(III)-porphyrins ranged on the order of 10^5–10^7. The Mn(II)-porphyrins reacted at least two orders of magnitude faster. A recent study showed the formation of a $[Mn^{III}OO]^+$ adduct as an intermediate in the reaction of Mn(II) complexes with superoxide [86]. Below is a brief description of the catalytic conversion of superoxide radical to oxygen and hydrogen peroxide by metalloenzymes.

4.1.4 Metalloenzymes

Superoxide dismutase (SOD) enzymes catalyze the dismutation of $O_2^{\bullet-}$ to O_2 and H_2O_2 through sequential reduction and oxidation of the metal centers (Eqs. 4.19–4.21):

$$M^{(n+1)+} + O_2^{\bullet-} \rightarrow M^{n+} + O_2 \tag{4.19}$$

$$M^{n+} + O_2^{\bullet-}(+2H^+) \rightarrow M^{(n+1)+} + H_2O_2 \tag{4.20}$$

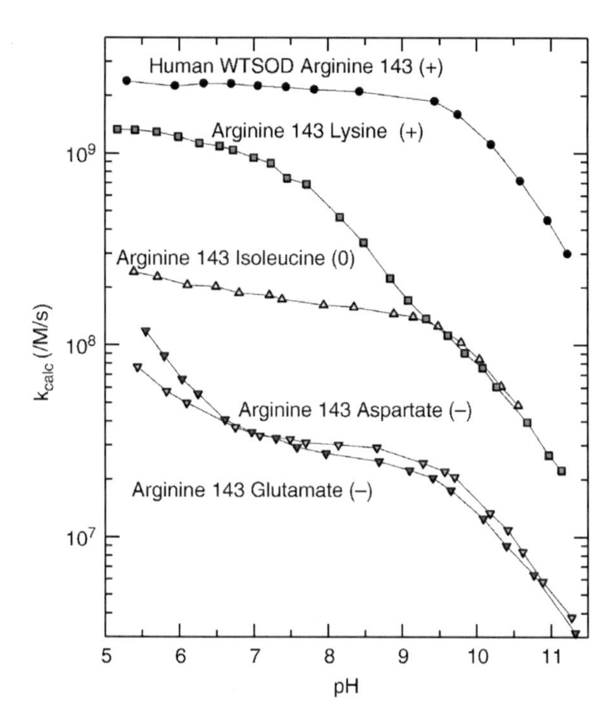

Figure 4.4. Rate constants for O_2^- dismutation by wild-type and various arginine 143 mutants of human sCuZnSOD as a function of pH (20 mM phosphate, 10 mM formate, 5 μM EDTA). The charge on the new residue is indicated in parentheses. WTSOD, wild-type superoxide dismutase (adapted from Abreu and Cabelli [31] with the permission of Elsevier Inc.). See color insert.

$$\text{Overall: } 2O_2^{\bullet-} + 2H^+ \rightarrow O_2 + H_2O_2. \tag{4.21}$$

The catalytically active metal of SOD can be iron, manganese, copper, or nickel [87]. The catalytic pH dependence activities of Cu, Zn-SOD, manganese superoxide dismutase (MnSOD), and Ni-SOD disproportionate to $O_2^{\bullet-}$ are shown in Figure 4.4 [31]. The diffusion-controlled rates were determined, which did not vary in the pH range of 5.0–9.5. Significantly, the catalytic rates were several orders of magnitude faster than the noncatalytic rates (Fig. 4.4). This influences the half-lives ($t_{1/2}$) of the dismutation of superoxide. For example, $t_{1/2}$ of the superoxide dismutation without SOD is 10 seconds at pH 7 for 10^{-6} M $[O_2^{\bullet-}/HO_2^{\bullet}]$. However, in the presence of the SOD enzyme concentration of 10^{-9} M, $t_{1/2}$ of the enzymatic dismutation of superoxide is over an order of magnitude faster than the spontaneous reaction.

Several studies have been carried out to understand the Cu-Zn-SOD-catalyzed dismutation of superoxide (Eqs. 4.22 and 4.23) [88–93]. Most studies were performed using pulse radiolysis as a result of its clean and rapid generation of $HO_2^{\bullet}/O_2^{\bullet-}$:

$$Cu^{2+}Zn^{2+}SOD + O_2^{\bullet-} \rightarrow Cu^{+}Zn^{2+}SOD + O_2 \quad k_{16} \approx 2 \times 10^9 /M/s \quad (4.22)$$

$$Cu^{+}Zn^{2+}SOD + O_2^{\bullet-}(+2H^{+}) \rightarrow Cu^{2+}Zn^{2+}SOD + H_2O_2 \quad k_{17} \approx 2 \times 10^9 /M/s. \quad (4.23)$$

Initial studies at very high concentrations of $O_2^{\bullet-}$ showed K_m should be greater than 5 mM $O_2^{\bullet-}$ if the saturation mechanism was involved [31]. It was also suggested the actual measured rate constants must be slower than 2×10^9/M/s, considering the surface area differential between $O_2^{\bullet-}$ and Cu-Zn-SOD (~1:150) [88, 93]. The measured "diffusion-controlled" rate constant may be from the electrostatic guidance of the negatively charged $O_2^{\bullet-}$ into the positively charged active site [88, 90, 93]. This suggestion was supported by pulse radiolysis experiments carried on various mutations of positively charged Arg at the entrance of the active site as a function of ionic strength and pH (Fig. 4.4) [89]. The rates were decreased by neutralization and reversal of the positive charge on Arg (Fig. 4.4). The effect of charge was also supported by the pH dependence behavior in which the protonation of anionic oxygen in glutamate/aspartate recovered the decrease in rate due to the negative charge. Furthermore, neutralization of the negative charge near the active site resulted in faster enzyme activity than the wild-type enzyme [91]. A detailed examination of reaction (4.22) using a pulse radiolysis technique demonstrated that enzymatic reduction and concomitant superoxide oxidation were pH independent. However, reaction (4.23) was pH dependent to yield the overall pH dependence of Cu-Zn-SOD on the dismutation of superoxide [92].

4.1.4.1 Manganese Superoxide Dismutase. MnSODs were first discovered in 1970 [94]. Dimer and tetramer forms of MnSODs with a single manganese atom per subunit have been determined. Bacteria and eukaryotes usually contain dimer and tetramer forms, respectively [31]. In the structure of MnSODs, manganese is bound in a trigonal bipyramidal geometry to four ligands from the protein (three histidines and an aspartic acid residue) and a fifth ligand from the solvent. It has been assumed that the solvent ligand is in hydroxide and protonated forms under oxidized and reduced conditions, respectively. The active sites and structures of three MnSODs (human, *Escherichia coli*, and *Deinococcus radiodurans*) have been studied in detail [95–97]. Due to the high sequence structure homology of iron superoxide dismutase (FeSOD) and MnSOD, iron instead of manganese in the active form can be incorporated into MnSOD. For example, the replacement of iron in *E. coli* MnSOD *in vivo* under anaerobic conditions has the ability to inactivate the enzyme [31, 98].

The mechanism of MnSOD dismutation is complex [99]. The kinetics results displayed a "burst phase" and a "zero-order" phase rather than a first-order phase at sufficiently high ratios of $[O_2^{\bullet-}]$:[MnSOD]. The first-order disappearance of $O_2^{\bullet-}$ has been seen in other SODs. Therefore, MnSOD in its reduced form has been suggested to react with two concomitant pathways [100–102].

Using pulse radiolysis studies, a simple mechanism expressed in Equations (4.24)–(4.27) has been proposed [98, 103, 104]:

$$Mn^{3+}SOD(OH^-) + O_2^{\bullet-}(+H^+) \rightarrow Mn^{2+}SOD(H_2O) + O_2 \quad k_{24} \qquad (4.24)$$

$$Mn^{2+}SOD(H_2O) + O_2^{\bullet-}(+H^+) \rightarrow Mn^{3+}SOD(OH^-) + H_2O_2 \quad k_{25} \qquad (4.25)$$

$$Mn^{2+}SOD(H_2O) + O_2^{\bullet-} \rightarrow Mn^{3+}SOD(H_2O)\text{-}O_2^{2-} \quad k_{26} \qquad (4.26)$$

$$Mn^{3+}SOD(H_2O)\text{-}O_2^{2-}(+H^+) \rightarrow Mn^{3+}SOD(OH^-) + H_2O_2. \quad k_{27} \qquad (4.27)$$

The mechanism includes the formation of both $Mn^{3+}SOD$ and the inhibited complex ($Mn^{3+}SOD(H_2O)\text{-}O_2^{2-}$).

The proposed mechanism was confirmed by conducting spectral studies in the visible region (350–600 nm) (Fig. 4.5) [31]. The growth of the absorption of $Mn^{3+}SOD$ was measured by reacting high concentrations of $Mn^{2+}SOD$ with rapid generated substoichiometric amounts of $O_2^{\bullet-}$ (Fig. 4.5). In the reaction of human $Mn^{2+}SOD$ with substoichiometric $O_2^{\bullet-}$, there was an initial growth corresponding to both $Mn^{3+}SOD(H_2O)\text{-}O_2^{2-}$ and $Mn^{3+}SOD$ with a rate constant of 4×10^9/M/s per tetramer. A slow process corresponds to the formation of $Mn^{3+}SOD$ (Eq. 4.27). However, only the growth of $Mn^{3+}SOD$ was observed ($k_{25} = 2 \times 10^9$/M/s per dimer) when *E. coli* MnSOD was used (Eq. 4.25).

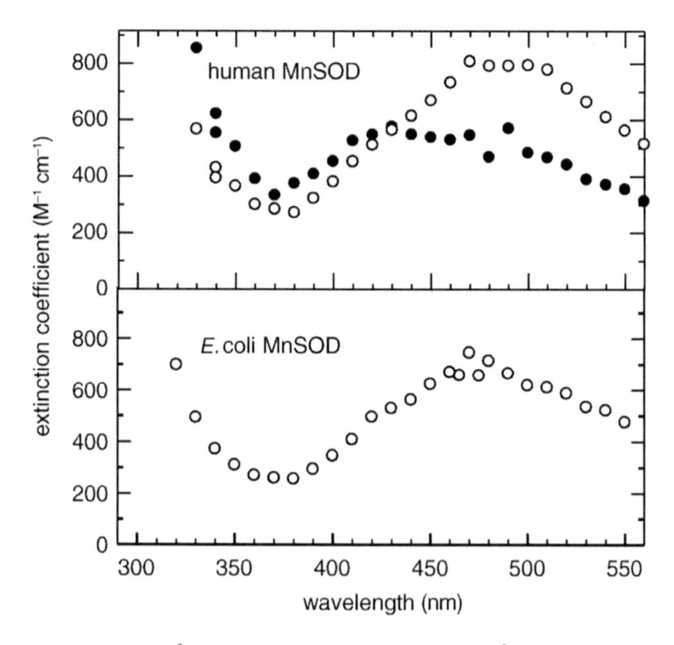

Figure 4.5. Spectra of Mn^{3+}SODs (open circles) and $Mn^{3+}SOD(H_2O) - O_2^{2-}$ (closed circles) as measured in pulse radiolysis experiments (adapted from Abreu and Cabelli [31] with the permission of Elsevier, Inc.).

TABLE 4.2. Individual Rate Constants in the Kinetic Mechanism of Different MnSODs and Their Mutants

MnSOD	k_{24} (/nM/s)	k_{25} (/nM/s)	k_{26} (/nM/s)	k_{26}' (/s)	k_{27} (/s)
Human	1.5	1.1	1.1	–	120
Y34A	0.25	<0.02	0.38	1600	330
Y34N	0.14	<0.02	0.15	850	200
Y34H	0.07	<0.02	0.04	250	61
Y34V	0.15	<0.02	0.15	–	1000
Y34F	0.55	<0.02	0.46	–	52
H30N	0.21	0.40	0.68	–	480
H30Q	0.57	0.79	0.79	–	200
H30V	0.005	0.03	0.16	–	0.7
E162D	0.36	0.13	0.21	–	40
E162A	0.06	0.05	0.09	–	30
F66A	0.6	0.5	0.7	–	82
F66L	0.7	0.8	0.2	–	40
W123F	0.76	<0.02	0.64	–	79
Y166F	0.2	0.2	0.2	–	270
W161A	0.08	<0.01	0.37	–	180
W161F	0.3	<0.01	0.46	–	33
W161V	na	na	0.27	–	265
W161Y	na	na	0.20	–	130
W161H	na	na	0.29	–	136
H30F/Y166F	0.1	0.1	0.1	–	440
Y34F/W123F	0.55	<0.22	0.46	–	52
E. coli	1.1	0.9	0.2	–	60
Deinococcus radiodurans	1.2	1.1	0.07	–	30
Y34F	0.9	0.9	0.5	–	30

Data were taken from Abreu and Cabelli [31] with the permission of the Elsevier Inc.
na, not available.

Table 4.2 displays the variation of rate constants of reactions (4.24)–(4.27) among the various species [31]. Reaction (4.25) represents an outer-sphere mechanism, while reactions (4.26) and (4.27) are of inner-sphere mechanism pathways. The ratios of k_{25}/k_{26} can thus distinguish outer-sphere and inner-sphere mechanisms for the reactions of $O_2^{\bullet -}$ with MnSODs. In reaction (4.27), dissociation of the complex resulted in H_2O_2, and hence, the value of k_{27} reflected the rate for the protonation of $O_2^{\bullet -}$ in the complex. Both reactions (4.25) and (4.27) were highly affected by the residue mutations involved in the hydrogen-bounded network (Table 4.2). The mutation of Y34 to histidine, glutamine, phenylalanine, and valine resulted in large variations in k_{27} (see Table 4.2). Both *E. coli* and *D. radiodurans* had two and four times slower k_{27} than for human enzymes, respectively. The change in histidine 30 in H30N and

H30Q did not show much variation in k_{25} and k_{26} but increased the rate significantly of the protonation off of the bound $O_2^{\bullet-}$ in human enzymes (Table 4.2). Other possible residues involved in the dissociation of the complex in human enzymes were E162 and E66. E162D, E162A, F66A, and F66L had a decreased k_{27} value in comparison with the wild-type (WT) MnSOD. Values of k_{27} for these mutants were similar to the values determined for prokaryotic enzymes. Overall, the rate constants determined in Table 4.2 assist in the understanding of the mechanistic differences in eukaryotic versus prokaryotic MnSODs [31, 100, 102, 105].

4.1.4.2 Iron Superoxide Dismutase. FeSOD, a prokaryotic enzyme, was discovered in some bacterial cells and in the cytosol area of plants [31]. Most of the structural and mechanistic studies have been performed on FeSOD obtained from *E. coli*. The structures of FeSODs are dimers in which each iron active site contains a single iron atom bonded to three histidines, one aspartate, and one water molecule [106]. The coordinated water molecule involves a hydrogen bond (H-bond) with an aspartate ligand and another with the conserved active site, glutamine 69 (Gln 69). The H-bonding network plays an important role to determine the reactivity of the Fe site.

A ping-pong mechanism is displayed in Equations (4.28) and (4.29), suggested in the dismutation of $O_2^{\bullet-}$ by FeSOD [107]:

$$Fe^{3+}SOD + O_2^{\bullet-} \rightarrow Fe^{2+}SOD + O_2 \tag{4.28}$$

$$Fe^{2+}SOD + O_2^{\bullet-}(+2H^+) \rightarrow Fe^{3+}SOD + H_2O_2. \tag{4.29}$$

In this mechanism, Fe^{III} is reduced by superoxide through the binding of $O_2^{\bullet-}$ to Fe (Eq. 4.28) [107, 108]. Reaction (4.29) indicates second-sphere binding of the next superoxide molecule, which results in the oxidation of Fe^{II} and the formation of hydrogen peroxide. Electron and proton transfers are coupled together during the activity of FeSOD, and in particular, the transfer of a proton to superoxide in Equation (4.29) determines the thermodynamic feasibility [108]. The involvement of the coordinated water molecule may be a redox-coupled proton acceptor and may possibly be donating one of the protons in peroxide [109].

The decay of $O_2^{\bullet-}$ in the presence of FeSOD, purified from marine bacterium *Photobacterium leiognathi*, was determined to be first order and was proportional to the concentration of FeSOD [107]. The second-order rate constant decreased from 6.1×10^8/M/s to 1.3×10^8/M/s with an increase in pH from 6.2 to 10.1. Similar results were also observed in FeSOD from the *filamentous cyanobacterium Nostoc* PCC 7120 [110]. However, the rate constants ranged from 5.3×10^9/M/s (pH 7) to 4.8×10^6 10^9/M/s (pH 10). The decay study using FeSOD from *E. coli* followed Michaelis–Menton kinetics (Eq. 4.30) [108]:

$$-d[O_2^{\bullet-}]/dt = k_{fir}[O_2^{\bullet-}] + 2k_{sec}[O_2^{\bullet-}]^2 + 2[FeSOD]TN[O_2^{\bullet-}]/(K_m + [O_2^{\bullet-}]). \tag{4.30}$$

The TN (turnover number) was determined to be independent of pH. Based on the concentration of Fe, the rate constant was ~2.6×10^4/s at 25°C. The values of K_m (Michaelis–Menton constant) were independent below pH 8 but increased by 10-fold above pH 10. K_m had a value of ~80 μM, which suggests the modification of Equation (4.28). Thus, reaction (4.28) may be followed by reaction (4.31):

$$Fe^{2+}SOD + O_2^{\bullet-} \rightleftharpoons [Fe^{2+}SOD\text{-}\ O_2^{\bullet-}] + (2H^+) \rightarrow Fe^{3+}SOD + H_2O_2. \quad (4.31)$$

The saturation process was not observed with all studied FeSODs [110]. FeSOD also goes through a parallel peroxidative mechanism (Eqs. 4.32 and 4.33). Reaction (4.32) is faster than reaction (4.33) [111]:

$$Fe^{3+}SOD + H_2O_2 \rightarrow Fe^{2+}SOD + HO_2 + H^+ \quad (4.32)$$

$$Fe^{2+}SOD + H_2O_2(+2H^+) \rightarrow Fe^{3+}SOD \cdots OH + OH^-. \quad (4.33)$$

A study on the activation of *E. coli* FeSOD was performed by the addition of primary amines [112]. The saturation kinetics was carefully observed for most of the studied amines. K_m was calculated as 100 ± 20 μM under the studied conditions. Activation of the proton transfer pathway was explored by seeking a linear relationship in the free energy plot of the apparent second-order rate constant of the activation and the pK_a of the amines (Fig. 4.6). The linearity in the plot confirmed the activation of the proton transfer. The slope of the plot was calculated as 0.50 ± 0.07, indicating the position of the transition state was between the reactants and products. The results suggest an occurrence of proton transfer in a ternary complex of amine with the enzyme-bound peroxide dianion. Furthermore, the activation by amines was uncompetitive with respect to superoxide [112].

The redox equilibria of FeSOD involve the inhibition by anions, N_3^- and F^-, which bind directly to the Fe^{3+} ion [108]. Complexes of FeSOD and MnSOD have thus been compared [113]. Crystal structures of N_3^- complexes with Fe^{3+}SOD and Mn^{3+}SOD showed N_3^- bound to the Fe^{3+}-to-coordinating-N vector in the Fe^{3+}SOD is more linear, while N_3^- engages in a H-bond with Tyr34 in the Mn^{3+}SOD [106]. The electron paramagnetic resonance (EPR) spectroscopy showed the fluoride ion binds to Fe^{2+}SOD outside the first coordination sphere. In the case of Mn^{2+}SOD, the fluoride ion joins the first coordination sphere [114].

4.1.4.3 Iron Superoxide Reductase (FeSOR). Based on the number of metal centers, two types of FeSORs have been classified: *neelaredoxins* (1Fe-SORs) and *desulfoferrodoxins* (2Fe-SORs) [115]. In FeSORs, both iron sites are near the molecular surfaces and are exposed to the solvent. Comparatively, metal centers in FeSODs are located inside the protein. Desulfoferrodoxins are a homodimeric nonheme iron protein found in some sulfate-reducing

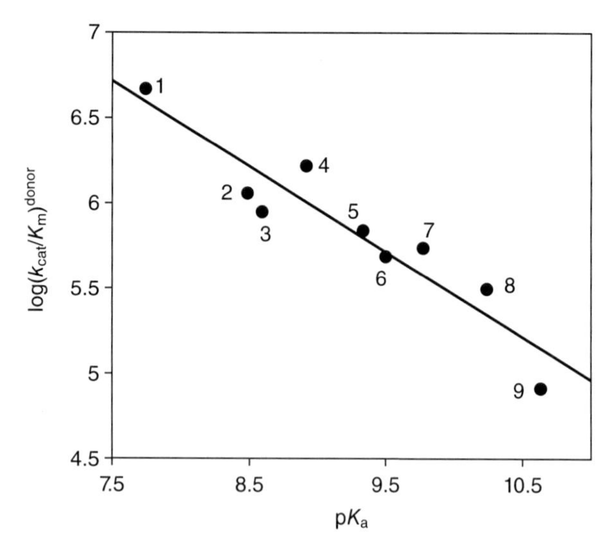

Figure 4.6. Variation with the pK_a of proton donors of $\log[(k_{cat}/K_m)^{donor}]$ for activation of *E.coli* FeSOD. $(k_{cat}/K_m)^{donor}$ is the contribution to k_{cat}/K_m^B due to the exogenous proton donors. Conditions are described in Greenleaf and Silverman [112]. The exogenous proton donors were glycine methyl ester (1), 2-bromoethylamine (2), 2-fluoroethylamine (3), 2-aminoethylsulfonic acid (4), benzylamine (5), ethanolamine (6), glycine (7), 3-aminopropionic acid (8), and ethylamine (9) (adapted from Greenleaf and Silverman [112] with the permission of the American Society for Biochemistry and Molecular Biology, Inc.).

bacteria and archaea [116, 117]. The crystal structure of desulfoferrodoxins obtained from *Desulfovibrio desulfuricans* [116] has center I ($[Fe(SCys)_4]$) and center II (square pyramidal $[Fe(NHis)_4(SCys)]$) sites. Center I and center II perform functions of electron transfer and react with superoxide, respectively [31, 115]. Neelaredoxins, from *Archaeoglobus fulgidus*, have biofunctional activities of both SOD and SOR. Desulfoferrodoxins and neelaredoxins have been mainly isolated in Fe^{2+} and Fe^{3+} forms, respectively [115]. The numbering method of amino acid ligands of the catalytic center of the 1Fe-SORs and 2Fe-SORs is given in Table 4.3 [115]. The iron in center I of the 2Fe-SORs is coordinated to four cysteins in a distorted tetrahedral geometry. The common site, center II, of both 1Fe-SORs and 2Fe-SORs, has a unique geometry. Iron in the +2 oxidation state (Fe^{2+}) is coordinated in a square-pyramidal geometry to four nitrogens from histidine-imidazoles and a fifth axial position is occupied by a cystein-sulfur (Table 4.3). Generally, it is assumed the sixth axial position is not occupied by the solvent molecule.

The UV–visible spectroscopy has been applied to study the catalytic activity of SORs (Table 4.4) [115, 118–125]. Center I of 2Fe-SORs has the characteristics of the $FeCys_4$ site in *desulforedoxin*, having maxima at 375 and 495 nm

TABLE 4.3. Numbering of Amino Acid Ligands of the Catalytic Center of the SORs Represented in the Sequence Alignment and in the Dendogram

	Fe Ligands						
	Glu	His	His	His	His	Cys	Lys
1Fe-SOR							
P. furiosus	E14	H16	H41	H47	H114	C111	K15
D. gigas	E15	H17	H45	H51	H118	C115	K16
A. fulgidus	E12	H14	H40	H46	H113	C110	K13
N. equitans	–	H10	H35	H41	H100	C97	K9
T. pendens	E24	H26	H51	H57	H113	C110	–
M. acetivorans	–	H27	H54	H60	H114	C111	K26
T. pallidum	E48	H50	H70	H76	H122	C119	K49
2Fe-SOR							
C. phytofermentans	E47	H49	H69	H75	H121	C118	K48
D. vulgaris	E47	H49	H69	H75	H119	C116	K48
D. desulfuricans	E47	H49	H69	H75	H119	C116	K48
D. baarsii	E47	H49	H69	H75	H119	C116	K48
G. uraniireducens	E47	H49	H69	H75	H119	C116	K48
A. fulgidus	E47	H49	H69	H75	H119	C116	K48
Rd SOR							
Uncultured bacterium	–	H53	H83	H89	H137	C134	K52

–, residues not conserved.
Adapted from Pinto et al. [115] with the permission of Elsevier Inc.

and a broad shoulder at 560 nm (Table 4.4). Center II has a broad absorption band at 660 nm, which represents a blue color of 1Fe-SORs or a gray color (a mixture of blue and pink) for the oxidized 2Fe-SORs and a shoulder at ca. 330 nm. A sulfur to iron charge transfer transition generates the 660-nm band [126]. In the reduced form, both types of SORs are colorless. In the half-reduced state (center I oxidized, center II reduced), both SOR types are pink. The UV–visible spectral characteristics were used to understand the catalytic mechanism of SORs using fast kinetic techniques (stopped-flow and pulse radiolysis).

The reduction potentials are 0 and 190–365 mV for center I and center II, respectively (Table 4.4). The resulting potentials are similar to reported values for SODs and sufficient for the reduction of superoxide, which has a potential of 890 mV (O_2^{-}/H_2O_2) at pH 7.0. The large difference between the two centers allowed a detailed study of the events at the catalytic site without interferences from the $FeCys_4$ center. Both 1Fe-SORs and 2Fe-SORs have pH-dependent equilibria (Table 4.4), important to understand the catalytic activities of SORs. Using an example of 1Fe-SOR of *A. fulgidus*, the catalytic cycle of SOR is represented in Figure 4.7. The first step of the reaction is the formation of the

TABLE 4.4. Spectroscopic and Redox Properties of SOR's Catalytic Center

	Oxidized λ_{max} (nm)		$E°$ (mV)		T1-Fe^{3+} "Hydroperoxo"		
	Low pH	High pH	Center II	pK_a	λ_{max} (nm)	k_1 ($\times 10^9$/M/s)	k_2 (Neutral pH) (s^{-1})
2Fe-SOR							
A. fulgidus	630	540	365	8.5	580	0.6	57
D. vulgaris	647	560	250	–	590	1.5	40
D. vulgaris E47A	–	–	–	–	600	1.5	65
D. vulgaris K48A	–	–	–	–	600	0.21	25
D. baarsii	644 (pH 7.6)	–	350 (pH 6–9)	9	610	1.1	500
D. baarsii E47A	580 (pH 7.6)	–	520 (pH5.5–6.5)	6.6	630	1.2	440
D. baarsii K48A	635 (pH 7.6)	–	520 (pH5.5–6.5)	7.6	600	0.38	300
1Fe-SOR							
A. fulgidus	666	590	250	9.6	620	1.2	400
A. fulgidus E12V	670	590	302	6.3	620	0.22	400
N. equitans	655	550	350	6.5	590	1	<10
T. pallidum	650	560	200	6	610	0.6	4800
T. pallidum E48A	650	560	–	6	600	0.6	2080
D. gigas	666	590	190	>9	–	–	–

Adapted from Pinto et al. [115] with the permission of Elsevier Inc.

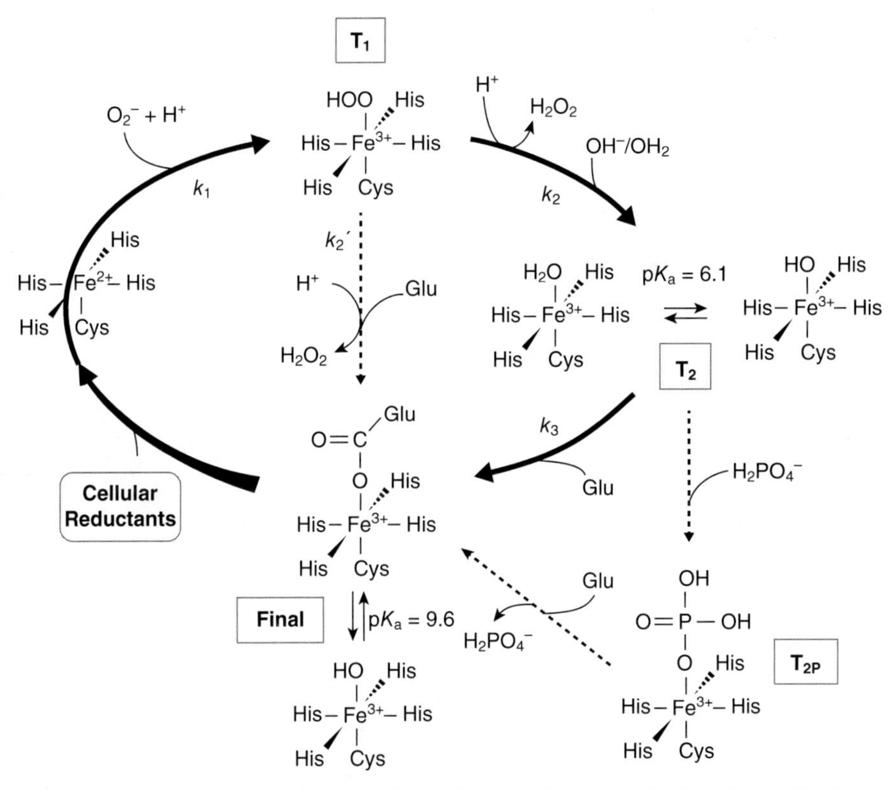

Figure 4.7. Catalytic cycle of superoxide reductase, showing only the observable intermediates. Large arrow: reductive cycle; narrow arrows: oxidative cycle (adapted from Pinto et al. [115] with the permission of Elsevier Inc.). See color insert.

intermediate, T1, with an absorption maximum at ca.620 nm, which occurs at a second-order rate constant of ~10^9/M/s (Table 4.4). TI intermediates have been suggested to be ferric-(hydro)peroxo species, which decay subsequently to another intermediate, T2, in the case of *A. fulgidus*. This step is a pseudo-first-order, unimolecular process (Table 4.4). The optical properties of T2 are identical to those reported for the basic form of ferric SOR [118]. T2 decays further to the resting oxidized state through a unimolecular process for the wild-type enzyme. This resting form was observed for the *D. vulgaris* enzyme T1 without the formation of T2. However, both intermediates, T1 and T2, were observed for the 1Fe-SOR from *Treponema pallidum* and 2Fe-SOR from *Desulfoarculus baarsii* [115]. The nature of intermediates T1 and T2 has been reviewed in detail [115]. Overall, progress has been made for the last few years on the elimination of superoxide by 1Fe- and 2Fe-SOR, but several issues still need to be resolved. These include the number of catalytic intermediates, reasons of low SOD activity of SOR, although thermodynamic properties suggest capability of oxidizing or reducing superoxide equally and the role of water in the catalytic activity of SOR [115]. A recent study on the catalytic mechanism of the reaction of $O_2^{\bullet-}$ with SOR suggests both ferrous iron-superoxo and ferric hydroperoxide species as intermediates of the reactions. Significantly, lysine 48 plays an important role in controlling the evolution of iron peroxide intermediate to yield H_2O_2 [127].

In summary, superpoxide is relatively unreactive with amino acids except Cys, Trp, His, and Phe. These amino acid residues as well as their locations play a significant role in the inactivation of selected dehydrogenases by $O_2^{\bullet-}$ [4]. Superoxide is reactive with complexes of transition metal ions such as Cu, Ni, Fe, and Mn. Several evidences have been given for possible targets of superoxide reactivity *in vivo*. In the research on the catalytic oxidation of superoxide, it was determined that the specificity for metals can differ with the type of SOD, and differences exist in the interaction between metals and proteins [128]. Research is in progress on understanding the mechanisms causing the change in the metal-specific activity. More spectroscopic and structural studies are thus needed on mutants of FeSOD and MnSOD. Mechanism studies on NiSOD are also required to unravel the inherent complexity of making nickel redox active with superoxide [31]. A study on the immobilization of SOD on organo-functionalized mesoporous silica nanoparticles has been conducted to determine the activity and structural changes of SOD upon immobilization [129]. Results showed that the immobilized SOD had a higher activity than the free enzymes and the structure of SOD did not change. More work in this area may provide information on designing drugs to reduce dangers from superoxide to the cells.

4.2 SINGLET OXYGEN

Formation of singlet oxygen (molecular oxygen in the $^1\Delta_g$ state; $^1O_2$) has been observed in dark- as well as in light-mediated processes in the presence of endogenous and exogenous sensitizers [13, 130–136]. The nonlight processes include biological systems such as peroxides (e.g., horseradish peroxidase, lactoperoxidase, MPO, and chloroperoxidase) and lipoxygenase-mediated reactions [137–141]. The generation of $^1O_2$ was also observed in several stimulated cell types (e.g., eosinophils and macrophases) [142, 143]. The glyoxal-peroxynitrite system generates $^1O_2$ [144]. Reactions of ozone with biological molecules and plant leaves produced $^1O_2$ [145, 146]. $^1O_2$ was observed during the reactions of H_2O_2 with sodium molybdate and HOCl [147, 148]. Steady-state irradiation of equilibrium mixtures of the retinal lipofuscins, 2-[2, 6-dimethyl-8-(2,6,6-trimethyl-1-cyclohexen-1-yl)-1E,3E,5E,7E-octatetraenyl]-1-(2-hydroxyethyl)-4-[4-methyl-6(2,6,6-trimethyl-1-cyclohexen-1-yl)-1E,3E, 5E-hexatrienyl]-pyridinium (A2E) and double-bond isomer of A2E (iso-A2E) also generates $^1O_2$ [140]. The self-reaction of sec-peroxy radicals has been shown to yield $^1O_2$ [149]. In photochemical reactions, the incidence of UV or visible light to aromatic compounds (e.g., naphathalene and anthracene) and conjugated alkenes (e.g., dye molecules and porphyrins) yielded excited states that went through rapid and efficient energy transfer to O_2 to generate $^1O_2$ [150, 151]. Generally, energy transfer to produce $^1O_2$ ($k = 1–3 \times 10^9$/M/s) competes with electron transfer ($k \leq 1 \times 10^7$/M/s), and thus, many photosensitizers form both $^1O_2$ and $O_2^{\bullet -}$ [152]. The overall yields of $^1O_2$ depend on the reaction conditions, excitation wavelength, and type of sensitizer [136, 152, 153]. The pH, solvent, and temperature can significantly affect the quantum yields [135, 154, 155]. The quantum yields of $^1O_2$ obtained from a number of biological molecules have been summarized [150]. Examples include quantitative characterization of $^1O_2$ in UVA excitation of 6-thioguanines in aqueous solution [156].

Recently, the generation of $^1O_2$ through the irradiation of air-saturated solutions of Phe, Tyr, and Trp in their zwitterionic forms and their methyl esters as well as of a few test proteins and immunoglobulins, in D_2O or MeCN using 266-nm pulse from a frequency quadrupled continuum Nd:YAG laser (8 ns, 3 mJ), was analyzed quantitatively [132]. This study reported quantum yields for generation of $^1O_2$ by biological macromolecules. The lifetime of $^1O_2$ in D_2O was longer than in H_2O, which was related to the role of high-frequency O–H vibrational modes in nonradiative electronic-to-vibrational energy transfer processes which lead to deactivation of $^1O_2$ [13]. The H_2O/D_2O test is thus a versatile tool to study $^1O_2$ in biological systems [157]. In most of the solvents, τ_Δ varied from 1 to 100 μs. The τ_Δ in water was determined to be 2–4 μs at neutral pH [158]. The following reactions (e.g., Phe) explain the formation of $^1O_2$ (Eqs. 4.34–4.37):

$$\text{Phe} + h\nu \rightarrow {}^1\text{Phe} \tag{4.34}$$

$$^1\text{Phe} \rightarrow {}^3\text{Phe} \tag{4.35}$$

$$^3\text{Phe} + {}^3\text{O}_2 \rightarrow \text{Phe} + {}^1\text{O}_2 \tag{4.36}$$

$$^1\text{O}_2 \rightarrow {}^1\text{O}_2 + h\nu\ (1270\ \text{nm}). \tag{4.37}$$

The quantum yields of $^1O_2$ were obtained by measuring the yields of its near-infrared (IR) emission (Eq. 4.37). A strong peak at ~1270 nm is the phosphorescence characteristic of a singlet oxygen. The kinetic traces in the inset of Figure 4.8 indicate a lifetime (τ_Δ) of 48 μs, similar to an earlier reported value of 53 ± 5 μs [158]. Table 4.5 reports the quantum yields of $^1O_2$ emission. The protonation state of the α-amino group in the aromatic amino acids had the influence on the quantum yields. For example, the quantum yields of N-acetylated amino acids were slightly greater than those of their free amino acids (Table 4.5). An equimolar mixture of free aromatic amino acids in solution had a weighted average $^1O_2$ quantum yield of 0.08, double that of the quantum yields for proteins and immunoglobulins (Table 4.5). An increase in viscosity of the matrix of protein could hinder and also limit the generation of $^1O_2$.

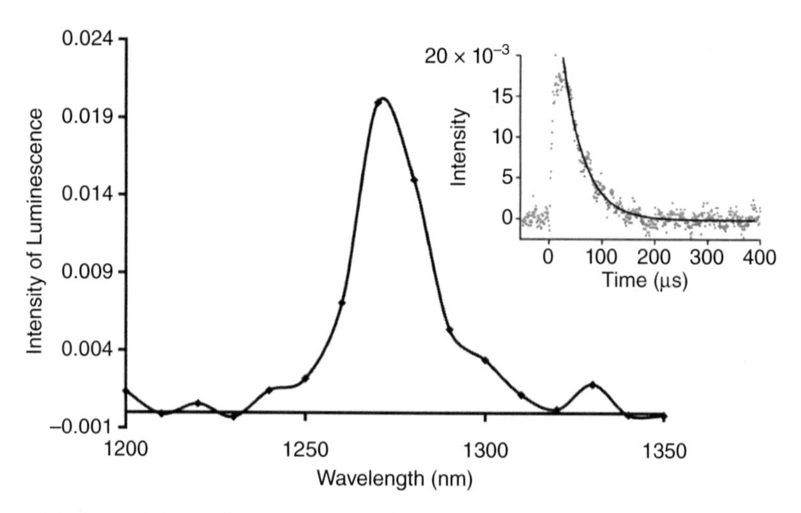

Figure 4.8. Near-infrared luminescence of singlet oxygen sensitized by UV irradiation of phenylalanine in D_2O. The emission spectrum shows a peak at about 1270 nm. Inset: decay of singlet oxygen phosphorescence as a function of time (in microsecond), sensitized by phenylalanine. The black line is a curve fit of the decay (adapted from Chin et al. [132] with the permission of the American Chemical Society).

TABLE 4.5. Quantum Yields of Singlet Oxygen Generated by Photosensitized Aromatic Amino Acids, N-Acetyl-Amino Acids, Proteins, and Immunoglobulins[a]

Compound	Solvent	Φ_Δ
Phe	D_2O	0.065 ± 0.004
Tyr	D_2O	0.138 ± 0.007
Trp	D_2O	0.062 ± 0.011
N-Acetyl-Phe	MeCN	0.083 ± 0.02
N-Acetyl-Tyr	MeCN	0.19 ± 0.05
N-Acetyl-Trp	MeCN	0.11 ± 0.02
BSA	D_2O	0.037 ± 0.008
Chick ovalbumin	D_2O	0.049 ± 0.005
Bovine IgG	D_2O	0.043 ± 0.008
Human IgG	D_2O	0.030 ± 0.010
Sheep IgG	D_2O	0.036 ± 0.007

[a] All experiments were conducted at room temperature and excited at 266 nm. Reported values are averages of five or more measurements.

Adapted from Chin et al. [132] with the permission of the American Chemical Society.

4.2.1 Reactivity

The rate constants for the reactions of $^1O_2$ with biological molecules such as sterols, lipids, DNA, RNA, proteins, ascorbic acid, Trolox, and amino acids have been determined to estimate potential consumption of $^1O_2$ by different cellular components [13, 130, 131, 151, 159, 160]. A number of reactions of $^1O_2$ with

TABLE 4.6. Rate Constants for Reactions of $^1O_2$ with Protein Side Chains

Side Chain	pH	k (/M/s)	Reference
Cys	7	0.9×10^7	[161]
Met	7	1.6×10^7	[161]
	8.5	2.2×10^7	[164]
Tyr	7	0.8×10^7	[161]
	8.5	0.9×10^7	[164]
Trp	7	3×10^{7a}	[161]
	7	$2–7 \times 10^{7b}$	[161]
	7.5	3.2×10^{8a}	[160]
	8.5	1.8×10^8	[164]
His	>8	1.0×10^8	[159]
	Low pH	0.5×10^7	[159]
	7.5	7.7×10^{7c}	[160]
	8.5	6.6×10^7	[164]
Carnosine	7.5	1.3×10^{8c}	[160]

[a] Chemical reactions.
[b] Physical quenching.
[c] Eu^{3+}-luminescence probe.

free amino acids have been studied to learn if proteins as cellular components/sites could be important targets of $^1O_2$ [151, 159, 161, 162]. Cys, Met, Tyr, Trp, and His were the only amino acids showing significant reactivity with $^1O_2$ at physiological pH (Table 4.6) [159–161, 163, 164]. Other aliphatic and aromatic amino acids had much lower reactivity ($k < 0.7 \times 10^7$/M/s). The predicted consumption of $^1O_2$ by intracellular targets based on rate constants of the reactions and the average concentration of each component in a typical leukocyte cell was ~68% [130]. Other factors such as the limited diffusion radius of $^1O_2$ from its site of generation and yields of $^1O_2$ would also determine proteins as potential target of $^1O_2$ in a cell.

The pH dependence for the quenching of $^1O_2$ by His and Trp is shown in Figure 4.9 [159]. Histidine had a pronounced effect of pH on the second-order rate constant, while Trp did not show significant variation of rate. The results for the rate constants of His were fitted nicely using a pK_a of 5.9, suggesting the protonation state of the imidazole ring in histidine governed the variation of rates with pH. The lower and upper limits of the rate constants were calculated as $\leq 10^4$ and 5.0×10^7/M/s, respectively. The unprotonated form of Arg and Lys also reacted rapidly with $^1O_2$ at higher pH values [162]. The lack of ionizable protons in the indole nucleus of Trp explained the trend in rates with pH. The ability of $^1O_2$ to damage critical targets in a cellular environment thus depends on the concentration of potential quencher species at a particular cellular location, which vary with pH.

Several studies have been performed to learn changes in proteins caused by $^1O_2$-mediated oxidation [13, 130, 134, 160, 163, 165–169]. The interaction of

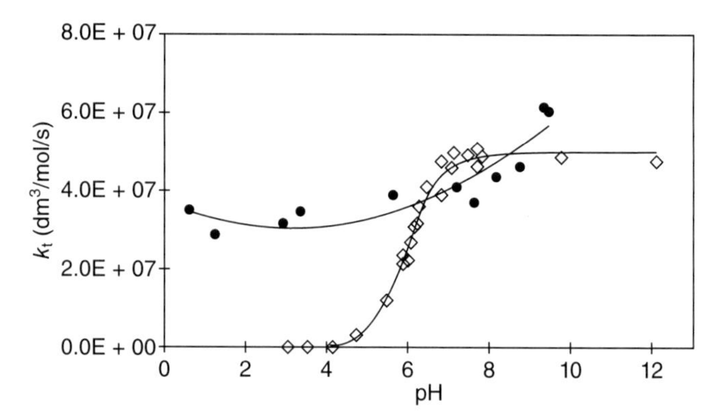

Figure 4.9. Effect of pH on the second-order rate constant for quenching of singlet oxygen (k_t) by His (◇) and Trp (•). The rates constants were measured in D_2O/acetonitrile (50:50 v/v) buffered with phosphate (40 mmol/dm^3) with the addition of NaOH or DCl as required (adapted from Bisby et al. [159] with the permission of the American Chemical Society).

protein side chains with $^1O_2$ could occur through both physical quenching and chemical reactions [13]. It appears Trp was the only amino acid to have a significant contribution to physical quenching with a k value of 10^7/M/s, similar to the chemical reaction rate constant (Table 4.6) [163]. The relative contribution of the two processes for Trp could be affected by several factors such as the environment of the residue and the presence of the substituent on the indole ring [170, 171]. The quenching of $^1O_2$ with N-acetyltryptophan amide (NATA), Trp$(CH_2)_{16}$, and melittin in D_2O and in solution containing liposome membrane using different sensitizers showed the quenching rate constants were greatly affected by the location of the sensitizer as well as the quencher in the liposome membrane or in the surrounding solution [165].

The products formed in the reactions of free Cys, cystine, and Met with $^1O_2$ have been characterized [152, 172–175]. In the case of Cys, products were pH dependent, and disulfide and oxyacids, such as cysteic acid (RSO_3H), were formed [152, 161]. The ratios of disulfide to oxyacids in the reactions of Cys residues in proteins with $^1O_2$ are expected to vary with the protein structure because electronic and steric barriers influence the formation of dimers. The quantitative analysis for the products of the reaction of free cystine with $^1O_2$ showed the formation of two molecules of RSS(=O)R through a zwitterionic-peroxy intermediate (Eqs. 4.38 and 4.39). Formation of this intermediate was suggested by conducting kinetics and trapping experiments and by carrying out low-temperature matrix-isolated Fourier transform infrared (FTIR) spectroscopy of sulfur–oxygen adducts:

$$R_2S + {}^1O_2 \rightarrow R_2S^+\text{-}OO^- \tag{4.38}$$

$$R_2S + R_2S\text{-}OO^- \rightarrow 2R_2S{=}O. \tag{4.39}$$

The zwitterionic-peroxy species was also observed in the reaction of free Met with $^1O_2$, which also reacted with another molecule of Met to produce 2 mol of sulfoxides [172]. However, the stoichiometry of $^1O_2$ to Met was complex and varied with pH. The stoichiometries were 1:2 and 1:1 below pH 5 and above pH 7, respectively. Several other intermediates such as nitrogen–sulfur cyclic species have been reported, which, on subsequent hydrolysis, formed sulfoxide [172].

The chemiluminescence in protein oxidation by $^1O_2$, generated by the irradiation of rose bengal (RB), was observed [166, 176]. The examination of chemiluminescence in the presence of several amino acids (Cys, Met, Trp, Tyr, and His) demonstrated the peroxides (and/or hydroperoxides) of the Trp residue of proteins were mainly responsible for chemiluminescence [166]. There is a significant amount of evidence for the formation of multiple endohydroperoxides in the reaction of tryptophan with $^1O_2$ has been provided by using different analytical techniques such as high-performance liquid chromatography (HPLC)/mass spectrometry (MS) and ^{18}O-labeled NMR [13, 133, 160, 166]. Scheme 4.1 in Figure 4.10 displays the pathway for the production of hydroperoxides. The initial reaction of $^1O_2$ with Trp yielded 3α-hydroperoxypyrroloindole (**1**) through the postulated precursor species. The subsequent decomposition of species produced N-formylkynurenine (NFK), whereas ring closure formed

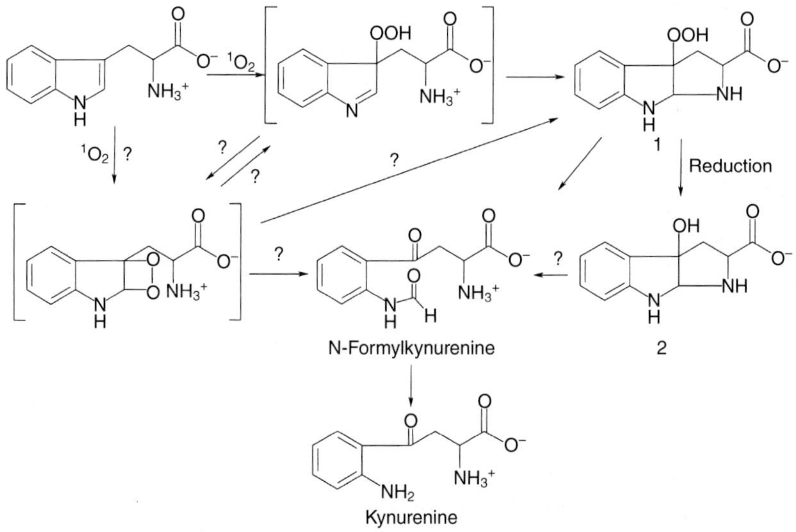

Scheme 4.1

Figure 4.10. Oxidation of Trp by $^1O_2$ (adapted from Davies [13] with the permission of the Royal Society of Chemistry).

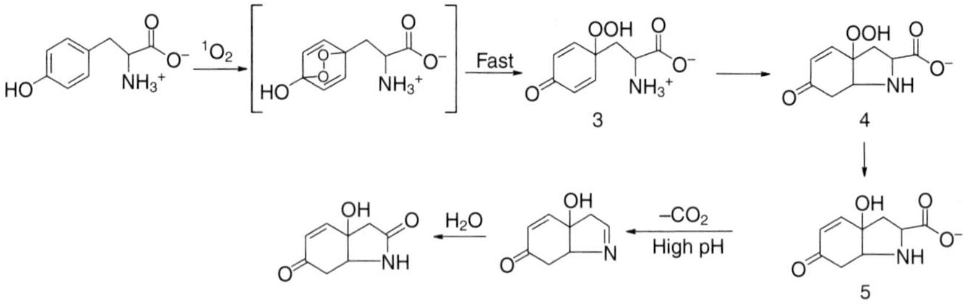

Scheme 4.2

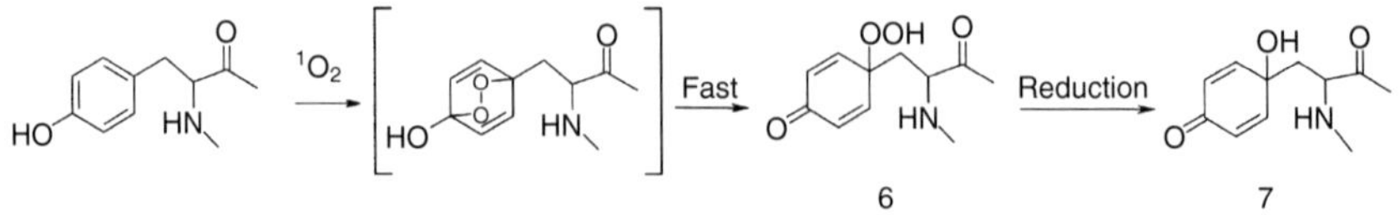

Scheme 4.3

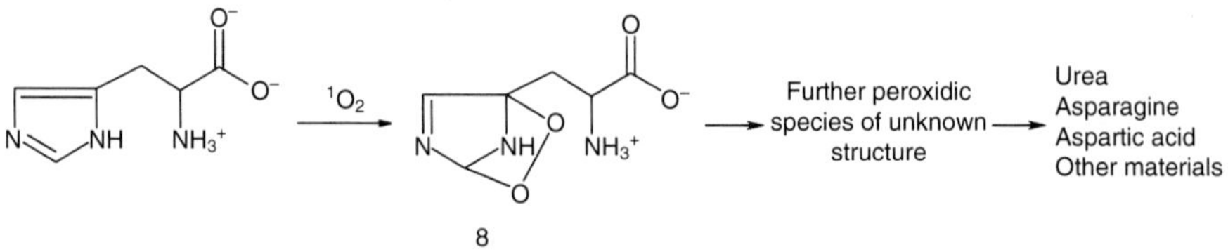

Scheme 4.4

Figure 4.11. Oxidation of Tyr by $^1O_2$ (adapted from Davies [13] with the permission of the Royal Society of Chemistry).

3α-hydrooxypyrroloindole (**2**). The formation of species **2** was supported by the Raman spectroscopic study in which a new band near 1090 cm^{-1}, ascribed to the stretching vibration of the O–O band in O_2^-, was observed following the reaction of Trp with $^1O_2$ [160]. The formation of kynurenine can occur through NFK [130]. The decomposition of hydroperoxypyrroloindole and hydroxypyrroloindole can also yield NFK [177, 178].

Figure 4.11 represents Schemes 4.2 and 4.3 for the formation of hydroperoxides in the reactions of $^1O_2$ with Tyr and peptide-bound Tyr. The formation of species **3** in Scheme 4.2 has been proven using NMR and MS analysis of the intermediate [179]. Other species related to species **3** have also been observed in the reactions of phenols with $^1O_2$ [180, 181]. Generation of the cyclized product, 3a-hydroperoxy-6-oxo-2,3,3a,6,7,7a-hexahydro-1*H*-indol-2-carboxylic acid (species **4**), involved a rapid ring closure at the unprotonated amine site of Tyr. This ring closure occurs somewhat freely but is slowed down in peptides, which do not contain Tyr residues at the N-terminus. The delocalization of the nitrogen lone pair in the amide bond of such peptides may result in less efficient nitrogen nucleophile in the ring-closing Michael reaction (Scheme 4.3). The C1 hydroperoxide (species **6**) was the major product in the reaction of the peptide-bound Tyr residue with $^1O_2$. Species **5** and **7** in Schemes 4.2 and 4.3 were formed via species **4** and **6**, respectively.

Oxidation of free His by $^1O_2$ was postulated to happen through the initial formation of one or more endoperoxides, similar to species **8** (Scheme 4.4, Fig. 4.11). One mole of His was consumed by 1 mol of $^1O_2$ [182]. At a very low temperature, peroxides such as species **8** have been seen in the reactions of $^1O_2$ with substituted His derivatives [183]. Such species rapidly decompose to poorly characterized intermediates, which ultimately yield a complex mixture of products including aspartic acid and aspartic derivatives and urea (Scheme 4.4). A dipeptide carosine, which contains His and β-amino alanine, showed damage to the imidazole ring when reacted with $^1O_2$ [160]. The stretching vibration of the C=C bond and the bending vibration of the NH bond in the imidazole ring were changed significantly in the Raman spectroscopy characterization of the products [160].

Exposure of $^1O_2$ to a wide range of proteins showed the formation of peroxide species regardless of the composition of proteins [13]. Peroxides were generated on Trp, Tyr, or His residues within the proteins, and the yields of peroxides were enhanced with D_2O as the solvent. In the case of tyrosine phosphotase-1B, inactivation of the proteins by $^1O_2$ resulted in oxidation of the active-site Cys 215 [168]. The capsid protein in the bacteriophage MS2 was also modified by $^1O_2$ [167]. The oxidized product detected was of the trypsin peptide (Ser84-Lys106). High yields of peroxides were formed on exposure of visible light to bulk cellular proteins, extracted from the cells, in the presence of sensitizers [13].

In a recent study, a detailed mass spectral analysis of cytochrome *c* and two model peptides P824 and tryptophan cage (Trp-cage) by $^1O_2$, generated by irradiation at 670- to 675-nm light in the presence of phthalocyanine Pc 4-malate salt, was conducted to see how $^1O_2$ oxidizes amino acid side chains of proteins and inactivates enzymes [169]. Cyt-*c* contains two Met (Met65 and Met80), three His (His 18, His 26, and His 33), one Trp, four Tyr, and four Phe (Phe10, Phe36, Phe46, and Phe82) residues. P824 is an eight-residue peptide (ASHLGLAR) with a single His residue. Trp-cage is a designed 20-residue miniprotein (NLYIQWLKDGGPSSGRPPPS) in which the Trp side chain is buried and more than 95% of it is folded in water at physiological pH [184]. The oxidized modification of Cyt-*c* was studied using matrix-assisted laser desorption/ionization-time of flight (MALDI-TOF) and tandem MS methods and is provided in Table 4.7 [169]. Modification of only one peptide, HKTGPNLHGLFG, was observed in H_2O, while several modifications occurred in the presence of D_2O (Table 4.7). The predominance of M + 16 and M + 32 modifications in D_2O suggests a single modification.

Modification as M + 16 in the case of Met65 and Met80 indicates sulfoxide as the oxidized product of this amino acid. Met80 showed an oxidized product of both +16 and +32. The product of the +32 adduct may be related to either

TABLE 4.7. List of Oxidized Peptides of Cyt-*c* in D_2O and H_2O

Parent		Modified (D_2O)			Modified (H_2O)	Found on MALDI[a]
Sequence	MW	Residue	δ	MW		
GITW^{59}K	604.3	W59	+16	620.3		
M^{80}IFAGIK	779.4	M80	+16	795.4	+	+
		M80	+32	811.4	+	
M^{80}IF^{82}AGIKK	907.5	F82	+16	923.5		
		M80	+16	923.5		
CAQCH^{18}TVEK	1018.4	H18[b]	+14	1032.4	+	
		H18[b]	+28	1046.4		
TGPNLH^{33}GLF^{36}GR	1168.6	H33	−22	1146.6		+
		H33	+14	1182.6		+
		H33	+26	1194.6		+
		H33	+32	1200.6		+
		H33	+34	1202.6	+	+
		F36	+16	1184.6		+
H^{26}KTGPNLHGLFGR	1433.8	H26	+14	1447.8		+
		H26	+32	1465.8	+	+
EETLM^{65}EYLENPK	1495.7	M65	+16	1511.7	+	

[a] Irradiated in D_2O.
[b] Modified residue is not certain.
Adapted from Kim et al. [169] with the permission of Elsevier Inc.

a peroxo species or sulfone. Of the three His residues, only His 26 and His 33 were modified. The incorporation of O was observed, which was also identified in a study using $H_2^{17}O$ [185]. A peptide containing His 18 was modified as +14 and +28. Significantly, the HKTGPNLHGLFGK + 32 Da peptide was continually the most intense modified peptide. The Tyr residue was not modified, resulting in observed unmodified peptides. Phe36 and Phe82 residues were modified as Phe + 16. Phe is not reactive with $^1O_2$; hence, observed hydroxylation of residues of Phe indicates either increased reactivity at the aromatic ring or the possibility of the involvement of a secondary oxidant (e.g., hyfroxyl radical) during the production of and exposue to $^1O_2$ [169]. Modification of these two Phe residues by the +16 was also seen independently in exposure to hydroxyl radicals [186].

Proteins are generally composed of 70% of dry mass of most cells in biological systems and have higher rate constants than other cellular components for the reactions with $^1O_2$ [151]. Therefore, proteins become a major target for $^1O_2$ as is shown in Figure 4.12. However, the potential localization and accumulation of sensitizers within particular cellular compartments (e.g., membranes) and the limiting diffusion radius of $^1O_2$ in the biological environment ultimately determine a preferential attack of $^1O_2$ on components of the cell [146, 167, 187].

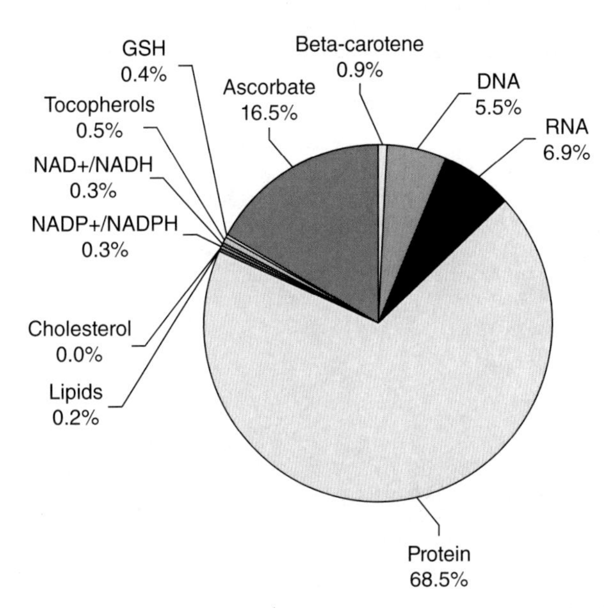

Figure 4.12. Predicted consumption of $^1O_2$ by intracellular targets calculated using the rate constant data given in Reference 151 and the average concentration of each component within a typical leukocyte cell [186] (adapted from Davies [130] with the permission of Elsevier Inc.). See color insert.

The hydroperoxides of peptides and proteins have been analyzed to demonstrate their production within proteins [133, 188, 189]. A wide range of sensitizers formed peroxides and the overlal yield was enhanced in using D_2O as a solvent. This is consistent with that $^1O_2$ was involved in forming peroxides. Research on the nature and exact position of these peroxides within the protein structure are in progress. Basically, interest in hydroperoxides is related to their high oxidation ability to induce damage by either directly inhibiting thiol-dependent proteins, cathepsins, and sarcoplasmic/endoplasmic reticulum Ca^{2+}-ATPase or by producing additional radical species in the presence of trace amounts of a transition metal ion [13, 130, 189–194]. Examples include the modification of DNA by protein hydroperoxides and the formation of DNA–protein cross-links [195, 196] and the decomposition of the hydroperoxides of Trp, His, and Try residues and a number of proteins at room temperature and in the presence of Fe^{2+}EDTA and $NaBH_4$ [133, 179, 188, 197].

The reactivity of hydroperoxides of amino acid, peptide, and protein hydroperoxides with peroxiredoxin s (Prx's) 2 and 3 has been conducted to quantitatively assess the importance of these reactions and also to know if reactions have significant rates to be involved in detoxification mechanism [198]. The second-order rate constants are presented in Table 4.8. Prx reacted faster than that of Prx2 and the rate constants varied from 2×10^3 for histidine

TABLE 4.8. Second-Order Rate Constants (/M/s) for the Reactions of Prx's with Various Hydroperoxides

Hydroperoxide	Prx2	Prx3
Lysozyme hydroperoxide	40	90
BSA hydroperoxide	160	360
Histidine hydroperoxide	2×10^3	3×10^3
N-Acetyl leucine hydroperoxide	4×10^4	ND

ND, not determined.

Adapted from Peskin et al. [198] with the permission of the Biochemical Society.

hydroperoxide to 4×10^4/M/s for *N*-acetyl leucine hydroperoxide. Tyrosine and Gly-Tyr-Gly hydroperoxides were slightly slower than the rate with *N*-acetyl leucine hydroperoxide, which suggests hydroperoxide on the α-carbon reacted slowly than those on the side chain. The rate constants provided in Table 4.8 are within close proximity to the measured values for the reactions of amino acids for small alkyl hydroperoxides [199]. However, Prx2 and Prx3 had much higher reactivity with H_2O_2 ($k > 10^7$/M/s) than their reactions with amino acid hydroperoxides [200, 201]. The reactivity of protein hydroperoxides was much slower with rate constants in the range of 4×10^1/M/s to 3.6×10^2/M/s (Table 4.8). Hydroperoxides of Leu and Ile as free amino acids and in peptides with either or both C- and N-termini blocked reacted rapidly with Prx2 and Prx3 (Table 4.9). Complete oxidation of Prx, except leucine hydroperoxide, was observed within 10–15 seconds. A significant Prx oxidation by histidine hydroperoxide was obtained, though the reaction was slow (Table 4.9). The results of Table 4.9 are supported by the rate studies [198]. Rate constants in the studies suggest that Prx2 and Prx3 may be able to remove peptide hydroperoxides with half-lives of 1–10 seconds, while a few minutes would be required to remove protein hydroperoxides [198]. Prx's would thus be able to prevent the accumulation of intracellularly generated such reactive hydroperoxides. Secondary damages to other targets by hydroperoxides may be prevented. The oxidation of Prx by hydroperoxides produced disulfides and alcohols (Eq. 4.40). The formation of alcohol was confirmed using mass analysis [198]:

$$2\text{PrxSH} + \text{ROOH} \rightarrow \text{PrxS-SPrx} + \text{ROH}. \tag{4.40}$$

The reactivity of hydroperoxides of amino acids (Tyr and Trp), peptides (*N*-Ac-Trp-Me and Gly-Tyr-Gly), and bovine serum albumin (BSA) with 26S proteasome has been examined to understand if oxidized proteins modulate 26S proteasome activity [189]. Impairment of proteasome has been reported in Alzheimer's and Parkinson's diseases. Results showed that studied hydroperoxides modified the activity of 26S proteasome activity. Hydroperoxides of

TABLE 4.9. Hydroperoxides of Amino Acids, Peptides, and Proteins Employed to React with Prx's

Hydroperoxide	Treatment to Produce Hydroperoxide	Relative Concentration of Hydroperoxide (mmol per mol of Treated Compound)	Reduced form of Prx's Remaining (%)
Leu	γ-Irradiation	24	16 ± 1
N-Acetyl Leu	γ-Irradiation	8	0
N-Acetyl Leu amide	γ-Irradiation	25	0
N-Acetyl Leu methyl ester	γ-Irradiation	23	0
Ile	γ-Irradiation	25	0
N-Acetyl Ile	γ-Irradiation	11	0
N-Acetyl Ile amide	γ-Irradiation	11	0
Gly-Leu	γ-Irradiation	22	0
Leu-Gly	γ-Irradiation	39	0
Gly-Leu-Leu-Gly	γ-Irradiation	20	0
Tyr	Photolysis	62	19 ± 25
Gly-Tyr-Gly	Photolysis	196	33 ± 28
His	Photolysis	16	27 ± 8
Lysozyme	Photolysis	45	55 ± 8
BSA	γ-Irradiation	759	30 ± 1
BSA	Photolysis	583	–

The level of the reduced form of Prx's remaining was determined after incubation for 10–30 seconds with typically 5 μM Prx2 or Prx3 and 7 μM hydroperoxide and is shown as means ± SD ($n = 2$–5) for pooled Prx2 and Prx3.

Adapted from Peskin et al. [198] with the permission of the Biochemical Society.

protein were efficient inhibitors of tryptic and chymotryptic activities of 26S proteasome. Loss of actcitivity may be related to the reaction with thiols in the proteasome. The inhibition of the proteasome activity by oxidized proteins may have implications in cell signaling processes because of the effect on the turnover of native proteins [189]. More recently, selenomethionine (SeMet) was shown to scavenge hydroperoxides present on amino acids, peptides, and proteins [202]. Significantly, Met could not remove these peroxides. The reaction of SeMet with peroxides resulted in selenoxide, confirmed by HPLC technique. In the presence of GSH, thioredoxin reductase (TrxR), thioredoxin (Trx), and NADPH system, selenoxide could be converted back to SeMet. The scheme of this recycling having removal of peroxides is presented in Figure 4.13 [202]. Reactions of selenoxide with GSH, TrxR/NADPH, and TrxR/Tx/NADPH were rapid, indicating enhanced removal rate of peroxide. A loss of peroxide was stoichiometric in consuming NADPH in the TrxR/Trx/NADPH system. This suggests efficient removal of peroxides. SeMet as an effective

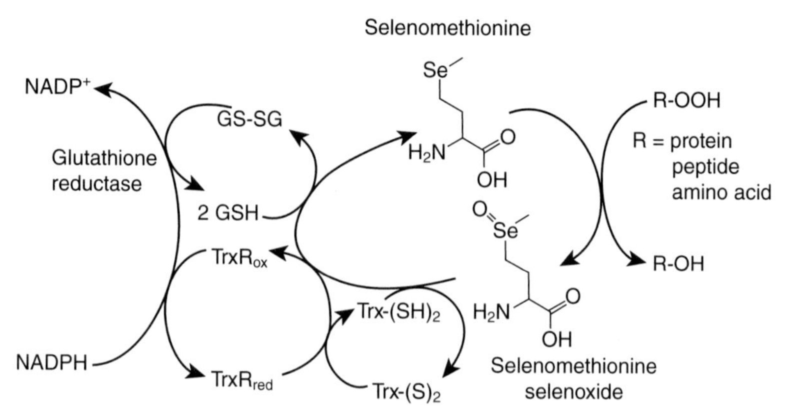

Figure 4.13. Proposed mechanisms for the enhanced peroxidatic activity detected in SeMet-supplemented cells exposed to Gly-His-Gly and *N*-Ac-Trp-OMe hydroperoxides (adapted from Rahmanto and Davies [202] with the permission of Elsevier, Inc.).

reducing agent was tested using murine macrophage-like J774A.1 cells (Fig. 4.13). A greater removal of two peptide peroxide hydroperoxides (Gly-His-Gly and *N*-Ac-Trp-OMe) was observed in the application of SeMet in these tests. SeMet may thus be involved in the catalytic activity of removing peptide hydroperoxides of peptide and proteins [202].

In summary, side chains of Cys, Met, cystine, Trp, Tyr, and His are most likely to be oxidized by $^1O_2$. Rates of modification of proteins by $^1O_2$ depend on other components present in solution. However, determined rates in *in vitro* studies should be used with caution in predicting intracellular phenomena because proteins tend to change structure and conformation and may also exist in form different from those *in vitro* experiments [203]. Studies on examining the influence of protein dynamics on the behavior of $^1O_2$ are forthcoming, but work on studying reactions of a wide range of proteins with $^1O_2$ should provide understanding the mechanism of $^1O_2$-induced cell death. A significant progress on identifying $^1O_2$-modified oxidized products of side chains of proteins, particularly Trp, has been made; however, much more research in this area using advanced analytical techniques is needed to fully characteriz the oxidized products. Furthermore, peroxide species have been determined to be the key species in protein damages induced by $^1O_2$, but limted information is currently known on their reactivity with biological molecules. More studies on protein peroxide and their effects on cellular enzymes may determine if these species are responsible in disease progression. A gas-phase approach using chemically generated "clean" $^1O_2$ using electrospray ionization (ESI) and ground-ion-beam scattering methods have been applied to gain insight on the mechanism of the reactions of protonated Cys and Trp with $^1O_2$ [204]. Such experimental studies combined with density functional theory (DFT) calculations on the reactions of $^1O_2$ with other amino acids may provide further information on the potential role of $^1O_2$ in the functioning of the cell.

4.3 OZONE

Ozone (O_3) is an allotrope of oxygen, which provides a protective shield to incoming UV radiation in the stratosphere. However, it is one of the most dangerous phytotoxic air pollutants in the troposphere. Thus, O_3 is associated with several health effects including exacerbation of asthma and airway inflammation and hyperreactivity [205, 206]. O_3 appears to be the source of degradation of self-assembled alkanethiol monolayers on silver and gold surfaces that are exposed to ambient air [207, 208]. O_3 is also damaging to plant life, which ranges from inhibition of photosynthesis and associated yield loss to premature senescence and visible tissue necrosis [209]. The formation of O_3 by human neutrophils may contribute to killing bacteria [210]. Met, Cys, His, and Trp were demonstrated to produce O_3 from $^1O_2$ in the water oxidation pathway [210]. In the aqueous phase, O_3 has applications in water treatment, whitening, and disinfection [211–213]. Ozone is unstable in the aqueous phase and decomposes through several initiation, propagation, and termination chain reactions, initiated by hydroxide ions [214–221]. The decomposition of O_3 depends on proton concentration, alkalinity, and inorganic and organic constituents of water. Studies have shown the acceleration of the decomposition of ozone with increasing pH and hydrogen peroxide [217] The following set of reactions are involved in the decomposition of O_3 [211]:

$$O_3 + OH^- \rightarrow HO_2^- + O_2 \quad k_{41} = 70/M/s \tag{4.41}$$

$$O_3 + HO_2^- \rightarrow {}^{\bullet}OH + O_2^{\bullet-} + O_2 \quad k_{42} = 2.8 \times 10^6/M/s \tag{4.42}$$

$$O_3 + O_2^{\bullet-} \rightarrow O_3^{\bullet-} + O_2 \quad k_{43} = 1.6 \times 10^9/M/s \tag{4.43}$$

$pH < \approx 8$:

$$O_3^{\bullet-} + H^+ \rightleftharpoons HO_3^{\bullet} \quad k_{44} = 5.0 \times 10^{10}/M/s \tag{4.44}$$

$$k_{-44} = 3.3 \times 10^2/s$$

$$HO_3^{\bullet} \rightarrow {}^{\bullet}OH + O_2 \quad k_{45} = 1.4 \times 10^5/s \tag{4.45}$$

$pH > \approx 8$:

$$O_3^{\bullet-} \rightleftharpoons O^{\bullet-} + O_2 \quad k_{46} = 2.1 \times 10^3/s \tag{4.46}$$

$$k_{-46} = 3.3 \times 10^9/s$$

$${}^{\bullet}OH + O_3 \rightarrow HO_2^{\bullet} + O_2 \quad k_{47} = 1.0 \times 10^8\text{–}2.0 \times 10^9/M/s. \tag{4.47}$$

Initiating reactions (4.41) and (4.42) suggest that the decomposition of ozone can be increased by raising the pH or by adding hydrogen peroxide. Recent studies on the decomposition of O_3 have estimated the value of k_{41} to be (1.7–1.8) × 10^2/M/s [220, 221]. The O_3 decomposition produces $^{\bullet}OH$ by a rapid reaction (4.45). Reaction (4.47) produces other oxidant species, superoxides. The generated $^{\bullet}OH$ reacts with carbonate species (HCO_3^- and CO_3^{2-}) and other constituents of water. A half-life of O_3 has thus a range of seconds to hours depending on the pH and concentrations of carbonate species and organics in water [211].

4.3.1 Reactivity

The reactivity of O_3 with a wide range of inorganic and organic compounds has been reported [222]. The second-order rate constants for the reactions vary widely between <0.1 and 7 × 10^9/M/s. The inorganic compounds such as Fe(II), Mn(II), H_2S, cyanide, and nitrite reacted rapidly with O_3 through an oxygen transfer mechanism. The reactions with organic compounds proceeded through well-defined mechanisms involving double bonds, activated aromatic systems, and neutral amines [213] In the water treatment process, the reactions of O_3 with components of water matrix compete with the reactions of ozone with desired pollutants. Therefore, only the pollutants resulting in high rate constants with O_3 can be removed by direct direction [212]. The kinetics of the oxidation of amino acids by O_3 in aqueous solution have been determined as a function of pH at 22–24°C [215, 223–229]. Second-order rate constants for the oxidation of selected amino acids by O_3 at different pH values are provided in Figure 4.14, which show an increase in rate constants with an increase in pH. Since carboxylic acids and protonated forms of amino acids reacted slowly, the variation in rates with pH was analyzed using the dissociation of the amino group of amino acids ($HA \rightleftarrows H^+ + A^-; K_{a2}$) [15]. The mathematical interpretation of the data considers the reactions of O_3 with undissociated (HA) and dissociated (A^-) forms of amino acids separately [223, 228]. The rate constants for the reaction of O_3 with HA (k_{HA}) and A^- (k_A) were obtained by plotting k versus ($K_{a2}/[H^+]$). The results of the oxidation of Gly and Pro by O_3 are shown in Figure 4.15. The slope of the plot gave the value of k_A while the intercept gave the value of k_{HA}. Values of k_{HA} near zero indicated the undissociated amino group of amino acids did not react significantly with O_3 in aqueous solutions. Therefore, the dependence of rate with pH was largely related to the reaction of the free amine form of amino acids with O_3.

The rate constants for various amino acids, except Cys, Met, and Trp, which reacted with rate constants, varied from 2.6 × 10^4/M/s to 4.4 × 10^6/M/s [223]. The order of reactivity was determined as Glu ≈ Gln ≈ Lys < Asp ≈ Asn ≈ Thr < Leu ≈ Ile ≈ Arg < Val < Ala < Ser < Gly ≈ His < Phe < Pro. The high reactivity of Cys with O_3 (Fig. 4.14) indicates the possible site of reaction was at the sulfhydryl functional group rather than the amino group. The measured rate constants shown in Figure 4.14 involved a change of the neutral form of

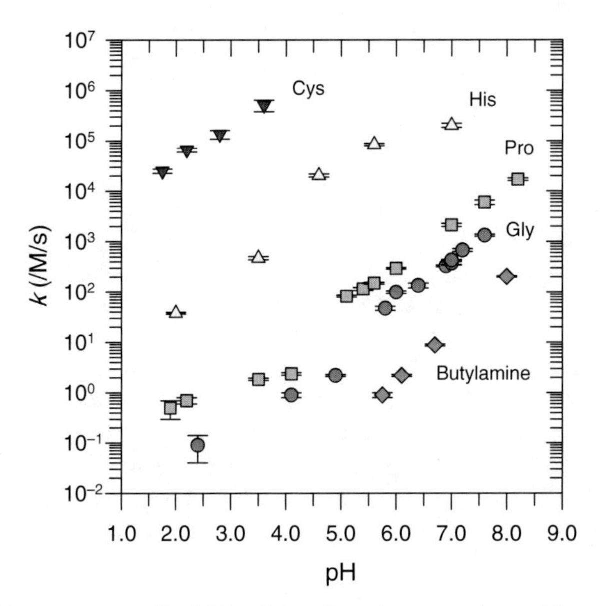

Figure 4.14. Rate constants (k, /M/s) of reaction of some amino acids and butylamine with O_3 in water (data were taken from Pryor et al. [223]). See color insert.

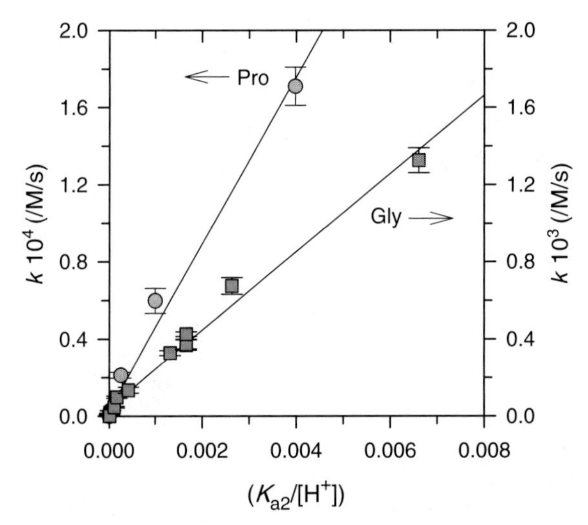

Figure 4.15. Plot of k (/M/s) versus (K_{a2}/[H^+]) for oxidation of Gly and Pro by O_3 (adapted from Sharma and Graham [15]). See color insert.

the sulfhydryl group (SH) to the thiolate ion (S^-) of cysteine. Thus, it could be concluded that the greater reactivity of O_3 with the thiolate ion was mostly responsible for the increase in the reaction rate with an increase in pH. The estimated rate constants were 2.0×10^4/M/s and 2.5×10^{10}/M/s for reactions of O_3 with Cys(SH) and Cys(S^-), respectively. The reactivity of O_3 with another sulfur-containing amino acid, Met, was also high [223].

Hydrocarbon-like amino acids (Glu and Asp) and amides (Asn, Gln, and Arg) have side chains that showed slow reactivity with O_3. Two hydroxyamino acids, Ser and Thr, also reacted slowly with O_3. Ala and Val contain alkyl groups and reacted slower than that of Gly, which has no alkyl group. Butylamine reacted much slower with O_3 than Gly and also showed the influence of the alkyl group on the rates (Fig. 4.14). Pro is a secondary amine and appears to react faster than primary amines, such as Gly (Fig. 4.14). It appears that the imidazole ring in His was responsible for its high reactivity with O_3. The aromatic amino acids, Phe and Trp, also had high reactivities with O_3 (Fig. 4.14). It seems that the aromatic ring and the benzylic hydrogens influenced the reactivity with O_3.

The rate constants at pH 8.0 are provided in Table 4.10. Amount of ozone required (i.e., ozone demand) for the oxidation of several amino acids were determined experimentally [230], which aided in the determination of the rate constants based on the decay rate of ozone, $k(O_3)$. The calculated values of k(AA) and half-lives are also provided in Table 4.10. The half-lives of most of the reactions of amino acids with O_3 were determined to be in tens of seconds. However, the half-lives for the oxidation of Met and Trp by O_3 were determined in milliseconds (Table 4.10).

Several studies on the products of the ozonation of amino acids, peptides, and proteins in aqueous solution have been performed using spectrophotometric and chromatographic techniques, and now ESI and MALDI methods [224, 231–250]. Ozonation of Gly by O_3 in the presence of HCO_3^- ions produced nitrate, oxamic, formic, and oxalic acids [251]. Ozonation of Met, Cys, and glutathionine formed sulfoxide and sulfone [236, 239, 252, 253]. Sulfoxide was the main oxidation product of Met (Fig. 4.16) [225, 236]. However, a mass analysis of the products showed both sulfoxide and sulfone as primary products of the oxidation reaction [239,254]. Cys was oxidized as a first step to cystine (CySSCy) (Fig. 4.16), but a further oxidation reaction by ozone yielded cysteic acid [236, 255]. Recently, an interfacial ozonolysis of Cys and GSH was carried out using online thermospray ionization MS [252, 253]. When the surface of the aqueous Cys microdroplets was exposed to O_3(g) for less than 1 millisecond, sulfenate ($CysSO^-$), sulfinate ($CysSO_2^-$), and sulfonate ($CysSO_3^-$) were simultaneously detected [252]. In similar experiments with GSH, sulfonates (GSO_3^-/GSO_3^{2-}) were formed [253].

The oxidation of Trp, Tyr, and His alone and as part of peptide/proteins residues by O_3 were carried out using the ESI-MS analysis [237, 239]. An earlier study suggested Pro was the oxidized product of His [225], but a later study reported an aspartyl residue, through a 2-oxohistidine intermediate, as

TABLE 4.10. Kinetics of the Oxidation of Amino Acids with O_3 at Ambient Temperature

Amino Acids	pK_{a2}	$k(O_3)$[a] (/M/s)	O_3 demand[b] (mol/mol)	$k(AA)$[c] (/M/s)	$t_{1/2}(AA)$[d]
Ala	9.87	1.0×10^3	6.7	1.5×10^2	217 seconds
Arg	9.04	5.3×10^3	9.1	5.8×10^2	58 seconds
Asn	8.72	8.0×10^3	7.3	1.1×10^3	30 seconds
Asp	9.82	6.2×10^2	4.3	1.4×10^2	230 seconds
Glu	9.47	8.8×10^2	3.8	2.3×10^2	144 seconds
Gln	9.13	1.9×10^3	–	–	–
Gly	9.78	3.5×10^3	5.4	6.5×10^2	52 seconds
His	9.18	1.7×10^4	9.7	1.7×10^3	19 seconds
Ile	9.76	9.8×10^2	–	–	–
Leu	9.74	9.6×10^2	5.7	1.7×10^2	197 seconds
Lys	8.95	3.5×10^3	9.3	3.7×10^2	89 seconds
Met	9.28	3.8×10^6	7.2	5.3×10^5	63 millisecond
Phe	9.31	1.9×10^4	9.2	2.1×10^3	16 seconds
Pro	10.6	1.1×10^4	8.3	1.3×10^3	25 seconds
Ser	9.21	8.0×10^3	9.7	8.2×10^2	41 seconds
Thr	9.1	3.6×10^3	9.8	3.7×10^2	91 seconds
Trp	9.39	7.0×10^6	12.9	5.4×10^5	61 millisecond
Tyr	9.11	–	13.4	–	–
Val	9.74	1.2×10^3	–	–	–

[a] Calculated at pH 8.0 from the values of k_{HA} and k_A given in Pryor et al. [223].
[b] Reported by Hureiki et al. [230] (pH 8).
[c] Calculated compound reaction rate constant at pH 8.
[d] Assuming $[O_3]$ = 1 mg/L (pH 8).
Adapted from Sharma and Graham [15].

the final product of ozonolysis [235]. However, a recent study yielded a compound in which three oxygen atoms were added to His. A proposed structure of a compound as the only oxidized product is shown in Figure 4.16 [239]. Two studies of the ozonolysis of Trp have indicated the formation of NFK as the major oxidized product (Fig. 4.16) [225, 239]. The ozonolysis of Tyr yielded 3,4-dihydroxyphenylalanine (DOPA) (Fig. 4.16) [225, 233]. Significantly, the ozonolysis of peptides containing Met (Gly-Gly-Phe-Met, Tyr-Gly-Gly-Phe-Met), His (His-Leu, Gly-His-Lys), and Trp (Lys-Trp-Lys) formed products that were in agreement with the expected products from the oxidation of free amino acids by O_3 [239]. A study on the degradation of Tyr-Gly-Gly (2.5 mg/L total organic carbon) by O_3 in the presence of bicarbonate buffer at pH 7.2 showed 0.26 mg of consumed O_3 achieved complete removal of the Tyr residue; however, only 23.3% of the Gly residue was removed [256].

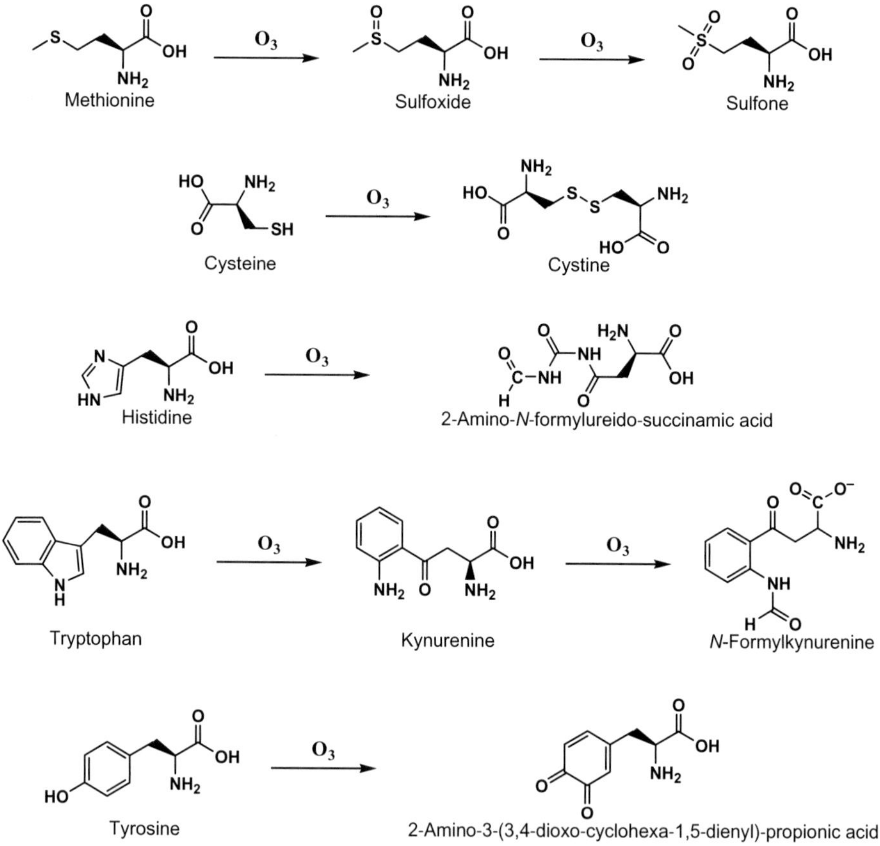

Figure 4.16. Ozone reactions with Met, Cys, His, Trp, and Tyr (adapted from Sharma and Graham [15]).

The ozonolysis of peptides containing residues of Tyr and His, angiotensin II (DRVYIHPF) and two analogues (DRVYIAPA and DRVAIHPA) has been carried out to characterize carbonyl products in order to understand the mechanisms of ozonolysis of biomolecules [231]. The second-order rate constants, k, for reactions with O_3 were determined as $(4.39 \pm 0.34) \times 10^2$/M/s, $(2.74 \pm 0.03) \times 10^2$/M/s, and $(3.79 \pm 0.03) \times 10^2$/M/s for angiotensin II, DRVYIAPA, and DRVAIHPA, respectively. Reaction rate constants suggest that the atmospheric lifetime of these peptides would be 20–30 days [231]. The major oxidized products identified were His + 3O, Tyr + O, and Phe + O. Two other adducts, Tyr + 3O and Tyr + 34, were also observed. The minor products include His + 5, His + 34, and His + 82, which suggests a continued oxidation of the products and of nonspecific sites of the parent peptide. Fragmentation mechanisms of the oxidized peptides in angiotensin II were interpreted using SID FT-ICR MS/MS experiments, Rice–Ramsperger–Kassel–Marcus (RRKM) modeling, and molecular dynamics [257]. Results indicate pathways of the reaction mechanism involve hydrogen bonding and close-ring structure and therefore ozonolysis was controlled by entropy effects. Calculated energies and entropies of activation allowed to proposed mechanisms for different pathways of ozonolysis.

There are few studies on the oxidation of proteins by O_3 [225, 237]. The oxidation of band 3 protein, cytochrome *c*, BSA, and human serum albumin (HSA) by O_3 has been carried out to examine the oxidation susceptibility of Trp residues present in different regions of proteins [225]. Isolation of band 3 protein from the ozone-treated ghosts followed by subjection of it to trypsin digestion yielded water-soluble peptides from the cytoplasmic N-terminal tail and the interhelical loops. Fluorescent peptides include WMEAAR from the N-terminal cytoplasmic tail and GWVIHPLGLR from the outer loop between helices 7 and 8. Trp residues in band 3 were monitored in three different environments: the N-terminal cytoplasmic tail, the transmembrane helices, and in one of the interhelical loops external to the lipid bilayer [225]. Cytochrome *c* contained the GITWK peptide in which Trp was at position 59 interacting with the heme moiety. Digestion by trypsin of BSA produced two fluorescent peptides (FWGK and AWSVAR) and two Trp residues, which were in different regions of the protein. HSA contained the AWAVAR peptide with one Trp residue within it.

The fluorescence of control- and O_3-treated proteins at pH 7.5 were measured (Fig. 4.17). As shown in Figure 4.17a,b, the peptides in band 3 and cytochrome *c* were not oxidized by O_3. The Trp in N-terminal cytoplasmic tail of band 3 is expected to be resistant to oxidation by O_3 because many proteins of the glycolytic sequence of the tail possibly react first before ozone could reach to Trp on the N-terminal cytoplasmic tail. A protection of Trp in the interhelical loop from the peripheral proteins on the outside of the membrane may be another cause of no oxidation observed in band 3. HPLC and amino acid analysis of O_3-treated cytochrome *c* demonstrated a modification of the protein and oxidation of Met. However, the Trp residue did not change, which suggests its protection by the porphyrin ring [225]. The results shown in Figure 4.17c,d clearly indicate the peptides in both BSA and HSA were oxidized by O_3. Trp residues in these proteins are involved in structures of membrane and hence are not protected from the attack of O_3. It is likely that the two Trp residues of BSA reacted differently with O_3 due to their positions in the three-dimensional structure. Overall, ozonolysis of tryptophans in studied proteins was affected by the position of the Trp residues within the protein.

The oxidation of five different proteins (invertase, pectinase, papain, trypsin, and gelatin) by O_3 has been examined to obtain clues on the action of ozone on more complex biological species such as virus and bacteria [237]. Measurements of spectroscopy and polarimetry were used to monitor the ozonolysis of the proteins. The results in Figure 4.18 showed two spectral maxima, one at 222 nm, attributed to the $n \rightarrow \pi^*$ transition of the amide bond, and the other at 275 nm, attributed to the aromatic acids, which are mainly Trp (ε_{277nm} = 5500/M/

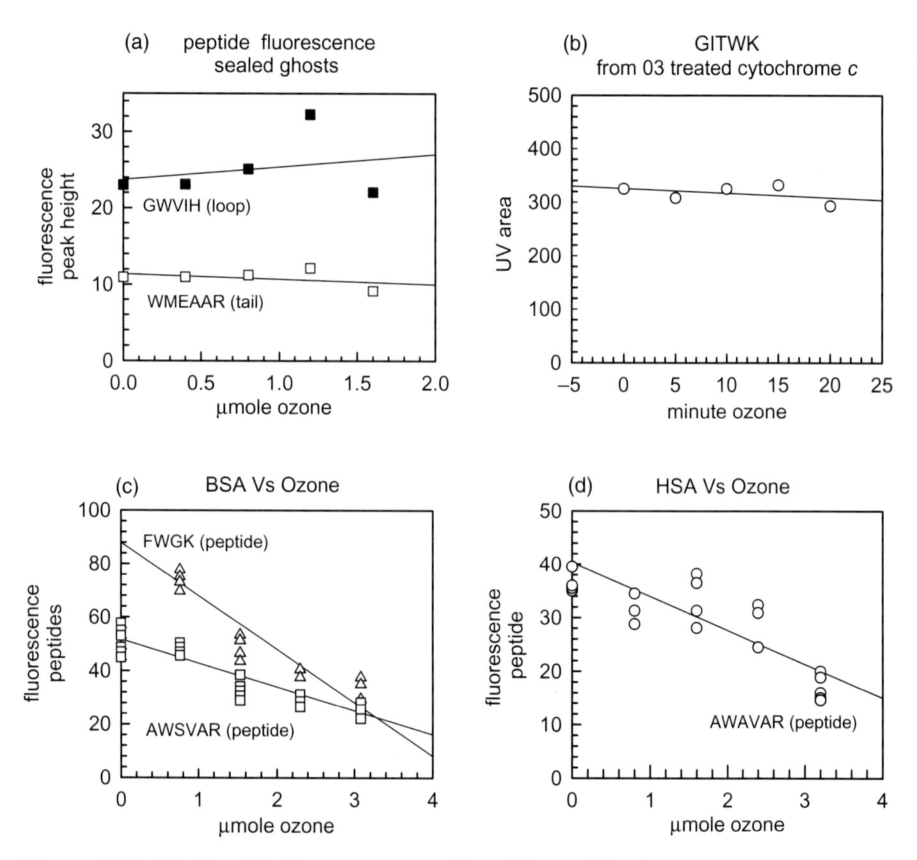

Figure 4.17. (a) Band 3 fluorescent peptides, (b) peptide GITWK from O_3-treated cytochrome *c*, (c) fluorescent peptides from O_3-treated BSA, and (d) fluorescent peptides from O_3-treated HSA (redrawn from Mudd et al. [232]).

cm), Tyr ($\varepsilon_{277\,nm}$ = 1500/M/cm), and Phe ($\varepsilon_{254\,nm}$ = 200/M/cm) [258]. His had a maximum at 210 nm and was buried by the amide $n \rightarrow \pi^*$ transition. Disulfide (S–S) was another bond in proteins, which absorbed at 250 nm ($\varepsilon_{277\,nm}$ = 300/M/cm). In the O_3-treated samples, the intensity of the peak at 275 nm decreased with a shift to shorter wavelengths and finally its elimination with an increase in the O_3 dosage. The intensity of the peaks at 222 nm remained similar, which indicated the attack on the amide group of the protein chain by O_3 was not significant.

In the ozonation of invertase and pectinase samples (Fig. 4.18a,b), a new peak at 327 nm appeared, which accounted for the formation of quinines and a ketone group attached to the benzene ring from the oxidation of aromatic amino acids. In the case of pectinase, an additional peak at 240 nm appeared from the oxidation of Cys and the formation of S–S bonds. Both papain and

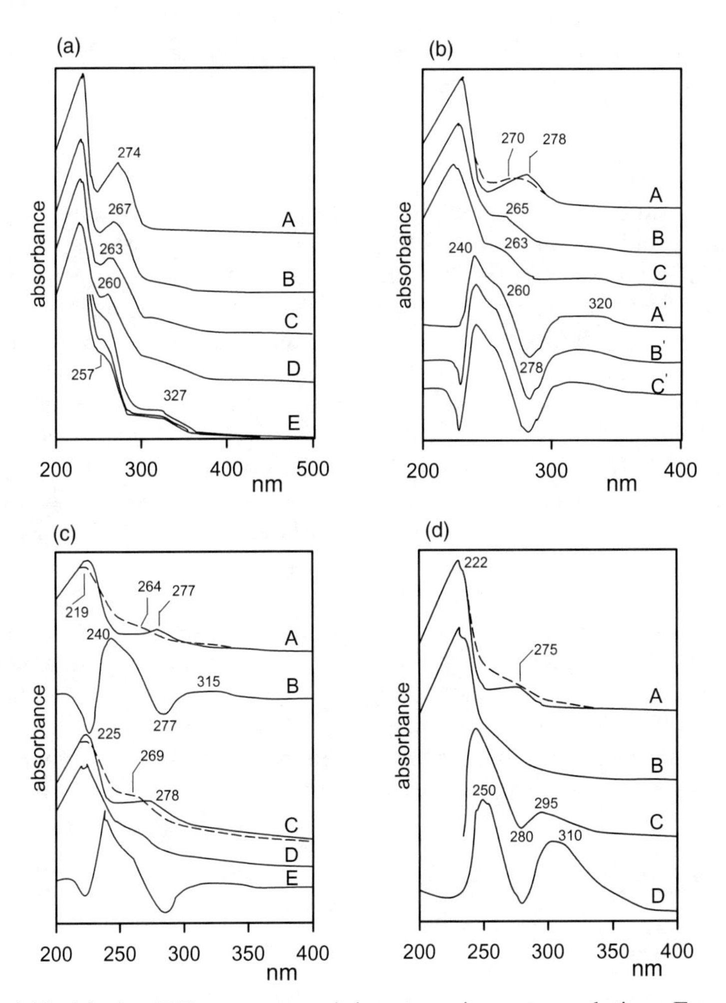

Figure 4.18. (a) A—UV spectrum of invertase in water solution. From B to E—increasing addition of O_3. (b) A—UV spectrum of pectinase in water solution (147 mg in 22 mL of H_2O). The dotted line is the spectrum after the addition of 5 mg; B—after the addition of 21 mg; C—after the addition of 56 mg. A'—differential spectrum of A; B'—differential spectrum of B; and C'—differential spectrum C. (c) A—UV spectrum of of papain before ozonation (solid line) and after addition of 10 mg O_3 (dotted line); B—differential spectrum of A; C—UV spectrum of trypsin (solid line) and after addition of 20 mg O_3; D—after addition of 45 mg O_3; E—differential spectrum of C. (d) UV spectrum of gelatin before ozonation (solid line) and after addition of 10 mg O_3; B—gelatin after addition of 45 mg O_3; C—differential spectrum of A; D—ratio of the spectrum B/A (redrawn from Cataldo [237]).

trypsin reacted in a similar manner to the oxidation of invertase by O_3 (Fig. 4.18c). It appears the thiol group, –SH, from cysteine, is enhanced by its interaction with a histidine group of another branch of the folded protein in papain, which was rapidly oxidized to disulfide (Fig. 4.18c). The gelatin molecule does not contain Trp and also had a low content of Phe and Tyr. This may result from a low intensity at 275 nm for gelatin (Fig. 4.18d). Similarly, a peak at 275 nm in gelatin disappeared with limited variation in the peak intensity at 222 nm. Other features at the 250- and 295-nm bands of the spectra were also similar to the O_3-treated protein samples. The polarimetric measurements of this study indicated the oxidation by O_3 treatment resulted in the denaturation of proteins by causing changes in their secondary and tertiary structures [237].

Interfacial reactions of O_3 with pulmonary surfactant protein B (SP-B) in a model surfactant (1-palmitoyl-2-oleoyl-*sn*-glycerol [POG]) system have been studied using field-induced droplet ionization (FIDI)/MS [259]. This study showed the structurally specific oxidative changes of $SP\text{-}B_{1-25}$ (a shortened version of human SP-B) at the air–liquid interface. The heterogeneous reaction of $SP\text{-}B_{1-25}$ at the interface was quite different from that in the solution phase. The homogeneous oxidation of $SP\text{-}B_{1-25}$ was nearly complete, while only a subset of the amino acids, which generally reacts with O_3, was oxidized in the hydrophobic interfacial environment, both with and without a surfactant layer [259]. A similar experimental approach was used to study the interfacial reactions of O_3 with pulmonary phospholipid surfactants [260]. Results of the interfacial studies may clarify the effect of smoking and airborne particles on the lung surfactant system.

4.4 HYDROXYL RADICAL

4.4.1 Generation

The production methods of $^{\bullet}OH$ include radiolysis, pulsed electron beams, photochemistry, electrochemistry, corona discharge, sonolysis, and Fenton chemistry [261–267]. In pulse radiolysis, the interaction of water with radiation initially produced $H_2O^{\bullet+}$, dry electron e^-_{dry}, and excited water H_2O^* (Eq. 4.48), which were then converted into highly reactive primary radical species, $^{\bullet}OH$, e^-_{aq}, and $H^{\bullet}$, within 10^{-12} seconds [268]:

$$H_2O + h\nu \rightarrow H_2O^{\bullet} + e^-_{dry} + H_2O^*. \tag{4.48}$$

The overall ionization produces freely diffusing species with the following stoichiometry [262]:

$$\begin{aligned} 4.14\ H_2O \xrightarrow{100\ eV} & 2.87^{\bullet}OH + 2.7\ e^-_{aq} + 0.61\ H^{\bullet} + 0.03\ HO_2^{\bullet} \\ & + 0.61\ H_2O_2 + 0.43\ H_2 + 2.7\ H^+. \end{aligned} \tag{4.49}$$

The stoichiometry coefficients are called *G* values, which indicate that every 100 eV of energy absorbed generates 2.8 $^{\bullet}OH$, 2.7 e^{-}_{aq}, 0.61 $H^{\bullet}$, and 0.03 $HO_2^{\bullet}$ (Eq. 4.49). H_2O_2 is the main stable oxidizing species in the radiolysis of water [269]. The hydrated electron is converted into $^{\bullet}OH$ in N_2O saturated solutions (24 mM N_2O, 1 atm, room temperature) to increase the yield of $^{\bullet}OH$, doubled through reaction (4.50). The results indicate a system contains 90% $^{\bullet}OH$ and 10% $H^{\bullet}$:

$$e^{-}_{aq} + N_2O \rightarrow {}^{\bullet}OH + N_2 + OH^{-}. \quad (4.50)$$

In synchrotron radiolysis of water, photons in the kilovolt X-ray range are used [270]. An electron pulse of 0.15-μs duration, using a Varian linear accelerator, provides doses of ≈4 Gy that could produce ≈2 μmol/L $^{\bullet}OH$.

The induced homolysis of H_2O_2 by UV light in aqueous solution forms two $^{\bullet}OH$ radicals (Eq. 4.51):

$$H_2O_2 + h\nu \rightarrow 2{}^{\bullet}OH. \quad (4.51)$$

The hemolytic fission of the peroxynitrous acid also yields $^{\bullet}OH$ (Eq. 4.52) [271]:

$$ONOOH \rightarrow NO_2^{\bullet} + {}^{\bullet}OH. \quad (4.52)$$

Reaction (4.52) has a half-life of 1 second at pH 7.0. The primary yield of reaction (4.53) is 0.4–0.5. Subsequently, $^{\bullet}OH$ proceeds through the Haber–Weiss chain reactions (Eqs. 4.53 and 4.54) [272]:

$$H_2O_2 + {}^{\bullet}OH \rightarrow HO_2^{\bullet} + H_2O \quad k_{53} = 2.7 \times 10^7\,/M/s \quad (4.53)$$

$$H_2O_2 + HO_2^{\bullet} \rightarrow H_2O + O_2 + {}^{\bullet}OH \quad k_{54} = 7 \times 10^9\,/M/s. \quad (4.54)$$

The Fenton reagent refers to a mixture of ferrous salts and H_2O_2, which generates $^{\bullet}OH$ through the reaction (Eq. 4.55)

$$Fe^{2+} + H_2O_2 \rightarrow Fe^{3+} + {}^{\bullet}OH + OH^{-}. \quad (4.55)$$

The $^{\bullet}OH$ spectrum has a broad absorption band with a maximum around 230 nm and a wavelength tail that extends beyond 320 nm [273]. The spectra of $^{\bullet}OH$, measured at different temperatures, are presented in Figure 4.19 [274]. The primary band at 230 nm may be due to hydrogen-bonded OH, while the band at 310 nm corresponds to "free" OH [274]. An increase in temperature in water-cooled nuclear power plants resulted in a decrease at 230 nm and a growth at 310 nm (Fig. 4.19). The isosbestic point appears at ~305 nm. The $^{\bullet}OH$ radicals self-combine at a nearly diffusion-controlled rate constant, yielding a hydrogen peroxide product (Eq. 4.56) [275]:

$${}^{\bullet}OH + {}^{\bullet}OH \rightarrow H_2O_2. \quad (4.56)$$

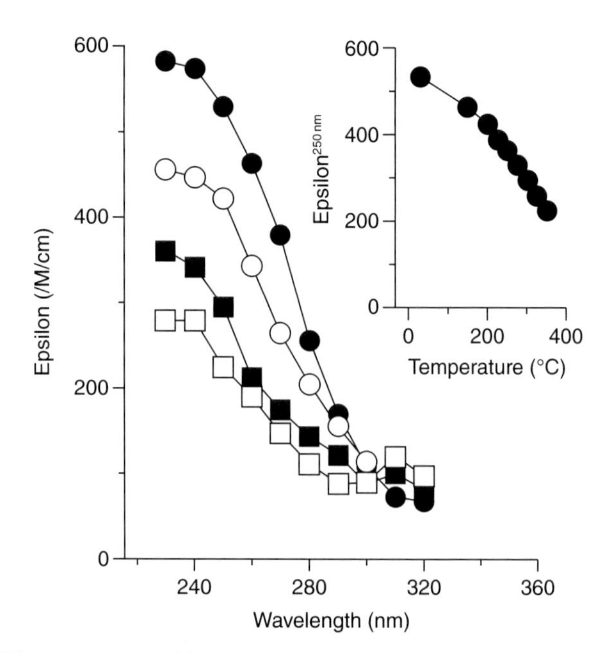

Figure 4.19. The spectrum of $^{\bullet}OH$ as a function of temperature; (●) 30°C; (○) 200°C; (■) 300°C, (□) 350°C. Inset: changes of Epsilon at 250 nm for $^{\bullet}OH$ versus temperature (adapted from Janik et al. [274] with the permission of the American Chemical Society).

The $^{\bullet}OH$ radical may also react with H_2O_2 to form $HO_2^{\bullet}$ (Eq. 4.53). This study of the $^{\bullet}OH$ radical at high temperature was able to model the radiation-induced chemistry in heat transport piping of the nuclear power plant reactor [274].

The acid dissociation constants of the $^{\bullet}OH$ radical (Eq. 4.57) have been measured over the temperature range 284–343 K [276]:

$$^{\bullet}OH \rightleftharpoons O^{\bullet-} + H^{+}. \quad K_a \tag{4.57}$$

At 298 K, pK_a and $\Delta_{ion}G^o$ were determined as 11.54 ± 0.04 kJ/mol and 65.9 ± 0.3 kJ/mol, respectively. The values of $\Delta_{ion}H^o_{298}$ and $\Delta_{ion}S^o_{298}$ for the $^{\bullet}OH$ radical were calculated as 24.85 ± 0.5 kJ/mol and −139 ± 2 J/mol, respectively [276].

4.4.2 Reactivity

The reactivity of $^{\bullet}OH$ has been studied using pulse radiolysis technique, γ radiation, and Fenton reactions [277, 278]. $^{\bullet}OH$ radical behaves as an electrophile and shows some selectivity in the kind of bonds with which it will react; however, $^{\bullet}OH$ generally reacts rapidly and nonselectively with most electron-rich sites of organic molecules [14]. The $^{\bullet}OH$ radical reacts via a hydroxyl addition to the carbon–carbon double bonds or aromatic rings or through hydrogen abstraction from saturated carbon sites of the molecules. These reactions produce transient radical species, which undergo further reactions depending on the radical and structural environment [261]. The factors that control both types of reactions include the possible number of sites available for the $^{\bullet}OH$ attack, the electronegativity of the substituents on the target sites, the strength of the C–H bond, the steric effects, and the nature of the produced organoradical [278]. For example, among the amino acids, H-abstraction from Cys was the most accessible because the average single bond energies for S–H, O–H, N–H, and C–H are 363, 459, 386, and 411 kJ/mol, respectively, at 25°C. The order of C–H reactivity for alkane functional groups was usually tertiary > secondary > primary. Based on several factors involved in the reactivity of $^{\bullet}OH$ with organic molecules, a kinetic model using a group contribution method was developed [279]. This model reasonably predicted the rate constants for several organic molecules. Quantum mechanical methods were also used to estimate aqueous-phase free energy of activation of reactions of $^{\bullet}OH$ with carboxylate ions [280].

The second-order rate constants for the reactions of $^{\bullet}OH$ with amino acids are provided in Table 4.11. The variation in rate constants ranged from 10^7 to 10^{10}/M/s. The relatively small variation indicates almost all side chains of the amino acids were oxidized by $^{\bullet}OH$. Sulfur- and aromatic-containing side chains had the highest reactivity (Table 4.11). Gly had higher reactivity than other aliphatic amino acids due to the influence of steric effects on the rates [281, 282]. Furthermore, the secondary α-carbon radical produced in the case of Gly appears to be more stable than the tertiary α-carbon radical, formed in other amino acids. Generally, less reactive radical species may have relatively higher reactivity toward Gly and other α-carbon sites of proteins [281]. Peptides reacted somewhat faster than the parent free amino acids, and the rate constants ranged from 10^8 to 10^9/M/s (Table 4.11). Proteins were highly reactive with diffusion-controlled rate constants (Table 4.11).

The rate constants of Table 4.11 suggest the $^{\bullet}OH$ radical can easily cause damage to both the side chain and backbone of proteins, causing fragmentation of the proteins. Cleavage of the main chains and oxidation of different residue side chains of proteins by $^{\bullet}OH$ are described below. The focus is on the progress made in the last few years, particularly the application of mass analysis of the products formed in the reactions.

4.4.2.1 Main-Chain Cleavage of Protein. Backbone cleavage through H-abstraction at the α-carbon site is shown in Figure 4.20. Two major pathways following the initiation of the formation of radicals occur [261]. One pathway involves the loss of $HO_2^{\bullet}$ from the peroxy radical and then hydrolysis of the newly produced imine species. In the other pathway, formation of the alkoxy radical at the α-carbon takes place, which ultimately results in further fragmentation to cleave the backbone of the protein. The scheme in Figure 4.20

TABLE 4.11. Rate Constants for Reactions of Amino Acids, Peptides, and Proteins with ˙OH Radical

Amino Acid	pH	*k* (/M/s)	Peptide	pH	*k* (/M/s)
Gly	5.9	1.7×10^7	*N*-Ac-Gly	6.6–8.7	4.0×10^8
Asn	6.6	4.9×10^7	Gly-Gly	5.5–7	2.4×10^8
Asp	6.9	7.5×10^7	Cyclo(Gly-Gly)	5.0	1.2×10^9
Ala	5.8	7.7×10^7	*N*-Ac-Gly-Gly	8.6	7.8×10^8
Glu	6.5	2.3×10^8	$(Gly)_3$	5.4	7.3×10^8
Ser	~6.0	3.2×10^8	$(Gly)_4$	5.5	4.5×10^8
Lys	6.6	3.5×10^8	*N*-Ac-Ala	6.6–9.2	4.7×10^8
Thr	6.6	5.1×10^8	Cyclo(Ala-Ala)	5.0	1.8×10^9
Gln	6.0	5.4×10^8	*N*-Ac-$(Ala)_3$	9.0	3.0×10^9
Pro	6.8	6.5×10^8			
Val	6.9	8.5×10^8			
Leu	~6.0	1.7×10^9			
Ile	6.6	1.8×10^9			
Cystine	6.5	2.1×10^9			
Arg	6.5–7.5	3.5×10^9			
His	7.5	4.8×10^9			
Phe	7–8	6.9×10^9			
Met	6–7	8.5×10^9			
Tyr	7.0	1.3×10^{10}			
Trp	6.5–8.5	1.3×10^{10}			
Cys	7.0	3.5×10^{10}			
			Lysozyme	5.6–7	$\approx 5 \times 10^{10}$
			HSA	7.0	7.8×10^{10}

Figure 4.20. H-abstraction at the α-carbon site of the protein (adapted from Xu and Chance [261] with the permission of the American Chemical Society).

shows the role of oxygen in achieving significant peptide fragmentation. In the absence of oxygen, less yield of fragmentation is expected.

Formation of the α-carbon radicals may also result from the •OH attack at the β-carbon (C-3) position (Fig. 4.21). The initial H-abstraction generates carbon-centered radicals, followed by the reaction with O_2 to yield peroxy radicals [283] (Fig. 4.21). The alkoxy radical may form from several different pathways using different precursors such as decomposition of intermediate hydroperoxides and termination reactions of the peroxy species with other radicals [284, 285]. The β-scission of the alkoxy radicals at a rate of $>10^7$/second may lead to the cleavage of the backbone [283]. The nature of the substituents, R and R′, determines the rate of the reactions [286]. The side chains of Asp, Val, and Leu usually go through such β-scission reactions to yield carbonyl products. The oxidation of Gly side chain by •OH via the abstraction of H from the γ-carbon site is shown in Figure 4.22. The γ-carbon peroxy radical, which forms in the presence of O_2, ultimately causes production of the modified chain products of different mass shifts [12] (Fig. 4.22). Decarboxylation with a mass shift of –30 was the major product, while mass shifts of +14 and +16 were minor products [287]. The dehydropeptide readily converted to keto form, which on hydrolysis formed a new amide and a keto acid-containing protein fragments [288].

The oxidation of Pro to the 2-pyrrolidone derivative by •OH is presented in Figure 4.23. Acid hydrolysis of the derivative to the γ-amino butyric acid leads to backbone cleavage of the protein [289, 290]. Oxidation of the C-5 site of Pro produced a pyroglutamyl residue with a mass shift of +14 Da, which on acid hydrolysis yielded glutamyl, causing backbone cleavage of the peptide. The products of mass shift +16 were also formed (Fig. 4.23). Glutamyl semialdehyde was the equilibrium product of 5-hydroxyproline.

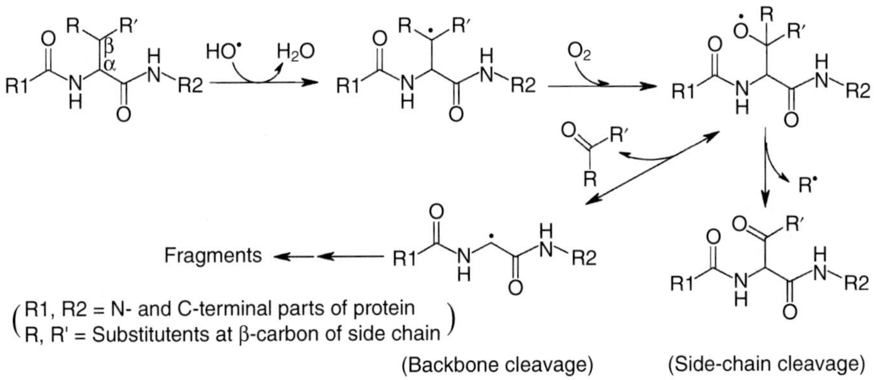

Figure 4.21. Backbone cleavage by radical transfer from the β-carbon at side chains (adapted from Xu and Chance [261] with the permission of the American Chemical Society).

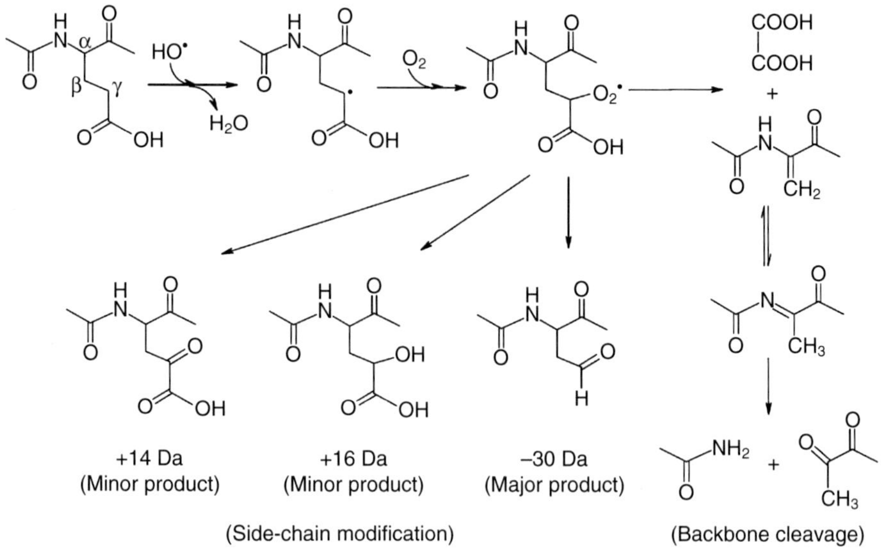

+14 Da (Minor product) +16 Da (Minor product) −30 Da (Major product)

(Side-chain modification) (Backbone cleavage)

Figure 4.22. Oxidation of Glu by the radical transfer from the γ-carbon at the side chains of the protein (adapted from Xu and Chance [261] with the permission of the American Chemical Society).

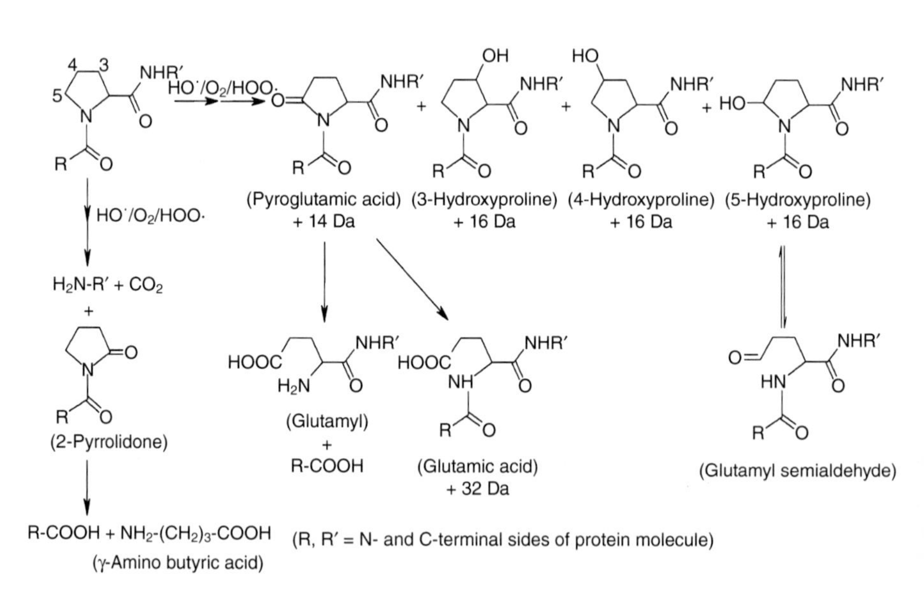

Figure 4.23. Modification of Pro (adapted from Xu and Chance [261] with the permission of the American Chemical Society).

4.4.2.2 Oxidation of Amino Acid Side Chains.

Aliphatic Side Chains. The effect of radiation on aliphatic amino acids has been performed [291–293]. Near-edge X-ray absorption fine structure (NEXAFS) measurements and DFT calculation showed a complex, a multi-step decomposition of Gly [291]. The reaction of $^{\bullet}$OH with Gly in solution was also found complex [294–296]. The reaction of $^{\bullet}$OH with the glycine anion (NH_2-CH_2-COO^-) immediately produced (~10-ns experimental time resolution) three radicals, glycyl (NH_2-$^{\bullet}$CH-COO^-), aminoethyl (NH_2-$^{\bullet}CH_2$), and aminyl ($^{\bullet}$NH-CH_2-COO^-), which were also confirmed by time-resolved EPR measurements [297–301]. A *scheme* has been proposed to demonstrate the pathways to the formation of these radicals (Fig. 4.24). Initially, the precursor species, aminium radical zwitterions ($^{+\bullet}NH_2$-CH_2-COO^- and HO:NH_2-CH_2-COO^-), were proposed to yield different radicals (reaction a) [297, 302]. Direct hydrogen atom transformation from the precursor species produced the glycyl and aminyl radicals (reactions b–d). The aminyl species could be formed from either decarboxylation or deprotonation (reaction e). Recently, the scheme given in Figure 4.24 was examined using the H/D kinetic isotope effects (KIE) to distinguish H abstraction and electron/proton transfer mechanisms [294]. The second-order rate constants for the reactions of $^{\bullet}$OH with various glycine anions are provided in Table 4.12 [294]. The partial rate constants, calculated by multiplying $k_{overall}$ with the relative yield of the each radical, for the formation of individual radicals are also presented in Table 4.12. The values of $k_{overall}$ were similar and decreased to some extent with an increasing degree of the Gly deuteration.

The reaction to yield the glycyl radical by a direct abstraction of hydrogen from –CH_2- occurred with a 37% probability, irrespective of $^{\bullet}$OH or $^{\bullet}$OD as reactive species and H_2O or D_2O solvent. The reaction was independent of the substitution of the H/D on the amino group of Gly (Table 4.12). The yields were 30% and 24% for the reactions of $^{\bullet}$OH or $^{\bullet}$OD with –CD_2-, respectively

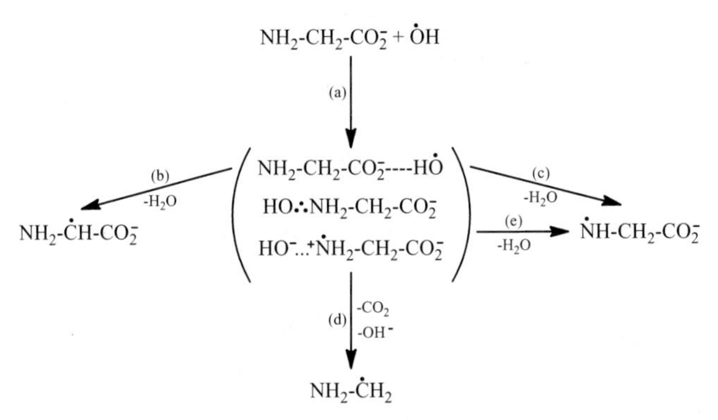

Figure 4.24. Radical products in the reaction of $^{\bullet}$OH with glycine (adapted from Štefanić et al. [294]).

TABLE 4.12. Relative Yields of Individual Radicals (G_{rel}), Overall, and Partial Rate Constants (k, /M/s) for the Reaction of $^{\bullet}OH$ in H_2O or $^{\bullet}OD$ in D_2O with Gly and Deuterated Gly Anions. Estimated Uncertainty ±10% for $k_{overall}$ and G_{rel} and ±14% for k Values

Solute		Glycyl	Aminoethyl	Aminyl	N-Attack
		G_{rel}	G_{rel}	G_{rel}	G_{rel}
(Solvent)	$k_{overall}$	k	k	k	k
NH_2-CH_2-COO^-		0.37	0.22	0.41	0.63
(H_2O)	2.8×10^9	1.0×10^9	0.62×10^9	1.2×10^9	1.8×10^9
NH_2-CD_2-COO^-		0.30	0.26	0.44	0.70
(H_2O)	2.8×10^9	0.63×10^9	0.55×10^9	0.92×10^9	1.5×10^9
ND_2-CH_2-COO^-		0.36	0.25	0.39	0.64
	2.1×10^9	0.76×10^9	0.53×10^9	0.82×10^9	1.4×10^9
ND_2-CD_2-COO^-		0.24	0.26	0.50	0.76
	1.6×10^9	0.38×10^9	0.42×10^9	0.80×10^9	1.2×10^9

Adapted from Štefanić et al. [294].

(Table 4.12). The results clearly demonstrate that majority of the reactions of $^{\bullet}OH$ and $^{\bullet}OD$ occurred by the interactions with the amino group of the glycine anion (Table 4.12). In this interaction, decarboxylation and aminomethyl radical formation contributed to ≈25%. The remaining percentage led to the formation of the aminyl radical (reactions c and e in Fig. 4.24).

The values of k^H/k^D for the appropriate couples of Gly anions are given in Table 4.13 [294]. The values varied from 1.6 to 2.0 in H_2O and D_2O, respectively, and were similar to the values determined for the abstraction of H/D from the α-C-H(D) bond in ethanol and methylamine by the $^{\bullet}OH$ radical in H_2O [303]. The secondary KIEs of the deuterated amino group for abstraction from the –CH_2(CD_2)- group were relatively high (1.3–1.7), suggesting $^{\bullet}OH$ and $^{\bullet}OD$ as reactants and/or H_2O and D_2O as solvents. The primary KIE increased to as high as 2.6 for Gly^- + $^{\bullet}OH$ in H_2O versus all-deuterated Gly^- + $^{\bullet}OD$ in D_2O. Comparatively, the KIE decreased to as low as 1.2 for ND_2-CH_2-COO^- + $^{\bullet}OD$ in D_2O versus NH_2-CD_2-COO^- + $^{\bullet}OH$ in H_2O.

Values for the formation of the aminyl radical were determined in the range of 1.1–1.5 for all effects (Table 4.13). These values were similar to the secondary KIE of 1.1, obtained for the H-atom abstraction by $^{\bullet}OH$ from the –NH_2 group of CH_3NH_2 or CD_3NH_2 in H_2O [303]. However, the aminoethyl radical formation showed a positive KIE for all, except one couple. Overall, it appears from the results of KIE that the process of heterogeneous C–C bond rupture to form CO_2 and the aminomethyl radicals may be sensitive to the electron density at the nitrogen, stability of the solvent cage, stability of the 3-e-bonded intermediate, and thermodynamic stabilities of the

TABLE 4.13. Kinetic Isotope Effects for Attacks of Hydroxyl Radical on Gly Anion. Estimated Uncertainty ±20%

		Attack at the Methylene-C		
		k^H/k^D	k^H/k^D	
Gly Anions	Solvent	Overall	Glycyl	
Primary effects				
$NH_2\text{-}CH_2\text{-}COO^-/NH_2\text{-}CD_2\text{-}COO^-$	H_2O/D_2O	1.3	1.6	
$ND_2\text{-}CH_2\text{-}COO^-/ND_2\text{-}CD_2\text{-}COO^-$	D_2O/D_2O	1.3	2.0	
Secondary effects				
$NH_2\text{-}CH_2\text{-}COO^-/ND_2\text{-}CH_2\text{-}COO^-$	H_2O/D_2O	1.3	1.3	
$NH_2\text{-}CD_2\text{-}COO^-/ND_2\text{-}CD_2\text{-}COO^-$	H_2O/D_2O	1.3	1.7	
Primary + secondary effects				
$NH_2\text{-}CH_2\text{-}COO^-/NH_2\text{-}CD_2\text{-}COO^-$	H_2O/D_2O	1.8	2.6	
$ND_2\text{-}CH_2\text{-}COO^-/ND_2\text{-}CD_2\text{-}COO^-$	D_2O/H_2O	1.0	1.2	
		Attack at the N-site		
		k^H/k^D	k^H/k^D	k^H/k^D
		N-Attack	Aminomethyl	Aminyl
Primary effects				
$NH_2\text{-}CH_2\text{-}COO^-/ND_2\text{-}CH_2\text{-}COO^-$	H_2O/D_2O	1.3	1.2	1.5
$ND_2\text{-}CD_2\text{-}COO^-/ND_2\text{-}CD_2\text{-}COO^-$	D_2O/D_2O	1.3	1.3	1.2
Secondary effects				
$NH_2\text{-}CH_2\text{-}COO^-/ND_2\text{-}CD_2\text{-}COO^-$	H_2O/D_2O	1.2	1.1	1.3
$ND_2\text{-}CH_2\text{-}COO^-/ND_2\text{-}CD_2\text{-}COO^-$	H_2O/D_2O	1.2	1.3	1.0
Primary + secondary effects				
$NH_2\text{-}CH_2\text{-}COO^-/ND_2\text{-}CD_2\text{-}COO^-$	H_2O/D_2O	1.5	1.5	1.5
$NH_2\text{-}CD_2\text{-}COO^-/ND_2\text{-}CH_2\text{-}COO^-$	D_2O/H_2O	1.1	1.0	1.1

Data are taken from Štefanić et al. [294].

radicals. Detailed theoretical examination for the reactions of Gly with $^{\bullet}OH$ has been carried out to further understand the simultaneous formation of three different primary radicals (see Fig. 4.24) [294, 295]. Calculations performed by previous studies also showed nitrogen and α-carbon sites are competitive for H-abstraction via an $^{\bullet}OH$ attack to generate different radicals [304]. The model amide system was also used to understand the reactions of $^{\bullet}OH$ radicals with peptide systems [305]. The $^{\bullet}OH$ radical attacked the α-carbon site of formamide; however, addition of a methyl group opened an alternative attack at the γ-carbon site. The γ-carbon H-abstraction was favorable compared to the β-carbon abstraction. Furthermore, H-abstraction of the nitrogen more likely occurred with the addition of a methyl group on the nitrogen. Geometric structures and stabilization of intermediates supported the preference of attacks by $^{\bullet}OH$ radicals.

The $^{\bullet}OH$ radical attacked the aliphatic hydrocarbon side chains (e.g., Ala, Val, Ile, and Leu) indiscriminately, and the reactivity increased with an increase in the number of C–H bonds and the length of the hydrocarbon side chains [306, 307]. The hydrogen abstraction by $^{\bullet}OH$ yielded a carbon-centered radical, which produced a peroxy radical upon reaction with O_2 under aerobic conditions. A number of reactions of the peroxy radical finally form carbonyl, hydroxide, and hydroperoxide products with mass shifts of +14, +16, and +32 Da, respectively. The alkoxy radical and O_2 may also be alternatively formed [307]. Hydroperoxides are generally unstable and decompose to produce further radicals or carbonyl and alcohol products.

In a recent study, oxidation of peptides containing Gly, Ala, Val, and Pro by $^{\bullet}OH/O_2$ was studied to quantify the nature and yields of oxidation products such as alcohols, carbonyls, hydroperoxides, and fragment species [308]. This study helped to learn the contribution of different pathways to the oxidation of peptides and proteins. The location of an amino acid in a peptide sequence influenced the proportions of the formation of different oxidation products. The hydroperoxides were formed nonrandomly (Pro > Val > Ala > Gly) and their concentrations were inversely related to the yields of both peptide-bound and released carbonyls. Both side chains and backbone site alcohols were produced. The overall yields of products were more than that of the initial $^{\bullet}OH$ generated, which suggest involvement of chain reactions in the oxidation of peptides.

Sulfur-Containing Side Chains. Met is highly reactive with the $^{\bullet}OH$ radical (see Table 4.11) and the reaction results in different intermediate radical species (Fig. 4.25). Methionine sulfoxide (+16 Da mass shift) was produced, which further oxidized to yield methionine sulfone (+32 Da mass shift). A formation of aldehyde at the γ-carbon gave a −32 Da mass shift [261, 309]. This mass shift product has been observed as the minor species in the oxidation of Met-containing peptides such as HDMNKVLDL [261]. Other mass shift products of −30 Da and (−30 + 16 Da) were due to the C-terminal decarboxylation from the original molecule and the formation of sulfone, respectively [287].

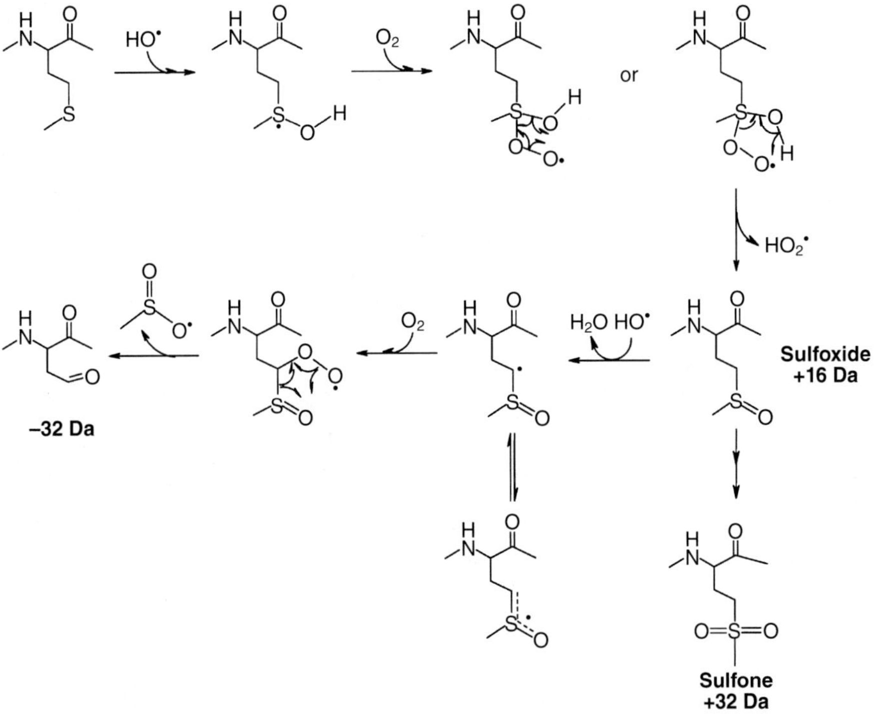

Figure 4.25. Oxidation of Met by $^{\bullet}OH$ radical (adapted from Xu and Chance [261] with the permission of the American Chemical Society).

In the radiolytic oxidation of Met, the addition of the $^{\bullet}OH$ radical to the sulfur atom and hydrogen abstraction was involved [310]. At pH > 3.0, the adduct released OH^- to yield a sulfur-centered radical cation $>S^{\bullet+}$. This step was catalyzed by an electron-rich heteroatom, such as N or S, on neighboring side chains [311]. The sulfur-centered cation may also be stabilized by the peptide bond through the formation of an intramolecular three-electron bonded cyclic transient [312, 313]. Theoretical calculations also showed the cyclic S–N bonded species in the one-electron oxidation of the Met peptide, in which the intramolecular three-electron interaction was between the S atom and the π orbital of the amide group [314]. The S–N bond radicals may convert into carbon-centered radicals located on the α-carbon moiety of the peptide backbone [312]. The dimeric radical cation has also been postulated as the intermediate of the reaction [315], which may deprotonate from $-CH_3$ or $-CH_2$- next to sulfur atom, followed by conversion into α-(alkylthio)alkyl radicals on the side chain of Met [312]. The factor analysis of the transient spectra obtained from the $^{\bullet}OH$ radical-induced oxidation of cyclo-(D-Met-L-Met) showed the formation of a radical cation via the stabilization of the oxidized sulfur through the formation of two-centered, three-electron bonds with the lone pairs of oxygen [316].

The formation of products related to Met in the oxidation of methionine-containing peptides by •OH is determined by the relative rates of the reactions of amino acids present in the peptides. When another amino acid had a higher reactivity than Met, the modification of Met was not observed [261, 309]. For example, the modification of Met with a –32 mass shift was observed in the γ-irradiation of peptides HDMNKVLDL and GSNKGAIIGLM(O) but not in peptides LWMRFA-NH_2 and YGGFM(O)R. The influence of other amino acids was also shown using DFT calculations for the reactions of Gly-Met-Gly and Gly-Nle-Gly tripeptides in which the former had 100 times faster reactivity than the latter [317]. The importance of protonated counterparts in relation to the parent molecule in the Met model peptide has also been explored [318]. ^{28}Met and ^{31}Met residues of CCK8 were shown to be susceptible to oxidation by •OH [319]. The •OH-induced oxidation of ^{27}Met to ^{27}Met-sulfoxide in the polypeptide hormone glucagon was also demonstrated [320].

As shown in Figure 4.26, the oxidation of Cys by •OH radical is complex. Numerous products were formed, depending on the reaction conditions and formation of the initial radical species [309, 321, 322]. In the initial step of the reaction of Cys with ROS, the thiyl radical species (RS•) was formed through

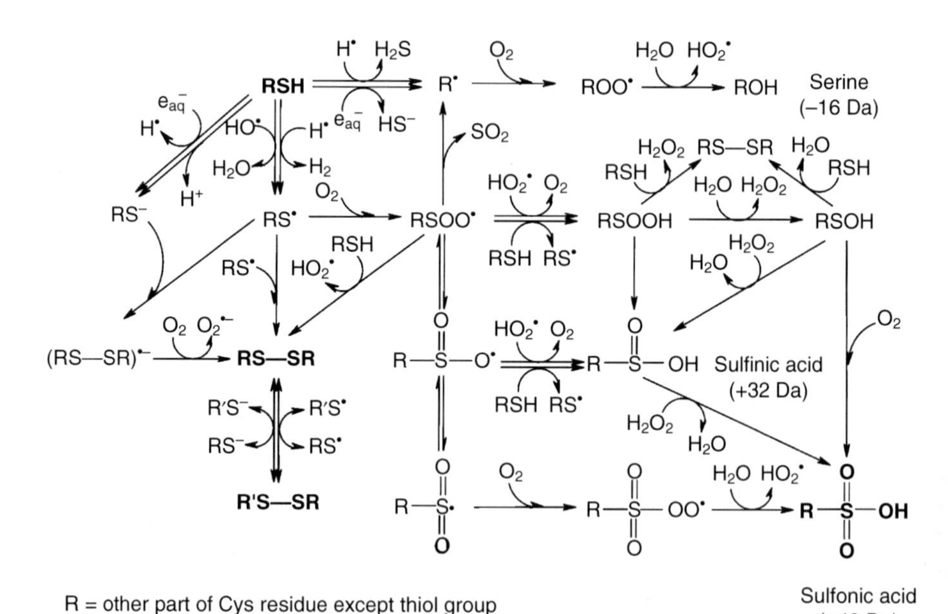

Figure 4.26. Oxidation of Cys by •OH radical (adapted from Xu and Chance [261] with the permission of the American Chemical Society).

hydrogen abstraction [51]. Further competitive reactions of $RS^{\bullet}$ with O_2 and RS^- generated the thiyl peroxy radical ($RSOO^{\bullet}$) and conjugated disulfide radical anion $(RSSR)^{\bullet-}$, respectively. $RSOO^{\bullet}$ could be the precursor species for the formation of sulfenic, sulfinic, and sulfonic acids, disulfide, and serine, possibly through thiyl hydroperoxide (RSOOH) as an intermediate [323]. Products of the reactions were confirmed by measuring the mass shifts of +32, +48, and −16 Da (Fig. 4.26). Formation of cysteine sulfinic and sulfonic acids was observed in the radiolysis of the cysteine-containing fibronectin peptide (RCDC) in water [324]. Other studies have shown the radiolysis of cysteine-containing peptides resulted in the formation of disulfide as one of the primary products under both aerobic and anaerobic conditions [323, 325]. The existence of cysteine sulfenic acid and cysteine sulfonamide has also been characterized in the $^{\bullet}OH$-mediated modification of the cysteine-containing synthetic peptide HCSAGIGRS [326]. Thiyl radicals, produced during the oxidation of Cys, may react with amino acids in peptides [327–329].

Reactions of seleno-derivatives of amino acids, selenomethionine (SeMet), selenocystine (SeCys), methyl selenocysteine (MeSeCys), and selenourea (SeU), selenocystamine (SeA), and diselenodipropionic acid (SeP) with $^{\bullet}OH$ have been studied to understand the biological activity of such compounds [330–332]. Second-order rate constants of the reactions were of the order of 10^9–10^{10}/M/s at pH 7.0. Selenium compounds are usually easier to oxidize and the oxidation chemistry of these compounds was similar to their sulfur analogues [331]. The reaction of MeSeCys with $^{\bullet}OH$ radicals in the pH range from 1 to 7 yielded monomer radicals with an absorption maximum at 350–370 nm. Diselenides reaction with $^{\bullet}OH$ radicals produced diselenide radical cations (λ_{max} = 560 nm) by the elimination of either OH^- or H_2O from the adduct. SeCys and SeA, which also contain amino functional groups, produced an additional transient (λ_{max} = 460 nm), which was likely a triselenide radical adduct in which a bridge may be formed having the sharing between the three Se stoms [333, 334].

Acidic Side Chains. C-Terminal decarboxylation of proteins and peptides has been reported in the generation of the alkoxy radical [307]. Figure 4.27 shows the mechanism of decarboxylation that produced the α-carbon alkoxy radical and $CO_2^{\bullet-}$. The decarboxylation product had a mass shift of –30, which agrees with the mechanism proposed in Figure 4.27. This type of decarboxylation product of Glu has been identified in the radiolysis of Glu-containing peptides and proteins [287, 309]. The mass shift of –30 was also observed in the radiolysis of Asp-containing peptides such as DSDPR, DRGDS, KQAGDV, ADSDGK, and DRGDS [287].

Basic Side Chains. The radiolytic oxidation of Lys hydroxylysine produced α-amino-adipyl-semialdehyde products [335]. However, different products were obtained in the presence of oxygen. In the radiolysis of the peptide DAHK, the mass shift products of +14 and +16 were observed [336]. If unstable

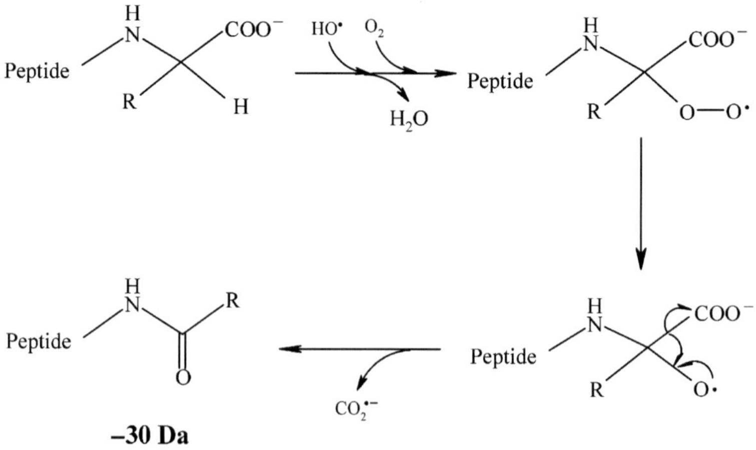

Figure 4.27. Decarboxylation of C-terminal carboxyl (adapted from Xu and Chance [261] with the permission of the American Chemical Society).

ε-hydroxylysine formed due to the attack of $^{\bullet}$OH on the ε position of Lys, it would convert into 2-amino-adipyl-semialdehyde with a loss of ammonia [337]. However, this product has not been identified in the radilolysis of Lys [261].

The deamination of Ser-containing peptides (Ala-Ser, Ser-Ala, Ala-Thr, and Ser-Ser-Ser) has been observed [338]. The presence of a hydroxyl group facilitated the degradation of the main chain of the N-terminal amino residue of the peptide, resulting in amides. However, the hydroxyl group at the C-terminal residue inhibited deamination. The deamination process decreased in oxygenated solutions of di- and tripeptides [338]. The study on the radiolysis oxidation of Thr and its derivative showed the presence of an α,β-aminoalcohol group in compounds was responsible for the C–C bond cleavage [338, 339]. Initially, nitrogen- and carbon-centered radicals were formed, which ultimately fragmented to cause degradation.

Products obtained in the oxidation of Arg by $^{\bullet}$OH are shown in Figure 4.28. A hydrogen abstraction from the δ-carbon of the side chain and O_2 addition produced a peroxy radical [340–342]. This initial attack on the δ-carbon has been confirmed by EPR and NMR studies [343, 344]. A similar approach has been used to identify radicals in reactions of $^{\bullet}$OH with aliphatic amines and polyamines as well as compounds containing one or two guanidine groups [343]. The loss of a guanidine group from the radical leads to the formation of γ-glutamyl semialdehyde.

A number of products were formed in the reaction of His with $^{\bullet}$OH in which some oxidation products have not been fully identified [14, 306]. Mass shift products of +16, −22, −23, +5, and −10 Da have been identified in the oxidation of the peptide DAHK and proteins (Fig. 4.29) [261]. Mass shifts of +16 Da and −22 Da correspond to 2-oxohistidine and aspartic acid, respectively [345]. In

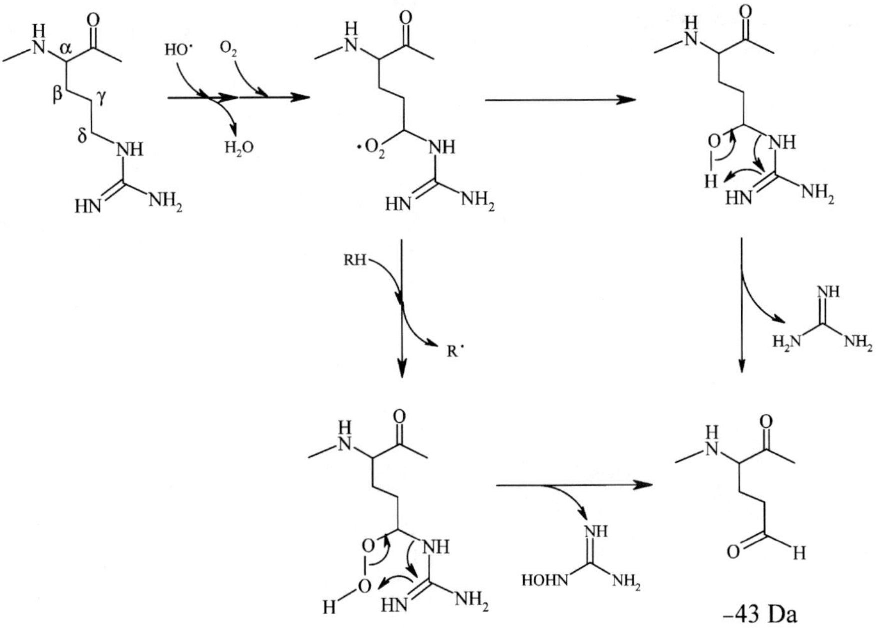

Figure 4.28. Oxidation of Arg by •OH radical (adapted from Xu and Chance [261] with the permission of the American Chemical Society).

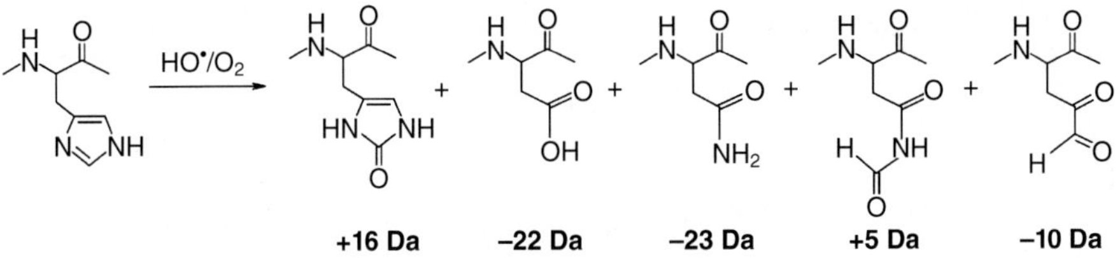

Figure 4.29. Oxidation of His by •OH radical (adapted from Xu and Chance [261] with the permission of the American Chemical Society).

the reactions, •OH initially attacks the imidazole ring of His at positions C-2, C-4, or C-5 and generates allyl-type radicals by OH addition. These radicals were confirmed by continuous-flow EPR experiments [346]. Significantly, allyl radicals can easily incorporate O_2 and yielded peroxy radicals. Further reaction of the peroxy radicals gave a mixture of products such as asparagines, aspartic acid, and 2-oxohistidine [343, 347–350]. Reaction conditions and the sequence of the local amino acid determine products of oxidation of His by •OH.

Aromatic Side Chains. Oxidation of Trp by •OH is complicated and oxidized products contain mass shifts of +16, +32, +4, and +20 Da, as determined in radiolysis and metal-catalyzed oxidation of Trp [14, 345, 351, 352]. The oxidation scheme in Figure 4.30 shows the addition of •OH to both the benzene ring and the pyrrole moiety at the C-2 and C-3 double bond with a relative ratio around 40:60 [353]. The oxidized products may further oxidize to yield dihydroxytryptophan and hydroxykynurenine. Such products were detected in the radiolysis of Trp-NH_2 and tripeptide GWG [261, 354].

Figure 4.31 presents the +16 oxidation product of Tyr [261]. In the oxidation, •OH adds to the sites next to the original hydroxyl substituent at the side chain.

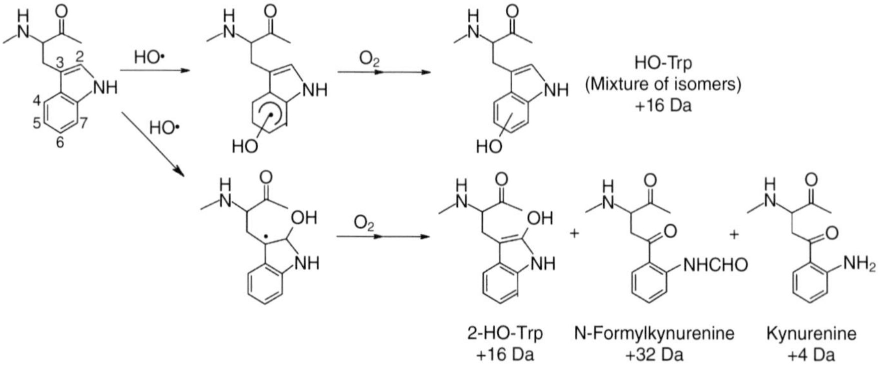

Figure 4.30. Oxidation of Trp by •OH radical (aapted from Xu and Chance [261] with the permission of the American Chemical Society).

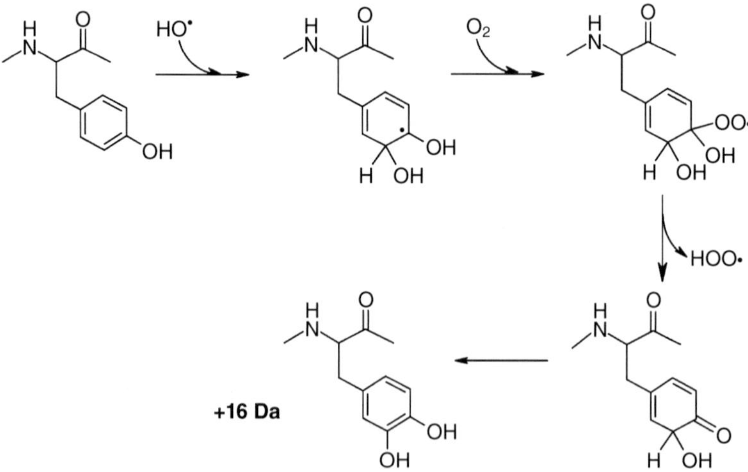

Figure 4.31. Oxidation of Tyr by •OH radical (adapted from Xu and Chance [261] with the permission of the American Chemical Society).

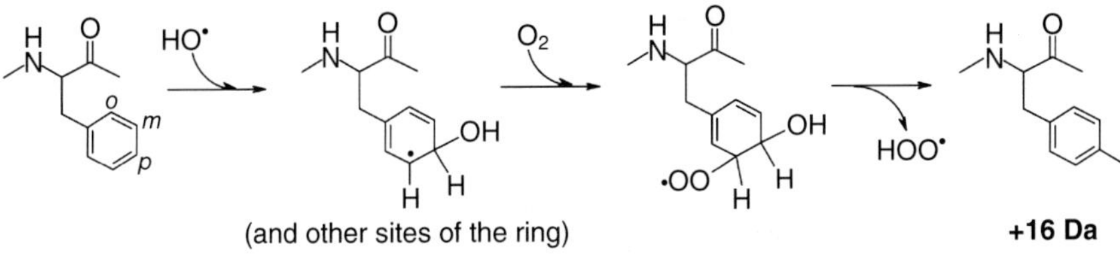

Figure 4.32. Oxidation of Phe by $^{\bullet}$OH radical (adapted from Xu and Chance [261] with the permission of the American Chemical Society).

In the presence of O_2, the addition of oxygen also takes place. The final product, DOPA, was formed with the elimination of HOO$^{\bullet}$, which on further hydroxylation yielded trihydroxyphenylalanine (TOPA). In the oxidation of peptides (Tyr-Leu and Leu-Tyr) by $^{\bullet}$OH, produced under the Fenton reaction, a wide variety of oxidation products were obtained, which include 1, 2, 3, and 4 oxygen atoms containing peptides [355]. Formation of the peroxy group occurred preferentially in the C-terminal residue. Other oxidation products with double bonds or keto groups and dimers were also identified. The oxidative damage to poly(glu, tyr) (4:1) peptides by $^{\bullet}$OH was also detected using the biosensor [319]. This biosensor thus may be used to study damage to the protein by $^{\bullet}$OH.

The oxidation of Phe by $^{\bullet}$OH is presented in Figure 4.32. The rapid addition of $^{\bullet}$OH to the aromatic ring with little selectivity yielded a hydroxycyclohexadienyl radical. In the presence of O_2, the fast reaction of the radical with O_2 and subsequent elimination of HOO$^{\bullet}$ formed a mixture of stereoisomers of Tyr with a ratio of *o*-/*m*-/*p*-Tyr being 2.0:1.0:1.5 or 1.3:1.0:2.1 under different pH and oxygen concentrations [14]. Further hydroxylation yielded DOPA and TOPA. The cross-link formation and less overall yield of these isomers occurred in the absence of O_2. In the oxidation of angiotension by $^{\bullet}$OH, generated by Fenton reaction, products related to the attack at side chains of Phe and Try were formed [356].

4.5 CONCLUSIONS

Cys, Met, Trp, Tyr, His, and Phe are among several amino acid residues of proteins, determined to be more likely prone to the attack of ROS. A comparison of their reactivity as free amino acids with different ROS is presented in Figure 4.33. The reactivity of $O_2^{\bullet-}$ was orders of magnitude lower than other ROS. $^1O_2$ and O_3 had significant reactivities (~10^4–10^8/M/s), while the $^{\bullet}$OH radical was the most reactive oxidant with nearly diffusion-controlled rate constants. However, the modification of proteins and the inactivation of enzymes must be interpreted by considering other factors such as the half-lives

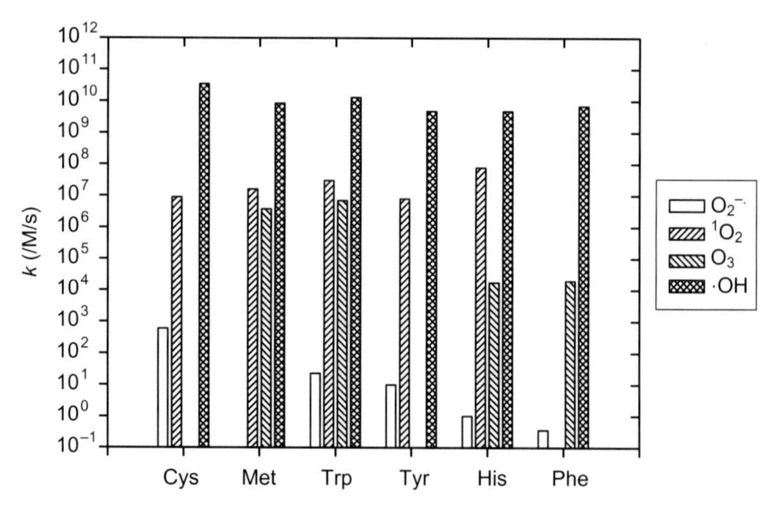

Figure 4.33. Comparison of the reactivity of some amino acids with ROS (data were taken from Tables 4.6, 4.10, and 4.11 and from Bielski and Shiue [48] for superoxide reactivity).

and the diffuse distance of ROS, as well as the location of residues in proteins or enzymes which control the efficiency of inactivation. For example, $O_2^{\bullet -}$ has a longer half-time than the $^\bullet$OH radical and therefore can diffuse at great distances to react with amino acid residues, even in locations including the protein's active site [4]. It has been determined that $O_2^{\bullet -}$ is efficient in the inactivation of protein A as is a $^\bullet$OH radical [357]. The rate constants of the reactions of glyceraldehydes-3-phosphate dehydrogenase (GAPDH) and alcohol dehydrogenase (ADH) with $O_2^{\bullet -}$ are at least three orders of magnitude lower than with the $^\bullet$OH radical [25, 277], but effectiveness results showed the inactivation of GAPDH and ADH by $O_2^{\bullet -}$ was only 2.4 and 2.8 times smaller, respectively, compared with the $^\bullet$OH radical [4]. The lifetime and diffusion distance of intracellular $^1O_2$ may also be significant in $^1O_2$-mediated cell death [157]. The conformation and structure of proteins *in vivo* would determine accessibility of the amino acid residues to ROS for their modifications. Comparative studies on proteins *in vivo* are very limited, and further studies are needed to establish the general criteria between the reactivity of ROS and the protein structure.

A comparison of the products obtained in the oxidation of Met, Trp, His, and Tyr in their free forms by different ROS is presented in Table 4.14 [358]. $^1O_2$ produced a sulfonate derivative in reaction with Met. However, a different product, hydroxymethione, was produced in reaction with $^\bullet$OH under similar conditions. Both NFK and hydroxyl-NFK were the products from the oxidation of Trp. The oxidation of His resulted in a number of compounds. A group

TABLE 4.14. Products and Their Yield from the Oxidation of Met, Trp, His, and Tyr by Singlet Oxygen

Amino Acid	Superoxide	Singlet Oxygen	Hydroxyl Radical
Met	–a	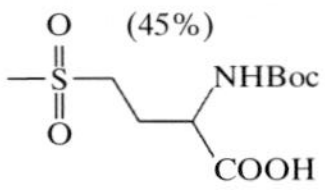 (45%) O, S, O, NHBoc, COOH	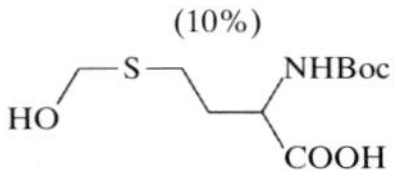 (10%) HO, S, NHBoc, COOH
Trp	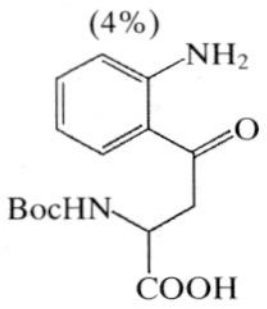 (4%) NH_2, O, BocHN, COOH	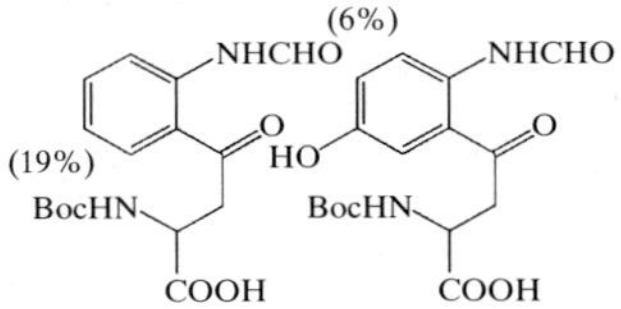 (19%) NHCHO, O, BocHN, COOH; (6%) HO, NHCHO, O, BocHN, COOH	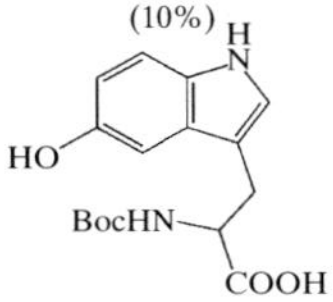 (10%) H, N, HO, BocHN, COOH
His	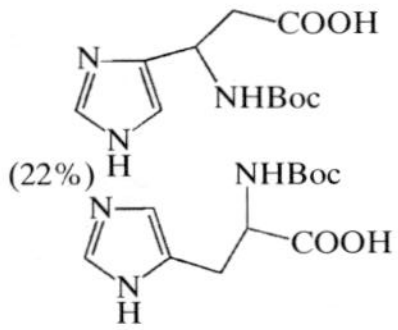 N, N, H, COOH, NHBoc; (22%) N, N, H, NHBoc, COOH	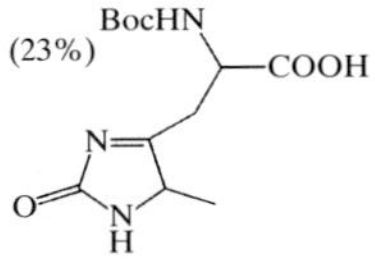 (23%) BocHN, COOH, N, O, N, H	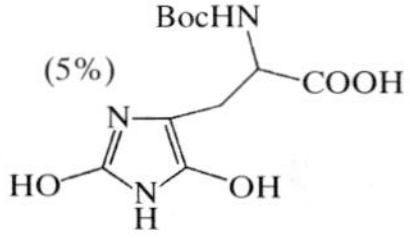 (5%) BocHN, COOH, N, HO, N, H, OH
Tyr	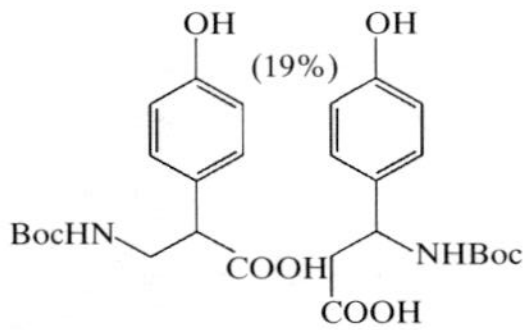 (19%) OH, BocHN, COOH; OH, NHBoc, COOH	–b	–c

a, not examined; b,c, not identified.

Adapted from Suto et al. [358].

translocation was observed in the reaction of $O_2^{\bullet-}$ with Tyr. Overall, there is progress in the identification and quantitative determination of amino acid residue products in their reaction with ROS in a complex biological environment, but further investigation is desired. The advancement of analytical techniques will help unravel many new oxidized products in future studies to understand the contribution of ROS in aging and diseases. Finally, product studies in combination with the ROS-specific detection method will enhance knowledge of the use of ROS by biological systems. This would not only require very low detection of ROS in biological matrices at nanomolar levels but it should also have high selectivity against ROS species such as $O_2^{\bullet-}$, $^1O_2$, and $^{\bullet}OH$.

REFERENCES

1 Sugamura, K. and Keaney, J.F., Jr. Reactive oxygen species in cardiovascular disease. *Free Radic. Biol. Med.* 2011, *51*, 978–992.

2 Salvemini, D., Little, J.W., Doyle, T., and Neumann, W.L. Roles of reactive oxygen and nitrogen species in pain. *Free Radic. Biol. Med.* 2011, *51*, 951–966.

3 Rains, J.L. and Jain, S.K. Oxidative stress, insulin signaling, and diabetes. *Free Radic. Biol. Med.* 2011, *50*, 567–575.

4 Rodacka, A., Serafin, E., and Puchala, M. Efficiency of superoxide anions in the inactivation of selected dehydrogenases. *Radiat. Phys. Chem.* 2010, *79*, 960–965.

5 Stadtman, E.R. Protein oxidation and aging. *Free Radic. Res.* 2006, *40*, 1250–1258.

6 Pattison, D.I., Rahmanto, A.S., and Davies, D.I. Photo-oxidation of proteins. *Photochem. Photobiol. Sci.* 2012, *11*, 38–53.

7 Dalsgaard, T.K., Nielsen, J.H., Brown, B.E., Stadler, N., and Davies, M.J. Dityrosine, 3,4-Dihydroxyphenylalanine (DOPA), and radical formation from tyrosine residues on milk proteins with globular and flexible structures as a result of riboflavin-mediated photo-oxidation. *J. Agric. Food Chem.* 2011, *59*, 7939–7947.

8 Stadtman, E.R. Free radical-mediated oxidation of proteins. In *Free Radical Technology*, D.B. Wallace, ed. Taylor and Francis, Washington, 1997, pp. 71–78.

9 Schuessler, H. and Schilling, K. Oxygen effect in the radiolysis of proteins. Part 2. Bovine serum albumin. *Int. J. Radiat. Biol.* 1984, *45*, 267–281.

10 Swallow, A.J. Effect of ionization radiation on proteins, RCO groups, peptide bond cleavage, inactivation, -SH oxidation. In *Radiation Chemistry of Organic Compounds*, A.J. Swallow, ed. Pergamon Press, New York, 1960, pp. 211–224.

11 Kopoldová, J. and Liebster, J. The mechanism of the radiation chemical degradation of amino acids-V. Radiolysis of norleucine, leucine and isoleucine in aqueous solution. *Int. J. Appl. Radiat. Isot.* 1963, *14*, 493–494, IN3–IN8, 495–498.

12 Garrison, W.M. Reaction mechanisms in the radiolysis of peptides, polypeptides, and proteins. *Chem. Rev.* 1987, *87*, 381–398.

13 Davies, M.J. Reactive species formed on proteins exposed to singlet oxygen. *Photochem. Photobiol.* 2004, *3*, 17–25.

14 Davies, M.J.D. and Dean, R.T. *Radical-Mediated Protein Oxidation*. Oxford University Press, Oxford, UK, 1997.

15 Sharma, V.K. and Graham, N.D. Oxidation of amino acids, peptides, and proteins by ozone. *Ozone Sci. Eng.* 2010, *32*, 81–90.

16 Bielski, B.H. Fast kinetic studies of dioxygen-derived species and their metal complexes. *Philos. Trans. R. Soc. Lond., B* 1985, *311*, 473–482.

17 Song, Y. and Buettner, G.R. Thermodynamic and kinetic considerations for the reaction of semiquinone radicals to form superoxide and hydrogen peroxide. *Free Radic. Biol. Med.* 2010, *49*, 919–962.

18 Zhang, C., Fan, F.-F., and Bard, A.J. Electrochemistry of oxygen in concentrated NaOH solutions: solubility, diffusion coefficients, and superoxide formation. *J. Am. Chem. Soc.* 2009, *131*, 177–181.

19 Bielski, B.H.J. and Cabelli, D.E. Highlights of current research involving superoxide and perhydroxyl radicals in aqueous solutions. *Int. J. Radiat. Biol.* 1991, *59*, 291–319.

20 Costentin, C., Evans, D.H., Robert, M., Savéant, J.M., and Singh, P.S. Electrochemical approach to concerted proton and electron transfers. Reduction of the water-superoxide ion complex. *J. Am. Chem. Soc.* 2005, *127*, 12490–12491.

21 René, A., Hauchard, D., Lagrost, C., and Hapiot, P. Superoxide protonation by weak acids in imidazolium based ionic liquids. *J. Phys. Chem. B* 2009, *113*, 2826–2831.

22 Rogers, E.I., Huang, X.-J., Dickinson, E.J.F., Hardacre, C., and Compton, R.G. Investigating the mechanism and electrode kinetics of the oxygenlsuperoxide (O_2IO_2-) couple in various room-temperature ionic liquids at gold and platinum electrodes in the temperature range 298–318 K. *J. Phys. Chem. C* 2009, *113*, 17811–17823.

23 Winterbourn, C.C. Reconciling the chemistry and biology of reactive oxygen species. *Nature Chem. Biol.* 2008, *4*, 278–286.

24 Benson, S.W. and Nangia, P.S. Electron affinity of HO_2 and HO_x radicals [5]. *J. Am. Chem. Soc.* 1980, *102*, 2843–2844.

25 Bielski, B.H.J., Cabelli, D.E., Arudi, R.L., and Ross, A.B. Reactivity of HO_2/O_2^- radicals in aqueous solution. *J. Phys. Chem. Ref. Data* 1985, *14*, 1041–1088.

26 Divišek, J. and Kastening, B. Electrochemical generation and reactivity of the superoxide ion in aqueous solutions. *J. Electroanal. Chem.* 1975, *65*, 603–621.

27 Saito, E. and Bielski, B.H.J. The electron paramagnetic resonance spectrum of the HO_2 radical in aqueous solution [2]. *J. Am. Chem. Soc.* 1961, *83*, 4467–4468.

28 Knowles, P.F., Gibson, J.F., Pick, F.M., and Bray, R.C. Electron-spin-resonance evidence for enzymic reduction of oxygen to a free radical, the superoxide ion. *Biochem. J.* 1969, *111*, 53–58.

29 Goldstein, S., Rosen, G.M., Russo, A., and Samuni, A. Kinetics of spin trapping superoxide, hydroxyl, and aliphatic radicals by cyclic nitrones. *J. Phys. Chem. A* 2004, *108*, 6679–6685.

30 Kim, S.U., Liu, Y., Nash, K.M., Zweier, J.L., Rockenbauer, A., and Villamena, F.A. Fast reactivity of a cyclic nitrone-calix[4]pyrrole conjugate with superoxide radical anion: theoretical and experimental studies. *J. Am. Chem. Soc.* 2010, *132*, 17157–17173.

31 Abreu, I.A. and Cabelli, D.E. Superoxide dismutases—a review of the metal-associated mechanistic variations. *Biochim. Biophys. Acta—Proteins Proteomics* 2010, *1804*, 263–274.

32 Legrini, O., Oliveros, E., and Braun, A.M. Photochemical processes for water treatment. *Chem. Rev.* 1993, *93*, 671–698.

33 Halliwell, B. Antioxidant characterization. Methodology and mechanism. *Biochem. Pharmacol.* 1995, *49*, 1341–1348.

34 Fridovich, I. Quantitative aspects of the production of superoxide anion radical by milk xanthine oxidase. *J. Biol. Chem.* 1970, *245*, 4053–4057.

35 Okado-Matsumoto, A. and Fridovich, I. Assay of superoxide dismutase: cautions relevant to the use of cytochrome c, a sulfonated tetrazolium, and cyanide. *Anal. Biochem.* 2001, *298*, 337–342.

36 Liu, R.H., Fu, S.Y., Zhan, H.Y., and Lucia, L.A. General spectroscopic protocol to obtain the concentration of the superoxide anion radical. *Ind. Eng. Chem. Res.* 2009, *48*, 9331–9334.

37 Kotani, H., Ohkubo, K., Crossley, M.J., and Fukuzumi, S. An efficient fluorescence sensor for superoxide with an acridinium ion-linked porphyrin triad. *J. Am. Chem. Soc.* 2011, *133*, 11092–11095.

38 Xu, X., Thompson, L.V., Navratil, M., and Arriaga, E.A. Analysis of superoxide production in single skeletal muscle fibers. *Anal. Chem.* 2010, *82*, 4570–4576.

39 Rose, A.L., Moffett, J.W., and Waite, T.D. Determination of superoxide in seawater using 2-methyl-6-(4-methoxyphenyl)-3,7-dihydroimidazo[1,2-a]pyrazin-3(7H)-one chemiluminescence. *Anal. Chem.* 2008, *80*, 1215–1227.

40 Bielski, B.H.J., Arudi, R.L., Cabelli, D.E., and Bors, W. Reevaluation of the reactivity of hydroxylamine with O_2^-/HO_2. *Anal. Biochem.* 1984, *142*, 207–209.

41 Hillard, E.A., De Abreu, F.C., Ferreira, D.C.M., Jaouen, G., Goulart, M.O.F., and Amatore, C. Electrochemical parameters and techniques in drug development, with an emphasis on quinones and related compounds. *Chem. Commun.* 2008, 2612–2628.

42 Wilshire, J. and Sawyer, D.T. Redox chemistry of dioxygen species. *Acc. Chem. Res.* 1979, *2*, 105–110.

43 Sawyer, D.T. and Valentine, J.S. How super is superoxide? *Acc. Chem. Res.* 1981, *14*, 393–400.

44 Zahir, K., Espenson, J.H., and Bakac, A. Reactions of polypyridylchromium(II) ions with oxygen: determination of the self-exchange rate constant of O_2/O_2^-. *J. Am. Chem. Soc.* 1988, *110*, 5059–5063.

45 Weinstock, I.A. Outer-sphere oxidation of the superoxide radical anion. *Inorg. Chem.* 2008, *47*, 404–406.

46 René, A., Abasq, M.L., Hauchard, D., and Hapiot, P. How do phenolic compounds react toward superoxide ion? A simple electrochemical method for evaluating antioxidant capacity. *Anal. Chem.* 2010, *82*, 8703–8710.

47 Chun, O.K., Kim, D.-O., and Lee, C.Y. Superoxide radical scavenging activity of the major polyphenols in fresh plums. *J. Agric. Food Chem.* 2003, *51*, 8067–8072.

48 Bielski, B.H. and Shiue, G.G. Reaction rates of superoxide radicals with the essential amino acids. *Ciba Found. Symp.* 1978, *65*, 43–56.

49 Benrahmoune, M., Thérond, P., and Abedinzadeh, Z. The reaction of superoxide radical with N-acetylcysteine. *Free Radic. Biol. Med.* 2000, *29*, 775–782.

50 Zhang, N., Schuchmann, H.P., and von Sonntag, C. The reaction of superoxide radical anion with dithiothreitol: a chain process. *J. Phys. Chem.* 1991, *95*, 4718–4722.

51 Winterbourn, C.C. and Metodiewa, D. Reactivity of biologically important thiol compounds with superoxide and hydrogen peroxide. *Free Radic. Biol. Med.* 1999, *27*, 322–328.

52 Cardey, B. and Enescu, M. Cysteine oxidation by the superoxide radical: a theoretical study. *Chemphyschem* 2009, *10*, 1642–1648.

53 Cardey, B., Foley, S., and Enescu, M. Mechanism of thiol oxidation by the superoxide radical. *J. Phys. Chem. A* 2007, *111*, 13046–13052.

54 Rush, J.D. and Cabelli, D.E. The reactions of a dinuclear ferric complex (oxo) di-iron(III) triethylenetetraamminehexaacetate, $Fe_2O(ttha)^{2-}$, with oxidizing and reducing free radicals. A pulse radiolysis study. *Radiat. Phys. Chem.* 1997, *49*, 661–674.

55 Rose, A.L. and Waite, T.D. Reduction of organically complexed ferric iron by superoxide in a simulated natural water. *Environ. Sci. Technol.* 2005, *39*, 2645–2650.

56 Fujii, M., Rose, A.L., Waite, T.D., and Omura, T. Oxygen and superoxide-mediated redox kinetics of iron complexed by humic substances in coastal seawater. *Environ. Sci. Technol.* 2010, *44*, 9337–9342.

57 Cabelli, D.E., Bielski, B.H.J., and Holcman, J. Interaction between copper(II)-arginine complexes and HO_2/O_2^- radicals, a pulse radiolysis study. *J. Am. Chem. Soc.* 1987, *109*, 3665–3669.

58 Weinstein, J. and Bielski, B.H.J. Reaction of superoxide radicals with copper(II)-histidine complexes. *J. Am. Chem. Soc.* 1980, *102*, 4916–4919.

59 Bull, C., McClune, G.J., and Fee, J.A. The mechanism of Fe-EDTA catalyzed superoxide dismutation. *J. Am. Chem. Soc.* 1983, *105*, 5290–5300.

60 Buettner, G.R., Doherty, T.P., and Patterson, L.K. The kinetics of the reaction of superoxide radical with Fe(III) complexes of EDTA, DETAPAC and HEDTA. *FEBS Lett.* 1983, *158*, 143–146.

61 Buettner, G.R. The reaction of superoxide, formate radical, and hydrated electron with transferrin and its model compound, Fe(III)-ethylenediamine-N,N′-bis[2-(2-hydroxyphenyl)acetic acid] as studied by pulse radiolysis. *J. Biol. Chem.* 1987, *262*, 11995–11998.

62 Cabelli, D.E., Rush, J.D., Thomas, M.J., and Bielski, B.H.J. Kinetics of the free-radical-induced reduction of Fe^{III}DTPA to Fe^{II}DTPA. A pulse radiolysis study. *J. Phys. Chem.* 1989, *93*, 3579–3586.

63 Borggaard, O.K., Faver, O., and Andersen, V.S. Polarographic study of the rate of oxidation of iron(II) chelates by hydrogen peroxide. *Acta Chem. Scand.* 1971, *25*, 3541–3545.

64 Borggaard, O.K. Polarographic determination of diffusion coefficients of hydrogen peroxide and iron chelates and rate constants of hydroxyl radical reactions. *Acta Chem. Scand.* 1972, *26*, 3393–3394.

65 Summers, J.S., Baker, J.B., Meyerstein, D., Mizrahi, A., Zilbermann, I., Cohen, H., Wilson, C.M., and Jones, J.R. Measured rates of fluoride/metal association correlate with rates of superoxide/metal reactions for $Fe^{III}EDTA(H_2O)$- and related complexes. *J. Am. Chem. Soc.* 2008, *130*, 1727–1734.

66 Brausam, A. and Van Eldik, R. Further mechanistic information on the reaction between Fe^{III}(EDTA) and hydrogen peroxide: observation of a second reaction step and importance of pH. *Inorg. Chem.* 2004, *43*, 5351–5359.

67 Brausam, A., Maigut, J., Meier, R., Szilágyi, P.A., Buschmann, H.-J., Massa, W., Homonnay, Z., and Van Eldik, R. Detailed spectroscopic, thermodynamic, and kinetic studies on the protolytic equilibria of Fe^{III}cydta and the activation of hydrogen peroxide. *Inorg. Chem.* 2009, *48*, 7864–7884.

68 Sharma, V.K., Szilágyi, P.A., Homonnay, Z., Kuzmann, E., and Vértes, A. Mössbauer investigation of peroxo species in Fe(III)-EDTA-H_2O_2 system. *Eur. J. Inorg. Chem.* 2005, *2005*, 4393–4400.

69 Sharma, V.K., Millero, F.J., and Homonnay, Z. Kinetics of the complex formation between iron(III)EDTA and hydrogen peroxide in aqueous solution. *Inorganica Chim. Acta* 2004, *357*, 3583–3586.

70 Neese, F. and Solomon, E.I. Detailed spectroscopic and theoretical studies on $[Fe(EDTA)(O_2)]^{3-}$: electronic structure of the side-on ferric-peroxide bond and its relevance to reactivity. *J. Am. Chem. Soc.* 1998, *120*, 12829–12848.

71 Kovacs, J.A. and Brines, L.M. Understanding how the thiolate sulfur contributes to the function of the non-heme iron enzyme superoxide reductase. *Acc. Chem. Res.* 2007, *40*, 501–509.

72 Namuswe, F., Kasper, G.D., Narducci Sarjeant, A.A., Hayashi, T., Krest, C.M., Green, M.T., Moënne-Loccoz, P., and Goldberg, D.P. Rational tuning of the thiolate donor in model complexes of superoxide reductase: direct evidence for a trans influence in Fe^{III}-OOR complexes. *J. Am. Chem. Soc.* 2008, *130*, 14189–14200.

73 Nam, E., Alokolaro, P.E., Swartz, R.D., Gleaves, M.C., Pikul, J., and Kovacs, J.A. Investigation of the mechanism of formation of a thiolate-ligated Fe(III)-OOH. *Inorg. Chem.* 2011, *50*, 1592–1602.

74 Rush, J.D. and Bielski, B.H.J. Pulse radiolytic studies of the reactions of HO_2/O_2^- with Fe(II)/Fe(III) ions. The reactivity of HO_2/O_2^- with ferric ions and its implication on the occurrence of the Haber-Weiss reaction. *J. Phys. Chem.* 1985, *89*, 5062–5066.

75 Groves, J.T. High-valent iron in chemical and biological oxidations. *J. Inorg. Biochem.* 2006, *100*, 434–447.

76 Walling, C. Fenton's reagent revisited. *Acc. Chem. Res.* 1975, *8*, 125–131.

77 Bielski, B.H.J. Reactivity of hypervalent iron with biological compounds. *Ann. Neurol.* 1992, *32*, 528–532.

78 Rush, J.D., Zhao, Z., and Bielski, B.H.J. Reaction of ferrate(VI)/ferrate(V) with hydrogen peroxide and superoxide anion—a stopped-flow and premix pulse radiolysis study. *Free Radic. Res.* 1996, *24*, 187–192.

79 Menton, J.D. and Bielski, B.H.J. Studies of the kinetics, spectral and chemical properties of Fe(IV) pyrophosphate by pulse radiolysis. *Radiat. Phys. Chem.* 1990, *36*, 725–733.

80 Egan, T.J., Barthakur, S.R., and Aisen, P. Catalysis of the Haber-Weiss reaction by iron-diethylenetriaminepentaacetate. *J. Inorg. Biochem.* 1992, *48*, 241–249.

81 Cyr, J.E. and Bielski, B.H.J. The reduction of ferrate(VI) to ferrate(V) by ascorbate. *Free Radic. Biol. Med.* 1991, *11*, 157–160.

82 Duerr, K., Olah, J., Davydov, R., Kleimann, M., Li, J., Lang, N., Puchta, R., Hübner, E., Drewello, T., Harvey, J.N., Jux, N., and Ivanović-Burmazović, I. Studies on an iron(III)-peroxo porphyrin. Iron(III)-peroxo or iron(II)-superoxo? *Dalton Trans.* 2010, *39*, 2049–2056.

83 Dürr, K., Jux, N., Zahl, A., Van Eldik, R., and Ivanović-Burmazović, I. Volume profile analysis for the reversible binding of superoxide to form iron(II)-superoxo/iron(III)-peroxo porphyrin complexes. *Inorg. Chem.* 2010, *49*, 11254–11260.

84 Peretz, P., Solomon, D., Weinraub, D., and Faraggi, M. Chemical properties of water-soluble porphyrins 3. The reaction of superoxide radicals with some metalloporphyrins. *Int. J. Radiat. Biol.* 1982, *42*, 449–456.

85 Solomon, D., Peretz, P., and Faraggi, M. Chemical properties of water-soluble porphyrins. 2. The reaction of iron(III) tetrakis(4-N-methylpyridyl)porphyrin with the superoxide radical dioxygen couple. *J. Phys. Chem.* 1982, *86*, 1842–1849.

86 Groni, S., Blain, G., Guillot, R., Policar, C., and Anxolabéhère-Mallart, E. Reactivity of Mn^{II} with superoxide. Evidence for a $[Mn^{III}OO]^+$ unit by low-temperature spectroscopies. *Inorg. Chem.* 2007, *46*, 1951–1953.

87 Ivanović-Burmazović, I. and Filipović, M.R. Reactivity of manganese superoxide dismutase mimics toward superoxide and nitric oxide. Selectivity versus cross-reactivity. *Adv. Inorg. Chem.* 2012, *64*, 53–95.

88 Cudd, A. and Fridovich, I. Electrostatic interactions in the reaction mechanism of bovine erythrocyte superoxide dismutase. *J. Biol. Chem.* 1982, *257*, 11443–11447.

89 Fisher, C.L., Cabelli, D.E., Tainer, J.A., Hallewell, R.A., and Getzoff, E.D. The role of arginine 143 in the electrostatics and mechanism of Cu,Zn superoxide dismutase: computational and experimental evaluation by mutational analysis. *Proteins Struct. Funct. Genet.* 1994, *19*, 24–34.

90 Getzoff, E.D., Tainer, J.A., Weiner, P.K., Kollman, P.A., Richardson, J.S., and Richardson, D.C. Electrostatic recognition between superoxide and copper, zinc superoxide dismutase. *Nature* 1983, *306*, 287–290.

91 Getzoff, E.D., Cabelli, D.E., Fisher, C.L., Parge, H.E., Viezzoli, M.S., Banci, L., and Hallewell, R.A. Faster superoxide dismutase mutants designed by enhancing electrostatic guidance. *Nature* 1992, *358*, 347–351.

92 Ellerby, L.M., Cabelli, D.E., Graden, J.A., and Valentine, J.S. Copper-zinc superoxide dismutase: why not pH-dependent? *J. Am. Chem. Soc.* 1996, *118*, 6556–6561.

93 Koppenol, W.H. Superoxide dismutase and oxygen toxicity. *Clin. Respir. Physiol.* 1981, *17*, 85–89.

94 Keele, B.B., Jr., McCord, J.M., and Fridovich, I. Superoxide dismutase from *Escherichia coli* B. A new manganese-containing enzyme. *J. Biol. Chem.* 1970, *245*, 6176–6181.

95 Borgstahl, G.E.O., Parge, H.E., Hickey, M.J., Beyer, W.F., Jr., Hallewell, R.A., and Tainer, J.A. The structure of human mitochondrial manganese superoxide dismutase reveals a novel tetrameric interface of two 4-helix bundles. *Cell* 1992, *71*, 107–118.

96 Dennis, R.J., Micossi, E., McCarthy, J., Moe, E., Gordon, E.J., Kozielski-Stuhrmann, S., Leonard, G.A., and McSweeney, S. Structure of the manganese superoxide dismutase from *Deinococcus radiodurans* in two crystal forms. *Acta Crystallograph. Sect. F Struct. Biol. Cryst. Commun.* 2006, *62*, 325–329.

97 Edwards, R.A., Baker, H.M., Whittaker, M.M., Whittaker, J.W., Jameson, G.B., and Baker, E.N. Crystal structure of *Escherichia coli* manganese superoxide dismutase at 2.1-Å resolution. *J. Biol. Inorg. Chem.* 1998, *3*, 161–171.

98 Grove, L.E. and Brunold, T.C. Second-sphere tuning of the metal ion reduction potentials in iron and manganese superoxide dismutases. *Comments Inorg. Chem.* 2008, *29*, 134–168.

99 Sheng, Y., Stich, T.A., Barnese, K., Gralla, E.B., Cascio, D., Britt, R.D., Cabelli, D.E., and Valentine, J.S. Comparison of two yeast mnsods: mitochondrial *Saccharomyces cerevisiae* versus cytosolic *Candida albicans*. *J. Am. Chem. Soc.* 2011, *133*, 20878–20889.

100 Bull, C., Niederhoffer, E.C., Yoshida, T., and Fee, J.A. Kinetic studies of superoxide dismutases: properties of the manganese-containing protein from *Thermus thermophilus*. *J. Am. Chem. Soc.* 1991, *113*, 4069–4076.

101 Pick, M., Rabani, J., Yost, F., and Fridovich, I. The catalytic mechanism of the manganese-containing superoxide dismutase of *Escherichia coli* studied by pulse radiolysis. *J. Am. Chem. Soc.* 1974, *96*, 7329–7333.

102 McAdam, M.E., Fox, R.A., Lavelle, F., and Fielden, E.M. A pulse radiolysis study of the manganese containing superoxide dismutase from *Bacillus stearothermophilus*. A kinetic model for the enzyme action. *Biochem. J.* 1977, *165*, 71–79.

103 Miller, A.-F., Padmakumar, K., Sorkin, D.L., Karapetian, A., and Vance, C.K. Proton-coupled electron transfer in Fe-superoxide dismutase and Mn-superoxide dismutase. *J. Inorg. Biochem.* 2003, *93*, 71–83.

104 Vance, C.K. and Miller, A.-F. Spectroscopic comparisons of the pH dependencies of Fe-substituted (Mn) superoxide dismutase and Fe-superoxide dismutase. *Biochemistry* 1998, *37*, 5518–5527.

105 Abreu, I.A., Hearn, A., An, H., Nick, H.S., Silverman, D.N., and Cabelli, D.E. The kinetic mechanism of manganese-containing superoxide dismutase from *Deinococcus radiodurans*: a specialized enzyme for the elimination of high superoxide concentrations. *Biochemistry* 2008, *47*, 2350–2356.

106 Lah, M.S., Dixon, M.M., Pattridge, K.A., Stallings, W.C., Fee, J.A., and Ludwig, M.L. Structure-function in *Escherichia coli* iron superoxide dismutase: comparisons with the manganese enzyme from *Thermus thermophilus*. *Biochemistry* 1995, *34*, 1646–1660.

107 Lavelle, F., McAdam, M.E., and Fielden, E.M. A pulse radiolysis study of the catalytic mechanism of the iron containing superoxide dismutase from *Photobacterium leiognathi*. *Biochem. J.* 1977, *161*, 3–11.

108 Bull, C. and Fee, J.A. Steady-state kinetic studies of superoxide dismutases: properties of the iron containing protein from *Escherichia coli*. *J. Am. Chem. Soc.* 1985, *107*, 3295–3304.

109 Miller, A.-F., Yikilmaz, E., and Vathyam, S. ^{15}N-NMR characterization of His residues in and around the active site of FeSOD. *Biochim. Biophys. Acta: Proteins Proteomics* 2010, *1804*, 275–284.

110 Regelsberger, G., Laaha, U., Dietmann, D., Rüker, F., Canini, A., Grilli-Caiola, M., Furtmüller, P.G., Jakopitsch, C., Peschek, G.A., and Obinger, C. The iron superoxide dismutase from the filamentous cyanobacterium Nostoc PCC 7120: localization, overexpression, and biochemical characterization. *J. Biol. Chem.* 2004, *279*, 44384–44393.

111 Yamakura, F. and Suzuki, K. Inactivation of *Pseudomonas* iron-superoxide dismutase by hydrogen peroxide. *Biochem. Biophys. Acta—Protein Struct. Mol. Enzym.* 1986, *874*, 23–29.

112 Greenleaf, W.B. and Silverman, D.N. Activation of the proton transfer pathway in catalysis by iron superoxide dismutase. *J. Biol. Chem.* 2002, *277*, 49282–49286.

113 Maliekal, J., Karapetian, A., Vance, C., Yikilmaz, E., Wu, Q., Jackson, T., Brunold, T.C., Spiro, T.G., and Miller, A. Comparison and contrasts between the active site pKs of Mn-superoxide dismutase and those of Fe-superoxide dismutase. *J. Am. Chem. Soc.* 2002, *124*, 15064–15075.

114 Whittaker, J.W. and Whittaker, M.M. Active site spectral studies on manganese superoxide dismutase. *J. Am. Chem. Soc.* 1991, *113*, 5528–5540.

115 Pinto, A.F., Rodrigues, J.V., and Teixeira, M. Reductive elimination of superoxide: structure and mechanism of superoxide reductases. *Biochim. Biophys. Acta: Proteins Proteomics* 2010, *1804*, 285–297.

116 Coelho, A.V., Matias, P., Fülöp, V., Thompson, A., Gonzalez, A., and Carrondo, M.A. Desulfoferrodoxin structure determined by MAD phasing and refinement to 1.9-Å resolution reveals a unique combination of a tetrahedral FeS4 centre with a square pyramidal FeSN4 centre. *J. Biol. Inorg. Chem.* 1997, *2*, 680–689.

117 Coulter, E.D., Emerson, J.P., Kurtz, D.M., Jr., and Cabelli, D.E. Superoxide reactivity of rubredoxin oxidoreductase (desulfoferrodoxin) from *Desulfovibrio vulgaris*: a pulse radiolysis study. *J. Am. Chem. Soc.* 2000, *122*, 11555–11556.

118 Rodrigues, J.V., Abreu, I.A., Cabelli, D., and Teixeira, M. Superoxide reduction mechanism of *Archaeoglobus fulgidus* one-iron superoxide reductase. *Biochemistry* 2006, *45*, 9266–9278.

119 Rodrigues, J.V., Saraiva, L.M., Abreu, I.A., Teixeira, M., and Cabelli, D.E. Superoxide reduction by *Archaeoglobus fulgidus* desulfoferrodoxin: comparison with neelaredoxin. *J. Biol. Inorg. Chem.* 2007, *12*, 248–256.

120 Rodrigues, J.V., Victor, B.L., Huber, H., Saraiva, L.M., Soares, C.M., Cabelli, D.E., and Teixeira, M. Superoxide reduction by *Nanoarchaeum equitans* neelaredoxin, an enzyme lacking the highly conserved glutamate iron ligand. *J. Biol. Inorg. Chem.* 2008, *13*, 219–228.

121 Mathé, C., Mattioli, T.A., Horner, O., Lombard, M., Latour, J.-M., Fontecave, M., and Nivière, V. Identification of iron(III) peroxo species in the active site of the superoxide reductase SOR from *Desulfoarculus baarsii*. *J. Am. Chem. Soc.* 2002, *124*, 4966–4967.

122 Lombard, M., Houée-Levin, C., Touati, D., Fontecave, M., and Nivière, V. Superoxide reductase from *Desulfoarculus baarsii*: reaction mechanism and role of glutamate 47 and lysine 48 in catalysis. *Biochemistry* 2001, *40*, 5032–5040.

123 Nivière, V., Asso, M., Weill, C.O., Lombard, M., Guigliarelli, B., Favaudon, V., and Houée-Levin, C. Superoxide reductase from *Desulfoarculus baarsii*: identification of protonation steps in the enzymatic mechanism. *Biochemistry* 2004, *43*, 808–818.

124 Chen, L., Sharma, P., Le Gall, J., Mariano, A.M., Teixeira, M., and Xavier, A.V. A blue non-heme iron protein from *Desulfovibrio gigas*. *Eur. J. Biochem.* 1994, *226*, 613–618.

125 Emerson, J.P., Coulter, E.D., Cabelli, D.E., Phillips, R.S., and Kurtz, D.M., Jr. Kinetics and mechanism of superoxide reduction by two-iron superoxide reductase from *Desulfovibrio vulgaris*. *Biochemistry* 2002, *41*, 4348–4357.

126 Clay, M.D., Jenney, F.E., Jr., Hagedoorn, P.L., George, G.N., Adams, M.W.W., and Johnson, M.K. Spectroscopic studies of *Pyrococcus furiosus* superoxide reductase: implications for active-site structures and the catalytic mechanism. *J. Am. Chem. Soc.* 2002, *124*, 788–805.

127 Bonnot, F., Molle, T., Ménage, S., Moreau, Y., Duval, S., Favaudon, V., Houée-Levin, C., and Nivière, V. Control of the evolution of iron peroxide intermediate in superoxide reductase from *Desulfoarculus baarsii*. Involvement of lysine 48 in protonation. *J. Am. Chem. Soc.* 2012, *134*, 5120–5130.

128 Osawa, M., Yamakura, F., Mihara, M., Okubo, Y., Yamada, K., and Hiraoka, B.Y. Conversion of the metal-specific activity of *Escherichia coli* Mn-SOD by site-directed mutagenesis of Gly165Thr. *Biochim. Biophys. Acta—Proteins Proteomics* 2010, *1804*, 1775–1779.

129 Falahati, M., Ma'Mani, L., Saboury, A.A., Shafiee, A., Foroumadi, A., and Badiei, A.R. Aminopropyl-functionalized cubic Ia3d mesoporous silica nanoparticle as an efficient support for immobilization of superoxide dismutase. *Biochim. Biophys. Acta—Proteins Proteomics* 2011, *1814*, 1195–1202.

130 Davies, M.J. Singlet oxygen-mediated damage to proteins and its consequences. *Biochem. Biophys. Res. Commun.* 2003, *305*, 761–770.

131 Ragàs, X., Cooper, L.P., White, J.H., Nonell, S., and Flors, C. Quantification of photosensitized singlet oxygen production by a fluorescent protein. *Chemphyschem* 2011, *12*, 161–165.

132 Chin, K.K., Trevithick-Sutton, C.C., McCallum, J., Jockusch, S., Turro, N.J., Scaiano, J.C., Foote, C.S., and Garcia-Garibay, M.A. Quantitative determination of singlet oxygen generated by excited state aromatic amino acids, proteins, and immunoglobulins. *J. Am. Chem. Soc.* 2008, *130*, 6912–6913.

133 Ronsein, G.E., Oliveira, M.C.B., Miyamoto, S., Medeiros, M.H.G., and Di Mascio, P. Tryptophan oxidation by singlet molecular oxygen [$O_2(^1\Delta_g)$]: mechanistic studies using ^{18}O-labeled hydroperoxides, mass spectrometry, and light emission measurements. *Chem. Res. Toxicol.* 2008, *21*, 1271–1283.

134 Reynoso, E., Biasutti, M.A., and García, N.A. Kinetics of the photosensitized oxidation of chymotrypsin in different media. *Amino Acids* 2008, *34*, 61–68.

135 Fang, Y. and Liu, J. Reaction of protonated tyrosine with electronically excited singlet molecular oxygen (a1Δg): an experimental and trajectory study. *J. Phys. Chem. A* 2009, *113*, 11250–11261.

136 Schweitzer, C. and Schmidt, R. Physical mechanisms of generation and deactivation of singlet oxygen. *Chem. Rev.* 2003, *103*, 1685–1757.

137 Kanofsky, J.R. Singlet oxygen production by biological systems. *Chem. Biol. Interact.* 1989, *70*, 1–28.

138 Fu, Y., Sima, P.D., and Kanofsky, J.R. Singlet-oxygen generation from liposomes: a comparison of 6β-cholesterol hydroperoxide formation with predictions from a one-dimensional model of singlet-oxygen diffusion and quenching. *Photochem. Photobiol.* 1996, *63*, 468–476.

139 Kanofsky, J.R. Measurement of singlet-oxygen *in vivo*: progress and pitfalls. *Photochem. Photobiol.* 2011, *87*, 14–17.

140 Kanofsky, J.R., Sima, P.D., and Richter, C. Singlet-oxygen Generation from A2E. *Photochem. Photobiol.* 2003, *77*, 235–242.

141 Kanofsky, J.R. and Sima, P.D. Assay for singlet oxygen generation by plant leaves exposed to ozone. *Method Enzymol.* 2000, *319*, 512–520.

142 Kanofsky, J.R., Hoogland, H., Wever, R., and Weiss, S.J. Singlet oxygen production by human eosinophils. *J. Biol. Chem.* 1988, *263*, 9692–9696.

143 Steinbeck, M.J., Khan, A.U., and Karnovsky, M.J. Extracellular production of singlet oxygen by stimulated macrophages quantified using 9,10-diphenylanthracene and perylene in a polystyrene film. *J. Biol. Chem.* 1993, *268*, 15649–15654.

144 Massari, J., Tokikawa, R., Medinas, D.B., Angeli, J.P.F., Di Mascio, P., Assunção, N.A., and Bechara, E.J.H. Generation of singlet oxygen by the glyoxal-peroxynitrite system. *J. Am. Chem. Soc.* 2011, *133*, 20761–20768.

145 Kanofsky, J.R. and Sima, P.D. Singlet oxygen generation from the reaction of ozone with plant leaves. *J. Biol. Chem.* 1995, *270*, 7850–7852.

146 Kanofsky, J.R. and Sima, P. Singlet oxygen production from the reactions of ozone with biological molecules. *J. Biol. Chem.* 1991, *266*, 9039–9042.

147 Aubry, J.M., Cazin, B., and Duprat, F. Chemical sources of singlet oxygen. 3. Peroxidation of water-soluble singlet oxygen carriers with the hydrogen peroxide-molybdate system. *J. Org. Chem.* 1989, *54*, 726–728.

148 Di Mascio, P., Briviba, K., Bechara, E.J.H., Medeiros, M.H.G., and Sies, H. Reaction of peroxynitrite and hydrogen peroxide to produce singlet molecular oxygen ($^1\Delta$(g)). *Method Enzymol.* 1996, *269*, 395–400.

149 Nakano, M., Takayama, K., Shimizu, Y., Tsuji, Y., Inaba, H., and Migita, T. Spectroscopic evidence for the generation of singlet oxygen in self-reaction of sec-peroxy radicals [9]. *J. Am. Chem. Soc.* 1976, *98*, 1974–1975.

150 Redmond, R.W. and Gamlin, J.N. A compilation of singlet oxygen yields from biologically relevant molecules. *Photochem. Photobiol.* 1999, *70*, 391–475.

151 Wilkinson, F., Helman, W.P., and Ross, A.B. Rate constants for the decay and reactions of the lowest electronically excited state of molecular oxygen in solution. *J. Phys. Chem. Ref. Data* 1995, *24*, 663–1021.

152 Straight, R.C. and Spikes, J.D. Photosensitized oxidation of biomolecules. In: *Singlet* O_2, Vol. 4, A.A. Frimer, ed. CRC Press, Boca Raton, FL, 1985, pp. 91–143.

153 Da Silva, E.F.F., Pedersen, B.W., Breitenbach, T., Toftegaard, R., Kuimova, M.K., Arnaut, L.G., and Ogilby, P.R. Irradiation- and sensitizer-dependent changes in the lifetime of intracellular singlet oxygen produced in a photosensitized process. *J. Phys. Chem. B* 2012, *116*, 445–461.

154 Arnbjerg, J., Johnsen, M., Nielsen, C.B., Jørgensen, M., and Ogilby, P.R. Effect of sensitizer protonation on singlet oxygen production in aqueous and nonaqueous media. *J. Phys. Chem. A* 2007, *111*, 4573–4583.

155 Arnbjerg, J., Paterson, M.J., Nielsen, C.B., Jørgensen, M., Christiansen, O., and Ogilby, P.R. One- and two-photon photosensitized singlet oxygen production: characterization of aromatic ketones as sensitizer standards. *J. Phys. Chem. A* 2007, *111*, 5756–5767.

156 Zhang, Y., Zhu, X., Smith, J., Haygood, M.T., and Gao, R. Direct observation and quantitative characterization of singlet oxygen in aqueous solution upon UVA excitation of 6-thioguanines. *J. Phys. Chem. B* 2011, *115*, 1889–1894.

157 Ogilby, P.R. Singlet oxygen: there is indeed something new under the sun. *Chem. Soc. Rev.* 2010, *39*, 3181–3209.

158 Lindig, B.A., Rodgers, M.A.J., and Schaap, A.P. Determination of the lifetime of singlet oxygen in D_2O using 9,10-anthracenedipropionic acid, a water-soluble probe. *J. Am. Chem. Soc.* 1980, *102*, 5590–5593.

159 Bisby, R.H., Morgan, C.G., Hamblett, I., and Gorman, A.A. Quenching of singlet oxygen by Trolox C, ascorbate, and amino acids: effects of pH and temperature. *J. Phys. Chem. A* 1999, *103*, 7454–7459.

160 Wei, C., Song, B., Yuan, J., Feng, Z., Jia, G., and Li, C. Luminescence and Raman spectroscopic studies on the damage of tryptophan, histidine and carnosine by singlet oxygen. *J. Photochem. Photobiol. A* 2007, *189*, 39–45.

161 Rougee, M., Bensasson, R.V., Land, E.J., and Pariente, R. Deactivation of singlet molecular oxygen by thiols and related compounds, possible protectors against skin photosensitivity. *Photochem. Photobiol.* 1988, *47*, 485–489.

162 Papeschi, G., Monici, M., and Pinzauti, S. pH effect on dye sensitized photooxidation of amino acids and albumins. *Med. Biol. Environ.* 1982, *10*, 245–250.

163 Matheson, I.B.C., Etheridge, R.D., Kratowich, N.R., and Lee, J. The quenching of singlet oxygen by amino acids and proteins. *Photochem. Photobiol.* 1975, *21*, 165–171.

164 Boreen, A.L., Edhlund, B.L., Cotner, J.B., and McNeill, K. Indirect photodegradation of dissolved free amino acids: the contribution of singlet oxygen and the differential reactivity of DOM from various sources. *Environ. Sci. Technol.* 2008, *42*, 5492–5498.

165 Vilensky, A. and Feitelson, J. Reactivity of singlet oxygen with tryptophan residues and with melittin in liposome systems. *Photochem. Photobiol.* 1999, *70*, 841–846.

166 Alarcón, E., Henríquez, C., Aspée, A., and Lissi, E.A. Chemiluminescence associated with singlet oxygen reactions with amino acids, peptides and proteins. *Photochem. Photobiol.* 2007, *83*, 475–480.

167 Rule Wigginton, K., Menin, L., Montoya, J.P., and Kohn, T. Oxidation of virus proteins during UV254 and singlet oxygen mediated inactivation. *Environ. Sci. Technol.* 2010, *44*, 5437–5443.

168 von Montfort, C., Sharov, V.S., Metzger, S., Schöneich, C., Sies, H., and Klotz, L.-O. Singlet oxygen inactivates protein tyrosine phosphatase-1B by oxidation of the active site cysteine. *Biol. Chem.* 2006, *387*, 1399–1404.

169 Kim, J., Rodriguez, M.E., Guo, M., Kenney, M.E., Oleinick, N.L., and Anderson, V.E. Oxidative modification of cytochrome c by singlet oxygen. *Free Radic. Biol. Med.* 2008, *44*, 1700–1711.

170 Michaeli, A. and Feitelson, J. Reactivity of singlet oxygen toward amino acids and peptides. *Photochem. Photobiol.* 1994, *59*, 284–289.

171 Palumbo, M.C., Garci′A, N.A., and Argüello, G.A. The interaction of singlet molecular oxygen $O_2(^1\Delta_g)$ with indolic derivatives. Distinction between physical and reactive quenching. *J. Photochem. Photobiol. B Biol* 1990, *7*, 33–42.

172 Sysak, P.K., Foote, C.S., and Chiang, T.Y. Chemistry of singlet oxygen-XXV. Photooxygenation of methionine. *Photochem. Photobiol.* 1993, *26*, 19–27.

173 Dillon, J., Chiesa, R., Wang, R.H., and McDermott, M. Molecular changes during the photooxidation of alpha-crystallin in the presence of uroporphyrin. *Photochem. Photobiol.* 1993, *57*, 526–530.

174 Finley, E.L., Busman, M., Dillon, J., Crouch, R.K., and Schey, K.L. Identification of photooxidation sites in bovine α-crystallin. *Photochem. Photobiol.* 1997, *66*, 635–641.

175 Foote, C.S. and Peters, J.W. Chemistry of singlet oxygen. XIV. A reactive intermediate in sulfide photooxidation. *J. Am. Chem. Soc.* 1971, *93*, 3795–3796.

176 Alarcón, E., Edwards, A.M., Aspée, A., Borsarelli, C.D., and Lissi, E.A. Photophysics and photochemistry of rose bengal bound to human serum albumin. *Photochem. Photobiol. Sci.* 2009, *8*, 933–943.

177 Wright, A., Hawkins, C.L., and Davies, M.J. Singlet oxygen-mediated protein oxidation: evidence for the formation of reactive peroxides. *Redox Rep.* 2000, *5*, 159–161.

178 Criado, S., Soltermann, A.T., Marioli, J.M., and García, N.A. Sensitized photooxidation of di- and tripeptides of tyrosine. *Photochem. Photobiol.* 1998, *68*, 453–458.

179 Wright, A., Bubb, W.A., Hawkins, C.L., and Davies, M.J. Singlet oxygen-mediated protein oxidation: evidence for the formation of reactive side chain peroxides on tyrosine residues. *Photochem. Photobiol.* 2002, *76*, 35–46.

180 Jensen, F. and Foote, C.S. Chemistry of singlet oxygen—48. Isolation and structure of the primary product of photooxygenation of 3,5-di-t-butyl catechol. *Photochem. Photobiol.* 1987, *46*, 325–330.

181 Saito, I., Kato, S., and Matsuura, T. Photoinduced reactions. XL addition of singlet oxygen to monocyclic aromatic ring. *Tetrahedron Lett.* 1970, *11*, 239–242.

182 Tomita, M., Irie, M., and Ukita, T. Sensitized photooxidation of histidine and its derivatives. Products and mechanism of the reaction. *Biochemistry* 1969, *8*, 5149–5160.

183 Kang, P. and Foote, C.S. Synthesis of a ^{13}C, ^{15}N labeled imidazole and characterization of the 2,5-endoperoxide and its decomposition. *Tetrahedron Lett.* 2000, *41*, 9623–9626.

184 Neidigh, J.W., Fesinmeyer, R.M., and Andersen, N.H. Designing a 20-residue protein. *Nat. Struct. Biol.* 2002, *9*, 425–430.

185 Au, V. and Madison, S.A. Effects of singlet oxygen on the extracellular matrix protein collagen: oxidation of the collagen crosslink histidinohydroxylysinonorleucine and histidine. *Arch. Biochem. Biophys.* 2000, *384*, 133–142.

186 Lentner, C. *Geigy Scientific Tables: Physical Chemistry, Composition of Blood, Hematology, Somatometric Data*. West Caldwell, Essex City, NJ, 1984.

187 Pedersen, B.W., Sinks, L.E., Breitenbach, T., Schack, N.B., Vinogradov, S.A., and Ogilby, P.R. Single cell responses to spatially controlled photosensitized production of extracellular singlet oxygen. *Photochem. Photobiol.* 2011, *87*, 1077–1091.

188 Gracanin, M., Hawkins, C.L., Pattison, D.I., and Davies, M.J. Singlet-oxygen-mediated amino acid and protein oxidation: formation of tryptophan peroxides and decomposition products. *Free Radic. Biol. Med.* 2009, *47*, 92–102.

189 Gracanin, M., Lam, M.A., Morgan, P.E., Rodgers, K.J., Hawkins, C.L., and Davies, M.J. Amino acid, peptide, and protein hydroperoxides and their decomposition products modify the activity of the 26S proteasome. *Free Radic. Biol. Med.* 2011, *50*, 389–399.

190 Dremina, E.S., Sharov, V.S., Davies, M.J., and Schöneich, C. Oxidation and inactivation of SERCA by selective reaction of cysteine residues with amino acid peroxides. *Chem. Res. Toxicol.* 2007, *20*, 1462–1469.

191 Davies, M.J., Fu, S., and Dean, R.T. Protein hydroperoxides can give rise to reactive free radicals. *Biochem. J.* 1995, *305*, 643–649.

192 Hampton, M.B., Morgan, P.E., and Davies, M.J. Inactivation of cellular caspases by peptide-derived tryptophan and tyrosine peroxides. *FEBS Lett.* 2002, *527*, 289–292.

193 Headlam, H.A., Gracanin, M., Rodgers, K.J., and Davies, M.J. Inhibition of cathepsins and related proteases by amino acid, peptide, and protein hydroperoxides. *Free Radic. Biol. Med.* 2006, *40*, 1539–1548.

194 Luxford, C., Morin, B., Dean, R.T., and Davies, M.J. Histone H1- and other protein- and amino acid-hydroperoxides can give rise to free radicals which oxidize DNA. *Biochem. J.* 1999, *344*, 125–134.

195 Luxford, C., Dean, R.T., and Davies, M.J. Induction of DNA damage by oxidized amino acids and proteins. *Biogerontology* 2002, *3*, 95–102.

196 Gebicki, S. and Gebicki, J.M. Crosslinking of DNA and proteins induced by protein hydroperoxides. *Biochem. J.* 1999, *338*, 629–636.

197 Agon, V.V., Bubb, W.A., Wright, A., Hawkins, C.L., and Davies, M.J. Sensitizer-mediated photooxidation of histidine residues: evidence for the formation of reactive side-chain peroxides. *Free Radic. Biol. Med.* 2006, *40*, 698–710.

198 Peskin, A.V., Cox, A.G., Nagy, P., Morgan, P.E., Hampton, M.B., Davies, M.J., and Winterbourn, C.C. Removal of amino acid, peptide and protein hydroperoxides by reaction with peroxiredoxins 2 and 3. *Biochem. J.* 2010, *432*, 312–321.

199 Parsonage, D., Karplus, P.A., and Poole, L.B. Substrate specificity and redox potential of AhpC, a bacterial peroxiredoxin. *Proc. Natl Acad. Sci. U.S.A.* 2008, *105*, 8209–8214.

200 Peskin, A.V., Low, F.M., Paton, L.N., Maghzal, G.J., Hampton, M.B., and Winterbourn, C.C. The high reactivity of peroxiredoxin 2 with H_2O_2 is not reflected in its reaction with other oxidants and thiol reagents. *J. Biol. Chem.* 2007, *282*, 11885–11892.

201 Cox, A.G., Peskin, A.V., Paton, L.N., Winterbourn, C.C., and Hampton, M.B. Redox potential and peroxide reactivity of human peroxiredoxin 3. *Biochemistry* 2009, *48*, 6495–6501.

202 Rahmanto, A.S. and Davies, M.J. Catalytic activity of selenomethionine in removing amino acid, peptide, and protein hydroperoxides. *Free Radic. Biol. Med.* 2011, *51*, 2288–2299.

203 Jensen, R.L., Arnbjerg, J., Birkedal, H., and Ogilby, P.R. Singlet oxygen's response to protein dynamics. *J. Am. Chem. Soc.* 2011, *133*, 7166–7173.

204 Liu, F., Fang, Y., Chen, Y., and Liu, J. Dissociative excitation energy transfer in the reactions of protonated cysteine and tryptophan with electronically excited singlet molecular oxygen ($a^1\Delta_g$). *J. Phys. Chem. B* 2011, *115*, 9898–9909.

205 Backus, G.S., Howden, R., Fostel, J., Bauer, A.K., Cho, H.-Y., Marzec, J., Peden, D.B., and Kleeberger, S.R. Protective role of interleukin-10 in ozone-induced pulmonary inflammation. *Environ. Health Perspect.* 2010, *118*, 1721–1727.

206 McWhinney, R.D., Gao, S.S., Zhou, S., and Abbatt, J.P.D. Evaluation of the effects of ozone oxidation on redox-cycling activity of two-stroke engine exhaust particles. *Environ. Sci. Technol.* 2011, *45*, 2131–2136.

207 Cohen, S.L. Ozone in ambient air as a source of adventitious oxidation. A mass spectrometric study. *Anal. Chem.* 2006, *78*, 4352–4362.

208 Schoenfisch, M.H. and Pemberton, J.E. Air stability of alkanethiol self-assembled monolayers on silver and gold surfaces. *J. Am. Chem. Soc.* 1998, *120*, 4502–5413.

209 Miles, G.P., Samuel, M.A., Ranish, J.A., Donohoe, S.M., Sperrazzo, G.M., and Ellis, B.E. Quantitative proteomics identifies oxidant-induced, AtMPK6-dependent changes in arabidopsis thaliana protein profiles. *Plant Signal. Behav.* 2009, *4*, 497–505.

210 Yamashita, K., Miyoshi, T., Arai, T., Endo, N., Itoh, H., Makino, K., Mizugishi, K., Uchiyama, T., and Sasad, M. Ozone production by amino acids contributes to killing of bacteria. *Proc. Natl Acad. Sci. U.S.A.* 2008, *105*, 16912–16917.

211 von Gunten, U. Ozonation of drinking water: part I. Oxidation kinetics and product formation. *Water Res.* 2003, *37*, 1443–1467.

212 Pocostales, J.P., Sein, M.M., Knolle, W., von Sonntag, C., and Schmidt, T.C. Degradation of ozone-refractory organic phosphates in wastewater by ozone and ozone/hydrogen peroxide (peroxone): the role of ozone consumption by dissolved organic matter. *Environ. Sci. Technol.* 2010, *44*, 8248–8253.

213 Ikehata, K., Naghashkar, N.J., and El-Din, M.G. Degradation of aqeous pharmaceuticals by ozonation and advanced oxidation processes: a review. *Ozone Sci. Eng.* 2006, *28*, 353–414.

214 Bader, H. and Hoigne, J. Determination of ozone in water by the indigo method; a submitted standard method. *Ozone Sci. Eng.* 1982, *4*, 169–176.

215 Hoigne, J., Bader, H., Haag, W.R., and Staehelin, J. Rate constants of reactions of ozone with organic and inorganic compounds in water—III: inorganic compounds and radicals. *Water Res.* 1985, *19*, 993–1004.

216 Sehested, K., Corfitzen, H., Holcman, J., Fischer, C.H., and Hart, E.J. The primary reaction in the decomposition of ozone in acidic aqueous solutions. *Environ. Sci. Technol.* 1991, *25*, 1589–1596.

217 Sehested, K., Corfitzen, H., Holcman, J., and Hart, E.J. On the mechanism of the decomposition of acidic O_3 solutions, thermally or H_2O_2-initiated. *J. Phys. Chem. A* 1998, *102*, 2667–2672.

218 Staehelin, J. and Hoigne, J. Decomposition of ozone in water: rate of initiation by hydroxide ions and hydrogen peroxide. *Environ. Sci. Technol.* 1982, *16*, 676–681.

219 Tomiyasu, H., Fukutomi, H., and Gordon, G. Kinetics and mechanism of ozone decomposition in basic aqueous solution. *Inorg. Chem.* 1985, *24*, 2962–2966.

220 Fábián, I. Reactive intermediates in aqueous ozone decomposition: a mechanistic approach. *Pure Appl. Chem.* 2006, *78*, 1559–1570.

221 Bezbarua, B.K. and Reckhow, D.A. Modification of the standard neutral ozone decomposition model. *Ozone Sci. Eng.* 2004, *26*, 345–357.

222 von Gunten, U. Ozonation of drinking water: part II. Disinfection and by-product formation in presence of bromide, iodide or chlorine. *Water Res.* 2003, *37*, 1469–1487.

223 Pryor, W.A., Giamalva, D.H., and Church, D.F. Kinetics of ozonation. 2. Amino acids and model compounds in water and comparisons to rates in nonpolar solvents. *J. Am. Chem. Soc.* 1984, *106*, 7094–7100.

224 Mumford, R.A., Lipke, H., Laufer, D.A., and Feder, W.A. Ozone-induced changes in corn pollen. *Environ. Sci. Technol.* 1972, *6*, 427–430.

225 Mudd, J.B., Leavitt, R., Ongun, A., and McManus, T.T. Reaction of ozone with amino acids and proteins. *Atmos. Environ.* 1969, *3*, 669–682.

226 Niki, E., Yamamoto, Y., Saito, T., Nagano, K., Yokoi, S., and Kamiya, Y. Ozonization of organic compounds. VII. Carboxylic acids, alcohols, and carbonyl compounds. *J. Chem. Soc. Jpn.* 1983, *56*, 223–228.

227 Bailey, P.S. Ozonation of nucleophiles. In *Ozonation in Organic Chemistry*, Vol. 2, H.H. Wasserman, ed. Academic Press Inc., New York, 1982, pp. 155–201.

228 Hoigne, J. and Bader, H. Rate constants of reactions of ozone with organic and inorganic compounds in water—II: dissociating organic compounds. *Water Res.* 1983, *17*, 185–194.

229 Yamamoto, Y., Niki, E., and Kamiya, Y. Ozonization of organic compounds. VI. Relative reactivity of protic solvents toward carbonyl oxide. *J. Chem. Soc. Jpn.* 1982, *55*, 2677–2678.

230 Hureiki, L., Croué, J.P., Legube, B., and Doré, M. Ozonation of amino acids: ozone demand and aldehyde formation. *Ozone Sci. Eng.* 1998, *20*, 381–402.

231 Lloyd, J.A., Spraggins, J.M., Johnston, M.V., and Laskin, J. Peptide ozonolysis: product structures and relative reactivities for oxidation of tyrosine and histidine residues. *J. Am. Soc. Mass Spectrochem.* 2006, *17*, 1289–1298.

232 Mudd, J.B., Dawson, P.J., Tseng, S., and Liu, F. Reaction of ozone with protein tryptophans: band III, serum albumin, and cytochrome c. *Arch. Biochem. Phys.* 1997, *338*, 143–149.

233 Verweij, H., Christianse, K., and Van Steveninck, J. Different pathways of tyrosine oxidation by ozone. *Chemosphere* 1982, *11*, 721–725.

234 Verweij, H., Christianse, K., and Van Steveninck, J. Ozone-induced formation of O,O′-dityrosine cross-links in proteins. *Biochim. Biophys. Acta* 1982, *701*, 180–184.

235 Berlett, B.S., Levine, R.L., and Stadtman, E.R. Comparison of the effects of ozone on the modification of amino acid residues in glutamine synthetase and bovine serum albumin. *J. Biol. Chem.* 1996, *271*, 4177–4182.

236 Cataldo, F. Ozone degradation of biological macromolecules: proteins, hemoglobin, RNA, and DNA. *Ozone Sci. Eng.* 2006, *28*, 317–328.

237 Cataldo, F. On the action of ozone on proteins. *Polym. Degrad. Stab.* 2003, *82*, 105–114.

238 Banerjee, S.K. and Mudd, J.B. Reaction of ozone with glycophorin in solution and in lipid vesicles. *Arch. Biochem. Phys.* 1992, *295*, 84–89.

239 Kotiaho, T., Eberlin, M.N., Vainiotalo, P., and Kostiainen, R. Electrospray mass and tandem mass spectrometry identification of ozone oxidation products of amino acids and small peptides. *J. Am. Soc. Mass Spectrochem.* 2000, *11*, 526–535.

240 Kuroda, M., Sakiyama, F., and Narita, K. Oxidation of tryptophan in lysozyme by ozone in aqueous solution. *J. Biochem.* 1975, *78*, 641–651.

241 Le Lacheur, R.M. and Glaze, W.H. Reactions of ozone and hydroxyl radicals with serine. *Environ. Sci. Technol.* 1996, *30*, 1072–1080.

242 Leh, F., Warr, T.A., and Mudd, J.B. Reaction of ozone with protease inhibitors from bovine pancreas, egg white, and human serum. *Environ. Res.* 1978, *16*, 179–190.

243 Knight, K.L. and Mudd, J.B. The reaction of ozone with glyceraldehyde-3-phosphate dehydrogenase. *Arch. Biochem. Biophys.* 1984, *229*, 259–269.

244 Meiners, B.A., Peters, R.E., and Mudd, J.B. Effects of ozone on indole compounds and rat lung monoamine oxidase. *Environ. Res.* 1977, *14*, 99–112.

245 Smith, C.E., Stack, M.S., and Johnson, D.A. Ozone effects on inhibitors of human neutrophil proteinases. *Arch. Biochem. Phys.* 1987, *253*, 146–155.

246 Spanggord, R.J. and McClurg, V.J. Ozone methods and ozone chemistry of selected organics in water. In *Ozone/Chlorine Dioxide Oxidation Products of Organic Materials*, R.G. Rice and J.A. Contruvo, eds. Ozone Press International, Cleveland, OH, 1978, pp. 115–123.

247 Uppu, R.M. and Pryor, W.A. The reactions of ozone with proteins and unsaturated fatty acids in reverse micelles. *Chem. Res. Toxicol.* 1994, *7*, 47–55.

248 Pryor, W.A. and Uppu, R.M. A kinetic model for the competitive reactions of ozone with amino acid residues in proteins in reverse micelles. *J. Biochem.* 1993, *265*, 3120–3126.

249 Dooley, M.M. and Mudd, J.B. Reaction of ozone with lysozyme under different exposure conditions. *Arch. Biochem. Phys.* 1982, *218*, 459–471.

250 Freeman, B.A. and Mudd, J.B. Reaction of ozone with sulfhydryls of human erythrocytes. *Arch. Biochem. Phys.* 1981, *208*, 212–220.

251 Berger, P., Leitner, N.K.V., Dore, M., and Legube, B. Ozone and hydroxyl radicals induced oxidation of glycine. *Water Res.* 1999, *33*, 433–441.

252 Enami, S., Hoffmann, M.R., and Colussi, A.J. Simultaneous detection of cysteine sulfenate, sulfinate, and sulfonate during cysteine interfacial ozonolysis. *J. Phys. Chem. B* 2009, *113*, 9356–9358.

253 Enami, S., Hoffmann, M.R., and Colussi, A.J. Ozone oxidizes glutathione to a sulfonic acid. *Chem. Res. Toxicol.* 2009, *22*, 35–40.

254 Lagerwerf, F.M., van de Weert, M., Heerma, W., and Haverkamp, J. Identification of oxidized methionine in peptides. *Rapid Commun. Mass Spectrom.* 1996, *10*, 1905–1910.

255 Nebel, C. Ozone. In *Kirk-Othmer Encyclopaedia of Chemical Technology*, Vol. 16, 3rd ed., Anonymous, ed. John Wiley & Sons, New York, 1988, pp. 683–713.

256 Karpel, N., Leitner, V., Berger, P., and Legube, B. Oxidation of amino groups by hydroxyl radicals in relation to the oxidation degree of the α-carbon. *Environ. Sci. Technol.* 2002, *36*, 3083–3089.

257 Spraggins, J.M., Lloyd, J.A., Johnston, M.V., Laskin, J., and Ridge, D.P. Fragmentation mechanisms of oxidized peptides elucidated by SID, RRKM modeling, and molecular dynamics. *J. Am. Soc. Mass Spectrom.* 2009, *20*, 1579–1592.

258 Havel, H.A. Derivative near-ultraviolet absorption techniques for investigating protein structure. In *Spectroscopic Methods for Determining Protein Structure in Solution*, H.A. Havel, ed. Wiley-VCH, New York, 1996, chapter 4.

259 Kim, H.I., Kim, H., Shin, Y.S., Beegle, L.W., Jang, S.S., Neidholdt, E.L., Goddard, W.A., Heath, J.R., Kanik, I., and Beauchamp, J.L. Interfacial reactions of ozone with surfactant protein B in a model lung surfactant system. *J. Am. Chem. Soc.* 2010, *132*, 2254–2263.

260 Kim, H.I., Kim, H., Shin, Y.S., Beegle, L.W., Goddard, W.A., Heath, J.R., Kanik, I., and Beauchamp, J.L. Time resolved studies of interfacial reactions of ozone with pulmonary phospholipid surfactants using field induced droplet ionization mass spectrometry. *J. Phys. Chem. B* 2010, *114*, 9496–9503.

261 Xu, G. and Chance, M.R. Hydroxyl radical-mediated modification of proteins as probes for structural proteomics. *Chem. Rev.* 2007, *107*, 3514–3543.

262 Buxton, G.V. An overview of the radiation chemistry of liquids. In *Radiation Chemistry: From Basics to Applications in Material and Life Science*, M. Spotheim-Maurizot, M. Mostafavi, and T.D. Jacquline, eds. EDP Sciences, L'Editeur, France, 2008, pp. 3–16.

263 Emmi, S.S. and Takacs, E. Water remediation by the electron beam treatment. In *Radiation Chemistry: from Basics to Applications in Material and Life Sciences*, M. Spotheim-Maurizot, M. Mostafavi, and T.D. Jacquline, eds. EDP Sciences, L' Editeur, France, 2008, pp. 79–95.

264 Shcherbakova, I., Mitra, S., Beer, R.H., and Brenowitz, M. Fast Fenton footprinting: a laboratory-based method for the time-resolved analysis of DNA, RNA and proteins. *Nucleic Acids Res.* 2006, *34*, document no. e48.

265 Watson, C., Janik, I., Zhuang, T., Charvátová, O., Woods, R.J., and Sharp, J.S. Pulsed electron beam water radiolysis for submicrosecond hydroxyl radical protein footprinting. *Anal. Chem.* 2009, *81*, 2496–2505.

266 Aye, T.T., Low, T.Y., and Sze, S.K. Nanosecond laser-induced photochemical oxidation method for protein surface mapping with mass spectrometry. *Anal. Chem.* 2005, *77*, 5814–5822.

267 Maleknia, S.D., Ralston, C.Y., Brenowitz, M.D., Downard, K.M., and Chance, M.R. Determination of macromolecular folding and structure by synchrotron X-ray radiolysis techniques. *Anal. Biochem.* 2001, *289*, 103–115.

268 Cooper, R. The history and development of radiation chemistry. *Aust. J. Chem.* 2011, *64*, 864–868.

269 Roth, O. and LaVerne, J.A. Effect of pH on H_2O_2 production in the radiolysis of water. *J. Phys. Chem. A* 2011, *115*, 700–708.

270 Rogers, F.M.A. *Radiation Chemistry: Principles and Applications*. VCH, New York, 1987.

271 Merényi, G., Lind, J., Goldstein, S., and Czapski, G. Peroxynitrous acid homolyzes into ·OH and $\cdot NO_2$ radicals. *Chem. Res. Toxicol.* 1998, *11*, 712–713.

272 Elliot, A.J. *AECL Report 11073, COG-95-167*, 1994.

273 Czapski, G. and Bielski, B.H.J. Absorption spectra of the ·OH and $O^{\cdot-}$ radicals in aqueous solutions. *Radiat. Phys. Chem.* 1993, *41*, 503–505.

274 Janik, I., Bartels, D.M., and Jonah, C.D. Hydroxyl radical self-recombination reaction and absorption spectrum in water up to 350°C. *J. Phys. Chem. A* 2007, *111*, 1835–1843.

275 Elliot, A.J., McCracken, D.R., Buxton, G.V., and Wood, N.D. Estimation of rate constants for near-diffusion-controlled reactions in water at high temperatures. *J. Chem. Soc. Faraday Trans.* 1990, *86*, 1539–1547.

276 Poskrebyshev, G.A., Neta, P., and Huie, R.E. Temperature dependence of the acid dissociation constant of the hydroxyl radical. *J. Phys. Chem. A* 2002, *106*, 11488–11491.

277 Buxton, G.V., Greenstock, C.L., Helman, W.P., and Ross, W.P. Critical review of rate constants for reactions of hydrated electrons, hydrogen atoms and hydroxyl radicals in aqueous solution. *J. Phys. Chem. Ref. Data* 1988, *17*, 513–886.

278 Pignatello, J.J., Oliveros, E., and MacKay, A. Advanced oxidation processes for organic contaminant destruction based on the Fenton reaction and related chemistry. *Crit. Rev. Environ. Sci. Technol.* 2006, *36*, 1–84.

279 Minakata, D., Li, K., Westerhoff, P., and Crittenden, J. Development of a group contribution method to predict aqueous phase hydroxyl radical ($HO^{\bullet}$) reaction rate constants. *Environ. Sci. Technol.* 2009, *43*, 6220–6227.

280 Minakata, D., Song, W., and Crittenden, J. Reactivity of aqueous phase hydroxyl radical with halogenated carboxylate anions: experimental and theoretical studies. *Environ. Sci. Technol.* 2011, *45*, 6057–6065.

281 Easton, C.J. Free-radical reactions in the synthesis of α-amino acids and derivatives. *Chem. Rev.* 1997, *97*, 53–82.

282 Easton, C.J. *Advances in Detailed Reaction Mechanism*. JAI Press, Greenwich, CT, 1991.

283 Headlam, H.A. and Davies, M.J. Beta-scission of side-chain alkoxyl radicals on peptides and proteins results in the loss of side-chains as aldehydes and ketones. *Free Radic. Biol. Med.* 2002, *32*, 1171–1184.

284 von Sonntag, C. and Schuchmann, H.P. The chemistry of free radicals: peroxy radicals in aqueous solutions. In *Peroxy Radicals*, Z.B. Alfassi, ed. John Wiley & Sons, Chichester, UK, 1997.

285 Steinmann, D., Ji, J.A, Wang, Y.J, Schöneich, C. Oxidation of human growth hormone by oxygen-centered radicals: formation of Leu-101 hydroperoxide and Tyr-103 oxidation products. *Mol. Pharmaceut.* 2012, *9*, 803–814.

286 Fossey, J., Lefort, D., and Sobra, J. *Free Radicals in Organic Chemistry*. Wiley, Chichester, UK, 1995.

287 Xu, G. and Chance, M.R. Radiolytic modification of acidic amino acid residues in peptides: probes for examining protein-protein interactions. *Anal. Chem.* 2004, *76*, 1213–1221.

288 Sokol, H.A., Bennett-Corniea, W., and Garrison, W.M. A marked effect of conformation in the radiolysis of poly-α-L-glutamic acid in aqueous solution. *J. Am. Chem. Soc.* 1965, *87*, 1391–1392.

289 Uchida, K., Kato, Y., and Kawakishi, S. A novel mechanism for oxidative cleavage of prolyl peptides induced by the hydroxyl radical. *Biochem. Biophys. Res. Commun.* 1990, *169*, 265–271.

290 Wolff, S.P., Garner, A., and Dean, R.T. Free radicals, lipids and protein degradation. *Trends Biochem. Sci.* 1986, *11*, 27–31.

291 Wilks, R.G., MacNaughton, J.B., Kraatz, H.B., Regier, T., Blyth, R.I.R., and Moewes, A. Comparative theoretical and experimental study of the radiation-induced decomposition of glycine. *J. Phys. Chem. A* 2009, *113*, 5360–5366.

292 Cataldo, F., Angelini, G., Iglesias-Groth, S., and Manchado, A. Solid state radiolysis of amino acids in an astrochemical perspective. *Radiat. Phys. Chem.* 2011, *80*, 57–65.

293 Cataldo, F., Ragni, P., Iglesias-Groth, S., and Manchado, A. Solid state radiolysis of sulphur-containing amino acids: cysteine, cystine and methionine. *J. Radioanal. Nucl. Chem.* 2011, *287*, 573–580.

294 Štefanić, I., Ljubić, I., Bonifačić, M., Sabljić, A., Asmus, K.-D., and Armstrong, D.A. A surprisingly complex aqueous chemistry of the simplest amino acid. A pulse radiolysis and theoretical study on H/D kinetic isotope effects in the reaction of glycine anions with hydroxyl radicals. *Phys. Chem. Chem. Phys.* 2009, *11*, 2256–2267.

295 Lin, R.J., Wu, C.C., Jang, S., and Li, F.Y. Variation of reaction dynamics for OH hydrogen abstraction from glycine between *ab initio* levels of theory. *J. Mol. Model.* 2010, *16*, 175–182.

296 Leitner, N.K.V., Berger, P., and Legube, B. Oxidation of amino groups by hydroxyl radicals in relation to the oxidation degree of carbon. *Environ. Sci. Technol.* 2002, *36*, 3083–3089.

297 Bonifačić, M., Štefanić, I., Hug, G.L., Armstrong, D.A., and Asmus, K.D. Glycine decarboxylation: the free radical mechanism. *J. Am. Chem. Soc.* 1998, *120*, 9930–9940.

298 Hug, G.L., Bonifačić, M., Asmus, K.D., and Armstrong, D.A. Fast decarboxylation of aliphatic amino acids induced by 4-carboxybenzophenone triplets in aqueous solutions. A nanosecond laser flash photolysis study. *J. Phys. Chem. B* 2000, *104*, 6674–6682.

299 Hug, G.L., Carmichael, I., and Fessenden, R.W. Direct EPR observation of the aminomethyl radical during the radiolysis of glycine. *J. Chem. Soc. Perkin Trans. 2* 2000, 907–908.

300 Stefanic, I., Bonifacic, M., Asmus, K., and Armstrong, D.A. Absolute rate constants and yields of transients from hydroxyl radical and H atom attack on glycine and methyl-substituted glycine anions. *J. Phys. Chem. A* 2001, *105*, 8681–8690.

301 Wisniowski, P., Carmichael, I., Fessenden, R.W., and Hug, G.L. Evidence for β scission in the oxidation of amino acids. *J. Phys. Chem. A* 2002, *106*, 4573–4580.

302 Mönig, J., Chapman, R., and Asmus, K.-D. Effect of the protonation state of the amino group on the ·OH radical induced decarboxylation of amino acids in aqueous solution. *J. Phys. Chem.* 1985, *89*, 3139–3144.

303 Bonifačić, M., Armstrong, D.A., Štefanić, I., and Asmus, K.D. Kinetic isotope effect for hydrogen abstraction by ·OH radicals from normal and carbon-deuterated ethyl alcohol and methylamine in aqueous solutions. *J. Phys. Chem. B* 2003, *107*, 7268–7276.

304 Liessmann, M., Hansmann, B., Blachly, P.G., Francisco, J.S., and Abel, B. Primary steps in the reaction of OH radicals with amino acids at low temperatures in laval nozzle expansions: perspectives from experiment and theory. *J. Phys. Chem. A* 2009, *113*, 7570–7575.

305 Doan, H.Q., Davis, A.C., and Francisco, J.S. Primary steps in the reaction of OH radicals with peptide systems: perspective from a study of model amides. *J. Phys. Chem. A* 2010, *114*, 5342–5357.

306 Hawkins, C.L. and Davies, M.J. EPR studies on the selectivity of hydroxyl radical attack on amino acids and peptides. *J. Chem. Soc. Perkin Trans. 2* 1998, 2617–2622.

307 Davies, M.J. The oxidative environment and protein damage. *Biochem. Biophys. Acta—Proteins Proteomics* 2005, *1703*, 93–109.

308 Morgan, P.E., Pattison, D.I., and Davies, M.J. Quantification of hydroxyl radical-derived oxidation products in peptides containing glycine, alanine, valine, and proline. *Free Radic. Biol. Med.* 2012, *52*, 328–339.

309 Xu, G. and Chance, M.R. Radiolytic modification and reactivity of amino acid residues serving as structural probes for protein footprinting. *Anal. Chem.* 2005, *77*, 4549–4555.

310 Hiller, K.-O., Masloch, B., Göbl, M., and Asmus, K.D. Mechanism of the OH˙ radical induced oxidation of methionine in aqueous solution. *J. Am. Chem. Soc.* 1981, *103*, 2734–2743.

311 Schöneich, C. and Yang, J. Oxidation of methionine peptides by Fenton systems: the importance of peptide sequence, neighbouring groups and EDTA. *J. Chem. Soc. Perkin Trans. 2* 1996, *5*, 915–924.

312 Schöneich, C., Pogocki, D., Hug, G.L., and Bobrowski, K. Free radical reactions of methionine in peptides: mechanisms relevant to β-amyloid oxidation and Alzheimer's disease. *J. Am. Chem. Soc.* 2003, *125*, 13700–13713.

313 Hong, J. and Schöneich, C. The metal-catalyzed oxidation of methionine in peptides by Fenton systems involves two consecutive one-electron oxidation processes. *Free Radic. Biol. Med.* 2001, *31*, 1432–1441.

314 Brunelle, P. and Rauk, A. One-electron oxidation of methionine in peptide environments: the effect of three-electron bonding on the reduction potential of the radical cation. *J. Phys. Chem. A* 2004, *108*, 11032–11041.

315 Asmus, K. *S-Centered Radicals*. John Wiley & Sons, New York, 1999.

316 Hug, G.L., Bobrowski, K., Pogocki, D., Marciniak, B., Schöneich, C., and Hörner, G. Factor analysis of transient spectra. Free radicals in cyclic dipeptides containing methionine. *Res. Chem. Intermed.* 2009, *35*, 431–442.

317 Francisco-Marquez, M. and Galano, A. Role of the sulfur atom on the reactivity of methionine toward OH radicals: comparison with norleucine. *J. Phys. Chem. B* 2009, *113*, 4947–4952.

318 Ji, W.-F., Li, Z.-L., Shen, L., Kong, D.-X., and Zhang, H.-Y. Density functional theory methods as powerful tools to elucidate amino acid oxidation mechanisms. A case study on methionine model peptide. *J. Phys. Chem. B* 2007, *111*, 485–489.

319 Qu, N., Guo, L.-H., and Zhu, B.-Z. An electrochemical biosensor for the detection of tyrosine oxidation induced by Fenton reaction. *Biosens. Bioelectron.* 2011, *26*, 2292–2296.

320 Ichiba, H., Ogawa, T., Yajima, T., and Fukushima, T. Analysis of hydroxyl radical-induced oxidation process of glucagon by reversed-phase HPLC and ESI-MS/MS. *Biomed. Chromatogr.* 2009, *23*, 1051–1058.

321 Prutz, W.A., Butler, J., Land, E.J., and Swallow, A.J. The role of sulphur peptide functions in free radical transfer: a pulse radiolysis study. *Int. J. Radiat. Biol.* 1989, *55*, 539–556.

322 Armstrong, D.A. *Sulfur-Centered Reactive Intermediates in Chemistry and Biology*. Plenum Press, New York, 1990.

323 Lal, M. Radiation induced oxidation of sulphydryl molecules in aqueous solutions. A comprehensive review. *Radiat. Phys. Chem.* 1994, *43*, 595–611.

324 Maleknia, S.D., Brenowitz, M., and Chance, M.R. Millisecond radiolytic modification of peptides by synchrotron X-rays identified by mass spectrometry. *Anal. Chem.* 1999, *71*, 3965–3973.

325 Dewey, D.L. and Beecher, J. Interconversion of cystine and cysteine induced by X-rays [30]. *Nature* 1965, *206*, 1369–1370.

326 Shetty, V. and Neubert, T.A. Characterization of novel oxidation products of cysteine in an active site motif peptide of PTP1B. *J. Am. Soc. Mass Spectrom.* 2009, *20*, 1540–1548.

327 Nauser, T., Casi, G., Koppenol, W.H., and Schöneich, C. Reversible intramolecular hydrogen transfer between cysteine thiyl radicals and glycine and alanine in model peptides: absolute rate constants derived from pulse radiolysis and laser flash photolysis. *J. Phys. Chem. B* 2008, *112*, 15034–15044.

328 Naumov, S. and Schöneich, C. Intramolecular addition of cysteine thiyl radical to phenylalanine and tyrosine in model peptides, Phe (CysS·) and Tyr(CysS·): a computational study. *J. Phys. Chem. A* 2009, *113*, 3560–3565.

329 Mozziconacci, O., Sharov, V., Williams, T.D., Kerwin, B.A., and Schöneich, C. Peptide cysteine thiyl radicals abstract hydrogen atoms from surrounding amino acids: the photolysis of a cystine containing model peptide. *J. Phys. Chem. B* 2008, *112*, 9250–9257.

330 Mishra, B., Sharma, A., Naumov, S., and Priyadarsini, K.I. Novel reactions of one-electron oxidized radicals of selenomethionine in comparison with methionine. *J. Phys. Chem. B* 2009, *113*, 7709–7715.

331 Mishra, B., Priyadarsini, K.I., and Mohan, H. Effect of pH on one-electron oxidation chemistry of organoselenium compounds in aqueous solutions. *J. Phys. Chem. A* 2006, *110*, 1894–1900.

332 Mishra, B., Kumbhare, L.B., Jain, V.K., and Priyadarsini, K.I. Pulse radiolysis studies on reactions of hydroxyl radicals with selenocystine derivatives. *J. Phys. Chem. B* 2008, *112*, 4441–4446.

333 Tobien, T., Bonifaić, M., Naumov, S., and Asmus, K.-D. Time-resolved study on the reactions of organic selenides with hydroxyl and oxide radicals, hydrated electrons, and H-atoms in aqueous solution, and DFT calculations of transients in comparison with sulfur analogues. *Phys. Chem. Chem. Phys.* 2010, *12*, 6750–6758.

334 Naumov, S., Bonifačić, M., Glass, R.S., and Asmus, K.D. Theoretical calculations and experimental data on spectral, kinetic and thermodynamic properties of Se∴N and S∴N three-electron-bonded, structurally stabilized $\sigma 2\sigma^*$ radicals. *Res. Chem. Intermed.* 2009, *35*, 479–496.

335 Morin, É., Bubb, W.A., Davies, M.J., Dean, R.T., and Fu, S. 3-Hydroxylysine, a potential marker for studying radical-induced protein oxidation. *Chem. Res. Toxicol.* 1998, *11*, 1265–1273.

336 Xu, G., Takamoto, K., and Chance, M.R. Radiolytic modification of basic amino acid residues in peptides: probes for examining protein-protein interactions. *Anal. Chem.* 2003, *75*, 6995–7007.

337 Requena, J.R. and Stadtman, E.R. Conversion of lysine to N(ε)-(carboxymethyl) lysine increases susceptibility of proteins to metal-catalyzed oxidation. *Biochem. Biophys. Res. Commun.* 1999, *264*, 207–211.

338 Shadyro, O.I., Sosnovskaya, A.A., and Vrublevskaya, O.N. Deamination and degradation of hydroxyl-containing dipeptides and tripeptides in the radiolysis of their aqueous solutions. *High Energy Chem.* 2000, *34*, 290–294.

339 Sosnovskaya, A.A., Sladkova, A.A., Dobridenev, I.S., and Shadyro, O.I. Radiation-induced transformations of threonine and its derivatives in aqueous solutions. *High Energy Chem.* 2009, *43*, 431–434.

340 Amici, A., Levine, R.L., Tsai, L., and Stadtman, E.R. Conversion of amino acid residues in proteins and amino acid homopolymers to carbonyl derivatives by metal-catalyzed oxidation reactions. *J. Biol. Chem.* 1989, *264*, 3341–3346.

341 Ayala, A. and Cutler, R.G. The utilization of 5-hydroxyl-2-amino valeric acid as a specific marker of oxidized arginine and proline residues in proteins. *Free Radic. Biol. Med.* 1996, *21*, 65–80.

342 Xu, G., Liu, R., Zak, O., Aisen, P., and Chance, M.R. Structural allostery and binding of the transferrin-receptor complex. *Mol. Cell. Proteomics* 2005, *4*, 1959–1967.

343 Mossoba, M.M., Rosenthal, I., and Riesz, P. E.s.r. and spin-trapping studies of dihydropyrimidines. γ-Radiolysis in the polycrystalline state and U.V. photolysis in aqueous solution. *Int. J. Radiat. Biol.* 1981, *40*, 541–552.

344 Nukuna, B.N., Goshe, M.B., and Anderson, V.E. Sites of hydroxyl radical reaction with amino acids identified by ^{2}H NMR detection of induced ^{1}H/^{2}H exchange. *J. Am. Chem. Soc.* 2001, *123*, 1208–1214.

345 Stadtman, E.R. Role of oxidized amino acids in protein breakdown and stability. *Method Enzymol.* 1995, *258*, 379–393.

346 Lassmann, G., Eriksson, L.A., Himo, F., Lendzian, F., and Lubitz, W. Electronic structure of a transient histidine radical in liquid aqueous solution: EPR continuous-flow studies and density functional calculations. *J. Phys. Chem. A* 1999, *103*, 1283–1290.

347 Kopoldova, J. and Hrncir, S. Gamma radiolysis of aqueous solution of histidine. *Z. Naturforsch., C, J. Biosci.* 1977, *32*, 482–487.

348 Schöneich, C. Mechanisms of metal-catalyzed oxidation of histidine to 2-oxo-histidine in peptides and proteins. *J. Pharm. Biomed. Anal.* 2000, *21*, 1093–1097.

349 Stadtman, E.R. Metal ion-catalyzed oxidation of proteins: biochemical mechanism and biological consequences. *Free Radic. Biol. Med.* 1990, *9*, 315–325.

350 Uchida, K. and Kawakishi, S. 2-Oxo-histidine as a novel biological marker for oxidatively modified proteins. *FEBS Lett.* 1993, *332*, 208–210.

351 Maskos, Z., Rush, J.D., and Koppenol, W.H. The hydroxylation of tryptophan. *Arch. Biochem. Biophys.* 1992, *296*, 514–520.

352 Finley, E.L., Dillon, J., Crouch, R.K., and Schey, K.L. Identification of tryptophan oxidation products in bovine α-crystallin. *Protein Sci.* 1998, *7*, 2391–2397.

353 Solar, S. Reaction of OH with phenylalanine in neutral aqueous solution. *Radiat. Phys. Chem.* 1985, *26*, 103–108.

354 Kiselar, J.G., Mahaffy, R., Pollard, T.D., Almo, S.C., and Chance, M.R. Visualizing Arp2/3 complex activation mediated by binding of ATP and WASp using structural mass spectrometry. *Proc. Natl Acad. Sci. U.S.A.* 2007, *104*, 1552–1557.

355 Fonseca, C., Domingues, M.R.M., Simões, C., Amado, F., and Domingues, P. Reactivity of Tyr-Leu and Leu-Tyr dipeptides: identification of oxidation products by liquid chromatography-tandem mass spectrometry. *J. Mass Spectrom.* 2009, *44*, 681–693.

356 Tian, Y., Liu, R., Zong, W., Sun, F., Wang, M., and Zhang, P. A new biomarker of protein oxidation degree and site using angiotensin as the target by MS. *Spectrochim. Acta A. Mol. Biomol. Spectrosc.* 2010, *75*, 908–911.

357 Moore, J.S., Sakhri, M., and Butler, J. The radiolysis of protein A. *Radiat. Phys. Chem.* 2000, *58*, 331–339.

358 Suto, D., Ikeda, Y., Fujii, J., and Ohba, Y. Structural analysis of amino acids, oxidized by reactive oxygen species and an antibody against N-formylkynurenine. *J. Clin. Biochem. Nutr.* 2006, *38*, 107–111.

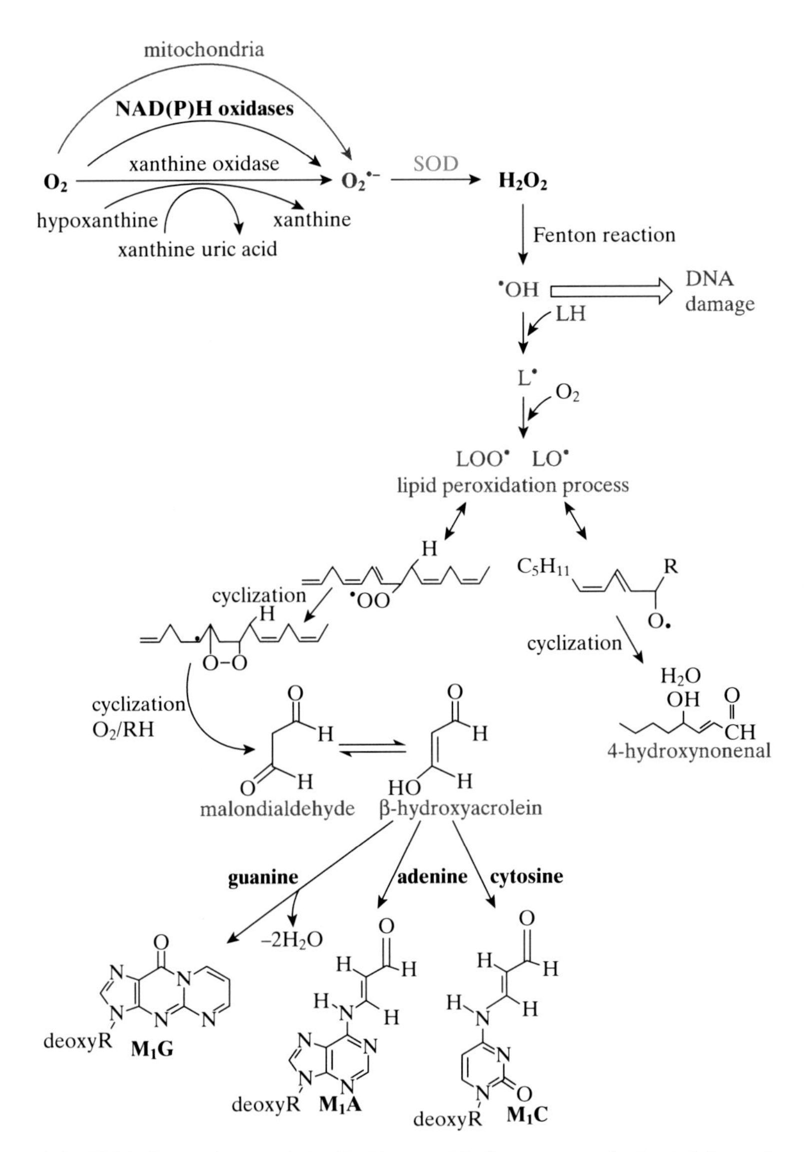

Figure 1.3. ROS formation and the lipid peroxidation process (adapted from Jomova and Valko [35] with the permission of the Elsevier Inc.).

Oxidation of Amino Acids, Peptides, and Proteins: Kinetics and Mechanism, First Edition.
Virender K. Sharma.

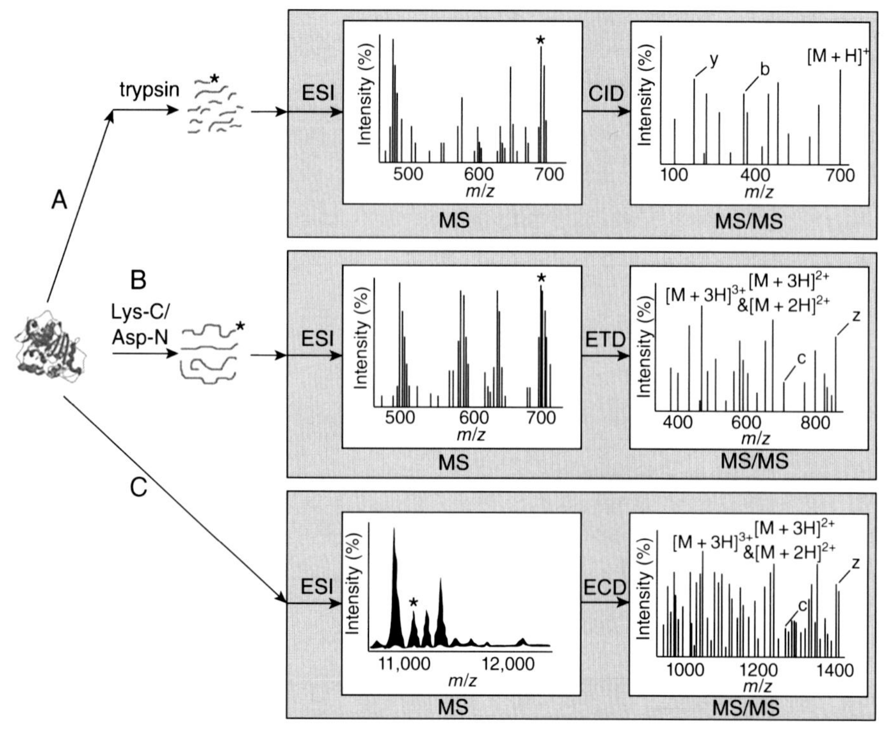

Figure 1.7. MS-based approaches A, B, and C. The gray boxes symbolize what takes place inside the MS equipment; the stars denote a certain peptide or protein (from an oxidized sample), chosen for CID, ETD, or ECD (adapted from Törnvall [90] with the permission of the Royal Society of Chemistry).

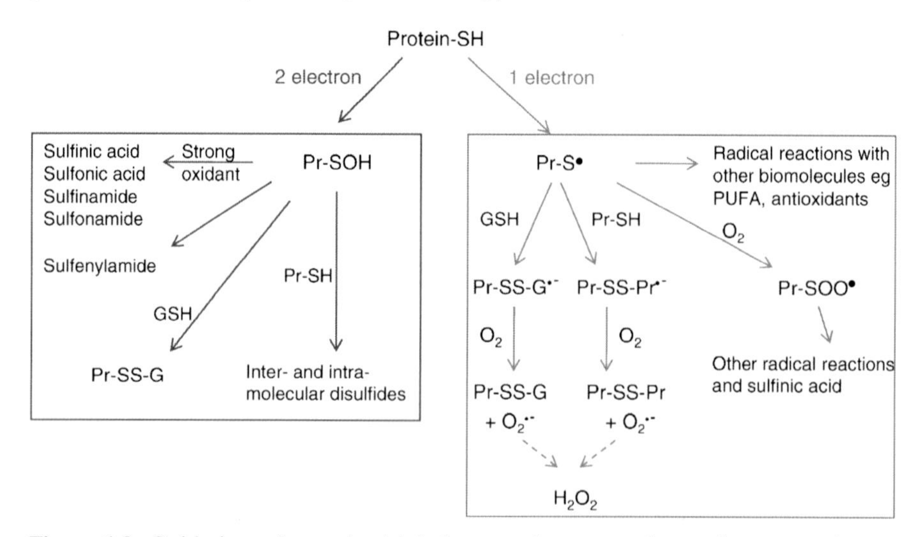

Figure 1.8. Oxidation of protein thiols by one-electron and two-electron pathways. PUFA, polyunsaturated fatty acids (adapted from Winterbourn and Hampton [230] with the permission of Elsevier Inc.)

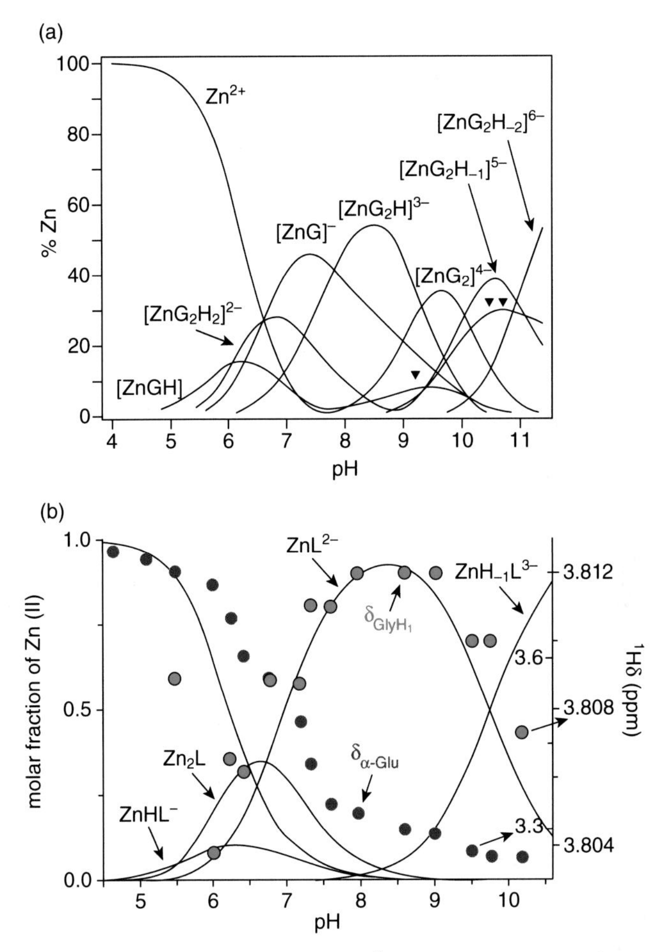

Figure 2.6. (a) Distribution diagram for the Zn^{2+}/GSH (H_3G) system (Zn : G = 1.0:1.81, $C_{Zn} = 1.6 \times 10^{-3}$ M): ▼ = $[Zn_2L_2H_{-1}]^{3-}$; ▼ ▼ = $[Zn_2L_2H_{-2}]^{4-}$ (adapted from Ferretti et al. [118] with the permission of the Elsevier Inc.). (See text for full caption)

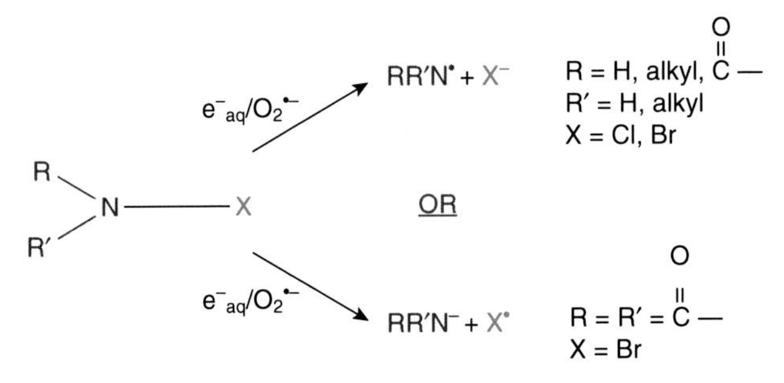

Figure 3.12. One-electron reduction of *N*-chlorinated and *N*-brominated species (adapted from Pattison et al. [125] with the permission of the American Chemical Society).

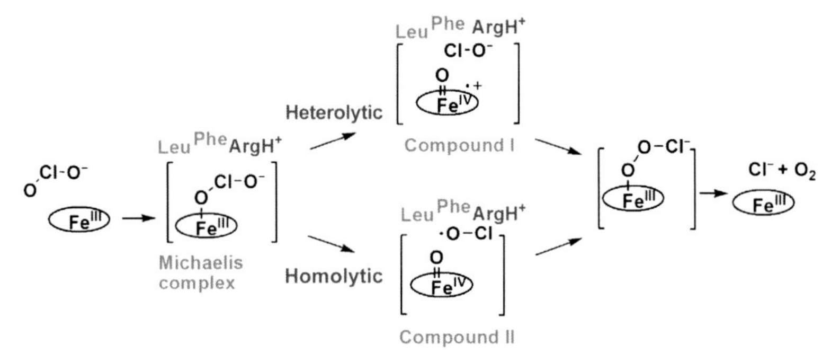

Figure 3.14. Proposed mechanisms for chlorite decomposition and O_2 evolution catalyzed by *Dechloromonas aromatica* (DA-Cld) (adapted from Goblirsch et al. [148] with the permission of Elsevier Inc.).

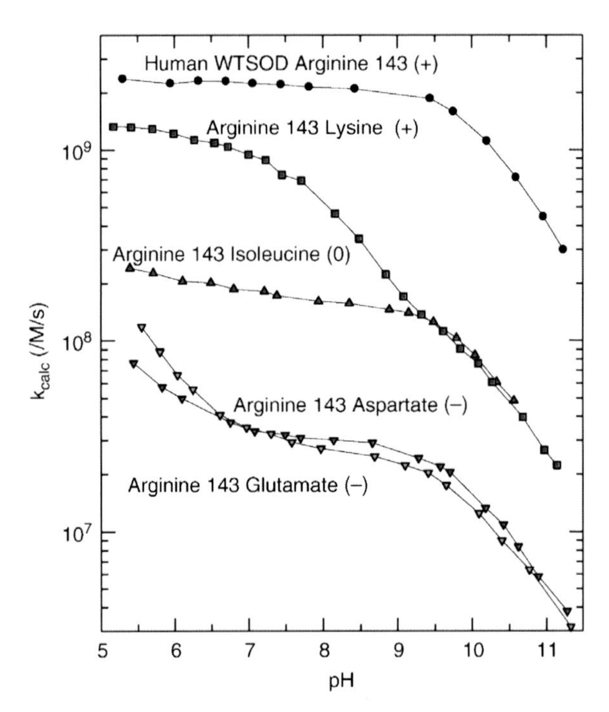

Figure 4.4. Rate constants for O_2^- dismutation by wild-type and various arginine 143 mutants of human sCuZnSOD as a function of pH (20 mM phosphate, 10 mM formate, 5 μM EDTA). The charge on the new residue is indicated in parentheses. WTSOD, wild-type superoxide dismutase (adapted from Abreu and Cabelli [31] with the permission of Elsevier Inc.).

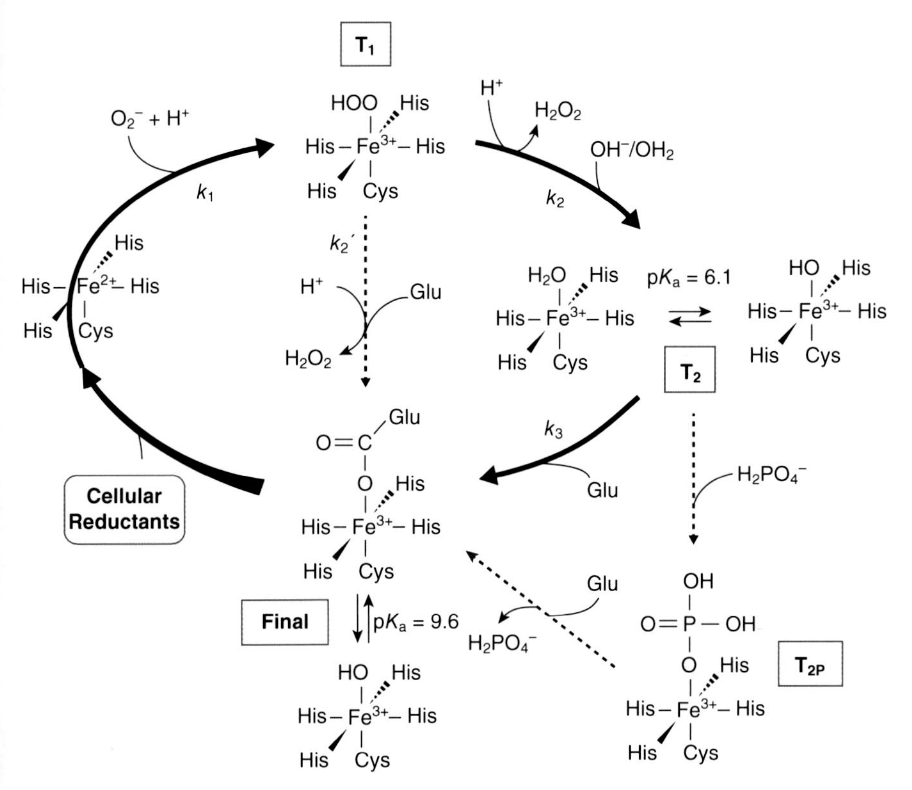

Figure 4.7. Catalytic cycle of superoxide reductase, showing only the observable intermediates. Large arrow: reductive cycle; narrow arrows: oxidative cycle (adapted from Pinto et al. [115] with the permission of Elsevier Inc.)

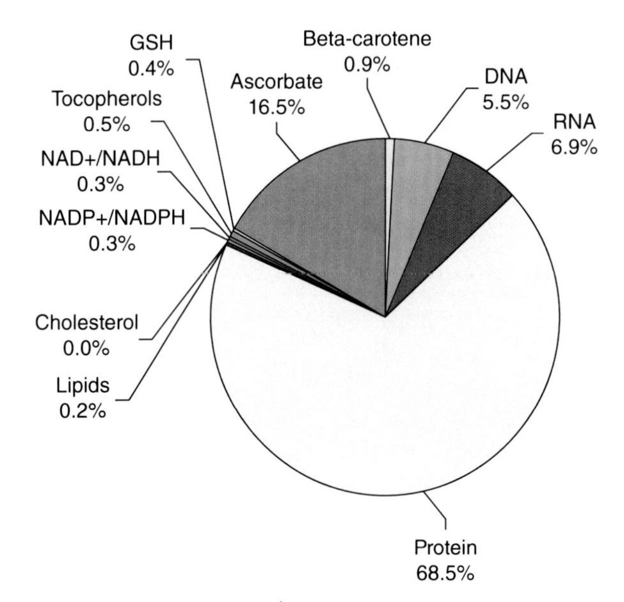

Figure 4.12. Predicted consumption of $^1O_2$ by intracellular targets calculated using the rate constant data given in Reference [151] and the average concentration of each component within a typical leukocyte cell [186] (adapted from Davies [130] with the permission of Elsevier Inc.).

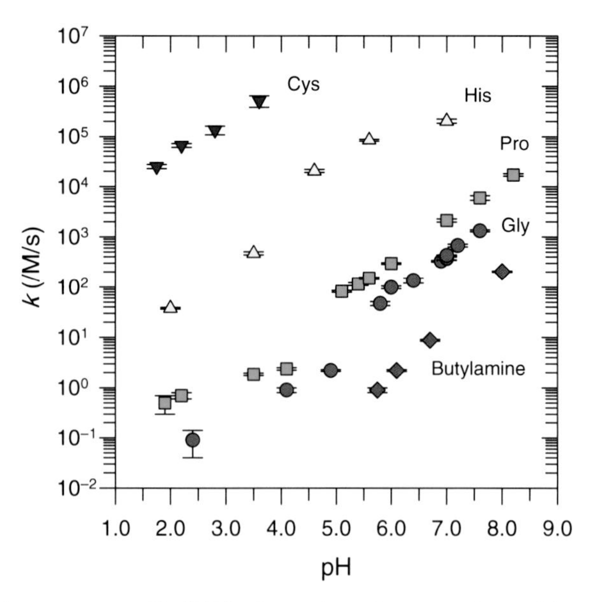

Figure 4.14. Rate constants (k, /M/s) of reaction of some amino acids and butylamine with O_3 in water (data were taken from Pryor et al. [223]).

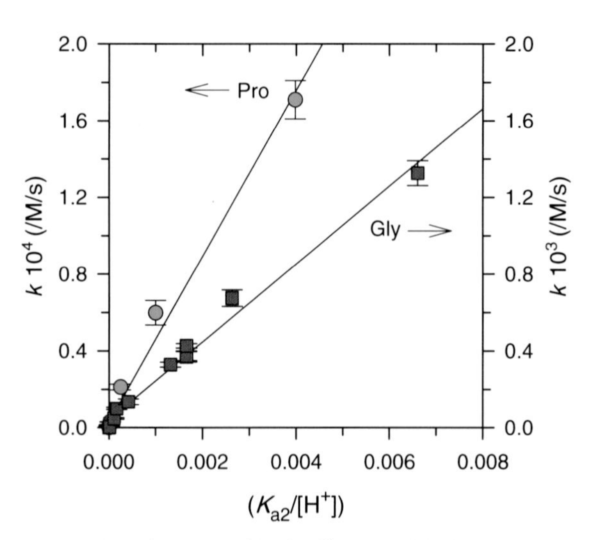

Figure 4.15. Plot of k (/M/s) versus (k_{a2}/[H^+]) for oxidation of Gly and Pro by O_3 (adapted from Sharma and Graham [15]).

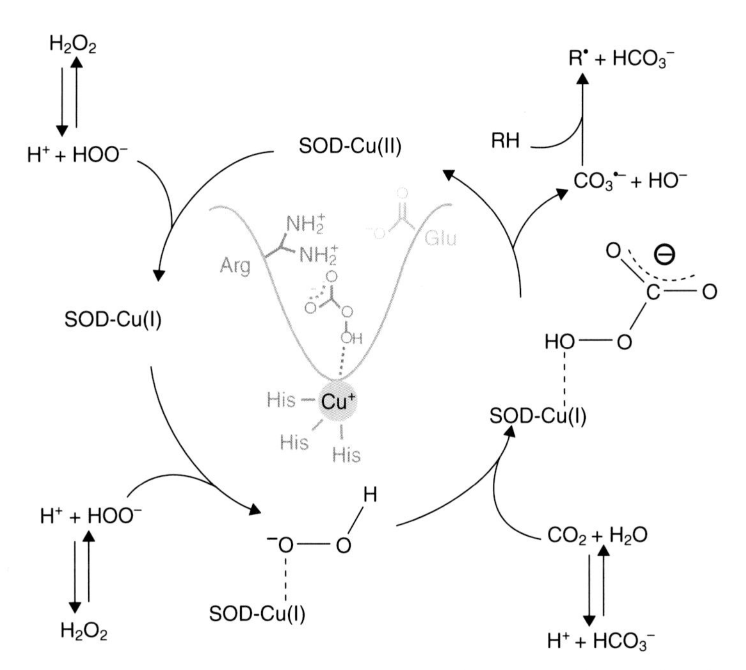

Figure 5.2. Mechanism proposed for carbonate radical production during the peroxidase activity of superoxide dismutase 1. The inset shows a pictorial view of the enzyme-Cu(I)-bound peroxymonocarbonate intermediate (adapted from Medinas et al. [2] with the permission of IUBMB).

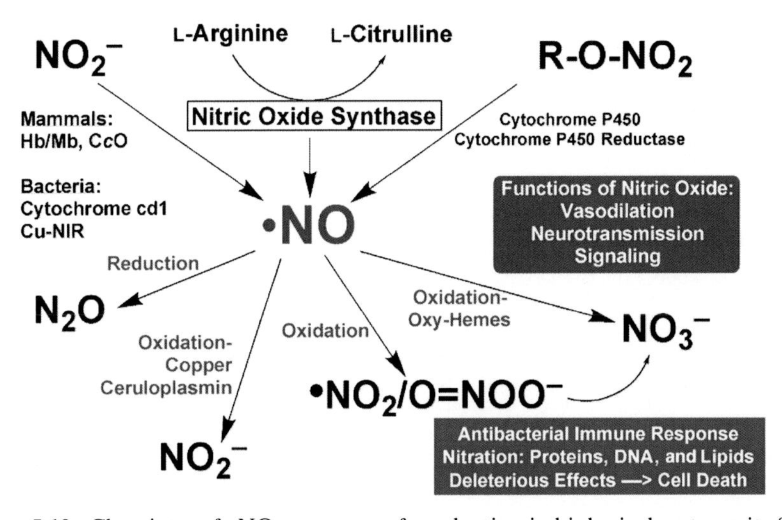

Figure 5.10. Chemistry of $\bullet NO_{(g)}$: sources of production in biological systems, its functions and roles, and multiple oxidation and reduction pathways, leading to the formation of a variety of species that may effect a diverse range of biomolecules and functions (adapted from Schopfer et al. [84] with the permission of the American Chemical Society).

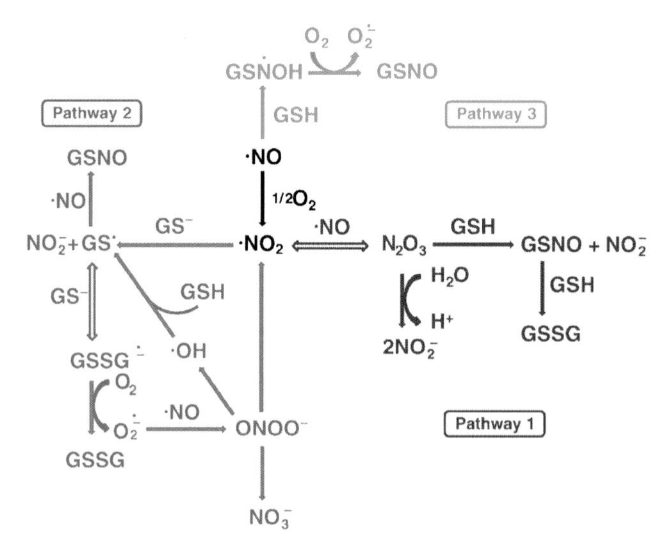

Figure 5.11. Proposed pathways of the reaction among GSH, NO, and oxygen (adapted from Keszler et al. [107] with the permission of Elsevier Inc.).

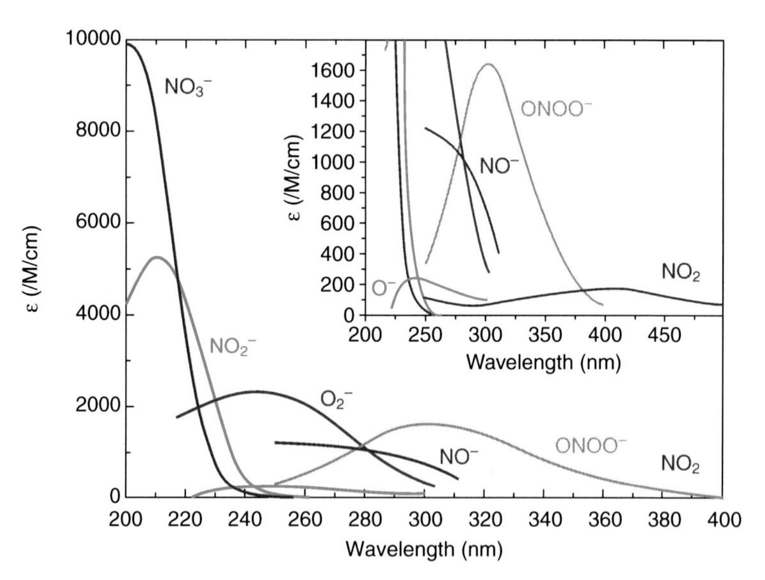

Figure 5.12. Absorption spectra of the species potentially involved in oxidation by peroxynitrite (adapted from Madsen et al. [131] with the permission of the American Chemical Society).

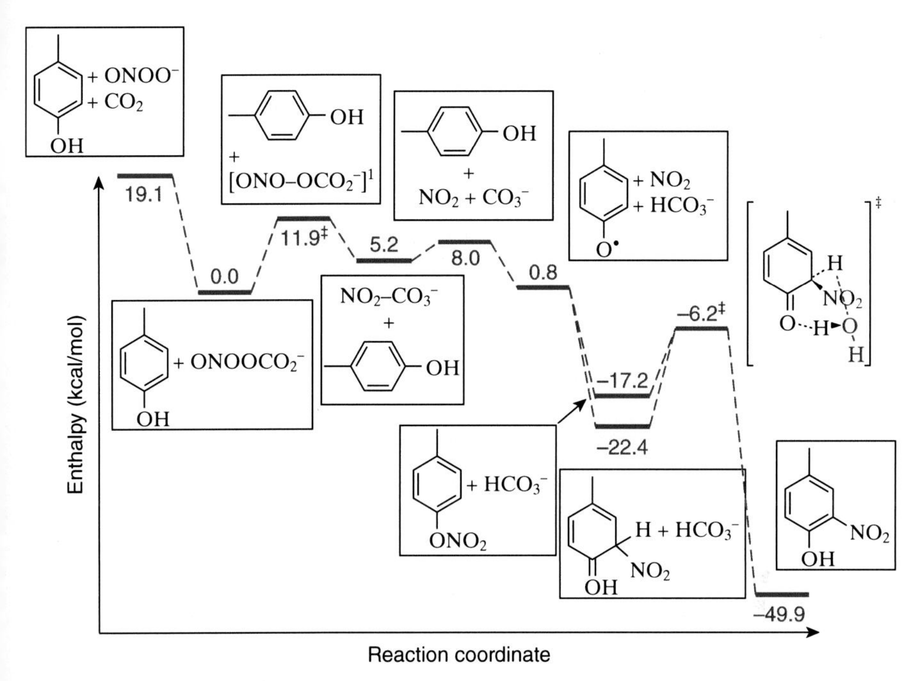

Figure 5.16. The mechanisms of tyrosine nitration by peroxynitrous acid investigated with the CBS-QB3 method (adapted from Gunaydin and Houk [288] with the permission of the American Chemical Society).

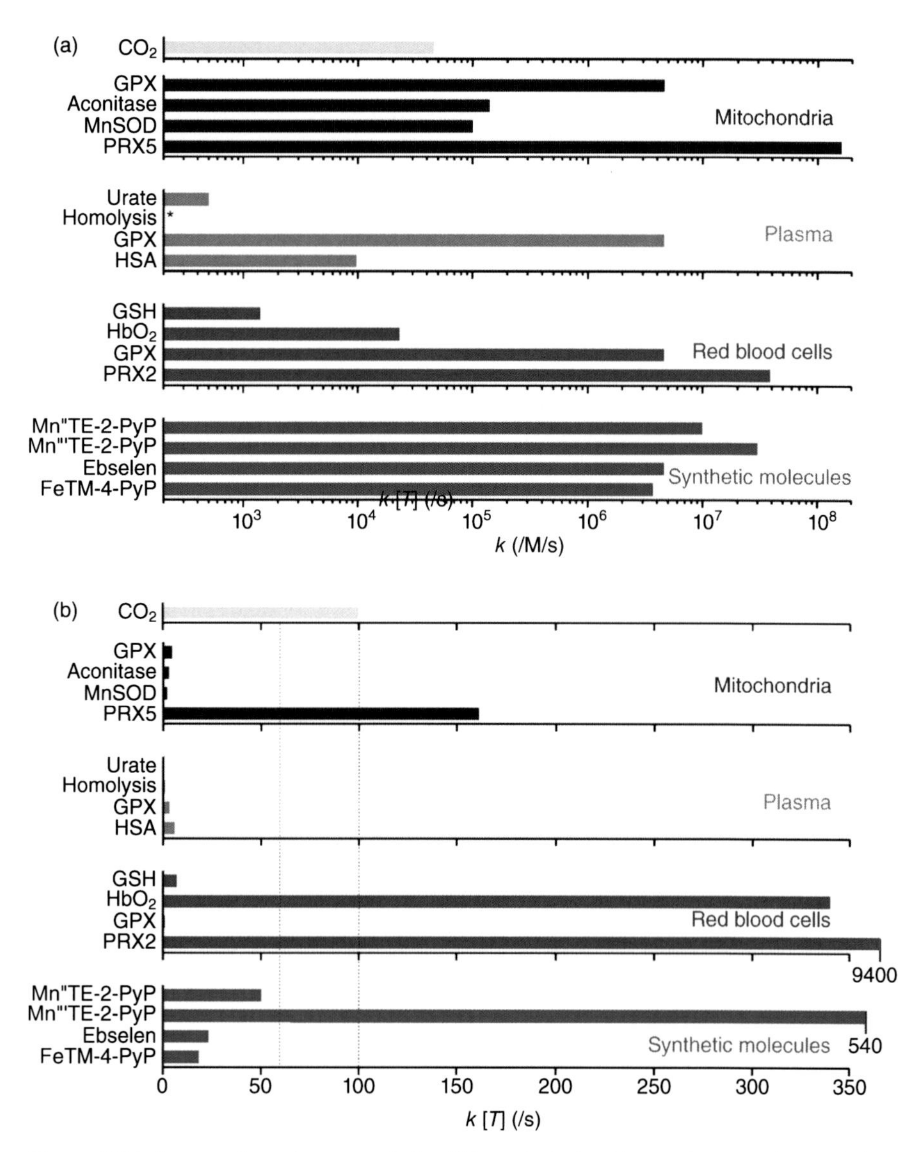

Figure 5.17. Scavenging of peroxynitrite. *Homolysis is a first-order reaction and thus cannot be plotted in this graph (adapted from Ferrer-Sueta and Radi [133] with the permission of the American Chemical Society).

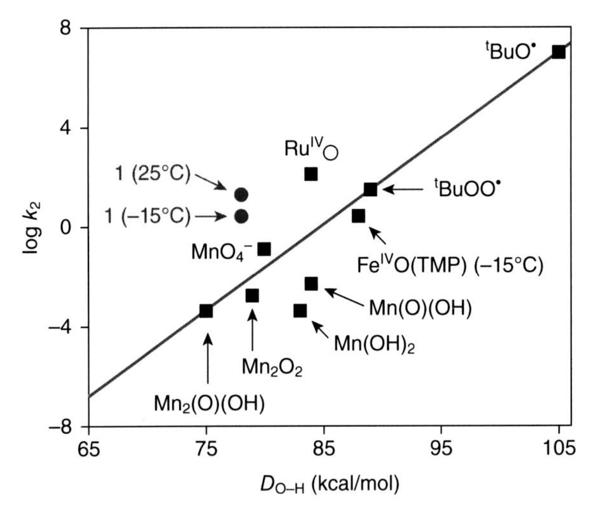

Figure 6.28. Plot of log k_2 of DHA oxidation and the strength of the O–H bond formed by the oxidants in CH_3CN at 25°C unless labeled otherwise. Data points shown in black filled squares were taken from References 32–37 and the blue filled circle was obtained in the present work of complex 1. The red straight line was drawn through the points belonging to the two oxygen radicals following Mayer's precedent. Mn(O)(OH), $[Mn^{IV}(Me2EBC)(O)(OH)]^+$; $Mn(OH)_2$, $[Mn^{IV}(Me_2EBC)(OH)_2]^{2+}$; Mn_2O_2, $[(phen)_2Mn^{IV}(O)_2Mn^{III}\text{-}(phen)_2]^{3+}$; $Mn_2(O)(OH)$, $[(phen)_2Mn^{III}(O)(OH)Mn^{III}(phen)_2]^{3+}$; $Ru^{IV}O$, $[Ru^{IV}(O)(bpy)_2(py)]^{2+}$ (adapted from Wang et al. [329] with the permission of the American Chemical Society).

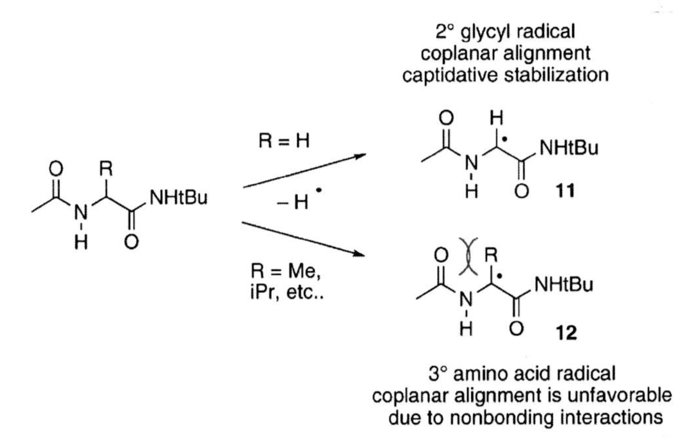

Figure 6.30. Carbon-centered radicals resulting from the abstraction of an α-hydrogen atom, stabilized in the case of Gly but destabilized with other amino acids (adapted from Abouelatta et al. [334] with the permission of the American Chemical Society).

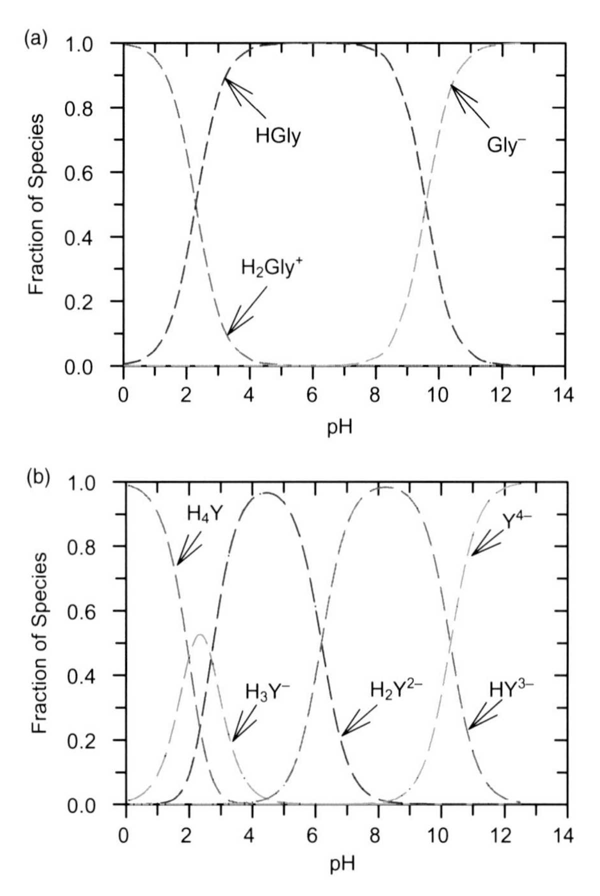

Figure 6.38. Speciation of Gly (a) and EDTA (b).

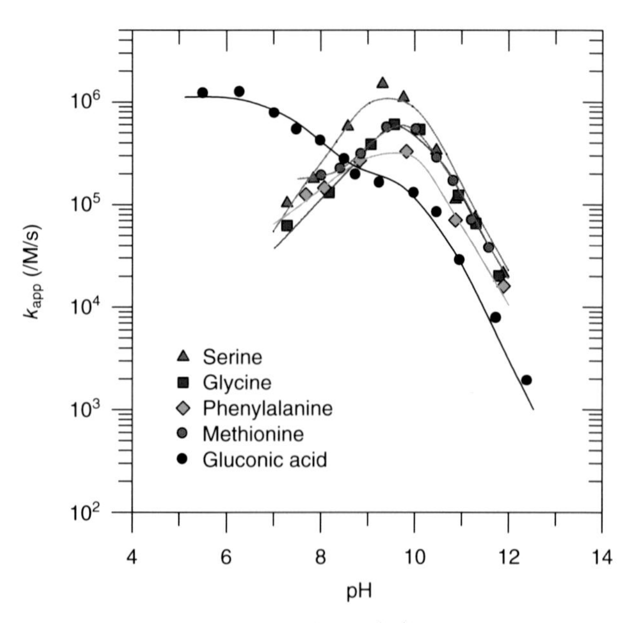

Figure 6.39. Second-order rate constants (k, $M^{-1}s^{-1}$) as a function of pH at 22°C for the oxidation of carboxylic acids by ferrate(V).

5

REACTIVE INORGANIC OXY-SPECIES OF C, N, S, AND P

In the last two decades, reactive species such as carbonate radical anion, nitrogen dioxide, and peroxynitrite have received attention due to their important role in cell homeostasis [1, 2]. The biological relevance of these species was observed in the efficiency of classical antioxidants, such as urate and nitroxides, in the protection of cells and the overall living organism against injuries related to an overproduction of nitric oxide [3, 4]. The potential physiological sources and biochemical reactions of the species are presented in Figure 5.1. Several rate studies supported the scheme shown in Figure 5.1 [5–7]. High rate constants for the reactions of the carbonate radical with biomolecules as the potential cell target at physiological pH were determined (discussed below). Thus, the carbonate radical can cause significant damage *in vivo* (Fig. 5.1). Among the different carbon species, the toxicity of CO_2 has also been reviewed [8]. Nitrogen dioxide can be produced from several physiological routes such as nitric oxide auto-oxidation and peroxidase-catalyzed reactions [2]. The selectivity of nitrogen dioxide with biomolecules is discussed below. The studies on sulfate and phosphate radicals are also presented because of their involvement in treating pollutants using advanced oxidation processes. Moreover, both sulfate and phosphate radicals are one-electron oxidants and are precursors of organic radicals of biologically relevant molecules, such as peptides. Additionally, the aqueous reactions of sulfate radicals are relevant to atmospheric chemistry [9, 10].

Oxidation of Amino Acids, Peptides, and Proteins: Kinetics and Mechanism, First Edition.
Virender K. Sharma.

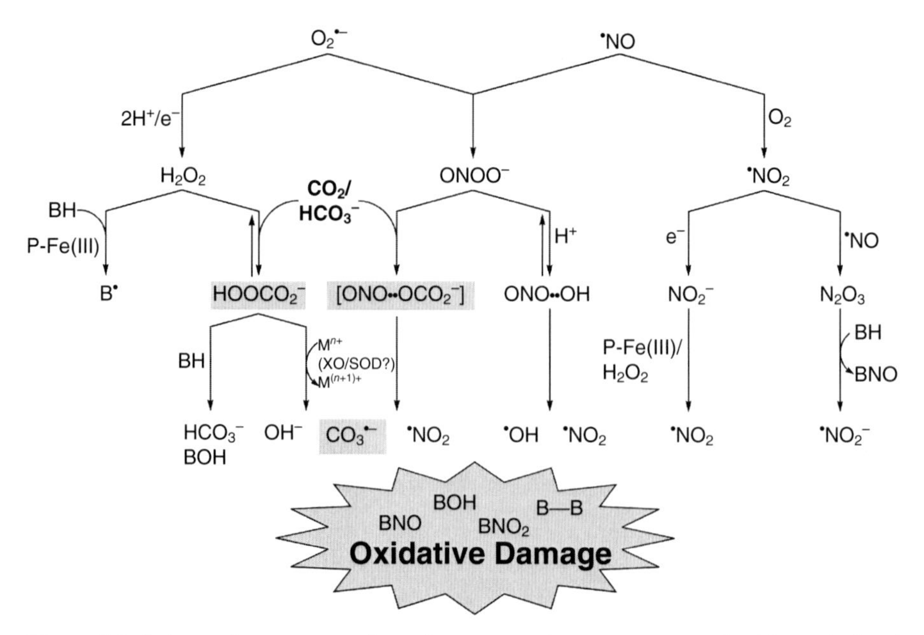

Figure 5.1. Schematic representation of sources and consequences of oxidants derived from bicarbonate buffer. The reactions were not balanced and some intermediates were omitted for clarity. XO, SOD and P-Fe(III), BH, and M^{n+} represent xanthine oxidase, superoxide dismutase 1, hemoproteins in the iron(III) state, general biomolecule, and transition metal ion, respectively (adapted from Medinas et al. [2] with the permission of IUBMB).

5.1 CARBON SPECIES

5.1.1 Carbonate Radical

5.1.1.1 Generation and Properties. The peroxidase activity of the mutated superoxide dismutase 1 (SOD-1) is one possibility of the production of the carbonate radical ($CO_3^{\bullet-}$). The proposed mechanism given in Figure 5.2 combines the results and suggestions of several studies [2, 11–14]. The reaction of SOD-1 with H_2O_2 proceeds in steps, which ultimately forms an enzyme-bound oxidant (e.g., $Cu(O) \leftrightarrow Cu(OH^{\bullet})^{2+} \leftrightarrow Cu^{3+}$) at the enzyme active site [15]. The oxidation of one or more histidine residues at the SOD-1 active site is promoted by this oxidant [16]. Formation of the hypothetical complex of Cu(I)-hydrogen peroxide in the mechanism reacts with CO_2 to form an enzyme-Cu(I)-bound peroxymonocarbonate (HCO_4^-), which reduces to produce a diffusible $CO_3^{\bullet-}$ (Fig. 5.2). The enzyme is recycled to the Cu(II) state. The formation of the complex is supported by the accelerating effects of the bicarbonate buffer on the oxidations of transition metal ions and enzyme–metal centers [17–20]. The formation of peroxymonocarbonate as a feasible

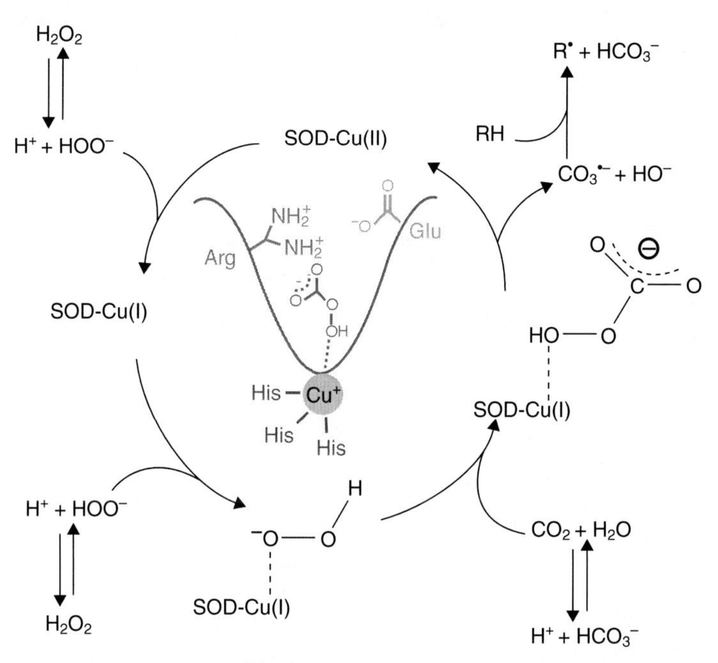

Figure 5.2. Mechanism proposed for carbonate radical production during the peroxidase activity of superoxide dismutase 1. The inset shows a pictorial view of the enzyme-Cu(I)-bound peroxymonocarbonate intermediate (adapted from Medinas et al. [2] with the permission of IUBMB). See color insert.

biological oxidant was suggested earlier [21, 22]. A xanthine oxidase (XO) turnover process involving peroxymonocarbonate as the precursor of $CO_3^{\bullet-}$ has also been proposed (see below).

XO, a complex enzyme, consists of two identical subunits each composed of one molybdenum atom, one of flavin adenine dinucleotide (FAD), and two nonidentical iron sulfur centers [2]. The generation of $CO_3^{\bullet-}$ by XO through the Fenton reaction was first proposed in the 1970s to explain the detection of low-level chemiluminescence from incubations of XO, acetaldehyde, and bicarbonate at pH 10.2 [23]. Recently, electron paramagnetic resonance (EPR) measurements provided strong evidence for the production of the $CO_3^{\bullet-}$ radical in the XO-mediated oxidation of xanthine, acetaldehyde, and hypoxanthine at pH 6.9–8.1 [1, 13, 22, 24]. A generalized scheme was proposed based on the many redox centers of XO involved in the electron flow from the oxidizing species to result in the production/escape of the intermediates, which are partially protonated/reduced, such as HO_2^- and $O_2^{\bullet-}$ (Fig. 5.3). The HO_2^- anion may react with the surrounding CO_2 to produce peroxymonocarbonate. This species is then reduced at the active site to $CO_3^{\bullet-}$ (Fig. 5.3). Another possibility is the peroxymonocarbonate leaves the active site and behaves as a two-electron oxidant.

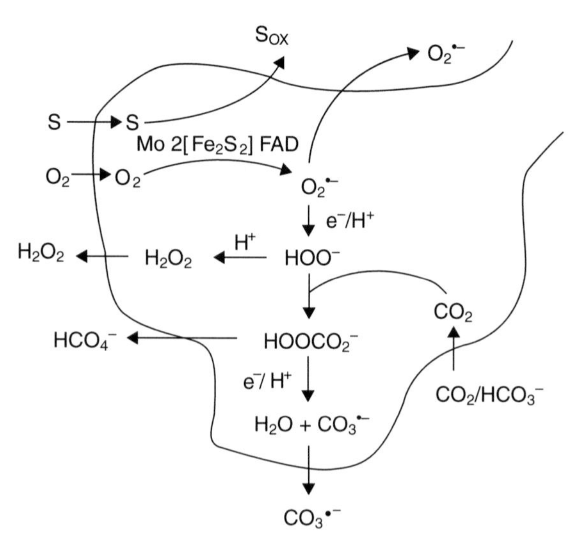

Figure 5.3. Schematic representation of a route proposed for carbonate radical anion production during XO turnover in biological environments (adapted from Bonini et al. [13] with the permission of the American Society for Biochemistry and Molecular Biology).

In the laboratory, $CO_3^{\bullet-}$ is produced by a one-electron oxidation of the carbonate or bicarbonate ion. The methods include the radiolysis of solid salt matrices and aqueous solutions, oxidation of HCO_3^- by the sulfate radical anion, photolysis of carbonatoamine complexes, and flash photolysis of carbonate solutions [25–31]. Generally, $CO_3^{\bullet-}$ is generated by the reactions of $^{\bullet}OH$ with either CO_3^{2-} or HCO_3^- (reactions 5.1 and 5.2):

$$^{\bullet}OH + CO_3^{2-} \rightarrow CO_3^{\bullet-} + OH^- \quad k_1 = 3.0 \times 10^8 /M/s\,(25°C)\,[32] \tag{5.1}$$

$$^{\bullet}OH + HCO_3^- \rightarrow CO_3^{\bullet-} + H_2O \quad k_2 = 8.5 \times 10^6 /M/s\,(25°C)\,[32]. \tag{5.2}$$

The reaction of $^{\bullet}OH$ with H_2CO_3 has also shown to produce $CO_3^{\bullet-}$ (Eq. 5.3):

$$^{\bullet}OH + H_2CO_3 \rightarrow CO_3^{\bullet-} + H^+ + H_2O \quad k_3 = 7.0 \times 10^4 /M/s\,(5°C)\,[33]. \tag{5.3}$$

The characteristic absorption spectrum of $CO_3^{\bullet-}$ ($\varepsilon_{600\,nm} = 1850 \pm 50\,M/cm$) at different pH values is shown in Figure 5.4 [33]. A similar spectrum in the pH range of the study suggests $HCO_3^{\bullet-}$ is acidic (Eq. 5.4):

$$HCO_3^{\bullet} \rightleftharpoons H^+ + CO_3^{\bullet-} \quad pK_4 < 0. \tag{5.4}$$

The reactions of $CO_3^{\bullet-}$ with thiocyanate, iodide, and ferrocyanide ions as a function of pH were conducted to support $HCO_3^{\bullet}$ with a $pK_4 < 0$ [34]. This estimated value was supported by the calculated pK_4 of –4.1 using high-level *ab initio* calculations [35]. The structure of $CO_3^{\bullet-}$ has been determined using

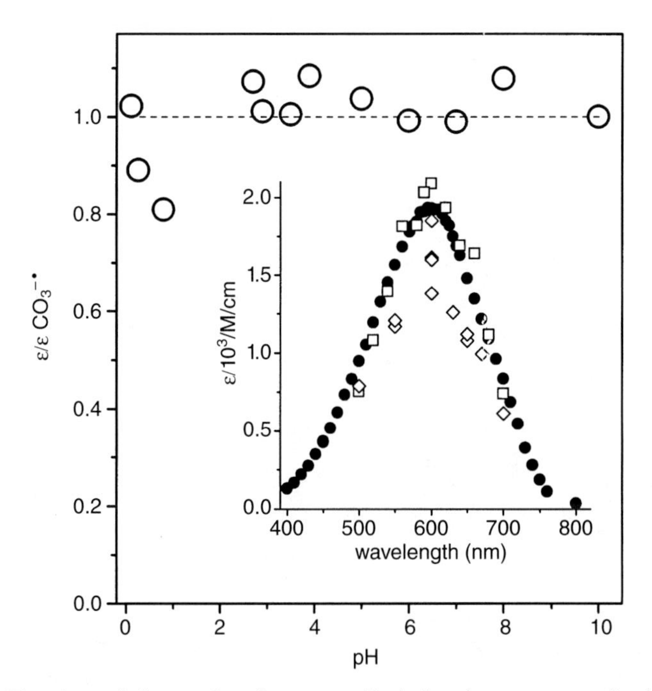

Figure 5.4. Fraction of the total carbonate radical that is present as the basic form, $CO_3^{\bullet-}$, as a function of the pH. Inset: the spectrum observed at pH 10 (filled circles), at pH 2.7 (open squares), and pH 0.8 (open diamonds) (adapted from [33] with the permission of the American Chemical Society).

resonance Raman spectroscopy [36]. The radical has C_{2v} symmetry, suggesting some distortion from the predicted D_{3h} structure. This study found no change in the Raman spectrum between pH 7.5 and 13.5, consistent with the absorption spectrum shown in Figure 5.4. The EPR spectrum with *g*-factor and line width values of 2.0113 and 5.0 G, respectively, has also been reported [26, 37]. The spectral studies concluded the carbonate radical anion is an electrophilic oxygen-centered radical.

The self-recombination of $CO_3^{\bullet-}$ is second order [38, 39]. The kinetics of the reaction was studied with temperatures of up to 250°C to understand the mechanism, which suggests oxygen transfer reactions [38]. Moreover, the chemistry of $CO_3^{\bullet-}$ at high temperature is relevant in designing an efficient nuclear reactor and in destroying pollutants under supercritical water conditions [39]. The mechanism involves a pre-equilibrium formation of the dimer, $C_2O_6^{2-}$, which dissociates to the peroxymonocarbonate anion and CO_2 (reaction 5.5). The $CO_3^{\bullet-}$ radical reacts with the peroxymonocarbonate anion, yielding a peroxymonocarbonate radical (reaction 5.6), which can further react with $CO_3^{\bullet-}$ to form $C_2O_7^{2-}$ (reaction 5.7):

$$CO_3^{\bullet-} + CO_3^{\bullet-} \rightleftharpoons C_2O_6^{2-} \rightarrow O_2COO^{2-} + CO_2 \quad k_5 = 4.25\times10^6/\text{M/s} \qquad (5.5)$$

$$CO_3^{\bullet-} + CO_4^{2-} \rightarrow CO_4^{\bullet-} + CO_4^{2-} \quad (5.6)$$

$$CO_3^{\bullet-} + CO_4^{\bullet-} \rightarrow C_2O_7^{2-} \quad k_7 = 1.0 \times 10^9 /M/s. \quad (5.7)$$

5.1.1.2 Reactivity. The reduction potential of $CO_3^{\bullet-}/CO_3^{2-}$ is high and the determined value at pH 12.5 was 1.58 V [34, 40]. The calculated value obtained using *ab initio* was in modest agreement with this experimental value [35]. The monomeric $CO_3^{\bullet-}$ form of the radical was also suggested [35]. A value of 1.78 V was determined for the $CO_3^{\bullet-},H^+/HCO_3^-$ pair at pH 7.0 [41]. This indicates rapid reactions with biomolecules, causing possible damage to protein and nucleic acids [1]. Literature shows determination of second-order rate constants for the reactions of $CO_3^{\bullet-}$ with a wide range of compounds [25, 31, 39, 42–50]. Table 5.1 gives the rate constants for some of the inorganic species. The measured rate constants ranged from $\leq 1.0 \times 10^6$/M/s to 8.0×10^9/M/s [25, 40, 42, 44, 47, 48, 51]. The rate constants for reactions of $CO_3^{\bullet-}$ with metal complexes increased with a decrease in the reduction potential of the metal complex. This is expected for electron transfer reactions. The activation energies of the studied compounds ranged from −11.4 to 18.8 kJ/mol [40]. The effect of temperature on the rate constants was also determined to know the mechanism of the reactions. The pre-exponential factors ranged from 6.4 to 9.7. The decrease in activation energies of the inorganic species with an increase in the driving force for the reactions suggests an outer-sphere electron transfer.

TABLE 5.1. Rate Constants and Arrhenius Parameters for the Reactions of $CO_3^{\bullet-}$ with Various Inorganic Species [25, 40, 42, 44, 47, 48, 51]

Species	pH	k_{298} (/M/s)	E_a	log A
O_2	11.0	$\leq 1.0 \times 10^6$	–	–
H_2O_2	–	8.0×10^5	–	–
HO_2^-	–	6.0×10^7	–	–
O_2^-	–	$0.4–1.5 \times 10^9$	–	–
O_3^-	–	6.0×10^7	–	–
I^-	11.4	2.5×10^8	-11.4 ± 0.6	6.4
$NH_2^{\bullet}$	11.0	8.0×10^9	–	–
NO_2^-	11.4	6.6×10^5	7.1 ± 1.1	7.1
ClO_2^-	11.4	3.4×10^7	6.8 ± 0.3	8.7
SO_3^{2-}	11.4	2.9×10^7	3.9 ± 1.3	8.2
SCN^-	11.4	1.6×10^5	18.8 ± 4.6	8.5
$Fe(CN)_6^{4-}$	11.4	3.6×10^8	6.5 ± 1.4	9.7
$Mo(CN)_8^{4-}$	11.2	3.5×10^7	8.4 ± 1.1	8.4
$W(CN)_8^{4-}$	11.2	2.4×10^8	3.3 ± 0.8	8.9

The rate constants for the reactions of $CO_3^{\bullet-}$ with organic compounds were used to derive the quantitative structure–activity relationships (QSARs) (Eqs. 5.8–5.10) [51]:

Monosubstituted benzene derivatives:

$$\log k = 3.4(\pm 0.8) - 3.6(\pm 0.8)\sigma_p^+ \quad n = 7, \quad r^2 = 0.96, \quad s = 0.45 \tag{5.8}$$

p-Substituted phenoxide ions:

$$\log k = 8.40(\pm 0.02) - 0.91(\pm 0.03)\sigma_p^+ \quad n = 8, \quad r^2 = 0.999, \quad s = 0.019 \tag{5.9}$$

p-Substituted anilines:

$$\log k = 8.710(\pm 0.06) - 0.98(\pm 0.15)\sigma_p^+ \quad n = 7, \quad r^2 = 0.98, \quad s = 0.052. \tag{5.10}$$

The relationships of the rate constants with molar Gibbs free energy change of electron transfer were obtained for p-substituted phenoxide ions and anilines (Fig. 5.5) [51]. The reaction with aniline was approximately one order of magnitude faster than the phenoxide ions, which is expected if reactions take place through inner-sphere mechanism [51]. This is supported by an earlier study in which low activation energies of reactions of $CO_3^{\bullet-}$ with organic compounds were determined [40]. This further suggests that Gibbs molar

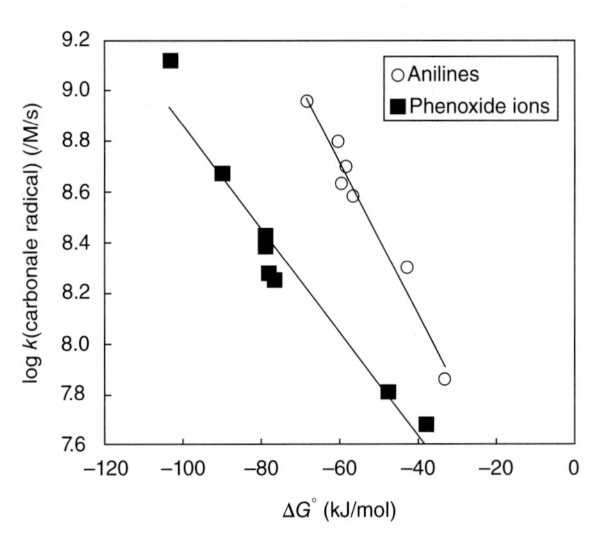

Figure 5.5. Linear correlations of rate constants for the oxidation of substituted phenoxide anions and anilines by the carbonate radical. Regression lines correspond to the following equations: (1) for phenoxide ions, $\log k = 6.8(\pm 0.4) - 0.021(\pm 0.005)\ \Delta G^{\circ\prime}$ ($n = 8$, $r^2 = 0.94$, $s = 0.12$), and (2) for anilines, $\log k = 6.9(\pm 0.3) - 0.030(\pm 0.006)\ \Delta G^{\circ\prime}$ ($n = 7$, $r^2 = 0.97$, $s = 0.07$) (adapted from Canonica and Tratnyek [51] with the permission of the Society of Environmental Toxicology and Chemistry).

free energy change does not fully controlled the reaction rates of the reactions with organic compounds. The QSARs for the effect of substituents on the rates for the same categories of substrates with different oxidants were developed [51]. The sensitivity to substituent groups increased in the order $^{\bullet}OH < CO_3^{\bullet-} \approx {}^1O_2 < ClO_2 \approx O_3$. Furthermore, this order was correlated approximately with the inverse order of reduction potentials for one-electron transfer. This suggests that the magnitude of second-order rate constants for oxidation is defined by the reaction mechanism.

The rate constants of the reactions of $CO_3^{\bullet-}$ with carbohydrates were determined in the range of 10^5–10^7/M/s and the rates increased with increasing pH from 8.0 to 13.0 [52]. This indicates that carbohydrates were activated toward oxidation due to deprotonation. At low pH, the mechanism involved hydrogen abstraction and/or electron/proton transfer [52]. However, deprotonated carbohydrates reacted via electron transfer at high pH [52]. Table 5.2 represents the rate constants of the reactions of $CO_3^{\bullet-}$ with compounds of biological interest. The rate constants vary from 4.0×10^5/M/s to 1.4×10^9/M/s [1, 27, 43, 44, 49, 53–61]. The activation energies for the reactions of $CO_3^{\bullet-}$ with organic compounds were generally low and also independent of the driving force, and hence, oxidation occurred via inner-sphere mechanisms [40].

Butyl amine had the lowest reactivity, while the reducing substrate, ascorbate, had the highest reactivity. Primary amines have been reported to react primarily by hydrogen abstraction [40, 42, 62]. The reactivity of the anionic form of Gly at pH 11 was higher than the neutral species at pH 7.5 (Table 5.2). Aromatic amines appear to react by electron transfer, suggested by their higher rate constants with $CO_3^{\bullet-}$ (Table 5.2). Cys reacted by electron transfer from the $-S^-$ group [40]. Reactions of Cys and Met most likely involve an addition to the sulfur [40]. Significantly, the sulfur-containing amino acids reacted similarly with $CO_3^{\bullet-}$; however, Met showed a negative temperature dependence, while the rates for the reaction of cystine with $CO_3^{\bullet-}$ increased with an increase in temperature. The observed results suggest the formation of the perthiyl radical through an intermediate complex in the cystine reaction with $CO_3^{\bullet-}$ (Eqs. 5.11, 5.12):

$$CO_3^{\bullet-} + RSSR \rightleftharpoons (RSSR)^{\bullet}CO_3^- \tag{5.11}$$

$$(RSSR)^{\bullet}CO_3^- + OH^- \rightarrow RSS^{\bullet} + ROH + CO_3^{2-}. \tag{5.12}$$

Comparatively, the reaction of $CO_3^{\bullet-}$ with Met initially resulted in an intermediate complex (Eq. 5.13). The decomposition of the complex requires charge separation (Eq. 5.14), a different pathway in comparison to the reaction of $CO_3^{\bullet-}$ with cystine. As the temperature increases, decomposition of the complex back to reactants is able to compete with the path of forming products. Thus, this causes negative temperature dependence for the overall reaction of $CO_3^{\bullet-}$ with Met:

$$CO_3^{\bullet-} + RSR \rightleftharpoons RSR^{\bullet}CO_3^- \tag{5.13}$$

TABLE 5.2. Second-Order Rate Constants for the Carbonate Radical Anion Reactions with Compounds [1, 27, 43, 44, 49, 53–61]

Compound	pH	k_{298} (/M/s)	E_a	log A
Ascorbate	11.4	1.4×10^9	8.3 ± 1.2	10.6
Butylamine	11.5	4.0×10^5	–	–
Triehthylamine	13.0	9.8×10^6	11.4 ± 0.9	9.0
Glycine	7.5	3.2×10^4	–	–
	11.0	1.3×10^5	–	–
Dopa-melanin	10.0	3.0×10^5	–	–
Cysteinyl-dopa-melanin	10.0	2.0×10^8	–	–
Methionine	7.0	3.6×10^7	–	–
	11.4	1.2×10^8	-9.2 ± 1.5	6.5
Cysteine	7.0	4.6×10^7	–	–
	11.4	1.8×10^8	6.4 ± 0.3	9.4
Cystine	11.4	1.2×10^7	10.5 ± 0.9	8.9
GS^-	Alkaline	7.1×10^8	–	–
Tryptophan	7.0	7.0×10^8	–	–
	11.4	4.9×10^8	-0.9 ± 0.2	8.5
Tyrosine	7.0	4.5×10^7	–	–
	11.0	1.4×10^8	–	–
Histidine	7.0	5.6×10^7	–	–
Horseradish peroxidase	8.4	1.2×10^8	–	–
	10.0	1.3×10^8	–	–
Fe(II) cytochrome *c*	8.4	1.1×10^9	–	–
	10.0	7.6×10^8	–	–
Fe(III) cytochrome *c*	8.4	5.1×10^7	–	–
	10.0	1.0×10^8	–	–
Catalase	8.4	3.7×10^9	–	–
Bovine serum albumin, native	8.8	1.7×10^8	–	–
Bovine serum albumin, denatured	8.6	2.1×10^8	–	–
$MbFe^{II}O_2$	10.0	5.7×10^7	–	–
$MbFe^{III}OH$	10.0	5.7×10^7	–	–
$MbFe^{II}NO$	10.0	5.2×10^8	–	–
$HbFe^{III}OH$	10.0	5.2×10^8	–	–
$HbFe^{II}O_2$	10.0	2.9×10^8	–	–
$HbFe^{II}NO$	10.0	5.2×10^8	–	–
Aconitase	7.4	3.0×10^8	–	–
d[G-oligo]	7.5	1.9×10^7	–	–
dGMP	7.5	6.6×10^7	–	–
dAMP	7.5	$\leq 2.8 \times 10^6$	–	–
dCMP	7.5	$\leq 1.9 \times 10^6$	–	–
dTMP	7.5	$\leq 1.7 \times 10^6$	–	–
8-Oxo-dGuo	7.5	7.9×10^8	–	–
Dichlorodihydrofluorescein ($DCFH_2$)	8.2	2.6×10^8	–	–
Dichlorofluorescein (DCF)	8.1	2.7×10^8	–	–
Dihydrorhodamine (DHR)	8.2	6.7×10^8	–	–
Rhodamine (DHR)	8.2	3.6×10^6	–	–

$$RSR^{\bullet}CO_3^- \rightarrow RSR^{\bullet+} + CO_3^{2-}. \quad (5.14)$$

The increased reactivity of cysteinyl-dopa-melanin compared to dopa-melanin with $CO_3^{\bullet-}$ (Table 5.2) is suggested to relate to the higher reactivity of the thiol molecule than the amine group. The reaction of glutathione (GSH) by $CO_3^{\bullet-}$ in alkaline solution ($k = 7.1 \times 10^8$/M/s) produced only the sulfur-centered $GS^{\bullet}$ ($GSSG^{\bullet-}$) radical [56].

The reactivity of $CO_3^{\bullet-}$ with aromatic amino acids (Trp, Tyr, and His) was ~10^8/M/s. A study of $CO_3^{\bullet-}$ with numerous tryptophan derivatives proposed an interaction of $CO_3^{\bullet-}$ with the aromatic system in the reaction mechanism [54,63]. In a biological environment, Trp and Tyr may likely be the target for $CO_3^{\bullet-}$. An example is the oxidative modification of Trp-32 in the peptide region (31–36) of hSOD1 to *N*-formylkynurenine (NFK)- and kynurenine (KYN)-containing peptides in the presence of SOD-1/H_2O_2/HCO_3^- or UV photolysis of a Co complex [64]. Based on electron spin resonance (ESR) spin trapping measurements as well as liquid chromatography–mass spectrometry (LC/MS) evidences, a mechanism shown in Figure 5.6 was proposed [64]. Initially, $CO_3^{\bullet-}$ forms a tryptophanyl radical intermediate, which then causes posttranslational modification of Trp-32 to NFK and KYN (Fig. 5.6).

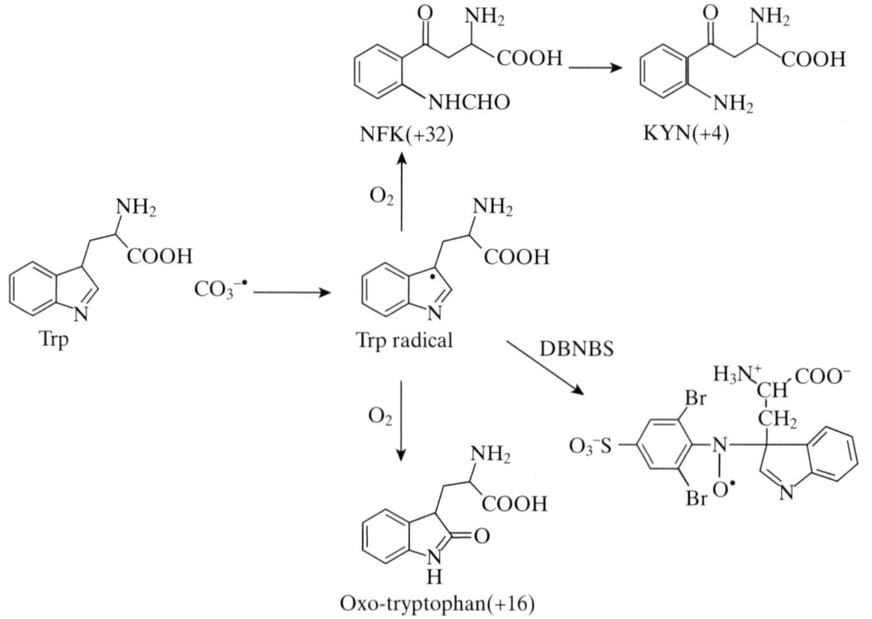

Figure 5.6. Proposed reaction pathway for oxidation modification of tryptophan by $CO_3^{\bullet-}$. N-Formylkynurenine is presumably formed from the tryptophanyl peroxy radical. The molecular weight of NFK is 32 mass units heavier than the parent peptide residue. Decarboxylation (-CO) of NFK yield a product containing kynurenine that is 4 mass units heavier. DBNBS, 3,5-dibromo-4-nitrosobenzenesulfonic acid (redrawn from Zhang et al. [64] with the permission of Elsevier Inc.).

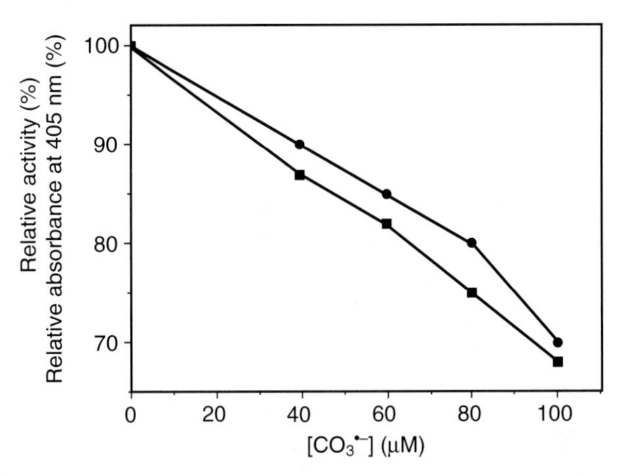

Figure 5.7. Relative activity (black squares) and relative absorbance at 405 nm (black circles) of HRP versus [$CO_3^{\bullet-}$]; [HRP] = 2.3 μM, pH = 8.4, dose rate = 1.1 Gy/min (adapted from Gebicka et al. [57] with the permission of Springer Americas).

The rate constants of the reactions of horseradish peroxidase (HRP) and cytochrome *c* are similar at pH 8.0–8.4 and at pH 10.0 (Table 5.2) [55, 57]. The rate constant for catalase has been determined only at pH 8.4 and was on the order of 10^9/M/s [57]. The pulse radiolysis experiments demonstrated $CO_3^{\bullet-}$ did not react directly with the heme iron of HRP, cytochrome *c*, and catalase. The loss of activity was 30% and 20% for HRP and catalase, respectively, after reactions with 100 μM of $CO_3^{\bullet-}$. A decrease in the absorbance of the Soret region of HRP (and catalase) with an increase in the cumulative concentration of $CO_3^{\bullet-}$ was observed (Fig. 5.7) [57]. However, the decrease was lower than the decrease of the enzyme activity. This suggests destruction of the heme center was the main, but not exclusive, cause for the loss of enzyme activity. Significantly, the changes in absorbances were not observed in the Soret region of HRP, cytochrome *c*, and catalase after pulse radiolysis experiments. This indicates the production of globin radicals decayed completely via dimerization/disproportionation processes under the studied conditions of the experiments (Fig. 5.7).

The reactivities of $CO_3^{\bullet-}$ with native and denatured bovine serum albumin (BSA) were similar (Table 5.2) and may likely be a result of the reactions with all of the solvent-exposed reactive residues, including TrpH and TyrOH (Eq. 5.15). The characteristic absorption maximum of the $Trp^{\bullet}$ radical in the visible region was observed (Fig. 5.8) [59]. The decay of $Trp^{\bullet}$ occurred as a result of electron transfer from TyrOH of either the same (intramolecular, Eq. 5.16a) or different albumin molecule (intermolecular, Eq. 5.16b), giving $TyrO^{\bullet}$. The decay rate of the absorption band at ~510 nm for $Trp^{\bullet}$ was equal to the formation rate of the absorption band at 410 nm for $TyrO^{\bullet}$:

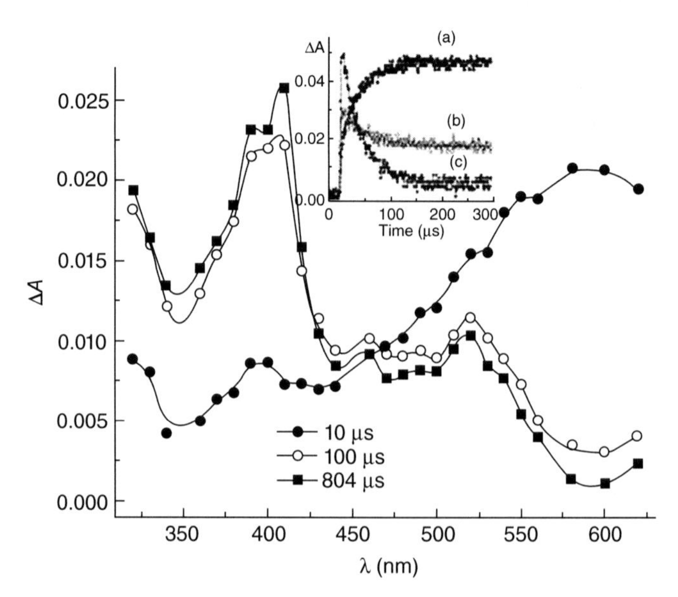

Figure 5.8. Transient absorption spectrum obtained on pulse radiolysis of N_2O-saturated aqueous solution containing native BSA (1×10^{-4} M) and HCO_3 (0.5 M) at pH 8.8. Inset: transient absorption profiles under identical conditions at (a) 410 nm, (b) 510 nm, and (c) 600 nm. Dose = 25 Gy (adapted from Joshi and Mukherjee [59] with the permission of Elsevier Inc.).

$$\text{BSA (TrpH, TyrOH ....)} + CO_3^{\bullet-} \text{ (600 nm)} \rightarrow \text{BSA}^{\bullet} + CO_3^{2-} + H^+ \qquad (5.15)$$

$$\text{BSA}_1 \text{ (Trp}^{\bullet}\text{, TyrOH, ...) (510 nm)} \rightarrow \text{BSA}_1 \text{ (TrpH, TyrO}^{\bullet}\text{, ...) (410 nm)} \qquad (5.16a)$$

$$\begin{aligned} &\text{BSA}_1 \text{ (Trp}^{\bullet}\text{, TyrOH, ...)} + \text{BSA}_2 \text{ (TrpH, TyrOH, ...)} \\ &\rightarrow \text{BSA}_1 \text{ (TrpH, TyrOH ....)} + \text{BSA}_2 \text{ (TrpH, TyrO}^{\bullet} \text{ ....).} \end{aligned} \qquad (5.16b)$$

The formation of $BSA^{\bullet}$ was also seen in the oxidation of oxymyoglobin and oxyhemoglobin by $CO_3^{\bullet-}$, which had rate constants in the range of 5.7×10^7/M/s to 5.2×10^7/M/s [58]. The $BSA^{\bullet}$ radical further oxidized the Fe^{II} center through both intra- and intermolecular forces. The reactions of $CO_3^{\bullet-}$ with nitrosyl(II) hemoglobin ($HbFe^{II}NO$) and nitrosyl(II)myoglobin ($MbFe^{II}NO$) have also been carried out (Table 5.2). Both reactions proceed through two steps (Eqs. 5.17 and 5.18). The outer-sphere oxidation of nitrosyl iron(II) proteins proceeds to their corresponding nitrosyl iron(III) forms (Eq. 5.17), followed by dissociation to $NO^{\bullet}$ (Eq. 5.18):

$$MbFe^{II}NO + CO_3^{\bullet-} \rightarrow MbFe^{III}NO + CO_3^{2-} \qquad (5.17)$$

$$MbFe^{III}NO + H_2O \rightarrow MbFe^{III}OH_2 + NO^{\bullet}. \qquad (5.18)$$

$CO_3^{\bullet-}$ is capable of selectively oxidizing guanines in the self-complementary oligonucleotide duplex d(AACGCGAATTCGCGTT), dissolved in air-equilibrated aqueous buffer solution (Table 5.2). Oxidative guanine base damage to DNA was demonstrated [60]. The oxidation of two common fluorescent probes, dichlorodihydrofluorescein ($DCFH_2$) and dihydrorhodamine (DHR), and their oxidized forms (DHR and Rh) by $CO_3^{\bullet-}$ have also been studied (Table 5.2) [61]. Oxidations of $DCFH_2$ and DHR were suggested to occur via a two-step process (e.g., Eqs. 5.19, 5.20):

$$DCFH_2 + CO_3^{\bullet-} \rightarrow DCFH^{\bullet} + HCO_3^- \qquad (5.19)$$

$$2DCFH^{\bullet} \rightarrow DCF + DCFH_2. \qquad (5.20)$$

The reactions of $CO_3^{\bullet-}$ with synthetic nitroxide antioxidants have been studied [65]. The reactivity of $CO_3^{\bullet-}$ radicals was independent of the ring size and side chain of cyclic nitroxides. The second-order rate constants were determined in the range of $(2–6) \times 10^8$/M/s at pH > 9.0. In consideration of the basic side chain of nitroxide, 4-amino-2,2,6,6-tetramethylpiperidine-1-oxyl, 3-(aminomethyl)-proxyl, the second-order rate constant was independent of the pH in the studied range from 9.1 to 11.7. The nitroxides were shown to be the most effective metal-independent scavengers of $CO_3^{\bullet-}$ radicals [65].

5.1.2 Peroxymonocarbonate

Peroxymonocarbonate (HCO_4^-) has been a known oxidant for many years [21]. The equilibrium reaction of bicarbonate with hydrogen peroxide rapidly forms HCO_4^- ions at near neutral pH ($t_{1/2} \approx 300$ seconds) (Eq. 5.21) [22]:

$$HCO_3^- + H_2O_2 \rightleftharpoons HCO_4^- + H_2O \quad k_{21} = 0.33. \qquad (5.21)$$

A shift in equilibrium to the right was determined when more soluble sources of bicarbonate (e.g., NH_4HCO_3) and an alcohol cosolvent were used [22]. HCO_4^- is a strong oxidant in aqueous solutions ($E^\circ(HCO_4^-/HCO_3^-) =$ 1.8 V vs. normal hydrogen electrode [NHE]). HCO_4^- has been characterized by vibrational spectroscopy and X-ray crystallography ($KHCO_4 \cdot H_2O_2$) [21, 66]. It can be classified as an anionic peracid having a structural formula of $HOOCO_2^-$ [66]. During the decomposition of HCO_4^-, chemiluminescence was observed [67–69].

5.1.2.1 Reactivity. The reactivities of HCO_4^- with organic compounds in aqueous and nonaqueous solvents have been studied [22, 24, 70–72]. The second-order rate constants for the oxidation of amines and thiols by H_2O_2 and HCO_4^- are given in Table 5.3 [72–74]. Generally, HCO_4^- reacts about twofold faster than H_2O_2. A detailed study on the oxidation of sulfides by HCO_4^- proposed a solvent-assisted S_N2 mechanism. Water as a solvent accelerates the oxygen transfer to the nucleophilic sulfide by assisting in the displacement of

TABLE 5.3. Second-Order Rate Constants for Oxidation of Compounds by H_2O_2 and HCO_4^- at 25°C [72–74]

Compound	$k(H_2O_2)$ (/M/s)	$k(HCO_4^-)$ (/M/s)	$k(HCO_4^-)/k(H_2O_2)$
N-Methylmorpholine[a]	3.5×10^{-5}	1.6×10^{-2}	457
N,N-Dimethylbenzylamine[b]	1.0×10^{-4}	4.2×10^{-2}	420
Cysteine[c]	1.7×10^{1}	9.9×10^{2}	58
Glutathione[c]	1.8×10^{1}	1.9×10^{3}	105
N-Acetylcysteine[c]	1.7×10^{1}	1.5×10^{3}	88
Methionine[a]	7.5×10^{-3}	4.8×10^{-1}	640

[a] Solvent: D_2O.
[b] Solvent: acetone-d_6/D_2O.
[c] Values are for thiolate anion.

the carbonate leaving group [22]. The oxidation of methyl disulfides by HCO_4^- resulted in the formation of thiosulfone and sulfonate products [73]. The mechanism of the oxidation of aryl sulfide by HCO_4^- has also been studied at the aqueous/cationic micellar interface [24]. The cetyltrimethylammonium carbonate ($CTAHCO_3$) significantly enhances the overall oxidation rate of sulfides compared to the addition to cetyltrimethylammonium chloride (CTACl) and bromide (CTABr) [24]. The oxidation of aliphatic amines by HCO_4^- was over 400-fold greater than H_2O_2 alone (Table 5.3) [72]. The enthalpy of activation was significantly lower for the oxidation by HCO_4^- than that of H_2O_2, resulting in an increased rate. The secondary aliphatic amines were oxidized to corresponding nitrones, while tertiary amines gave corresponding *N*-oxides in high yields.

The scheme for the activation of H_2O_2 by HCO_3^-/CO_2 in the oxidation of Cys and Met is presented in Figure 5.9 [73, 74]. In the scheme, k_1 is the second-order rate constant for the oxidations by HCO_4^- (see Table 5.3). The left side of the scheme is the proposed perhydration of CO_2 to form percarbonic acid in equilibrium with its conjugate base, HCO_4^-. The right side of the scheme in Figure 5.9 is the equilibrium of CO_2/HCO_3^-, which may be catalyzed by carbonic anhydrase and various model complexes. The scheme demonstrated the reaction of HCO_4^- with substrate S via O atom transfer formed back to HCO_3^- (top portion). This supports the role of HCO_3^- as a catalyst in the oxidation of Met and Cys. The possible involvement of HCO_4^- in protein oxidation has also been explored [73, 74]. For example, *in vitro* and *in vivo* studies have demonstrated the enhancement of the oxidation of the human neutrophil elastase inhibitor αI-PI in the presence of HCO_3^- [74]. The loss of inhibitory activity was essentially complete at the longest reaction time when the highest concentration of HCO_3^- was used. Evidence was provided for the oxidation of two methionines (Met351 and Met358) in the reactive center loop of αI-PI. A

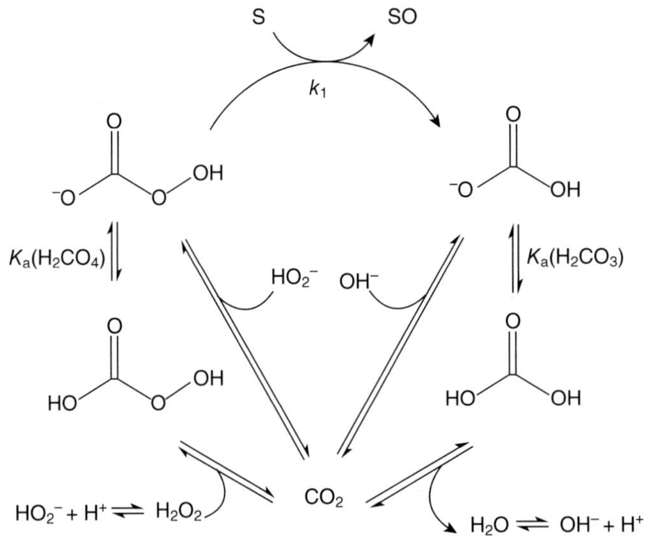

Figure 5.9. Scheme for the activation of H_2O_2 by HCO_3^-/CO_2 in the oxidation of cysteine and methionine. Hydration (right side) and the analogous perhydration (left side) pathways are shown. At physiological pH values, the pathway involving direct reaction of HO_2^- with CO_2 is likely to be significant (center) (adapted from Richardson et al. [74] with the permission of Elsevier Inc.).

significant oxidation of a third methionine (Met226) was also observed. An additional example is the catalytic decomposition of HCO_4^- by catalase [73].

5.1.3 Carboxyl Radical

The production of the carboxyl radical ($CO_2^{\bullet-}$) during hydralazine and carbon tetrachloride metabolism has been reported [75, 76]. The α-keto acid decarboxylation promoted by hydrogen peroxide/transition metal ions or peroxynitrite is the likely source of $CO_2^{\bullet-}$ at neutral pH [77]. The carboxyl radical is a powerful one-electron reductant with a redox potential of –1.85 V ($CO_2/CO_2^{\bullet-}$, pH 7.0) [78, 79]. The reduction of the disulfide bond in aponeocarzinostatin, the aporiboflavin-binding protein, and bovine immunoglobin has been investigated in detail [80]. These proteins do not contain free Cys in the native form. The reactions yielded protein-bound cysteine-free thiols under γ-ray irradiation, which were pH and protein concentration dependent. The efficiency of the chain increased upon acidification of the solution and decreased sharply below pH 3.6. The major protein radical species formed was protonated disulfide radical anion under acidic conditions. This radical decayed to thiyl radical, which could react with formate to propagate the chain reaction with the generation of $CO_2^{\bullet-}$. The decay of the disulfide radical anion at pH 8 was through competing intermolecular and/or intramolecular pathways. Overall, the pulse radiolysis experiments showed $CO_2^{\bullet-}$ was also capable of inducing a rapid one-electron oxidation of thiols and tyrosine phenolic groups in addition to reducing exposed disulfide bonds of the protein [80].

5.2 NITROGEN SPECIES

5.2.1 Nitrogen Monoxide

Nitrogen monoxide (nitric oxide, $^{\bullet}NO$) was previously considered a toxic to the environment, destroying ozone and causing acid rain [81, 82]. In 1987, $^{\bullet}NO$ was discovered to be formed by enzymatic reactions in a variety of mammalian cells [83]. This has led to numerous biological chemistry studies on this molecule [82, 84–91]. The generation of $^{\bullet}NO$ at low physiological levels is involved in processes such as blood pressure modulation, immune system control, peristalsis, and neurotransmission [92–96]. Sources of the production of $^{\bullet}NO$ in biological systems are given in Figure 5.10. These include (1) reduction of NO_2^- with various metalloprotiens, (2) reduction of organic nitrates, (3) nonenzymatic reduction of NO_2^-, (4) oxidation of L-arginine to L-citrulline by nitric oxide synthase (NOS), and (5) nitrite reduction by nitrite reductase (NOR). $^{\bullet}NO$ may be transformed to NO_3^-, peroxynitrite, and N_2O by the reactions mediated by the metal ion center of proteins (Eqs. 5.22–5.24) [84, 97, 98]:

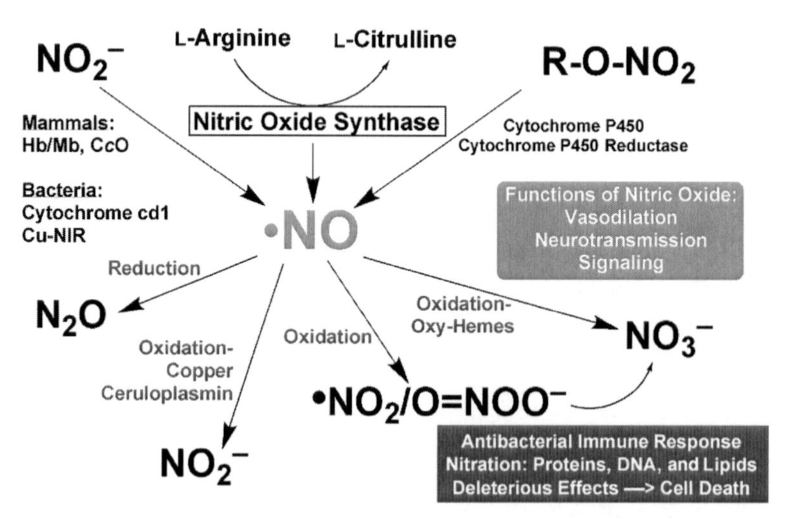

Figure 5.10. Chemistry of $\bullet NO_{(g)}$: sources of production in biological systems, its functions and roles, and multiple oxidation and reduction pathways, leading to the formation of a variety of species that may effect a diverse range of biomolecules and functions (adapted from Schopfer et al. [84] with the permission of the American Chemical Society). See color insert.

$$M^{n+}(O_2) + {}^{\bullet}NO \rightarrow M^{n+1} + NO_3^- \tag{5.22}$$

$$M^{n+}(O_2) + {}^{\bullet}NO \rightarrow M^{n+1}(O = NOO^-) \tag{5.23}$$

$$2{}^{\bullet}NO + 2e^- \text{ (from metal ions)} + 2H^+ \rightarrow N_2O + H_2O. \tag{5.24}$$

An example of reaction (5.22) is the rapid formation of $Fe^{III}OON{=}O$ from the reaction of oxyheme ($Fe^{III}O_2^{\bullet -}$) with ${}^{\bullet}NO_{(g)}$, which then decomposes immediately to form the NO_3^- anion [92]. A study on the binding of NO in Fe(II) and Fe(III) heme proteins using spectroscopic and density functional theory (DFT) calculations has recently been reported [99–103]. The structure, chemical properties, and biological action of dinitrosyl iron complexes have recently been reviewed [104].

The half-life of ${}^{\bullet}NO$ is determined by the concentration of oxygen [105]. The reaction between ${}^{\bullet}NO$ and O_2 is more rapid within membranes than in the surrounding aqueous medium [106]. Half-lives of ${}^{\bullet}NO$ have been estimated in the range of seconds to minutes [105]. ${}^{\bullet}NO$ is difficult to oxidize to NO^+ ($E°(NO^+/{}^{\bullet}NO) = 1.2V$). However, NO^- is more easily formed, which is of biological importance. Recently, the reaction between thiols, ${}^{\bullet}NO$, and O_2 has been studied in detail *in vitro* [107]. Three pathways from the reaction, given in Figure 5.11, were proposed for the formation of S-nitrosothiols and thiol disulfides. These pathways were based on kinetic, stoichiometric, and scavenger studies [108–113]. Pathway 1 involves the reaction of ${}^{\bullet}NO$ and O_2 to yield N_2O_3. Pathway 2 associated with the radical–radical combination of the ${}^{\bullet}NO$

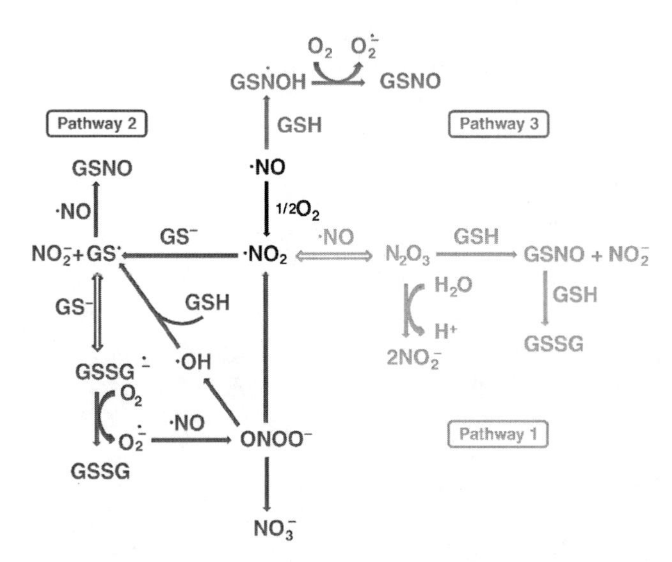

Figure 5.11. Proposed pathways of the reaction among GSH, NO, and oxygen (adapted from Keszler et al. [107] with the permission of Elsevier Inc.). See color insert.

and thiyl radicals, which are formed from the one-electron oxidation of thiol by NO_2. Pathway 3 involves the direct combination of $^{\bullet}NO$ and thiol to form a putative intermediate radical (RSNOH). The following conclusions were drawn using both experiment and kinetic simulation: (1) the simultaneous occurrence of radical and nonradical pathways for S-nitrosation; (2) the direct addition of $^{\bullet}NO$ to thiol for S-nitrosation is not significant; and (3) S-nitrosation of GSH is not catalyzed or enhanced by the hydrophobic environment of protein [107]. Therefore, any possible reaction that scavenges the thiyl radical would favor N_2O_3-mediated nitrosation over $GS^{\bullet}$-mediated nitrosation. Furthermore, significant levels of SOD, GSH, other thiols, oxygen, and any other targets for $GS^{\bullet}$ in cells may facilitate the nitrosation of GSH by N_2O_3. Also, any mechanism associated with enhancing the formation of the thiyl radical, such as action of peroxidases and increased formation of superoxide, may support nitrosation through thiyl radicals. Conclusions (2) and (3) may not be valid for some individual proteins containing thiol residues in hydrophobic protein environments due to the possible reactivity with nitrogen oxides to cause S-nitrosation.

A recent study on the reactivity of thiyl radicals derived from cysteine, GSH, and penicillamine with $^{\bullet}NO$ showed a near diffusion-controlled rate constant ($k = (2\text{–}3) \times 10^9$/M/s) [114]. The high rate constants suggest the formation of *S*-nitrosothiols in biological systems. Reactions of thiyl radicals with cyclic nitroxides have also been studied [115]. The rate constants of their reactions were determined to be $(5\text{–}7) \times 10^8$/M/s at pH 5–7 and were independent of the structure of the thiyl radical and the nitroxide. The main products of the reactions were identified as the corresponding amines [115]. The proposed mechanism could account for the protective effect of nitroxides against thiyl radicals.

In recent years, the formation of $^{\bullet}NO$ in natural waters has been explored [116, 117]. The NO_2^- ion, a trace compound in natural waters, is photochemically unstable because it can absorb in the sea-level UV region to yield $^{\bullet}NO$ [109, 117–122]:

$$NO_2^- + H_2O + h\nu \rightarrow {}^{\bullet}NO + {}^{\bullet}OH + OH^-. \tag{5.25}$$

In natural waters, the reactions of alkyl peroxy radicals and $^{\bullet}NO$ may be one of the sources of alkyl nitrates [123]. The photoformation rates of $^{\bullet}NO$ in river water and seawater have been measured as $(9.4\text{–}300) \times 10^{-12}$/M/s and $(5.3\text{–}39) \times 10^{-12}$/M/s, respectively [116, 117]. The scavenging rate constants for $^{\bullet}NO$ in the surface seawater samples were measured as 0.0–0.33/second [116]. The concentrations of $^{\bullet}NO$ were determined as $(2.4 - 32) \times 10^{-11}$ M [116]. Nitrification and denitrification processes in natural waters can also produce $^{\bullet}NO$ [124]. However, a recent study demonstrated there is a negligible contribution of photobiological processes to produce $^{\bullet}NO$ in natural waters [116].

5.2.2 Nitrogen Dioxide Radical

Nitrogen dioxide ($^{\bullet}NO_2$) produces and participates in a variety of biological reactions [65, 115, 125, 126]. The kinetics and mechanism of the reactions of $^{\bullet}NO_2$ with nitroxides, thiols, peptides, and proteins have been studied [65, 125, 127]. Reactions of piperidine, pyrrolidine, and oxozolidine nitroxides ($RNO^{\bullet}$) by $^{\bullet}NO_2$ radicals have been performed [65]. The kinetics of the reactions demonstrated nitroxides are the most efficient scavengers of $^{\bullet}NO_2$ at physiological pH ($k = (3–9) \times 10^8$/M/s). Reactions were dependent on the ring size and nature of the side chain of nitroxides, similar to the reactions of nitroxides with $CO_3^{\bullet -}$ [65]. The reactions produced the respective oxoammonium cation, which could be efficiently scavenged by $ABTS^{2-}$, forming $ABTS^{\bullet -}$ [65].

The reactivity of $^{\bullet}NO_2$ with GSH and Cys has been performed using pulse radiolysis [127]. Formation of the thiyl radical was observed (Eqs. 5.26, 5.27):

$$GSH + {}^{\bullet}NO_2 \rightarrow GS^{\bullet} + NO_2^- + H^+ \qquad (5.26)$$

$$CysSH + {}^{\bullet}NO_2 \rightarrow GysS^{\bullet} + NO_2^- + H^+. \qquad (5.27)$$

The rate constants of reactions (5.26) and (5.27) increased linearly with an increase in pH between ~5.7 and 8.0. The slopes of the plot of log$k$ versus pH were determined as 0.63 ± 0.02 and 0.57 ± 0.01 for the reactivity of $^{\bullet}NO_2$ with GSH and cysteine, respectively. At pH 7.4, the rate constants for reactions (5.25) and (5.26) were estimated as $\sim 2 \times 10^7$/M/s and 5×10^7/M/s, respectively.

Nitrosylation of peptide and proteins by $^{\bullet}NO_2$ has been carried out [128–130]. The reactivity of Gly-Tyr with $^{\bullet}NO_2$ in aqueous solution was strongly pH dependent with rate constants varying from 3.2×10^5/M/s at pH 7.5 to 2.0×10^7/M/s at pH 11.3. The reaction generated phenoxyl radicals, which could further react with $^{\bullet}NO_2$ to yield nitrotyrosine ($k \sim 3.0 \times 10^9$/M/s), the predominant final product in the neutral solution. The reaction of $^{\bullet}NO_2$ with cysteine-thiolate was faster ($k \sim 2.4 \times 10^8$/M/s at pH 9.2), which involved the transient formation of cystinyl radical anions. The reactivity of $^{\bullet}NO_2$ with Gly-Trp was relatively slow ($k \sim 10^6$/M/s at pH 9.0). Reactivity was not observed for $^{\bullet}NO_2$ with Met-Gly and $(Cys\text{-}Gly)_2$ in the pulse radiolysis experiments. At pH 7.0–9.0, the selective nitration of Tyr in the interaction of $^{\bullet}NO_2$ with histone, lysozyme, and ribonuclease A was found [129]. Nitration of tyrosine in hydrophilic and hydrophobic environments has recently been performed, which suggested the involvement of free radical mechanisms [128]. The involvement of radical was supported by the restrictions imposed on obtaining a large yield of protein 3-nitrotyrosine formation *in vivo* due to the presence of strong reducing systems such as GSH. Elimination of important soluble reductants allowed nitration of Tyr [128].

Recently, the reactivity of $^{\bullet}NO_2$ with various oxidation states of myoglobin was studied [125]. The interconversion of various oxidation states of myoglobin was observed (Eqs. 5.28–5.32):

$$\mathrm{MbFe^{II}O_2 + {}^{\bullet}NO_2 \rightarrow MbFe^{III}OONO_2} \quad (5.28)$$

$$\mathrm{MbFe^{III}OONO_2 \rightarrow MbFe^{V}=O + NO_3^-} \quad (5.29)$$

$$\mathrm{MbFe^{V}=O \rightarrow {}^{\bullet}MbFe^{IV}=O} \quad (5.30)$$

$$\mathrm{{}^{\bullet}MbFe^{IV}=O + MbFe^{II}O_2 + H_2O \rightarrow MbFe^{IV}=O + MbFe^{III}OH_2 + O_2} \quad (5.31)$$

$$\mathrm{MbFe^{IV}=O + MbFe^{II}O_2 + H_2O + 2H^+ \rightarrow 2MbFe^{III}OH_2 + O_2.} \quad (5.32)$$

The reaction of $MbFe^{II}O_2$ with $^{\bullet}NO_2$ yielded $MbFe^{III}OONO_2$ ($k_{28} = 4.5 \pm 0.3 \times 10^7$/M/s; pH 7.4), which rapidly transformed to perferryl species ($MbFe^V$=O) and NO_3^- through heterolysis along the O–O bond (reaction 5.29). The perferryl species converted rapidly to the ferryl species with a radical site on the globin (reaction 5.30), which oxidizes another oxymyoglobin to form equal amounts of ferrylmyoglobin and metmyoglobin (reaction 5.31; $10^4/M/s < k_{31} < 10^7/M/s$). At a longer timescale of minutes, a slow buildup of ferrylmyoglobin from the comproportionation of ferryl species with oxymyoglobin occurred (reaction 5.32; $k_{32} = (2.13 \pm 0.53) \times 10^1$/M/s). Overall, three oxymyoglobin molecules converted to metmyoglobin, consumed by each $^{\bullet}NO_2$ radical.

The intermediate, $MbFe^{III}OONO_2$ species, produced in reaction (5.28), was also observed in the reaction of peroxynitrite with metmyoglobin ($k = (4.6 \pm 0.3) \times 10^4$/M/s). The chemistry and reactions of peroxynitrite with biological molecules are discussed in the next section. The ferryl species could also be reduced by $^{\bullet}NO_2$ to yield transient species, $MbFe^{III}ONO_2$ ($k_{33} = (1.2 \pm 0.2) \times 10^7$/M/s), which dissociated into metmyoglobin and nitrate ($k_{34} = (1.9 \pm 0.2) \times 10^2$/M/s) [125]:

$$\mathrm{MbFe^{IV}=O + {}^{\bullet}NO_2 \rightarrow MbFe^{III}ONO_2} \quad (5.33)$$

$$\mathrm{MbFe^{III}ONO_2 + OH^- \rightarrow MbFe^{III}OH + NO_3^-.} \quad (5.34)$$

5.2.3 Peroxynitrite

5.2.3.1 Generation. Numerous studies have been performed to suggest the formation of $ONOO^-$ through photochemical and biological systems [131–134]. Photolysis of the aqueous nitrate ion in the UV region produces $ONOO^-$ (Eq. 5.35) [131, 135–139]. A spectrum of NO_3^- exhibits peaks at 200 nm ($\varepsilon = 9900$/M/cm) (Fig. 5.12) [131]. With the excitation of the $\pi \rightarrow \pi^*$ band, the nitrate ion can isomerize to peroxynitrite (Eq. 5.35) [140–143]:

$$\mathrm{NO_3^- + h\nu \rightarrow ONOO^-.} \quad (5.35)$$

The spectra of other nitrogen and oxygen species involved in the photodynamics are also depicted in Figure 5.12. Absorption of $^{\bullet}NO$ is weak, while its

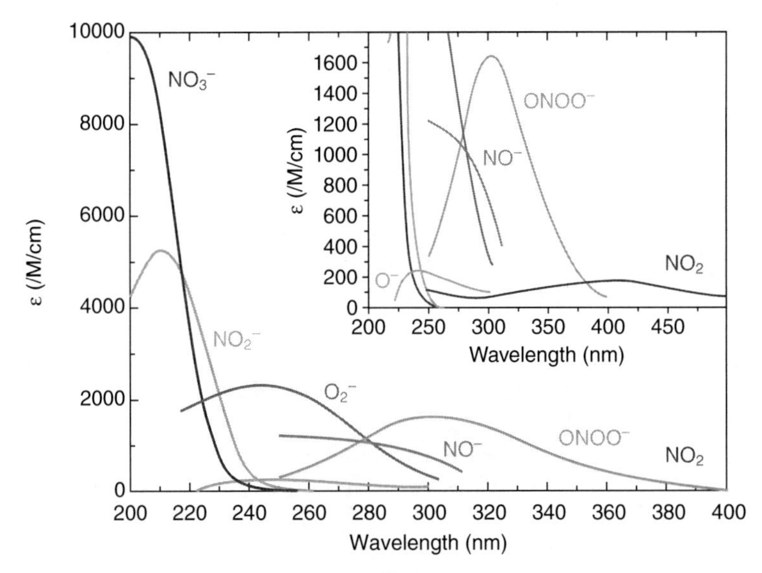

Figure 5.12. Absorption spectra of the species potentially involved in oxidation by peroxynitrite (adapted from Madsen et al. [131] with the permission of the American Chemical Society). See color insert.

anion, NO^- (aq), absorbs below 300 nm. Both $^•O^-$ (aq) and $^•NO_2$ have weak absorptions ($\varepsilon < 250$/M/cm). Peroxynitrite ion has an absorption band at 302 nm ($\varepsilon = 1670$/M/cm). Peroxynitrous acid exhibits an absorption band at 240 nm ($\varepsilon = 770$/M/cm) with no significant absorption greater than 300 nm [120, 144].

In biological systems, the diffusion-controlled reaction of $O_2^{•-}$ and $^•NO$ has been suggested to be the most probable source of peroxynitrite (Eq. 5.36) [143, 145]:

$$O_2^{•-} + {}^•NO \rightarrow ONOO^-. \tag{5.36}$$

Reaction (5.36) has been studied in aqueous solutions [143, 146–150]. Reported rate constants are in the range of $(3.8–20.0) \times 10^9$/M/s. The large variation in rate constants was suggested to relate to the reaction studied under irreversible conditions [150]. A study made under reversible conditions at pH > 12 explained the dependence of the rate constant on the distance between $ONOO^-$ particles (sites of generation of $O_2^{•-}$ and $^•NO$) [150].

$ONOO^-$ can be synthesized by several methods [151–155]. The methods include the reaction of ozone with azide, the reaction of hydrogen peroxide and nitrite (or alkyl nitrite) under acidic conditions, the reaction of hydroxylamine and oxygen in alkaline solution, and the mixed-phase reaction of solid KO_2 with gaseous $^•NO$. The pure peroxynitrite was synthesized by the reaction of $(Me_4N)ONOO$ with $^•NO$ gas in liquid ammonia [155, 156].

5.2.3.2 Decomposition in Aqueous Solution. The pK_a for deprotonation of peroxynitrous acid (ONOOH) was determined to be 6.5–6.8 [144, 149, 157, 158]. Thus, both ONOOH and its anion, $ONOO^-$, exist at physiological pH. The value for standard Gibbs energy for formation, $\Delta_f G°$, of ONOOH from the reaction of HNO_2 with H_2O_2 was determined as 29.7 kJ/mol. Using the pK_a(ONOOH) = 6.6, $\Delta_f G°(ONOO^-)$ was estimated as 67.4 kJ/mol. The bond strength of O–O bond in HOONO is 90 kJ/mol, weaker than that of hydrogen peroxide (170 kJ/mol) [88]. ONOOH decomposes relatively fast ($k_0 = 1.25 \pm 0.05$/second at 25°C) [149, 158, 159]. Comparatively, $ONOO^-$ is a stable species. The lifetime of peroxynitrite is ~1 second at physiological pH [88]. The homolysis of ONOOH produces $^{\bullet}NO_2$ and $^{\bullet}OH$ radicals with a free radical yield of 28 ± 4% [132, 143, 160, 161]. The activation parameters, E_a and frequency factor (A), at pH 4.0 were determined as 86.6–92.0 kcal/mol and $(0.18–1.0) \times 10^{16}$/second, respectively [132, 143]. The value of A is of the same order of magnitude as has been determined for hemolytic reactions of peroxides in gas phase and in nonpolar organic solvents. The activation parameters suggest the homolysis of ONOOH. The value of E_a was similar at pH 14 (90.8–100.8 kJ/mol), while a lesser value of A was determined ($(0.08–4.9) \times 10^{12}$/second) [159, 162]. The activation volume, ΔV^*, at pH 4.0 was evaluated as 6.0–14.0 cm^3/mol [163–166], which also indicates a process of bond breakage.

Figure 5.13 presents the reactions involving the decomposition of ONOOH and $ONOO^-$ [132]. The rate constants of the involved reactions are given in Table 5.4 [108, 120, 132, 143, 167–173]. The reaction of $^{\bullet}OH$ with ONOOH is much slower than the reaction with NO_2^-, forming NO_3^- as a final product (Fig. 5.13). However, the rate constants of $^{\bullet}OH$ with $ONOO^-$ and NO_2^- are similar (see Table 5.4). The ratio of $[ONOO^-]/[NO_2^-]$ thus determines the contribution of the reaction of $^{\bullet}OH$ with $ONOO^-$. Overall, the change in the reaction pattern of peroxynitrite was observed when the pH approaches the exceeded pK_a of ONOOH. The results were described using the homolysis of ONOOH with the formation of $^{\bullet}NO_2$ and $^{\bullet}OH$. The homolytic equilibrium between $ONOO^-$ and $^{\bullet}NO + O_2$ and other subsequent reactions of these radicals [170] are depicted in Fig. 5.13. The formation of the powerful electrophile, N_2O_3, also occurred (see Fig. 5.13). The mechanism for the homolysis of the species into free radicals in aqueous solution was also simulated by applying metadynamics (MT) [143]. The formation of $^{\bullet}NO_2$ and $^{\bullet}OH$ through a bond breakage of peroxide was obtained in the simulation. MT further suggests the formation of a short-lived cage radical pair with an average lifetime of less than 1 ns, which could diffuse into the bulk of the solution to result in free radicals [143]. The molecular dynamics of the decomposition of HONOOH in water demonstrated nitrate formed 86% of the time from the water-caged radical pair [174]. Remaining 14% of the time was the solvent-separated radical pair from the water-caged radical pair [174]. However, NO_3^- did not form with an increase in pH. Formation of O_2 and NO_2^- at ratios of 1:2 occurred at higher pH. The yield of NO_2^- reached to ~80% at pH 9–10 [132, 162, 175–177].

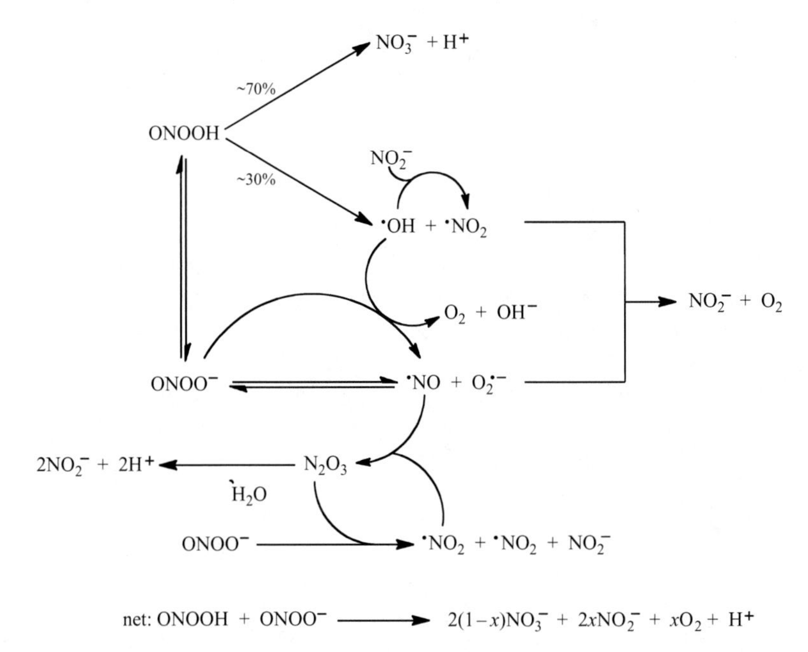

Figure 5.13. Decomposition of peroxynitrite (adapted from Goldstein and Merényi [132] with the permission of Elsevier Inc.).

TABLE 5.4. Summary of the Reactions and Their Rate Constants Involved in the Decomposition Mechanism of Peroxynitrite at 25°

Reaction	K	Reference
$ONOOH \rightarrow NO_3^- + H^+$	0.90 ± 0.05/second	[a]
$ONOOH \rightarrow NO_2 + OH$	0.35 ± 0.03/second	[a]
$2NO_2 + H_2O \rightarrow NO_2^- + NO_3^- + 2H^+$	$2k = 1.3 \times 10^8$/M/s	[167]
$OH + NO_2^- \rightarrow OH^- + NO_2$	5.3×10^9/M/s	[168]
$OH + ONOO^- \rightarrow NO + O_2 + OH^-$	4.8×10^9/M/s	[169]
$NO + O_2^- \rightarrow ONOO^-$	$(5 \pm 1) \times 10^9$/M/s	[143]
$ONOO^- \rightarrow NO + O_2^-$	0.020 ± 0.003/second	[168, 170]
$NO_2 + NO_{\text{(ReversReact)}} N_2O_3$	$k_7 = 1.1 \times 10^9$/M/ s$k_{-7} = 8.4 \times 10^4$/second	[171]
$N_2O_3 + H_2O \rightarrow 2NO_2^- + 2H^+$	$2 \times 10^3 + 10^8[OH^-] + (6.4 - 9.4) \times 10^5$[phosphate]/ second	[108, 172]
$ONOO^- + N_2O_3 \rightarrow 2NO_2 + NO_2^-$	3.1×10^8/M/s	[173]
$NO_2 + O_2^- \rightarrow NO_2^- + O_2$	4.5×10^8/M/s	[120]

[a] ONOOH decomposes with $k_d = 1.25$/second, yielding 72% $NO_3^- + H^+$ and 28% $NO_2 + OH$.
Adapted from Goldstein and Merényi [132].

More recently, the formation of an intermediate, peroxynitrate, was observed in the decomposition of peroxynitrite at a near neutral pH according to the following proposed reaction sequence (Eq. 5.37) [178]:

$$ONOOH + ONOO^- \rightarrow NO_2^- + O_2NOO^- + H^+ \rightarrow 2NO_2^- + O_2. \quad (5.37)$$

The formation of $^1\Delta_gO_2$ from the decay of $ONOO^-$ was used as an indicator for the decomposition of O_2NOO^- [179]. A study of the isomerization of $ONOO^-$ using *ab initio* also showed the formation of O_2NOO^- from the reaction of $ONOO^-$ and ONOOH [180].

A study on the photon-initiated homolysis of peroxynitrous acid has been performed in detail [181]. The homolysis of ONOOH at pH 4.0–5.5, studied using laser flash photolysis at 355 nm, occurred exclusively at the N–O bond rather than the O–O bond (Eq. 5.38):

$$ONOOH \rightleftharpoons {}^{\bullet}NO + HO_2^{\bullet}. \quad (5.38)$$

The radicals formed in reaction (5.38) recombined with a rate constant of $(1.2 \pm 0.2) \times 10^{10}$/M/s [181]. The quantum yields, resulting from excitation of ONOOH and $ONOO^-$ at 266 or 355 nm, were similar (15%) within experimental errors. Hence, it was concluded that the bond dissociation energies of ON–OOH and ON–OO^- bonds are similar (100 kJ/mol).

5.2.3.3 Reactivity with CO_2. The reaction between peroxynitrite and CO_2 was observed almost four decades ago when peroxynitrite was determined to be unstable in carbonate buffered solution [182]. Now decades later, this reaction was recognized that the bicarbonate buffer inhibits *Escherichia coli* killing by peroxynitrite [183, 184]. In later years, studies were conducted for the reaction between $ONOO^-$ and CO_2, and the reaction mechanism can be presented by Equations (5.39) and (5.40) [185]:

$$ONOO^- + CO_2 \rightleftharpoons ONOOC(O)O^- \quad k_{39} = (2.9 \pm 0.3) \times 10^4\,/\text{M/s [186]} \quad (5.39)$$

$$k_{-39} = 1.5 \times 10^6\,/\text{second [185]}$$

$$ONOOC(O)O^- \rightleftharpoons {}^{\bullet}NO_2 + CO_3^{\bullet-} \quad k_{40} = 1.9 \times 10^9\,/\text{M/s [185]} \quad (5.40)$$

$$k_{-40} = 5.0 \times 10^8\,/\text{M/s [187]}.$$

The formation of the carbonate radical was directly detected using EPR [37, 188]. Both mixtures of peroxynitrite and bicarbonate at pH 6–9 and peroxynitrite and CO_2-saturated solution in alkaline media yielded carbonated radicals ranging from a few percent [188] to 35% [37]. In the presence of excess CO_2, the oxidation or nitration of several substrates by peroxynitrite did not exceed more than 33% [143]. Significantly, the rates of decay of $ONOO^-$ in these reactions were similar to the rates of formation of oxidized or nitrated substrates [161, 189–195]. The observations indicate $ONOOC(O)O^-$, formed in reaction (5.39), did not accumulate; however, it decomposed readily to form ~33% $^{\bullet}NO_2$ and $CO_3^{\bullet-}$ radicals. The radicals in the absence of potentially reducing substrates yielded nitrate, suggesting nitrate is produced through the formation of an intermediate nitrocarbonate, $O_2NO\text{-}C(O)O^-$ (Eq. 5.41):

$$^{\bullet}NO_2 + CO_3^{\bullet-} \rightarrow O_2NO\text{-}C(O)O^- \rightarrow NO_3^- + CO_2. \quad (5.41)$$

The rate of diffusion of radicals out of solvent cages decreases with increasing solvent viscosity; therefore, a study was conducted to learn the effect of solvent viscosity on product yields and rate constants for the decomposition of free radical initiators [196]. The ~33% yield of $Fe(CN)_6^{3-}$ and $ABTS^{\bullet-}$ from the reactions of $ONOO^-$ with $Fe(CN)_6^{3-}$ and $ABTS^{2-}$ in the presence of excess CO_2^-, respectively, decreased with increasing concentration of added glycerol [196]. Moreover, yields of the oxidized products were independent of the initial concentrations of $Fe(CN)_6^{3-}$ and $ABTS^{2-}$ within experimental error, an indication that the decreased yields in the presence of glycerol were not due to a competing reaction of glycerol with $Fe(CN)_6^{3-}$ and $ABTS^{2-}$, but instead, in the increased viscosity in the reaction mixtures. Thus, oxidizing radicals, $^{\bullet}NO_2$ and $CO_3^{\bullet-}$, were in a water cage and only ~3% could escape from the cage to the bulk of the solution.

The changes in Gibbs energies for the reactions involving peroxynitrite and CO_2 have been evaluated [185]. An energy diagram for the reaction of peroxynitrite and CO_2 is shown in Figure 5.14. The relative low value of $\Delta G°$ for reaction (5.39) suggests the O–O bond in $ONOOC(O)O^-$ is weak. This conclusion was also supported with theoretical calculations [197]. Therefore, an adduct, $ONOOC(O)O^-$, is short-lived and a rate constant for the decay of the adduct was estimated between 10^7 and 10^9/second [185, 198]. Furthermore, the lifetime for the decay of the adduct below ca. 0.1 μs clearly indicates an insufficient time to react with any biological matter prior to homolyzation into $^{\bullet}NO_2$ and $CO_3^{\bullet-}$ radicals. Hence, $^{\bullet}NO_2$ and $CO_3^{\bullet-}$ radicals are responsible species for any biological damage related to the simultaneous presence of $ONOO^-$ and CO_2.

5.2.3.4 Reactivity with Inorganic and Organic Substrates. Peroxynitrite is a strong oxidant and has a reduction potential of 1.4 V at pH 7.0 [199]. The reactivity of peroxynitrite with inorganic species showed a descending order of Sn(II) > Sb(II) > As(III) > S(IV) >> P(I) > P(III) [200]. This order was not related to the formal potentials of the reducing centers and reactions involving oxygen transfer via O-bridged precursor complexes [200]. Other studied inorganic compounds include NH_2OH, iodide, cyanide, $Fe(CN)_6^{4-}$, and $Mo(CN)_8^{3-}$ [164, 200–203]. Among the studied organic substrates were amino polycarboxylates, 1,2-glycols, pyruvate, thioethers, thiols, sulfides, sulfite, dimethylsulfide (DMS), dimethylselenide (DMSe), carbonyls, and phenols [145, 204–216]. Common chelators such as desferrioxamine, salicylaldehyde isonicotinoyl hydrazone (SIH), ethylenediaminetetraacetate (EDTA), diethylenetraminepentaacetate

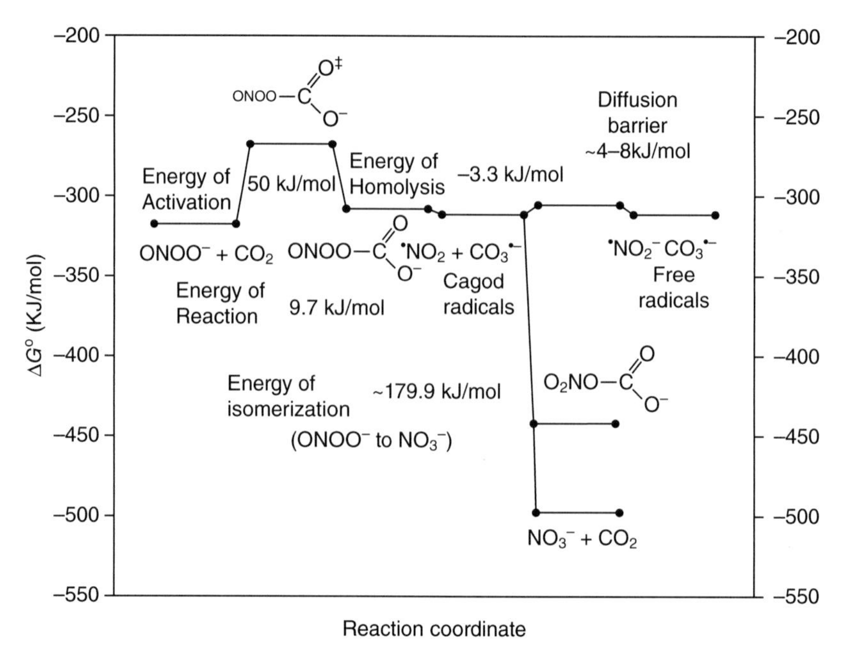

Figure 5.14. Standard Gibbs energy diagram of the reaction of peroxynitrite with CO_2 showing the energies of reactive species (adapted from Squadrito and Pryor [185] with the permission of the American Chemical Society).

(DTPA), and nitriloacetate (NTA) had significant reactivity with peroxynitrite [217]. Peroxynitrite reacted with pyruvate with a second-order rate constant of $(8.8 \pm 0.7) \times 10^1$/M/s at pH 7.4 and 37°C [77]. Pyruvate was decarboxylated by peroxynitrite at pH 7.4 and 5.5 through the formation of a carbon dioxide radical anion [77]. The formation of radicals was also observed in the oxidation of thiols by peroxynitrite [213].

In the reactions of peroxynitrite with DMS and DMSe, the rates were first order with respect to the concentrations of both peroxynitrite and DMS (or DMSe) [208]. The rate constants for DMS decreased from $(2.1 \pm 0.2) \times 10^4$/M/s to $(2.3 \pm 0.2) \times 10^{-1}$/M/s in the pH range of 4.6–13.0 [208]. In the same pH range, the rate constants for DMSe varied from $(2.7 \pm 0.3) \times 10^4$/M/s to (6.0 ± 0.5)/M/s. This indicates a higher reactivity of ONOOH than $ONOO^-$ with both DMA and DMSe. A partial replacement of water with ethanol decreased the rate of the reaction of $ONOO^-$ but had no effect on the rate of the reaction with ONOOH. Both species of peroxynitrite ($ONOO^-$ and ONOOH) reacted with DMS (or DMSe) to yield NO_2^- and DMS=O (or DMSe=O). The postulated mechanism involved the formation of a prereaction complex, a radical pair ({(DMSOH)$^•$ . . . $^•$(ONO)}), followed by O–O bond cleavage in a rate-limiting step. Theoretical studies on the system also supported the postulated mechanism [218]. A recent study performed one-electron oxidation of acetohydroxamic acid by $^{\bullet}OH$, $O_2^{\bullet-}$, $^{\bullet}N_3$, $^{\bullet}NO_2$, and $CO_3^{\bullet-}$ has been studied [219]. The aceto hydroxamic acid has an importance in metal chelation and enzyme inhibition as well as in scavenging of radicals. The effective reactivity was observed only with $^{\bullet}OH$, $^{\bullet}N_3$, and $CO_3^{\bullet-}$, resulting respective transient nitroxide radical through a complex mechanism [219].

The addition of $ONOO^-$ to carbonyl compounds represents an example of nucleophilic addition [143]. Initially, a fast equilibration between the reactants and the corresponding tetrahedral adduct anion is formed with a strong tendency of an equilibrium shift toward the reactant side. A fast protonation of the adduct anion by water and added buffers occurs [220, 221]. The rate of nucleophilic addition to carbonyls (e.g., amides) is much slower than that of the reaction of $ONOO^-$ with CO_2 [77, 161, 221–223]; therefore, the oxidation of carbonyls by $ONOO^-$ may not be of biological significance.

Recently, the reactions of $ONOO^-$ with phenyl and coumarin boronates have been studied in order to develop boronate-based fluorescent detection [224]. The reactions were rapid ($k \sim 10^6$/M/s at pH 7.4). Comparatively, the reactions of $ONOO^-$ with aryl boronates were ~10^6 and 2×10^2 times faster than H_2O_2 and HOCl [224]. Both H_2O_2 and HOCl reacted with boronates stoichiometrically to yield a 100% phenolic product through a two-electron process. However, in the case of reaction of $ONOO^-$ with boronates, the nitrated species as the minor products (10–15%) were also obtained in addition to phenols as the major products (~85 to 100%) [224]. The positive effects of oxygen and inhibitory effects of hydrogen atom donor (acetonitrile and 2-propanol) and general radical scavengers (Tyr, GSH, NADH, and ascorbic acid) on the yield of minor product indicates the intermediate of the reaction was the phenoxy radical [225]. Results of this study may have relevance in mitigating peroxynitrite-mediated cytotoxicity and therefore warrant further understanding of the reaction mechanism between peroxynitrite and boronates.

5.2.3.5 Reactivity with Proteins and Nonprotein Metal Centers. The second-order rate constants for the reactions of peroxynitrite with protein and nonprotein metal centers are given in Table 5.5 [199, 201, 226–243]. The rate constants vary from 10^3 to 10^7/M/s. The reactions of peroxynitrite with water-soluble iron porphyrins at physiological pH have been studied [227, 244–247]. The rate constants vary from 10^5 to 10^7/M/s. For example, 5,10,15,20-tetrakis (*N*-methyl-4′-pyridyl)porphinatoiron(III) [Fe(III)TMPyP] rapidly reacts with $ONOO^-$ with a rate constant $k = 5 \times 10^7$/M/s [245]. Comparatively, the rate constant of Fe(III)TMPyP with O_2^- under the same conditions was 1.9×10^7/M/s [245].

Reactions with metal-centers can be rationalized by Equation (5.42):

$$M^{n+} + ONOO^- \rightarrow M^{n+}\text{-}ONOO^{n+} \rightarrow M^{n+}\text{-}O^{\bullet-} + {}^{\bullet}NO_2 \rightarrow = M^{(n+1)+} = O + {}^{\bullet}NO_2. \quad (5.42)$$

TABLE 5.5. Second-Order Rate Constants of Peroxynitrite Reactions with Protein and Nonprotein Metal Centers

Metal Center	pH[a]	*t* (°C)	*k* (/M/s)	Reference
Mn(III)-tm-2-pyp[b]	7.4	37	1.85×10^7	[226]
Mn(III)-tm-3-pyp[b]	7.4	37	3.82×10^6	[226]
Mn(III)-tm-4-pyp[b]	7.4	37	4.33×10^6	[226]
Fe(III)-tm-4-pyp[b]	7.4	37	2.2×10^6	[227]
Fe(III)-tmps[c]	7.4	37	6.45×10^5	[227]
Fe(III)EDTA[d]	7.5	37	5.7×10^3	[199]
Fe(III)-myeloperoxidase	7.0	25	6.8×10^6	[228]
Fe(II)-myeloperoxidase	7.0	25	1.3×10^6	[228]
Compound I	7.0	25	7.6×10^6	[228]
Ni(II)-cyclam[e]	NR	27	3.25×10^4	[201]
Mn(III)-tbap[f]	7.2	37	6.8×10^4	[229]
Zn(II)-tbap[f]	7.2	37	4.9×10^5	[229]
Myeloperoxidase (heme)	7.2	12	6.2×10^6	[230, 242]
Lactoperoxidase (heme)	7.4	12	3.3×10^5	[230, 242]
Horseradish peroxidase (heme)	Ind[g]	25	3.2×10^6	[230, 242]
Alcohol dehydrogenase (zinc sulfur cluster)	7.4	23	2.6×10^5	[231]
Aconitase (iron sulfur cluster)	7.6	25	1.4×10^5	[232]
Cytochrome c^{2+} (heme)	7.4	25	1.3×10^4	[233]
	7.4	37	2.5×10^4	[233]
Metmyoglobin (metMb)	7.4	20	$1.0\text{–}1.4 \times 10^4$	[234, 243]
H64A metMb	7.5	20	5.8×10^6	[235]
H64D metMb	7.5	20	4.8×10^6	[235]
H64L metMb	7.5	20	5.7×10^4	[235]
F43W/H64L metMb	7.5	20	5.2×10^4	[235]
H64Y/H93G metMb	7.5	20	0.9×10^4	[235]
sw WT-metMb	7.5	20	1.6×10^4	[235]
hh metMb	7.5	20	1.2×10^4	[236]
Oxyhemoglobin (monomer)	7.4	25	1.04×10^4	[237]
	NR	20	$2\text{–}3 \times 10^4$	[238]
Mn superoxide dismutase (monomer)	7.4	37	2.5×10^4	[229]
Mn(II) superoxide dismutase	8.0	13	3.0×10^2	[239]
Mn(III) superoxide dismutase	8.0	13	1.0×10^2	[239]
Cu,Zn superoxide dismutase (monomer) 7.5	37	9.4×10^3	[240]	
Tyrosine hydroxylase	7.4	25	3.8×10^3	[241]

[a] pH: This column shows the pH at which the rate constant was determined.
[b] tmpyp: 5,10,15,20,-tetrakis(N-metil-40-pyridyl)porphyrin.
[c] tmps: 5,10,15,20,-tetrakis(2,4,6-trimetil-3,5-sulfonatofenil)porphyrin.
[d] EDTA: ethylenediaminetetraacetic acid.
[e] cyclam: 1,4,8,11-tetraazacyclotetradecane.
[f] tbap: tetrakis-(4-benzoic acid) porphyrin.
[g] ind: pH-independent value.
NR, not reported.

The type of substrate determines the rate-limiting step and the yield of radicals diffusing out of the solvent cage [248]. The reactivity of peroxynitrite with transition metal centers are complex and several different mechanisms are involved [228, 236, 248–255]. Significantly, the ferryl intermediate has been detected in the reaction of peroxynitrite with chloroperoxidase [256]. Therefore, reactions of peroxynitrite can generate a secondary oxidizing species in addition to $^{\bullet}NO_2$. However, the oxidizing species may be reversed by available reductants such as GSH, ascorbic acid, and even excess peroxynitrite.

Recently, the kinetics and spectral measurements on the Fe(III)EDTA/peroxynitrite system as a function of pH (10.4–12.3) at 25°C in aqueous solution were made [257]. The reaction between Fe(III)EDTA and peroxynitrite immediately formed a purple-colored species with an absorption maximum at 520 nm, similar to species obtained in the Fe(III)EDTA/H_2O_2 system in an alkaline medium [257–260]. However, the molar extinction coefficient $\varepsilon_{520\,nm}$ = 13/M/cm was much lower than $\varepsilon_{520\,nm}$ = 520/M/cm for the Fe^{III}(EDTA)$O_2]^{3-}$, a purple high-spin Fe^{III} side-on-bound peroxo complex, observed in the Fe(III)EDTA/H_2O_2 system. The rate of the reaction between Fe(III)EDTA and peroxynitrite increased rapidly and slowed down with increasing concentration of peroxynitrite. This indicates peroxynitrite first binds to Fe(III), the initial complex, which then formed the colored species. This species may be an end-on peroxynitrite complex because of much lower absorptivity compared to the peroxo-complex species [257].

5.2.3.6 Reactivity with Amino Acids, Peptides, and Proteins. The kinetics of the reactions of peroxynitrite with essential amino acids have been determined [261]. Met, Cys, and Trp were the only amino acids to undergo a direct reaction with peroxynitrite. The second-order rate constants of these amino acids are given in Table 5.6 [205, 206, 210, 216, 248, 261, 262, 263, 264, 265, 266, 267, 268, 269, 273, 274]. Other amino acids were modified indirectly by oxidizing species (e.g., $^{\bullet}NO_2$, $CO_3^{\bullet-}$, ferryl), suggesting their reactions with peroxynitrite in the absence or presence of transition metals [261]. Direct and indirect reactions lead to the formation of oxidized, nitrated, and nitrosated products, discussed below.

Methionine reacted with peroxynitrite with a second-order rate constant on the order of 10^2/M/s (Table 5.6). The nucleophilic sulfur atom in the side chain of Met reacted with peroxynitrite to form methionine sulfoxide and nitrite (Table 5.7) [248]. Similar oxidations of Met in proteins, such as α1-antitrypsin inhibitor and glutamine synthetase, *in vitro* by peroxynitrite, have also been observed. Other sulfur-containing amino acids, such as Cys, reacted the fastest with peroxynitrite (Table 5.6). The second-order rate constants of the reaction of peroxynitrite with homocysteine, GSH, and the thiol of albumin were on the order of ~10^3/M/s (Table 5.6). Comparatively, thiols in glyceraldehyde 3-phosphate dehydrogenase, creatine kinase, tyrosine phosphatase, and perooxiredoxin reacted faster with second-order rate constant values in the range of ~10^5–10^7/M/s (Table 5.6). Plots of apparent second-order

TABLE 5.6. Second-Order Rate Constants of Peroxynitrite Reactions with Free Amino Acids, Peptides, and Nonmetal Containing Proteins [248]

Amino Acid, Peptide, or Protein	pH[a]	t (°C)	k (/M/s)	Reference
Glutathione peroxidase (selenocysteine, reduced)[b]	7.4	25	8×10^6	[262]
Glutathione peroxidase (selenocysteine, oxidized)	7.4	25	7.4×10^5	[262]
Peroxiredoxin alkylhydroperoxide reductase(cysteine)	7.0	NR	1.51×10^6	[263]
Protein tyrosine phosphatases (cysteine)	7.4	37	$2–20 \times 10^7$	[264]
Creatine quinase (cysteine)	6.9	NR	8.85×10^5	[265]
Glyceraldehyde 3-phosphate dehydrogenase(cysteine)	7.4	25	2.5×10^5	[266]
Human serum albumin (whole protein)	7.4	37	9.7×10^3	[261]
Human serum albumin (cysteine)	7.4	37	3.8×10^3	[261]
Cysteine	7.4	37	4.5×10^3	[205]
Glutathione	7.4	37	1.36×10^3	[267, 268]
Homocysteine	7.4	37	7.0×10^2	[267]
N-Acetylcysteine	7.4	37	4.15×10^2	[267]
Cysteine sulfinic acid	7.4	25	7.6×10^1	[216]
Hypotaurine	7.4	25	7.7×10^1	[216]
Lipoic acid (disulfide)	7.4	37	1.4×10^3	[267]
Selenomethionine	7.8	25	1.48×10^3	[210]
Methionine	7.4	25	$1.7–1.8 \times 10^2$	[206, 269]
	7.4	37	3.64×10^2	[261]
N-Acetylmethionine	6.3	25	1.6×10^3	[269]
Threonylmethionine	7.4	27	2.83×10^2	[273]
Glycylmethionine	7.4	27	2.80×10^2	[273]
Lysozyme	7.4	37	7.0×10^2	[248]
Tryptophan	7.4	37	3.7×10^1	[274]

[a] pH: This column shows the pH at which the rate constant was determined.
[b] In the case of proteins, the critical residue is shown in parentheses.
NR, not reported.

rate constants as a function of pH for thiols were a bell-shaped curve. This indicates the protonated form of one species reacted with the anionic species of the other. However, the rate constants for a number of thiols increased as the pK_a of the thiol compound decreased, suggesting ONOOH was the reactive species with the anionic thiolate (RS^-) [267].

The reactions of peroxynitrite with RS^- form corresponding disulfides and nitrites (Table 5.7). Disulfide is formed through an intermediate of sulfenic acid, RSOH (Eq. 5.43), which reacts further with another molecule of thiol to form the corresponding disulfide (Eq. 5.44) [270]:

TABLE 5.7. Products and Intermediates Detected in Peroxynitrite-Damaged Amino Acids

Amino Acid	Product	Reference
Cysteine	Disulfide (RSSR)	[205]
Sulfenic acid	(RSOH)	[205, 263, 270]
	Sulfinic acid (RSO_2H)	[205]
	Sulfonic acid (RSO_3H)	[205]
	Nitrosocysteine (RSNO)	[271, 275]
	Nitrocysteine ($RSNO_2$)	[271]
	Thiyl radical ($RS^•$)	[276]
	(Disulfide radical anion ($RSSR^{-•}$)	[41]
	Sulfinyl radical ($RSO^•$)	[41]
	Sulfinic acid (RSO_2H)	[205]
Cysteine sulfinic acid	Sulfonates	[216]
Hypotaurine	Sulfonates	[216]
Methionine	Methionine sulfoxide	[206, 269, 273]
Tryptophan	5- and 6-Nitrotryptophan	[274, 405]
	Hydroxytryptophan	[274]
	N-Formylkynurenine	[274, 277]
	Hydropyrroloindole	[277]
	Oxindole	[277]
	Tryptophanyl radical	[278]
Tyrosine	3-Nitrotyrosine	[199, 207, 212, 279, 280]
	3-Hydroxytyrosine	[212, 279]
	Dityrosine	[279, 280]
	3,5-Dinitrotyrosine	[406]
	Tyrosyl radical	[287]
Phenylalanine	*o*-, *m*- and *p*-Tyrosine	[279]
	Nitrophenylalanine	[279]
	Nitrotyrosine	[279]
	Dityrosine	[279]
Histidine	Oxo-histidine	[240]
	Nitrohistidine	[240]
	Histidinyl radical	[240]

Adapted from Alvarez and Radi [248].

$$ONOOH + RS^- \rightarrow NO_2^- + RSOH \quad (5.43)$$

$$RSOH + RS^- \rightarrow RSSR + OH^-. \quad (5.44)$$

Disulfide does not usually react with peroxynitrite [267]. However, reactive disulfide in lipoic acid displayed reactivity with peroxynitrite (1.4×10^3/M/s), forming thiosulfinate or disulfide-oxide [267]. Sulfinate, such as taurine and cysteine sulfinic acid, also had reactivity with peroxynitrite, but at least an order of magnitude slower than that of sulfur-containing amino acids (Table

5.6) to yield sulfonate (Table 5.7) [216]. In addition to disulfide and sulfonate, the disulfide radical anion ($RSSR^{\bullet-}$), sulfinyl radical ($RSO^{\bullet}$), and nitroso- and nitrothiol have been detected (Table 5.7) [41, 271].

The oxidation of selenium-containing compounds, selenomethionine, selenocysteine, and ebselen (2-phenyl-11,2-benzisoselenazol-3(2H)-one) by peroxynitrite has been performed (Table 5.6). The selenium compounds reacted much faster than their sulfur analogues (Table 5.6). Peroxynitrite reacts with ebselen at a rate constant of 2×10^6/M/s, producing selenoxide and nitrite [272].

Among the aromatic amino acids, Trp had a direct reaction (Table 5.6), while His, Phe, and Tyr did not react directly with peroxynitrite. The formation of the tryptophan radical and nitrate products were observed, with increasing concentrations in the presence of CO_2 (Table 5.7) [41, 199, 205, 206, 207, 212, 216, 240, 248, 263, 269, 273, 274, 270, 271, 275, 276, 277, 278, 279, 280, 209]. Nitrotryptophan also formed *in vitro* when the protein, human Cu,Zn superoxide dismutase, was exposed to peroxynitrite/CO_2 [281]. A recent study examined the modification of Trp in a number of proteins by peroxynitrite [282]. Peroxynitrate resulted in nitration of residues of Trp to form nitrotryptophan (NO_2-Trp), which may be significant in nitrosative stress [282]. The indirect reaction of peroxynitrite with His formed a product with a nitro addition, while Phe resulted in the formation of o-, p-, and m-tyrosine and nitrophenylalanine (Table 5.7). Formations of oxo- and nitro-derivatives were observed in the reaction of peroxynitrite with histidine-containing peptides.

Nitration of tyrosine residues by peroxynitrite under physiological conditions has been studied in detail [132, 283]. The mechanism of the reaction in the presence of excess CO_2 at pH7.5 is presented in Figure 5.15. As stated above, peroxynitrite decomposed to ~33% yield of $CO_3^{\bullet-}$ and $^{\bullet}NO_2$, and therefore, ~33% of peroxynitrite formed the tyrosyl radical ($TyrO^{\bullet}$) and $^{\bullet}NO_2$ due to the fast reaction of $CO_3^{\bullet-}$ with Tyr. The cross recombination of radicals resulted in 3-nitrotyrosine. The expected yield of 3-nitrotyrosine from the reaction of $TyrO^{\bullet}$ with $^{\bullet}NO_2$ is 45%, consisting of 15% peroxynitrite. The experimental results agreed with the expected yield [192, 284–286]. The formation of 3-nitrotyrosine in peroxynitrite-treated oxyHb, metHb, and Co-Hb have also been reported [287]. The individual steps of the mechanism, displayed in Figure 5.15, were recently examined theoretically and the calculated enthalpy changes are given in Figure 5.16 [288]. The calculations were based on the nitration of *p*-methylphenol [288]. Initially, the unimolecular decomposition of peroxynitrous acid or $ONOOCOO^-$ yielded $CO_3^{\bullet-}$ and $^{\bullet}NO_2$. The reaction proceeded through the phenoxy radical intermediate from the reaction of Tyr with $CO_3^{\bullet-}$. The reaction of this intermediate with $^{\bullet}NO_2$, followed by tautomerization, formed nitrated tyrosine. The yield of nitration depended on the concentration of peroxynitrite and the dosage rate of peroxynitrite [132]. Dityrosine was the dimerization product of $TyrO^{\bullet}$ (Fig. 5.15). Both *in vitro* and *in vivo* studies have shown that the peroxynitrate and other nitrating species can cause nitration of Tyr to 3-nitrotyrosine in mitochondrial

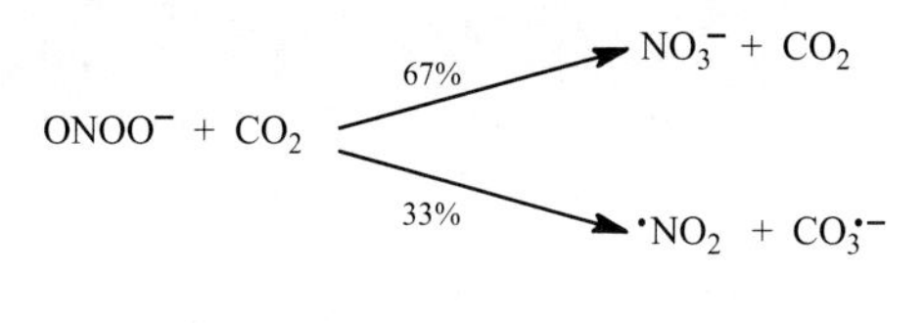

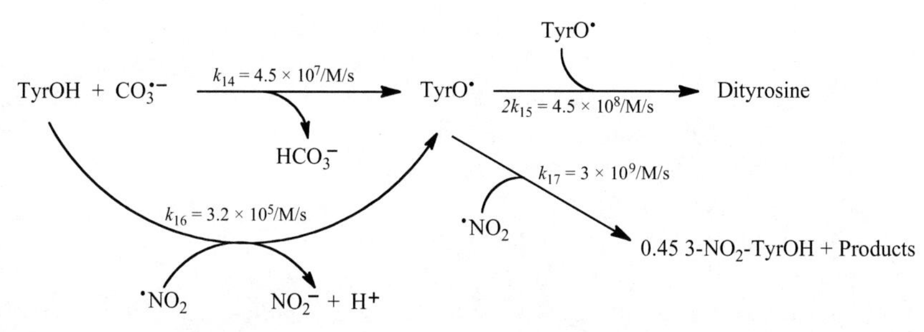

Figure 5.15. Mechanism of nitration and oxidation of tyrosine by peroynitrite in the presence of CO_2 at pH 7.5 (adapted from Goldstein and Merényi [132] with the permission of Elsevier Inc.).

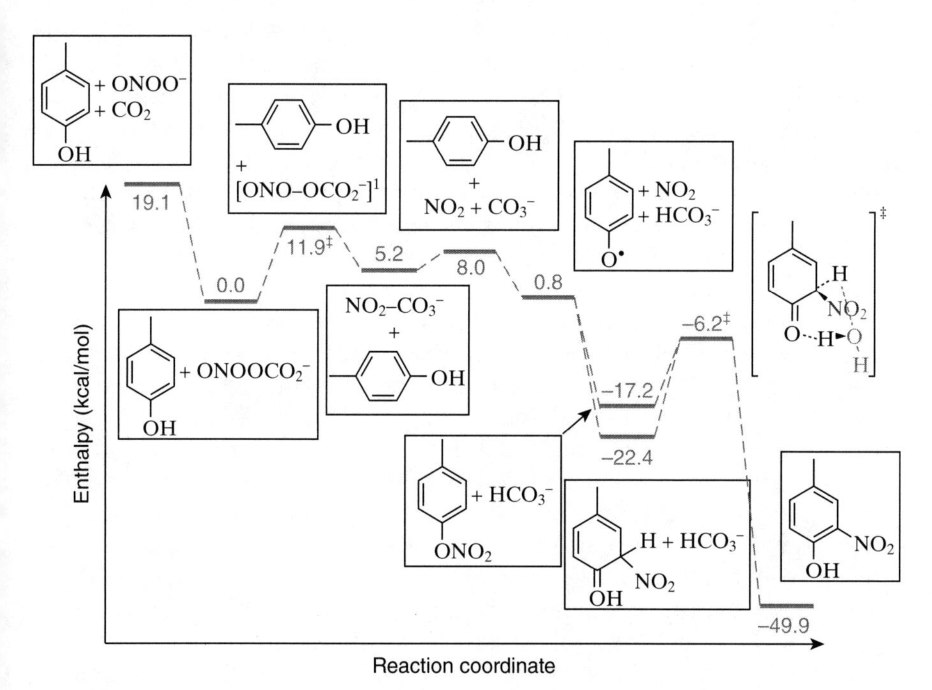

Figure 5.16. The mechanisms of tyrosine nitration by peroxynitrous acid investigated with the CBS-QB3 method (adapted from Gunaydin and Houk [288] with the permission of the American Chemical Society). See color insert.

proteins [289]. An enhancement of 3-nitrotyrosine has been reported in numerous diseases [289]. A study has been performed to demonstrate that imidazole-based thiourea and selenourea derivatives inhibited protein tyrosine nitration, mediated by peroxynitrite and peroxidase [290].

Recently, acetylation of amino acids L-His and L-Lys by acetyl radicals, formed from the reaction of diacetyl and methylglyoxal, has been studied [291, 292]. Peroxynitrite reacted with diacetyl and methylglyoxal with second-order rate constants of 1.0×10^4/M/s and 1.0×10^3/M/s, respectively, at pH 7.2 and 25°C. The mechanism of acylation of amino acids mediated by acetyl radicals was confirmed by EPR measurements [292].

In recent years, studies have started appearing on the reactivity of peroxynitrite with peroxiredoxins (Prx's) [263, 293–304]. Generally, Prx's have shown high reactivity with peroxynitrite ($k \sim 10^6$–10^7/M/s at pH 7.4 and 25°C) [293]. For example, tryparedoxin peroxidases from *Trypanosoma cruzi* had high efficiency in the catalytic elimination of peroxynitrite and hydrogen peroxide. Studied tryparedoxin peroxidases were cytosolic (c-TXNPx) and mitochondrial (m-TXNPx). Two cysteine residues in both TXNPxs (Cys52 and Cys173 in c-TXNPx and Cys81 and Cys204 in m-TXNPx) were identified that played the catalytic role in reducing peroxynitrite and hydrogen peroxide [293]. The HRP compound I also showed efficient reaction of its protein-cysteine residues with peroxynitrite ($k = 3.7 \times 10^5$/M/s) and hydrogen peroxide ($k = 3.4 \times 10^7$/M/s) at pH 7.4 and 25°C [305]. Significantly, alkyl hydroperoxide reductase (AhpE) showed ~10^3 times faster reactivity with peroxynitrite than with hydrogen peroxide [306, 307]. AhpE is a novel subgroup of the Prx family and contains *Mycobacterium tuberculosis AhpE* (*Mt*AhpE) and AhpE-like proteins. In the mechanism study of the reactions, peroxidative thiol oxidation and sulfenic acid overoxidation through sulfenate anion were proposed. Besides peroxynitrite reaction, oxidation ($k \sim 10^8$/M/s) and overoxidation ($k \sim 10^8$/M/s) of *Mt*AhpE by fatty acid-derived hydroperoxides were also suggested.

The biological chemistry of peroxynitrite was recently evaluated [133]. The concentrations of target molecules (R) and rate constants of their reactions with peroxynitrite (k) ultimately determine the relevance of peroxynitrite targets in the biological system. Based on these two parameters, only a few targets such as CO_2, Prx's, other thiol proteins (e.g., GSH peroxidase), and some heme proteins need to be considered (Fig. 5.17) [133]. A range of $[R] \times k = 60$–100/second has been estimated for the reaction of peroxynitrite with CO_2, considering the concentration of CO_2 is high (≥1.3 mM) and the rate constant is pH dependent with a value of 5.8×10^4/M/s at 37°C. This range is the established benchmark to assess the relative importance of other biological targets. For example, the rate constants of peroxynitrite with Prx5 is 7.0×10^7/M/s at pH 7.4 and 37°C and the concentration is >1 μM, giving $[R] \times k$ >70/second, which is higher than that of CO_2. Similarly, the estimated concentration of 2 μM of the selenium-containing protein, GSH peroxidase, in hepatocytes and the rapid reaction with peroxynitrite ($k = 8 \times 10^6$/M/s; pH 7.4, 25°C) gives $[R]k > 16$/M/s. This value is below that of CO_2 but is still significant. Displayed

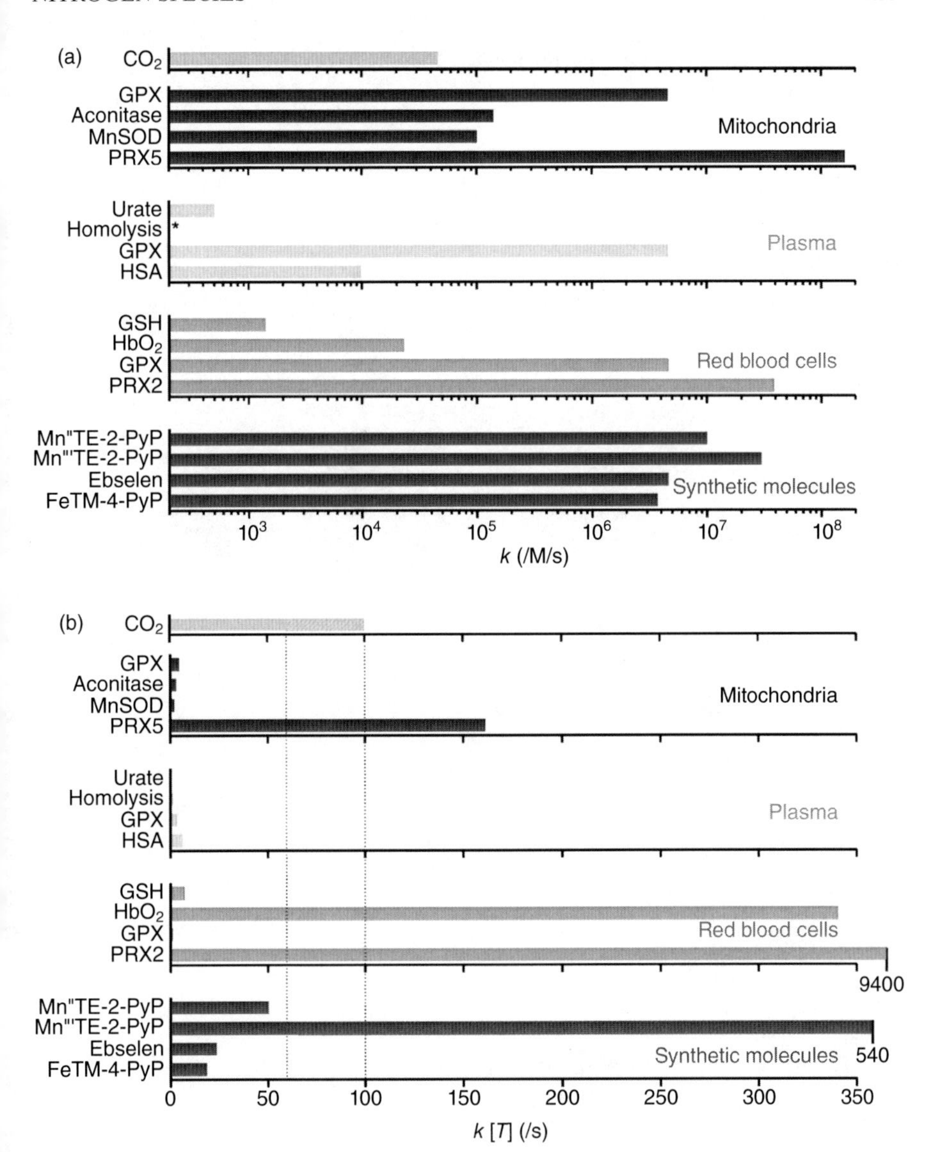

Figure 5.17. Scavenging of peroxynitrite. *Homolysis is a first-order reaction and thus cannot be plotted in this graph (adapted from Ferrer-Sueta and Radi [133] with the permission of the American Chemical Society). See color insert.

in Figure 5.17, only a few biological targets have greater values than the range of 60–100/second for CO_2. These include HbO_2, Prx2, and Mn^{III}TE-2-PyP. Other heme proteins such as Mn^{II}TE-2-PyP, ebselen, and FeTM-4-Pyp have lower values of [R]k than that of CO_2 but are significant [133].

Studies have reported the inactivation of several enzymes by peroxynitrite (Table 5.8) [49, 211, 232, 240, 241, 247, 248, 256, 262, 264–266, 308–346]. It

TABLE 5.8. Enzymes Reported to Become Inactivated upon Exposure to Peroxynitrite [248]

Enzyme	Modified Residue	Reference
ATPase[a]	ND	[308]
Succinate dehydrogenase[a]	ND	[308]
Fumarate reductase (*Trypanosoma cruzi*)[a]	ND	[309]
NADH: ubiquinone oxidoreductase[a]	ND	[308]
Cytochrome P450 BM-3[b]	Cys, Tyr	[256]
Cytochrome P450 2B1[b]	Tyr	[310]
Prostacyclin synthase[a,b]	Tyr	[311, 312]
Inducible nitric oxide synthase[b]	Heme	[313]
Glutathione peroxidase[a,b]	Selenocysteine	[211, 262, 314, 315]
Alcohol dehydrogenase[b]	Zinc sulfur cluster	[231]
Aconitase[a,b]	Iron sulfur cluster	[232, 316–318]
6-Phosphogluconate dehydratase[a]	Iron sulfur cluster	[316]
Fumarase A[a]	Iron sulfur cluster	[316]
Creatine kinase[a,b]	Cys	[265]
Glyceraldehyde 3-phosphate dehydrogenase[a,b]	Cys	[266, 316]
Glutamine synthetase[b]	Tyr, Met	[319]
Succinyl-CoA:3-oxoacid CoA transferase[a]	Tyr	[320]
Mn superoxide dismutase[a,b]	Tyr	[321]
Cu,Zn superoxide dismutase[b]	His B.	[240]
Tyrosine hydroxylase[a,b]	Tyr, Cys	[241, 322]
Tryptophan hydroxylase[b]	Cys	[323]
Ca2þ-ATPase[a]	Cys	[324, 325]
Caspase 3[a,b]	Cys	[326]
Protein tyrosine phosphatase[b]	Cys	[264]
Nicotinamide nucleotide transhydrogenase[a]	Tyr	[327]
Ribonucleotide reductase[b]	Tyr	[328]
Zn^{2+}-glycerophosphocolin cholinephos phodiesterase[b]	Tyr	[329]
NADPH-cytochrome P450 reductase[b]	ND	[330]
Glutathione reductase[b]	Tyr	[331, 332]
Glutathione S-transferase[a,b]	ND	[333]
Glutaredoxin[a]	ND	[334]
Protein kinase C[a,b]	Tyr	[335]
Ornithine decarboxylase[a,b]	Tyr	[336]
Xanthine oxidase[b]	Molybdenum center	[337, 338]
Lysozyme[b]	Trp	[248]
Cystathionine β-synthase (CBS)	Trp, Tyr	[339]
Histone deacetylases	Tyr	[340]
Mitochondrial Lon protease	Glutathione	[341]
Paraoxonase-1	Trp	[342]

TABLE 5.8. (*Continued*)

Enzyme	Modified Residue	Reference
Metalloproteinase-4	Trp	[343]
Muscle glycogen phosphorylase	Tyr	[344]
Mitochondrial aconitase	Glutathione	[49]
Mammalian glutamine synthetase	Tyr	[345]
Microsomal glutathione S-transferase 1	Tyr	[346]

[a] Enzyme in cell extracts, *ex vivo* or *in vivo* systems.
[b] Purified enzyme.
ND, not determined.

appears the modification of a critical amino acid residue or prosthetic group resulted from the irreversible inactivation of enzymes (Table 5.8). In some instances, the loss of activity and modification of amino acids were observed *in vitro* and *in vivo*. For example, peroxynitrite-treated chronically rejected human renal allograft showed nitration and inactivation of manganese superoxide dismutase [321]. Overall, both the reaction rate constants and studies conducted with cells, extracts, or other biological systems must be taken into account to learn about the inactivation of enzymes by peroxynitrite.

5.3 SULFUR SPECIES

5.3.1 Oxysulfur Radicals

Sulfur–oxygen radicals $SO_n^{\bullet -}$ (n = 3, 4, and 5) are involved in the conversion of $SO_{2(g)}$ to acid rain (H_2SO_4). Several studies have therefore been performed to examine the pathways that result in the $SO_n^{\bullet -}$ formation and their oxidation mechanisms [347, 348]. The oxysulfur radicals are also reactive toward biomolecules [347]. Hence, their reactivity toward proteins, DNA, and enzymes has been carried out [349–351]. In this section, the generation and oxidation mechanism of the oxysulfur radicals are described.

5.3.1.1 Generation. The $^{\bullet}SO_3^-$ radical has been generated by different methods: (1) oxidation of sulfite with Ce^{4+} in acidic solution (Eq. 5.45); (2) reaction of sulfite produced by Fenton-type reagents, for example, $^{\bullet}OH$ and $^{\bullet}NH_2$; (3) photoionization of sulfite directly (Eq. 5.46) or through photosensitizers or photolysis of dithionite and thiosulfate; and (4) reaction of sulfite with radiolytically produced $^{\bullet}OH$ radicals (Eq. 5.47) or other oxidizing radicals [352]:

$$Ce^{4+} + SO_3^{2-} \rightarrow [Ce^{3+} \ldots SO_3^-] \rightleftharpoons Ce^{3+} + {}^{\bullet}SO_3^- \quad (5.45)$$

$$SO_3^{2-} + h\nu \rightarrow {}^\bullet SO_3^- \tag{5.46}$$

$$SO_3^{2-} + {}^\bullet OH \rightarrow {}^\bullet SO_3^- + OH^- \quad k_{47} = 5.5 \times 10^9 /M/s. \tag{5.47}$$

The formation of ${}^\bullet SO_3^-$ in the solution was confirmed by a single-line spectrum [352]. The optical absorption of ${}^\bullet SO_3^-$ has an absorption at 255 nm ($\varepsilon_{255\,nm}$ = 1000/M/cm). The decay of ${}^\bullet SO_3^-$ followed second-order kinetics (Eq. 5.48) [352]:

$$2{}^\bullet SO_3^- \rightarrow S_2O_6^{2-} \text{ or } SO_3 + SO_3^{2-} \quad 2k_{48} = 1.1 \times 10^9 /M/s. \tag{5.48}$$

Generally, the kinetics of ${}^\bullet SO_3^-$ with different substrates is studied by monitoring the radicals produced from the reaction [352].

The sulfate radical ($SO_4^{\bullet -}$) has been generated by hemolytic scission of the peroxide of persulfate (PS) or peroxymonosulfate (PMS) by transition metal activation, thermal activation, and photolytic and radiolytic methods [353–357]. The catalytic decomposition of PMS in a homogeneous pathway uses transition metals as an aid (M: Ag^+, Fe^{2+}, Co^{2+}, Ni^{2+}, and Mn^{2+}) (Eq. 5.49):

$$M^{2+} + HSO_5^- \rightarrow M^{3+} + SO_4^{\bullet -} + OH^-. \tag{5.49}$$

An approach of using Fe–Co mixed oxide nanocatalysts for the heterogeneous activation of PMS has also been suggested to produce the $SO_4^{\bullet -}$ radical [358]. Thermal activation of PS yields the $SO_4^{\bullet -}$ radical (Eq. 5.50):

$$S_2O_8^{2-} + \text{heat} \rightarrow 2SO_4^{\bullet -}. \tag{5.50}$$

Reaction (5.50) was recently studied in detail using an EPR spin trapping technique [359]. A mechanism of the reaction (5.50) has also been studied in the temperature range of 60–80°C in the presence and absence of sodium formate [360]. The rate constant was measured as (0.7–1.0) × 10^{-4}/second at 80°C [360]. A chemical probe to differentiate $HO^\bullet$ and $SO_4^{\bullet -}$ radicals in thermal activation of PS has been developed [357].

In the radiolytic method, the reduction of PS by a hydrated electron produces sulfate radicals (Eq. 5.51):

$$S_2O_8^{2-} + e^-_{(aq)} \rightarrow SO_4^{\bullet -} + SO_4^{2-}. \tag{5.51}$$

The spectrum obtained at pH 7.0 using this method is shown in Figure 5.18 [353]. The absorption spectrum has a peak at 450 nm. The sulfate radical has a weak absorption ($\varepsilon_{450\,nm}$ = 1100/M/cm). The photolysis of PS generates the sulfate radical (Eq. 5.52) [354, 361]:

$$S_2O_8^{2-} + h\nu\,(\lambda < 300\text{ nm}) \rightarrow SO_4^{\bullet -}. \tag{5.52}$$

Similarly, sulfate and hydroxyl radicals are produced in the degradation of PMS upon UV irradiation (λ < 260 nm) [354]. The decay kinetics of $SO_4^{\bullet -}$ in

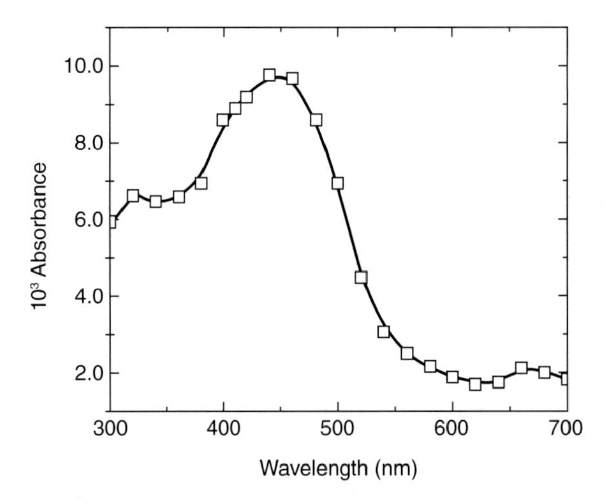

Figure 5.18. Transient absorption spectrum obtained for the $SO_4^{-\bullet}$ radical in N_2-saturated pH 7.4 water containing 5.0 mM phosphate buffer, 5.0 mM $K_2S_2O_8$, and 0.50 M *tert*-butanol (adapted from Rickman and Mezyk [353] with the permission of Elsevier).

the photolytic system of PS (Eq. 5.50) was described by mixed first- and second-order kinetics [361]. The mechanism of the decay was given by Equations (5.53)–(5.55) [362]:

$$SO_4^{\bullet-} + SO_4^{\bullet-} \rightarrow S_2O_8^{2-} \quad (5.53)$$

$$SO_4^{\bullet-} + H_2O \rightarrow {}^{\bullet}OH + HSO_4^{\bullet-} \quad (5.54)$$

$$SO_4^{\bullet-} + S_2O_8^{2-} \rightarrow SO_4^{2-} + S_2O_8^{2-} \quad k_{55} \leq 10^4 /M/s. \quad (5.55)$$

At 296 K, the rate constants of reactions (5.53) and (5.54) were determined as $2k_{53}/\varepsilon = (2.5 \pm 0.2) \times 105$ cm/s (ε is the absorption coefficient of SO4$^{\bullet-}$ at 454 nm) and $k_{54}[H_2O] = 440 \pm 50$/second, respectively. Rate constants k_{53} and k_{54} were found to increase strongly and nonlinearly with increasing ionic strength [361]. Reaction (5.52) was studied as a function of temperature (11.8–74.4°C) and the rate constants were expressed by Equations (5.56) and (5.57) [361]:

$$2k_{53}/\varepsilon = (4.8 \pm 2.0) \times 10^5 \exp(-1.7 \pm 1.1 \text{ kJ/mol/RT}) \text{ cm/s} \quad (5.56)$$

$$k_{54} = (4.7 \pm 0.1) \times 10^3 \exp(-15.5 \pm 0.6 \text{ kJ/mol/RT})/\text{M/s}. \quad (5.57)$$

The reaction of ${}^{\bullet}SO_3^-$ with oxygen generates the peroxysulfate radical ($SO_5^{\bullet-}$) (Eq. 5.58) [352]:

$${}^{\bullet}SO_3^- + O_2 \rightarrow SO_5^{\bullet-} \quad k_{58} = 2.1 \times 10^9 /M/s. \quad (5.58)$$

The reactions of $^{\bullet}OH$ with HSO_5^- (acidic medium) and SO_5^{2-} (alkaline medium) also produce $SO_5^{\bullet-}$ (Eqs. 5.59, 5.60) [363]:

$$^{\bullet}OH + HSO_5^- \rightarrow SO_5^{\bullet-} + H_2O \quad k_{59} = 1.7 \times 10^7\,/M/s \tag{5.59}$$

$$^{\bullet}OH + SO_5^{2-} \rightarrow SO_5^{\bullet-} + OH^- \quad k_{60} = 2.1 \times 10^9\,/M/s. \tag{5.60}$$

The absorption band of $SO_5^{\bullet-}$ is in the UV region (λ_{max} = 260–265 nm) with $\varepsilon = 1065 \pm 80$/M/cm [364].

5.3.1.2 Reactivity. The $^{\bullet}SO_3^-$ radical can act as a reductant and an oxidant. For example, $^{\bullet}SO_3^-$ can reduce $Fe^{VI}O_4^{2-}$ to $Fe^{V}O_4^{3-}$ ($k = (1.9 \pm 0.3) \times 10^8$/M/s) [365]. Reactions of $^{\bullet}SO_3^-$ with organic compounds involve hydrogen abstraction or addition to double bonds [352]. The $^{\bullet}SO_3^-$ oxidizes chlorpromazine, hydroquinine, phenylenediamines, ascorbate, and methoxyphenol with rate constants varying from 10^6 to 10^9/M/s [352]. The rate constant of the $^{\bullet}SO_3^-$ radical with Trp at pH 3.2 is ~8×10^4/M/s, while the reactivity with Gly is sluggish ($k < 10^3$/M/s). The $SO_5^{\bullet-}$ radical is a stronger oxidant than the $^{\bullet}SO_3^-$ radical, and it oxidizes biological molecules (e.g., ascorbate) more rapidly than $^{\bullet}SO_3^-$ [352]. Comparatively, the $SO_4^{\bullet-}$ radical is very reactive toward organic compounds in which it is capable of oxidizing by electron transfer, abstracting H atoms, or adding to double bonds.

In recent years, several studies of the $SO_4^{\bullet-}$ radical with compounds of environmental interest have been carried out [366]. The compounds studied include sulfur-containing compounds (dimethylsulfoxide, dimethylsulfone, and methane sulfone), pyrimidines, pyridines, gallic acid, fulvic acid, humic acids, dodecylbenzene sulfonate, iopromide, triclosan, sulfamethoxazole, acetaminophen, and microcystin-LR [354, 356, 360, 367–375]. Among the sulfur-containing compounds, dimethylsulfoxide reacted the fastest at 298 K (Eqs. 5.61–5.63) [373]:

$$CH_3S(O)CH_3 + SO_4^{\bullet-} \rightarrow \text{products} \quad k_{61} = (3.0 \pm 0.4) \times 10^9\,/M/s \tag{5.61}$$

$$CH_3(O)S(O)CH_3 + SO_4^{\bullet-} \rightarrow \text{products} \quad k_{62} < (3.9 \pm 0.5) \times 10^6\,/M/s \tag{5.62}$$

$$CH_3(O)S(O)O^- + SO_4^{\bullet-} \rightarrow \text{products} \quad k_{63} = (1.1 \pm 0.3) \times 10^4\,/M/s. \tag{5.63}$$

The activation energies were determined as 12.0 ± 0.4, 11.3 ± 1.3, and 20.7 ± 4.3 kJ/mol for reactions (5.61)–(5.63), respectively. One-electron transfer mechanism for reaction (5.61) has been demonstrated (Eq. 5.64) [376]:

$$CH_3S(O)CH_3 + SO_4^{\bullet-} \rightarrow CH_3S(O^+)CH_3 + SO_4^{2-}. \tag{5.64}$$

Gallic acid and the gallate ion reacted with $SO_4^{\bullet-}$ with bimolecular rate constants, $(6.3 \pm 0.7) \times 10^8$/M/s and $(2.9 \pm 0.2) \times 10^9$/M/s, respectively [368]. The mechanism involves a formation of the phenoxy radical by the hydrogen atom abstraction from the phenol to the $SO_4^{\bullet-}$ radical. This mechanism was supported by DFT calculations in the gas phase and in the aqueous solution [368]. Studies on the oxidation of fulvic and humic acids by $SO_4^{\bullet-}$ showed the formation of H-bonds to $SO_4^{\bullet-}$ to yield adducts [371, 372]. Formation of radical adducts has been suggested in the reactions of pyrimidine and pyridine [367, 374]. The rate constants for the reactions of the $SO_4^{\bullet-}$ radical with a wide range of β-lactum antibiotics in water were determined as $(2.1 \pm 0.6) \times 10^9$/M/s and $(1.6 \pm 0.9) \times 10^9$/M/s for five- and six-member antibiotics at zero ionic strength and 20–22°C [353]. The transient spectra obtained during the reactions showed the $SO_4^{\bullet-}$ radicals reacted at the five- or six-member rings adjacent to the β-lactum moiety, mainly at the sulfur atom and the double bond, respectively [353].

The kinetics for the oxidation of amino acids and peptides by $SO_4^{\bullet-}$ radicals have been performed (Table 5.9) [352, 377–380]. Peptides were found to be more reactive than their parent free amino acids (Table 5.9). Aromatic amino acids had a higher reactivity than aliphatic amino acids. The aromatic amino acids confer a higher reactivity than those of aliphatic acids in the peptides (Table 5.9). For example, Tyr and Trp amino acids containing peptides reacted at diffusion-controlled rate constants, which are much faster than those containing only Gly and/or Ala [377].

The spectra of transients during the oxidation of amino acids and peptides by $SO_4^{\bullet-}$ radicals were used to understand the mechanisms of the reactions

TABLE 5.9. Rate Constants for Reactions of Amino Acids and Peptides with $SO_4^{\bullet-}$

Amino Acids	k (/M/s)	Peptide	k (/M/s)
Gly	$(3.7 \pm 0.1) \times 10$ [6, 377]	Gly-Gly	$(8.3 \pm 1.0) \times 10^7$ [377]
	~9×10 [6, 352]	Gly-Ala	$(9.8 \pm 0.8) \times 10^7$ [377]
Ala	$(4.9 \pm 0.1) \times 10$ [6, 377]	Ala-Ala	$(4.8 \pm 0.4) \times 10^7$ [377]
Val	2.1×10^7 [378]	Gly-Trp	$(1.1 \pm 0.1) \times 10^{10}$ [377]
Lys	1.6×10^7 [379]	Trp-Gly	$(1.5 \pm 0.2) \times 10^{10}$ [377]
Arg	7.5×10^7 [380]	Gly-Tyr	$(1.2 \pm 0.4) \times 10$ [10, 377]
Trp	2.9×10 [9, 377]	Tyr-Gly	$(2.7 \pm 0.2) \times 10$ [10, 377]
	~2×10 [9, 352]	Tyr-Val	8.2×10^8 [378]
Tyr	~3×10 [9, 352]	Val-Tyr	1.2×10^9 [378]
	5.8×10 [9, 378]	Gly-Gly-Gly	$(1.2 \pm 0.2) \times 10^8$ [377]
His	~2.5×10 [9, 352]	Ala-Gly-Gly	$(8.4 \pm 0.8) \times 10$ [7, 377]
		Val-Tyr-Val	4.4×10^8 [378]
		Gly-Gly-Gly-Gly	$(1.3 \pm 0.2) \times 10^8$ [377]

[377, 378, 380]. The electron transfer–decarboxylation of Gly and Ala by $SO_4^{\bullet-}$ radicals lead to reducing α-amino radicals (Eqs. 5.65, 5.66) [377]:

$$XCYHCOO^- + SO_4^{\bullet-} \rightarrow XCYHCOO^\bullet + SO_4^{2-} \quad (5.65)$$

$$XCYHCOO^\bullet \rightarrow CO_2 + {}^\bullet CYHX. \quad (5.66)$$

The formation of radicals via hydrogen abstraction occurs in the oxidation of Leu and Ser by $SO_4^{\bullet-}$ radicals [350]. A kinetic study of the reaction of L-Arg with $SO_4^{\bullet-}$ radicals indicated the formation of C-centered radicals as the transient species [380]. However, generation of N-centered radicals was suggested in the reaction of $SO_4^{\bullet-}$ radicals with L-Lys [379]. Carbon- and nitrogen-centered radicals have been suggested in the reaction of $SO_4^{\bullet-}$ radicals with the guanidine derivatives [381].

Because dipeptides Ala-Ala and Gly-Gly reacted much faster than either Gly or Ala, a different initial step involving the peptide bond in the mechanism was suggested (Eq. 5.67):

$$\begin{aligned} &{}^+H_3NCHXCONHCHYCOO^- + SO_4^{\bullet-} \\ &\rightarrow [{}^+H_3NCHXCONHCHYCOO^-]^{\bullet+} + SO_4^{2-}. \end{aligned} \quad (5.67)$$

The interaction of $SO_4^{\bullet-}$ with the aromatic amino acid of the dipeptides, Trp-Gly, Gly-Trp, Gly-Tyr, and Tyr-Gly is expected to occur [377]. Figure 5.19 shows the oxidation of Tyr-Gly and Gly-Tyr by $SO_4^{\bullet-}$ radicals [377]. An adduct was initially formed, which then yielded the radical cation of the aromatics and the formation of the hydroxycyclohexadienyl radical. The elimination of sulfite yields the phenoxy radicals, which was detected experimentally [377]. The phenoxyl radicals were also observed in the reactions of flavonoids with $SO_4^{\bullet-}$ ($k \approx 10^9$/M/s) [382]. Recently, $SO_4^{\bullet-}$ was suggested to be a new reagent for protein footprinting due to its potential as an oxidant for proteins [351]. Examples include the formation of protein cross-linkage through the interaction with $SO_4^{\bullet-}$ and occurrence of the metal-catalyzed oxidation of proteins by $SO_4^{\bullet-}$ [383, 384]. The inactivation of enzymes, ribonucleases, and lysozymes has also been reported [349].

5.4 PHOSPHOROUS SPECIES

5.4.1 Phosphate Radicals

Photochemical and/or thermal processes in natural and atmospheric water resulted in reactive species such as ${}^\bullet OH$ and $SO_4^{-\bullet}$ radicals, which have a capability of generating secondary radicals by reacting with inorganic anions. Phosphate ions can react with reactive species to form phosphate radicals that have unexpected consequences in overall chemical processes [385]. For example,

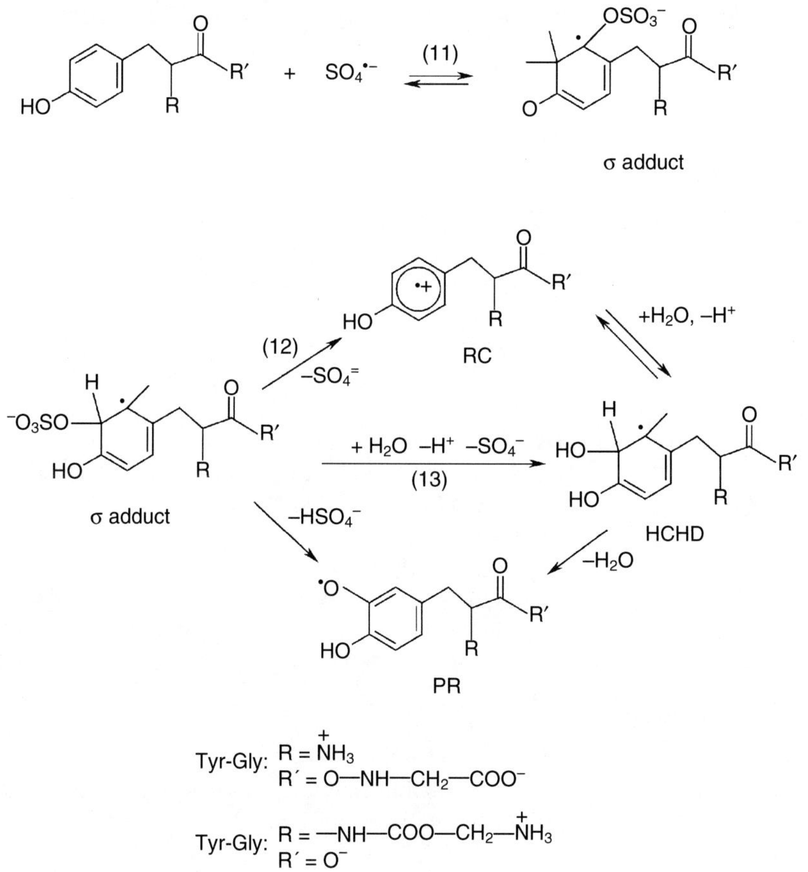

Figure 5.19. Scheme proposed for the reactions of Tyr-Gly and Gly-Tyr with sulfate radicals. (adapted from [377] with the permission of the Royal Society of Chemistry).

the phosphate radicals are formed in sunlight irradiated natural waters containing phosphate ions or during advanced oxidation treatments of contaminated waters. The fate of the pollutants in both cases was affected by the presence of the phosphate radicals. The chemistry of phosphate radicals is briefly described below.

5.4.1.1 Generation, Equilibria, and Spectral Characteristics. The reaction of $^{\bullet}OH$ and $SO_4^{\bullet-}$ with phosphates results in the formation of phosphate radicals (Eqs. 5.66–5.69) [347, 386]:

$$H_3PO_4 + {}^{\bullet}OH \rightarrow H_2PO_4^{\bullet} + H_2O \quad k_{66} = 2.6 \times 10^6\,/M/s \tag{5.66}$$

$$H_2PO_4^- + {}^{\bullet}OH \rightarrow H_2PO_4^{\bullet}/HPO_4^{\bullet-} + OH^-/H_2O \quad k_{67} = 2.2 \times 10^6\,/M/s \tag{5.67}$$

$$HPO_4^{2-} + {}^{\bullet}OH \rightarrow HPO_4^{\bullet-}/PO_4^{\bullet 2-} + OH^-/H_2O \quad k_{68} = 8.0 \times 10^5/M/s \tag{5.68}$$

$$HPO_4^{2-} + SO_4^{\bullet-} \rightarrow HPO_4^{\bullet-} + SO_4^{2-} \quad k_{69} = 1.2 \times 10^6/M/s. \tag{5.69}$$

The photolysis of the phosphate ion also yields phosphate radicals (Eq. 5.70) [387, 388]:

$$HPO_4^{2-} + h\nu(\lambda_{exc} < 200\ nm) \rightarrow HPO_4^{\bullet-} + e_{aq}^-. \tag{5.70}$$

Phosphate radicals exist in three acid–base forms (Eqs. 5.71, 5.72) [386]:

$$H_2PO_4^{\bullet} \rightleftharpoons H^+ + HPO_4^{\bullet-} \quad pK_{a1} = 5.7 \tag{5.71}$$

$$HPO_4^{\bullet-} \rightleftharpoons H^+ + PO_4^{\bullet 2-}. \quad pK_{a2} = 8.9 \tag{5.72}$$

Phosphate radicals have broad absorptions in the 400- to 600-nm range: $H_2PO_4^{\bullet}$ (λ_{max} 520 nm, ε_{max} 1850/M/cm), $HPO_4^{\bullet-}$ (λ_{max} 510 nm, ε_{max} 1550/M/cm), and $PO_4^{\bullet 2-}$ (λ_{max} 530 nm, ε_{max} 2150/M/cm) [386].

The decay kinetics of the three acid–base forms of phosphate radicals have been studied [389]. The kinetics results were independent of the dissolved oxygen in the solution. Bimolecular rate constants ($2k/\varepsilon_{500\,nm}$) for the decay of $HPO_4^{\bullet-}$ and $PO_4^{\bullet 2-}$ radical ions were determined as 1.3×10^5 and 2.8×10^5 cm/s, respectively [388, 389]. The kinetics of $H_2PO_4^{\bullet}$ was determined to be mixed first and second orders with rate constants of 5×10^3/second and 1.0×10^6 cm/s, respectively.

5.4.1.2 Reactivity. The kinetics of the reactions of phosphate radicals with a number of organic compounds has been performed [363]. The reactivity of $H_2PO_4^{\bullet}$ radicals were determined to be higher than that of $HPO_4^{\bullet-}$ and $PO_4^{\bullet 2-}$ by a factor of ~4 to 10. The reactivities of $HPO_4^{\bullet-}$ and $PO_4^{\bullet 2-}$ were similar. Phosphate radicals reacted with compounds by the abstraction of hydrogen atoms, and the rate constants ranged from ~10^5/M/s for acetic acid and 2-methyl-2-propanol to ~10^8/M/s for 2-propanol and formate. The rate constants for the reactivity of phosphate radicals with aromatic compounds were in the range of ~10^8 to 10^9/M/s [382, 385, 386, 390]. Similar to $SO_4^{\bullet-}$ radicals, the $H_2PO_4^{\bullet}$ radical reacted with the aromatic ring by a one-electron transfer to the inorganic radical and also produced phenoxy radicals in reactions with phenolic derivatives of α, α, and α-trifluorotoluene and flavanoid [382, 390]. Comparatively, the singlet oxygen ($^1O_2$) involved the charge transfer mechanism and was found to be much less reactive with undissociated trifluoromethylphenols than the $HPO_4^{\bullet-}$. However, products obtained using $^1O_2$ were more oxidized than those observed (e.g., biphenyl) in the oxidation by $HPO_4^{\bullet-}$. These results are significant in wastewater treatment of phenol in which advanced oxidation process is combined with biological mineralization [385].

The rates of hydrogen abstraction reactions of $SO_4^{\bullet-}$ and $H_2PO_4^{\bullet}$ radicals with aromatic compounds were ~10 to 100 times lower than those of the ${}^{\bullet}OH$ radical. The general reactivity of radicals can be expressed as

$^{\bullet}OH > SO_4^{\bullet-} > H_2PO_4^{\bullet} > HPO_4^{\bullet-} > PO_4^{\bullet 2-}$. This general trend does not necessarily represent one-electron oxidizing capability as the $SO_4^{\bullet-}$ radical is a stronger oxidant (2.5–3.1 V vs. NHE) than the $^{\bullet}OH$ (2.3 V vs. NHE in neutral solution). Therefore, the relative reactivity of the sulfate and phosphate radicals with inorganic and compounds predict the reduction potential tendencies as $E(SO_4^{\bullet-}) > E(H_2PO_4^{\bullet}) > E(HPO_4^{\bullet-}) > E(PO_4^{\bullet 2-})$ [386]. The selectivity of $SO_4^{\bullet-}$ and $H_2PO_4^{\bullet}$ radicals were higher than the $^{\bullet}OH$ radical. Reactions of substituted benzenes go through addition and hydrogen atom abstractions, while sulfate radicals involved electron transfer reactions. The effect of the substituents has been demonstrated in a Hammett-type plot for substituted benzenes [389, 391]. The Hammett parameters (ρ^+) were determined as −1.5, −1.2, −1.3, and −1.7 for reactions of $SO_4^{\bullet-}$, $H_2PO_4^{\bullet}$, $HPO_4^{\bullet-}$, and $PO_4^{\bullet 2-}$, respectively [387, 392]. The negative values of ρ^+ suggests that the reactions were favored by high electron density at the site of the reaction [387]. Comparatively, the parameter, ρ, for the reaction of $^{\bullet}OH$ with aromatic compounds was determined between −0.5 and −0.4 [393].

The reactivity of phosphate radicals with amino acids and peptides is presented in Table 5.10 [349, 377, 378]. Similar to $SO_4^{\bullet-}$ radical, $HPO_4^{\bullet-}$ and $PO_4^{\bullet 2-}$ radicals reacted faster with peptides than free amino acids. Also, aromatic amino acids had a higher reactivity than aliphatic amino acids. Significantly,

TABLE 5.10. Second-Order Rate Constants for the Rate Constants of $HPO_4^{\bullet-}$ and $PO_4^{\bullet 2-}$ with Amino Acids and Peptides [349, 377, 378]

	k (/M/s)	
Amino Acids/Peptides	$HPO_4^{\bullet-}$	$PO_4^{\bullet 2-}$
Gly	$(5.2 \pm 0.6) \times 10^5$	–
Ala	$(9.4 \pm 0.6) \times 10^5$	–
Val	4.1×10^7	1.6×10^7
Trp	$(2.7 \pm 0.2) \times 10^8$	–
Tyr	4.0×10^8	1.2×10^8
Ala-Ala	$(2.1 \pm 0.2) \times 10^6$	–
Gly-Ala	$(5.0 \pm 0.2) \times 10^6$	–
Gly-Trp	$(5 \pm 1) \times 10^9$	–
Trp-Gly	$(2.6 \pm 0.5) \times 10^9$	–
Gly-Tyr	$(2.0 \pm 0.9) \times 10^9$	–
Tyr-Gly	$(2.1 \pm 0.1) \times 10^9$	–
Val-Tyr	2.2×10^8	1.0×10^8
Tyr-Val	5.9×10^8	1.2×10^8
Gly-Gly-Gly	$(1.6 \pm 0.1) \times 10^7$	–
Ala-Gly-Gly	$(1.3 \pm 0.1) \times 10^7$	–
Val-Tyr-Val	4.7×10^8	1.4×10^8
Gly-Gly-Gly-Gly	$(1.3 \pm 0.1) \times 10^7$	–

the reactivity of the $HPO_4^{\bullet-}$ radical was higher than that of the $PO_4^{\bullet 2-}$ radical [377, 378]. The reaction mechanisms for the oxidation of Trp and Tyr peptides by phosphate radicals were studied by observing the spectra of transients. The organic radicals were formed from reaction (5.73):

$$HPO_4^{\bullet-} + \text{Gly-Trp (or Tyr-Gly)} \rightarrow \text{organic radicals.} \qquad (5.73)$$

The phenoxy radical of Tyr as one of the organic radicals was detected as an intermediate involved in the oxidation of Tyr by $HPO_4^{\bullet-}$ [378].

5.5 CONCLUSIONS

Carbonate, nitrogen dioxide, sulfate, and phosphate radicals are one-electron oxidants, while peroxynitrite is considered a two-electron oxidant. Peroxynitrite can be derived through $^{\bullet}OH$, $^{\bullet}NO_2$, and $CO_3^{\bullet-}$ in the absence and presence of CO_2. Rate constants for the reactions of one- and two-electron oxidants of biological importance with sulfur-containing molecules are compared in Table 5.11 [127, 267, 268, 394–400]. Hydrogen sulfide is an endogenously generated gaseous molecule and can mediate a wide range of biological responses and is therefore included as a substrate in comparing reactivities [397, 401]. Peroxynitrite reacted much faster than hydrogen peroxide (Table 5.11). This may be related to the leaving-group tendency of these oxidants (HNOOH; $pK_a = 3.15$ vs. $pK_a = 15.7$). The calculation for the intrinsic reactivity of HS^- from the data in Table 5.11 showed that the reactivity of HS^- with oxidants was faster than Cys and GSH. Hydrogen sulfide is a less favorable target by oxidants *in vivo* unless its local concentration is high. One-electron oxidation of thiols yielded thiyl radicals ($RS^{\bullet}$), which could (1) recombine to form disulfides, (2) react with oxygen to yield secondary radicals (e.g., $RSOO^{\bullet}$), and (3) react with a thiolate to yield a reductive disulfide radical anion ($RSSR^{\bullet-}$). The radical anion can react with oxygen to form superoxide, which may also react with thiols to initiate the oxygen-dependent chain reaction.

Peroxynitrite is very reactive against the Cys, Met, Trp, Tyr, Phe, and His residues in proteins. The significance of peroxynitrite in the biological environment is also related to the rates of its formation and decay and to the diffusion across membranes of the involved species [133]. The levels of CO_2 present in biological systems also need to be considered to assess the role of peroxynitrite. While performing studies on reactions of peroxynitrite with heme proteins, simple spectroscopic methods such as UV–visible spectroscopy have been used. Future studies may examine intermediates of the reactions by using additional analytical methods such as Mössbauer, EPR, and X-ray absorption spectroscopies [402, 403]. Further studies are needed to understand the importance of nitrated protein modifications in normal cellular function and in

TABLE 5.11. Rate Constants for Reactions of Hydrogen Sulfide and the Low-Molecular-Weight Thiols Cysteine and Glutathione with One- and Two-Electron Oxidants

	Hydrogen Sulfide (pK_a 7.0)	Cysteine (pK_a 8.1–8.4)	Glutathione (pK_a 8.6–9.2)
Two-electron oxidants			
Peroxynitrite	4.8×10^3/M/s (pH 7.4, 37°C) This work	4.5×10^3/M/s (pH 7.4, 37°C) [400]	1.3×10^3/M/s (pH 7.4, 37°C) [267, 268]
Hydrogen peroxide	0.73/M/s (pH 7.4, 37°C) This work	2.9/M/s (pH 7.4, 37°C) [267]	0.87/M/s (pH 7.4, 37°C) [267]
Hypochlorite	8×10^7/M/s (pH 7.4, 37°C) This work	3.2×10^7/M/s (pH 7.4, 25°C) [399]	$\geq 10^7$/M/s (pH 7.4, 25°C) [398]
TauCl	303/M/s (pH 7.4, 37°C) This work	205/M/s (pH 7.4, 24°C) [394]	115/M/s (pH 7.4, 24°C) [394]
One-electron oxidants			
$^{\bullet}NO_2$	1.2×10^7/M/s (pH 7.5, 25°C) This work	5.0×10^7/M/s (pH 7.0) [127]	2.0×10^7/M/s (pH 7.0) [127]
$HO^{\bullet}$	1.5×10^{10}/M/s (pH 6) [395]	3.5×10^{10}/M/s (pH 7.0) [395]	2.3×10^{10}/M/s (pH 6.8) [395]
$CO_3^{\bullet-}$	2.0×10^8/M/s (pH 7.0) [396]	4.6×10^7/M/s (pH 7.0) [395]	5.3×10^6/M/s (pH 7.0) [395]

Adapted from Carballal et al. [397] with the permission of Elsevier Inc.

diseases [133]. The significance of the nitration of protein-Tyr and protein-Trp has been demonstrated. Advancement made in the proteomic and immunological approaches will help to enhance understanding of the site of modification and mechanisms *in vivo* [282]. Kinetics of the reaction between Prx's and reactive species such as peroxynitrite and peroxides are forthcoming to understand how *M. tuberculosis* can survive even in the presence of reactive oxygen and nitrogen species. However, mechanisms of these reactions are still far from being understood, and future work in this field may allow development of new processes for tuberculosis treatment. Reactive nitrogen species have shown physiological relevance in enhancing nucleotide exchange in Ras proteins, but future studies are needed to understand how redox species such as reactive nitrogen species (e.g., $^{\bullet}NO$) regulate Ras proteins [404].

Sulfate and phosphate radicals appear to play no significant direct role in biological systems. However, these radicals are important in atmospheric and advanced oxidation processes. Sulfate radicals can efficiently oxidize nitrogenous compounds (e.g., amino acids, proteins, and microcystins) in water. The product determination of the reactions with sulfate and phosphate radicals remains largely elusive. Future studies may include applications of analytical methods such as LC/MS/MS to identify products in order to learn reaction pathways of the reactions.

REFERENCES

1 Augusto, O., Bonini, M.G., Amanso, A.M., Linares, E., Santos, C.C.X., and De Menezes, S.L. Nitrogen dioxide and carbonate radical anion: two emerging radicals in biology. *Free Radic. Biol. Med.* 2002, *32*, 841–859.

2 Medinas, D.B., Cerchiaro, G., Trindade, D.F., and Augusto, O. The carbonate radical and related oxidants derived from bicarbonate buffer. *IUBMB Life* 2007, *59*, 255–262.

3 Finkel, T. and Holbrook, N.J. Oxidants, oxidative stress and the biology of ageing. *Nature* 2000, *408*, 239–247.

4 Radi, R. Nitric oxide, oxidants, and protein tyrosine nitration. *Proc. Natl. Acad. Sci. U.S.A.* 2004, *101*, 4003–4008.

5 Zeldes, H. and Livingston, R. A paramagnetic species in irradiated $NaNO_2$. *J. Chem. Phys.* 1961, *35*, 563–567.

6 Lttz, Z., Retjveni, A., Holmberg, R.W., and Silver, B.L. ESR of ^{17}O-labeled nitrogen dioxide trapped in a single crystal of sodium nitrite. *J. Chem. Phys.* 1969, *51*, 4017–4024.

7 Brailsford, J.R. and Morton, J.R. Paramagnetic resonance spectra of NO_2 trapped in alkali halides. *J. Magn. Reson.* 1969, *1*, 575–583.

8 Guais, A., Brand, G., Jacquot, L., Karrer, M., Dukan, S., Grévillot, G., Molina, T.J., Bonte, J., Regnier, M., and Schwartz, L. Toxicity of carbon dioxide: a review. *Chem. Res. Toxicol.* 2011, *24*, 2061–2070.

9 Wine, P.H. Editorial: atmospheric and environmental physical chemistry: pollutants without borders. *J. Phys. Chem. Lett.* 2010, *1*, 1749–1751.

10 Zhu, L., Nicovich, J.M., and Wine, P.H. Temperature-dependent kinetics studies of aqueous phase reactions of SO_4—radicals with dimethylsulfoxide, dimethylsulfone, and methanesulfonate. *J. Photochem. Photobiol. A.* 2003, *157*, 311–319.

11 Liochev, S.I. and Fridovich, I. Copper, zinc superoxide dismutase and H_2O_2: effects of bicarbonate on inactivation and oxidations of NADPH and urate, and on consumption of H_2O_2. *J. Biol. Chem.* 2002, *277*, 34674–34678.

12 Liochev, S.I. and Fridovich, I. CO_2, not HCO_3^-, facilitates oxidations by Cu,Zn superoxide dismutase plus H_2O_2. *Proc. Natl. Acad. Sci. U.S.A.* 2004, *101*, 743–744.

13 Bonini, M.G., Miyamoto, S., Mascio, P.D., and Augusto, O. Production of the carbonate radical anion during xanthine oxidase turnover in the presence of bicarbonate. *J. Biol. Chem.* 2004, *279*, 51836–51843.

14 Elam, J.S., Malek, K., Rodriguez, J.A., Doucette, P.A., Taylor, A.B., Hayward, L.J., Cabelli, D.E., Valentine, J.S., and Hart, P.J. An alternative mechanism of

bicarbonate-mediated peroxidation by copper-zinc superoxide dismutase. Rates enhanced via proposed enzyme-associated peroxycarbonate intermediate. *J. Biol. Chem.* 2003, *278*, 21032–21039.

15 Ramirez, D.C., Gomez-Mejiba, S.E., Corbett, J.T., Deterding, L.J., Tomer, K.B., and Mason, R.P. Cu,Zn-superoxide dismutase-driven free radical modifications: copper and carbonate radical anion-initiated protein radical chemistry. *Biochem.* 2009, *417*, 341–353.

16 Kurahashi, T., Miyazaki, A., Suwan, S., and Isobe, M. Extensive investigations on oxidized amino acid residues in H_2O_2-treated Cu,Zn-SOD protein with LC-ESI-Q-TOF-MS, MS/MS for the determination of the copper-binding site. *J. Am. Chem. Soc.* 2001, *123*, 9268–9278.

17 Ramirez, D.C., Mejiba, S.E.G., and Mason, R.P. Copper-catalyzed protein oxidation and its modulation by carbon dioxide: enhancement of protein radicals in cells. *J. Biol. Chem.* 2005, *280*, 27402–27411.

18 Liochev, S.I. and Fridovich, I. The role of CO_2 in cobalt-catalyzed peroxidations. *Arch. Biochem. Biophys.* 2005, *439*, 99–104.

19 Liochev, S.I. and Fridovich, I. Carbon dioxide mediates Mn(II)-catalyzed decomposition of hydrogen peroxide and peroxidation reactions. *Proc. Natl. Acad. Sci. U.S.A.* 2004, *101*, 12485–12490.

20 Arai, H., Berlett, B.S., Chock, P.B., and Stadtman, E.R. Effect of bicarbonate on iron-mediated oxidation of low-density lipoprotein. *Proc. Natl. Acad. Sci. U.S.A.* 2005, *102*, 10472–10477.

21 Flanagan, J., Jones, D.P., Griffith, W.P., Skapski, A.C., and West, A.P. On the existence of peroxocarbonates in aqueous solution. *J. Chem. Soc. Chem. Commun.* 1986, *1*, 20–21.

22 Richardson, D.E., Yao, H., Frank, K.M., and Bennett, D.A. Equilibria, kinetics, and mechanism in the bicarbonate activation of hydrogen peroxide: oxidation of sulfides by peroxymonocarbonate. *J. Am. Chem. Soc.* 2000, *122*, 1729–1739.

23 Hodgson, E.K. and Fridovich, I. The mechanism of the activity dependent luminescence of xanthine oxidase. *Arch. Biochem. Biophys* 1976, *172*, 202–205.

24 Yao, H. and Richardson, D.E. Bicarbonate surfoxidants: micellar oxidations of aryl sulfides with bicarbonate-activated hydrogen peroxide. *J. Am. Chem. Soc.* 2003, *125*, 6211–6221.

25 Behar, D., Czapski, G., and Duchovny, I. Carbonate radical in flash photolysis and pulse radiolysis of aqueous carbonate solutions. *J. Phys. Chem.* 1970, *74*, 2206–2210.

26 Chawla, O.P. and Fessenden, R.W. Electron spin resonance and pulse radiolysis studies of some reactions of $SO_4^{\cdot-}$. *J. Phys. Chem.* 1975, *79*, 2693–2700.

27 Chen, S.N., Cope, V.W., and Hoffman, M.Z. Behavior of CO_3—radicals generated in the flash photolysis of carbonatoamine complexes of cobalt(III) in aqueous solution. *J. Phys. Chem.* 1973, *77*, 1111–1116.

28 Hayon, E. Flash photolysis in the vacuum ultraviolet region of SO_4^{2-}, CO_3^{2-} and OH^- ions in aqueous solutions. *J. Phys. Chem.* 1967, *71*, 1472–1477.

29 Hisatsune, I.C., Adl, T., Beahm, E.C., and Kempf, R.J. Matrix isolation and decay kinetics of carbon dioxide and carbonate anion free radicals. *J. Phys. Chem.* 1970, *74*, 3225–3231.

30 Serway, R.A. and Marshall, S.A. Electron spin resonance absorption spectra of CO_3^- and CO_3^{3-} molecule-ions in irradiated single-crystal calcite. *J. Chem. Phys.* 1967, *46*, 1949–1952.

31 Busset, C., Mazellier, P., Sarakha, M., and De Laat, J. Photochemical generation of carbonate radicals and their reactivity with phenol. *J. Photochem. Photobiol. A*. 2007, *185*, 127–132.

32 Buxton, G.V., Greenstock, C.L., Helman, W.P., and Ross, W.P. Critical review of rate constants for reactions of hydrated electrons, hydrogen atoms and hydroxyl radicals in aqueous solution. *J. Phys. Chem. Ref. Data* 1988, *17*, 513–886.

33 Czapski, G. Acidity of the carbonate radical. *J. Phys. Chem. A* 1999, *103*, 3447–3450.

34 Lymar, S.V., Schwarz, H.A., and Czapski, G. Medium effects on reactions of the carbonate radical with thiocyanate, iodide, and ferrocyanide ions. *Radiat. Phys. Chem.* 2000, *59*, 387–392.

35 Armstrong, D.A., Waltz, W.L., and Rauk, A. Carbonate radical anion—thermochemistry. *Can. J. Chem.* 2006, *84*, 1614–1619.

36 Bisby, R.H., Johnson, S.A., Parker, A.W., and Tavender, S.M. Time-resolved resonance Raman spectroscopy of the carbonate radical. *J. Chem. Soc. Faraday Trans.* 1998, *94*, 2069–2072.

37 Bonini, M.G., Radi, R., Ferrer-Sueta, G., Ferreira, A.M.D., and Augusto, O. Direct EPR detection of the carbonate radical anion produced from peroxynitrite and carbon dioxide. *J. Biol. Chem.* 1999, *274*, 10802–10806.

38 Haygarth, K.S., Marin, T.W., Janik, I., Kanjana, K., Stanisky, C.M., and Bartels, D.M. Carbonate radical formation in radiolysis of sodium carbonate and bicarbonate solutions up to 250°C and the mechanism of its second order decay. *J. Phys. Chem. A* 2010, *114*, 2142–2150.

39 Wu, G., Katsumura, Y., Muroya, Y., Lin, M., and Morioka, T. Temperature dependence of carbonate radical in $NaHCO_3$ and Na_2CO_3 solutions: is the radical a single anion? *J. Phys. Chem. A* 2002, *106*, 2430–2437.

40 Huie, R.E., Shoute, L.C.T., and Neta, P. Temperature dependence of the rate constants for reactions of the carbonate radical with organic and inorganic reductants. *Int. J. Chem. Kinet.* 1991, *23*, 541–552.

41 Bonini, M.G. and Augusto, O. Carbon dioxide stimulates the production of thiyl, sulfinyl, and disulfide radical anion from thiol oxidation by peroxynitrite. *J. Biol. Chem.* 2001, *276*, 9749–9754.

42 Clifton, C.L. and Huie, R.E. Rate constants for some hydrogen abstraction reactions of the carbonate radical. *Int. J. Chem. Kinet.* 1993, *25*, 199–203.

43 Alvarez, M.N., Peluffo, G., Folkes, L., Wardman, P., and Radi, R. Reaction of the carbonate radical with the spin-trap 5,5-dimethyl-1-pyrroline-N-oxide in chemical and cellular systems: pulse radiolysis, electron paramagnetic resonance, and kinetic-competition studies. *Free Radic. Biol. Med.* 2007, *43*, 1523–1533.

44 Clarke, K., Edge, R., Johnson, V., Land, E.J., Navaratnam, S., and Truscott, T.G. The carbonate radical: its reactivity with oxygen, ammonia, amino acids, and melanins. *J. Phys. Chem. A* 2008, *112*, 10147–10151.

45 Draganic, Z.D., Negron-Mendoza, A., Sehested, K., Vujosevic, S.I., Navarro-Gonzales, R., Albarran-Sanchez, M.G., and Draganic, I.G. Radiolysis of aqueous

solutions of ammonium bicarbonate over a large dose range. *Radiat. Phys. Chem.* 1991, *38*, 317–321.

46 Huie, R.E., Clifton, C.L., and Neta, P. Electron transfer reaction rates and equilibria of the carbonate and sulfate radical anions. *Radiat. Phys. Chem.* 1991, *38*, 477–481.

47 Larson, R.A. and Zepp, R.G. Environmental chemistry. Reactivity of the carbonate radical with aniline derivatives. *Environ. Toxicol. Chem.* 1988, *7*, 265–274.

48 Nemes, A., Fábián, I., and Van Eldik, R. Kinetics and mechanism of the carbonate ion inhibited aqueous ozone decomposition. *J. Phys. Chem. A* 2000, *104*, 7995–8000.

49 Tórtora, V., Quijano, C., Freeman, B., Radi, R., and Castro, L. Mitochondrial aconitase reaction with nitric oxide, S-nitrosoglutathione, and peroxynitrite: mechanisms and relative contributions to aconitase inactivation. *Free Radic. Biol. Med.* 2007, *42*, 1075–1088.

50 Wu, C. and Linden, K.G. Phototransformation of selected organophosphorus pesticides: roles of hydroxyl and carbonate radicals. *Water Res.* 2010, *44*, 3585–3594.

51 Canonica, S. and Tratnyek, P.G. Quantitative structure-activity relationships for oxidation reactions of organic chemicals in water. *Environ. Toxicol. Chem.* 2003, *22*, 1743–1754.

52 Stenman, D., Carlsson, M., Jonsson, M., and Reitberger, T. Reactivity of the carbonate radical anion towards carbohydrate and lignin model compounds. *J. Wood Chem. Technol.* 2003, *23*, 47–69.

53 Boccini, F., Domazou, A.S., and Herold, S. Pulse radiolysis studies of the reactions of $CO_3^{\bullet-}$ and NO_2 with nitrosyl(II)myoglobin and nitrosyl(II)hemoglobin. *J. Phys. Chem. A* 2006, *110*, 3927–3932.

54 Chen, S.N. and Hoffman, M.Z. Rate constants for the reaction of the carbonate radical with compounds of biochemical interest in neutral aqueous solution. *Radiat. Res.* 1973, *56*, 40–47.

55 Domazou, A.S. and Koppenol, W.H. Oxidation-state-dependent reactions of cytochrome c with the trioxidocarbonate(•1-) radical: a pulse radiolysis study. *J. Biol. Inorg. Chem.* 2007, *12*, 118–125.

56 Eriksen, T.E. and Fransson, G. Radical-induced oxidation of glutathione in alkaline aqueous solution. *Radiat. Phys. Chem.* 1988, *32*, 163–167.

57 Gebicka, L., Didik, J., and Gebicki, J.L. Reactions of heme proteins with carbonate radical anion. *Res. Chem. Inter.* 2009, *35*, 401–409.

58 Goldstein, S. and Samuni, A. Intra- and intermolecular oxidation of oxymyoglobin and oxyhemoglobin induced by hydroxyl and carbonate radicals. *Free Radic. Biol. Med.* 2005, *39*, 511–519.

59 Joshi, R. and Mukherjee, T. Carbonate radical anion-induced electron transfer in bovine serum albumin. *Radiat. Phys. Chem.* 2006, *75*, 760–767.

60 Shafirovich, V., Dourandin, A., Huang, W., and Geacintov, N.E. The Carbonate radical is a site-selective oxidizing agent of guanine in double-stranded oligonucleotides. *J. Biol. Chem.* 2001, *276*, 24621–24626.

61 Wrona, M., Patel, K., and Wardman, P. Reactivity of 2′,7′-dichlorodihydrofluorescein and dihydrorhodamine 123 and their oxidized forms toward carbonate,

nitrogen dioxide, and hydroxyl radicals. *Free Radic. Biol. Med.* 2005, *38*, 262–270.

62 Elango, T.P., Ramakrishnan, V., Vancheesan, S., and Kuriacose, J.C. Reactions of the carbonate radical with aliphatic amines. *Tetrahedron* 1985, *41*, 3837–3843.

63 Chen, S.-N., Hoffman, M.Z., and Parsons, G.H., Jr. Reactivity of the carbonate radical toward aromatic compounds in aqueous solution. *J. Phys. Chem.* 1975, *79*, 1911–1912.

64 Zhang, H., Joseph, J., Crow, J., and Kalyanaraman, B. Mass spectral evidence for carbonate-anion-radical-induced posttranslational modification of tryptophan to kynurenine in human Cu,Zn superoxide dismutase. *Free Radic. Biol. Med.* 2004, *37*, 2018–2026.

65 Goldstein, S., Samuni, A., Hideg, K., and Merenyi, G. Structure—activity relationship of cyclic nitroxides as SOD mimics and scavengers of nitrogen dioxide and carbonate radicals. *J. Phys. Chem. A* 2006, *110*, 3679–3685.

66 Adam, A. and Mehta, M. $KH(O_2)CO_2{\cdot}H_2O_2$—an oxygen-rich salt of monoperoxocarbonic acid. *Ang. Chem. Int. Ed.* 1998, *37*, 1387–1388.

67 Chen, H., Lin, L., Lin, Z., Guo, G., and Lin, J.-M. Chemiluminescence arising from the decomposition of peroxymonocarbonate and enhanced by CdTe quantum dots. *J. Phys. Chem. A* 2010, *114*, 10049–10058.

68 Liu, M., Cheng, X., Zhao, L., and Lin, J.-M. On-line preparation of peroxymonocarbonate and its application for the study of energy transfer chemiluminescence to lanthanide inorganic coordinate complexes. *Luminescence* 2006, *21*, 179–185.

69 Liu, M., Zhao, L., and Lin, J.-M. Chemiluminescence energy transfer reaction for the on-line preparation of peroxymonocarbonate and Eu(II)-dipicolinate complex. *J. Phys. Chem. A* 2006, *110*, 7509–7514.

70 Attiogbe, F.K., Bose, S.K., Wang, W., McNeillie, A., and Francis, R.C. The peroxymonocarbonate anions as pulp bleaching agents. Part 1. Results with lignin model compounds and chemical pulps. *BioResources* 2010, *5*, 2208–2220.

71 Attiogbe, F.K., Wang, W., McNeillie, A., and Francis, R.C. The peroxymonocarbonate anions as pulp bleaching agents. Part 2. Mechanical pulp brightening and effects of metal ions. *BioResources* 2010, *5*, 2221–2231.

72 Balagam, B. and Richardson, D.E. The mechanism of carbon dioxide catalysis in the hydrogen peroxide N-oxidation of amines. *Inorg. Chem.* 2008, *47*, 1173–1178.

73 Regino, C.A.S. and Richardson, D.E. Bicarbonate-catalyzed hydrogen peroxide oxidation of cysteine and related thiols. *Inorg. Chim. Acta* 2007, *360*, 3971–3977.

74 Richardson, D.E., Regino, C.A.S., Yao, H., and Johnson, J.V. Methionine oxidation by peroxymonocarbonate, a reactive oxygen species formed from CO_2/bicarbonate and hydrogen peroxide. *Free Radic. Biol. Med.* 2003, *35*, 1538–1550.

75 Connor, H.D., Thurman, R.G., Galizi, M.D., and Mason, R.P. The formation of a novel free radical metabolite from CCl4 in the perfused rat liver and in vivo. *J. Biol. Chem.* 1986, *261*, 4542–4548.

76 Wong, P.K., Poyer, J.L., DuBose, C.M., and Floyd, R.A. Hydralazine-dependent carbon dioxide free radical formation by metabolizing mitochondria. *J. Biol. Chem.* 1988, *263*, 11296–11301.

77 Vásquez-Vivar, J., Denicola, A., Radi, R., and Augusto, O. Peroxynitrite-mediated decarboxylation of pyruvate to both carbon dioxide and carbon dioxide radical anion. *Chem. Res. Toxicol.* 1997, *10*, 786–794.

78 Surdhar, P.S. and Armstrong, D.A. Reduction potentials and exchange reactions of thiyl radicals and disulfide anion radicals. *J. Phys. Chem.* 1987, *91*, 6532–6537.

79 Surdhar, P.S. and Armstrong, D.A. Redox potentials of some sulfur-containing radicals. *J. Phys. Chem.* 1986, *90*, 5915–5917.

80 Favaudon, V., Tourbez, H., Houée-Levin, C., and Lhoste, J.M. $CO_2^{\bullet-}$ radical induced cleavage of disulfide bonds in proteins. A γ-ray and pulse radiolysis mechanistic investigation. *Biochemistry* 1990, *29*, 10978–10989.

81 Li, S., Matthews, J., and Sinha, A. Atmospheric hydroxyl radical production from electronically excited NO_2 and H_2O. *Science* 2008, *319*, 1657–1660.

82 Bauer, G., Chatgilialoglu, C., Gebicki, J.L., Gebicka, L., Gescheidt, G., Golding, B.T., Goldstein, S., Kaizer, J., Merenyi, G., Speien, G., and Wardman, P. Biologically relevant small radicals. *Chimia* 2008, *62*, 704–712.

83 Palmer, R.M.J., Ferrige, A.G., and Moncada, S. Nitric oxide release accounts for the biological activity of endothelium-derived relaxing factor. *Nature* 1987, *327*, 524–526.

84 Schopfer, M.P., Wang, J., and Karlin, K.D. Bioinspired heme, heme/nonheme diiron, heme/copper, and inorganic NO_x chemistry: $^{\bullet}$NO(g) oxidation, peroxynitrite-metal chemistry, and $^{\bullet}$NO(g) reductive coupling. *Inorg. Chem.* 2010, *49*, 6267–6282.

85 Ullrich, V. and Kissner, R. Redox signaling: bioinorganic chemistry at its best. *J. Inorg. Biochem.* 2006, *100*, 2079–2086.

86 Rafikova, O., Rafikov, R., and Nudler, E. Catalysis of S-nitrosothiols formation by serum albumin: the mechanism and implication in vascular control. *Proc. Natl. Acad. Sci. U.S.A.* 2002, *99*, 5913–5918.

87 Nedospasov, A., Rafikov, R., Beda, N., and Nudler, E. An autocatalytic mechanism of protein nitrosylation. *Proc. Natl. Acad. Sci. U.S.A.* 2000, *97*, 13543–13548.

88 Hughes, M.N. Chemistry of nitric oxide and related species. *Method Enzymol.* 2008, *436*, 3–19.

89 Abbas, K., Breton, J., Planson, A.G., Bouton, C., Bignon, J., Seguin, C., Riquier, S., Toledano, M.B., and Drapier, J.C. Nitric oxide activates an Nrf2/sulfiredoxin antioxidant pathway in macrophages. *Free Radic. Biol. Med.* 2011, *51*, 107–114.

90 Santolini, J. The molecular mechanism of mammalian NO-synthases: a story of electrons and protons. *J. Inorg. Biochem.* 2011, *105*, 127–141.

91 Arikawa, Y. and Onishi, M. Reductive N-N coupling of NO molecules on transition metal complexes leading to N_2O. *Coord. Chem. Rev.* 2012, *256*, 468–478.

92 Moncada, S., Palmer, R.M.J., and Higgs, E.A. Nitric oxide: physiology, pathophysiology, and pharmacology. *Pharmacol. Rev.* 1991, *43*, 109–142.

93 Dedon, P.C. and Tannenbaum, S.R. Reactive nitrogen species in the chemical biology of inflammation. *Arch. Biochem. Biophys* 2004, *423*, 12–22.

94 Favaloro, J.L. and Kemp-Harper, B.K. The nitroxyl anion (HNO) is a potent dilator of rat coronary vasculature. *Cardiovasc. Res.* 2007, *73*, 587–596.

95 Andrews, K.L., Irvine, J.C., Tare, M., Apostolopoulos, J., Favaloro, J.L., Triggle, C.R., and Kemp-Harper, B.K. A role for nitroxyl (HNO) as an

endothelium-derived relaxing and hyperpolarizing factor in resistance arteries. *Br. J. Pharmacol.* 2009, *157*, 540–550.

96 Lok, H.C., Rahmanto, Y.S., Hawkins, C.L., Kalinowski, D.S., Morrow, C.S., Townsend, A.J., Ponka, P., and Richardson, D.R. Nitric oxide storage and transport in cells are mediated by glutathione S-transferase P1-1 and multidrug resistance protein 1 via dinitrosyl iron complexes. *J. Biol. Chem.* 2012, *287*, 607–618.

97 Kurtikyan, T.S. and Ford, P.C. FTIR and optical spectroscopic studies of the reactions of heme models with nitric oxide and other NO_x in porous layered solids. *Coord. Chem. Rev.* 2008, *252*, 1486–1496.

98 Basu, S., Keszler, A., Azarova, N.A., Nwanze, N., Perlegas, A., Shiva, S., Broniowska, K.A., Hogg, N., and Kim-Shapiro, D.B. A novel role for cytochrome c: efficient catalysis of S-nitrosothiol formation. *Free Radic. Biol. Med* 2010, *48*, 255–263.

99 Ascenzi, P., Cao, Y., Di Masi, A., Gullotta, F., De Sanctis, G., Fanali, G., Fasano, M., and Coletta, M. Reductive nitrosylation of ferric human serum heme-albumin. *FEBS J.* 2010, *277*, 2474–2485.

100 Ascenzi, P., Di Masi, A., Gullotta, F., Mattu, M., Ciaccio, C., and Coletta, M. Reductive nitrosylation of ferric cyanide horse heart myoglobin is limited by cyanide dissociation. *Biochem. Biophys. Res. Commun.* 2010, *393*, 196–200.

101 Ascenzi, P., Santucci, R., Coletta, M., and Polticelli, F. Cytochromes: reactivity of the “dark side” of the heme. *Biophys. Chem.* 2010, *152*, 21–27.

102 De Marinis, E., Casella, L., Ciaccio, C., Coletta, M., Visca, P., and Ascenzi, P. Catalytic peroxidation of nitrogen monoxide and peroxynitrite by globins. *IUBMB Life* 2009, *61*, 62–73.

103 Soldatova, A.V., Ibrahim, M., Olson, J.S., Czernuszewicz, R.S., and Spiro, T.G. New light on NO bonding in Fe(III) heme proteins from resonance Raman spectroscopy and DFT modeling. *J. Am. Chem. Soc.* 2010, *132*, 4614–4625.

104 Lewandowska, H., Kalinowska, M., Brzóska, K., Wójciuk, K., Wójciuk, G., and Kruszewski, M. Nitrosyl iron complexes—synthesis, structure and biology. *Dalton Trans.* 2011, *40*, 8273–8289.

105 Di Simplicio, P., Franconi, F., Frosalí, S., and Di Giuseppe, D. Thiolation and nitrosation of cysteines in biological fluids and cells. *Amino Acids* 2003, *25*, 323–339.

106 Liu, X., Miller, M.J.S., Joshi, M.S., Thomas, D.D., and Lancaster, J.R., Jr. Accelerated reaction of nitric oxide with O_2 within the hydrophobic interior of biological membranes. *Proc. Natl. Acad. Sci. U.S.A.* 1998, *95*, 2175–2179.

107 Keszler, A., Zhang, Y., and Hogg, N. Reaction between nitric oxide, glutathione, and oxygen in the presence and absence of protein: how are S-nitrosothiols formed? *Free Radic. Biol. Med.* 2010, *48*, 55–64.

108 Goldstein, S. and Czapski, G. Mechanism of the nitrosation of thiols and amines by oxygenated $^{\bullet}NO$ solutions: the nature of the nitrosating intermediates. *J. Am. Chem. Soc.* 1996, *118*, 3419–3425.

109 Jourd’heuil, D., Jourd’heuil, F.L., and Feelisch, M. Oxidation and nitrosation of thiols at low micromolar exposure to nitric oxide: evidence for a free radical mechanism. *J. Biol. Chem.* 2003, *278*, 15720–15726.

110 Gow, A.J., Buerk, D.G., and Ischiropoulos, H. A novel reaction mechanism for the formation of S-nitrosothiol in vivo. *J. Biol. Chem.* 1997, *272*, 2841–2845.

111 Schrammel, A., Gorren, A.C.F., Schmidt, K., Pfeiffer, S., and Mayer, B. S-nitrosation of glutathione by nitric oxide, peroxynitrite, and •NO/$O_2^{\bullet-}$. *Free Radic. Biol. Med.* 2003, *34*, 1078–1088.

112 Wink, D.A., Nims, R.W., Derbyshire, J.F., Christodoulou, D., Hanbauer, I., Cox, G.W., Laval, F., Laval, J., Cook, J.A., Krishna, M.C., DeGraff, W.G., and Mitchell, J.B. Reaction kinetics for nitrosation of cysteine and glutathione in aerobic nitric oxide solutions at neutral pH. Insights into the fate and physiological effects of intermediates generated in the NO/O_2 reaction. *Chem. Res. Toxicol.* 1994, *7*, 519–525.

113 Keshive, M., Singh, S., Wishnok, J.S., Tannenbaum, S.R., and Deen, W.M. Kinetics of S-nitrosation of thiols in nitric oxide solutions. *Chem. Res. Toxicol.* 1996, *9*, 988–993.

114 Madej, E., Folkes, L.K., Wardman, P., Czapski, G., and Goldstein, S. Thiyl radicals react with nitric oxide to form S-nitrosothiols with rate constants near the diffusion-controlled limit. *Free Radic. Biol. Med.* 2008, *44*, 2013–2018.

115 Goldstein, S., Samuni, A., and Merenyi, G. Kinetics of the reaction between nitroxide and thiyl radicals: nitroxides as antioxidants in the presence of thiols. *J. Phys. Chem. A* 2008, *112*, 8600–8605.

116 Olasehinde, E.F., Takeda, K., and Sakugawa, H. Photochemical production and consumption mechanisms of nitric oxide in seawater. *Environ. Sci. Technol.* 2010, *44*, 8403–8408.

117 Olasehinde, E.F., Takeda, K., and Sakugawa, H. Development of an analytical method for nitric oxide radical determination in natural waters. *Anal. Chem.* 2009, *81*, 6843–6850.

118 Arakaki, T., Miyake, T., Shibata, M., and Sakugawa, H. Photochemical formation and scavenging of hydroxyl radical in rain and dew waters. *Chem. Soc. Jpn.-Chem. Indus. Chem. J.* 1998, *9*, 339–340.

119 Gomes, A., Fernandes, E., and Lima, J.L.F. C. Use of fluorescence probes for detection of reactive nitrogen species: a review. *J. Fluoresc.* 2006, *16*, 119–139.

120 Løgager, T. and Sehested, K. Formation and decay of peroxynitric acid: a pulse radiolysis study. *J. Phys. Chem.* 1993, *97*, 10047–10052.

121 Smith, R.L. and Yoshinari, T. Occurrence and turnover of nitric oxide in a nitrogen-impacted sand and gravel aquifer. *Environ. Sci. Technol.* 2008, *42*, 8245–8251.

122 Zafiriou, O.C. and McFarland, M. Determination of trace levels of nitric oxide in aqueous solution. *Anal. Chem.* 1980, *52*, 1662–1667.

123 Mopper, K. and Zhou, X. Hydroxyl radical photoproduction in the sea and its potential impact on marine processes. *Science* 1990, *250*, 661–664.

124 Zhou, X. and Mopper, K. Determination of photochemically produced hydroxyl radicals in seawater and freshwater. *Mar. Chem.* 1990, *30*, 71–88.

125 Goldstein, S., Merenyi, G., and Samuni, A. Kinetics and mechanism of $^{\bullet}NO_2$ reacting with various oxidation states of myoglobin. *J. Am. Chem. Soc.* 2004, *126*, 15694–15701.

126 Ford, P.C., Laverman, L.E., and Lorkovic, I.M. Reaction mechanisms of nitric oxide with biologically relevant metal centers. *Adv. Inorg. Chem.* 2003, *54*, 203–257.

127 Ford, E., Hughes, M.N., and Wardman, P. Kinetics of the reactions of nitrogen dioxide with glutathione, cysteine, and uric acid at physiological pH. *Free Radic. Biol. Med.* 2002, *32*, 1314–1323.

128 Bartesaghi, S., Ferrer-Sueta, G., Peluffo, G., Valez, V., Zhang, H., Kalyanaraman, B., and Radi, R. Protein tyrosine nitration in hydrophilic and hydrophobic environments. *Amino Acids* 2007, *32*, 501–515.

129 Prütz, W.A., Mönig, H., Butler, J., and Land, E.J. Reactions of nitrogen dioxide in aqueous model systems: oxidation of tyrosine units in peptides and proteins. *Arch. Biochem. Biophys.* 1985, *243*, 125–134.

130 Prutz, W.A., Butler, J., Land, E.J., and Swallow, A.J. The role of sulphur peptide functions in free radical transfer: a pulse radiolysis study. *Int. J. Radiat. Biol.* 1989, *55*, 539–556.

131 Madsen, D., Larsen, J., Jensen, S.K., Keiding, S.R., and Thøgersen, J. The primary photodynamics of aqueous nitrate: formation of peroxynitrite. *J. Am. Chem. Soc.* 2003, *125*, 15571–15576.

132 Goldstein, S. and Merényi, G. The chemistry of peroxynitrite: implications for biological activity. *Method Enzymol.* 2008, *436*, 49–61.

133 Ferrer-Sueta, G. and Radi, R. Chemical biology of peroxynitrite: kinetics, diffusion, and radicals. *ACS Chem. Biol.* 2009, *4*, 161–177.

134 Ieda, N., Nakagawa, H., Peng, T., Yang, D., Suzuki, T., and Miyata, N. Photocontrollable peroxynitrite generator based on N-methyl-N- nitrosoaminophenol for cellular application. *J. Am. Chem. Soc.* 2012, *134*, 2563–2568.

135 Anan'ev, V. and Miklin, M. The peroxynitrite formation under photolysis of alkali nitrates. *J. Photochem. Photobiol. A.* 2005, *172*, 289–292.

136 Warneck, P. and Wurzinger, C. Product quantum yields for the 305-nm photodecomposition of NO_3^- in aqueous solution. *J. Phys. Chem.* 1988, *92*, 6278–6283.

137 Mark, G., Korth, H.-G., Schuchmann, H.-P., and von Sonntag, C. The photochemistry of aqueous nitrate ion revisited. *J. Photochem. Photobiol. A.* 1996, *101*, 89–103.

138 Bradley, R.A., Jr. Lanzendorf, E., McCarthy, M.I., Orlando, T.M., and Hess, W.P. Molecular NO desorption from crystalline sodium nitrate by resonant excitation of the NO_3^- π-π^* transition. *J. Phys. Chem.* 1995, *99*, 11715–11721.

139 Goldstein, S. and Rabani, J. Mechanism of nitrite formation by nitrate photolysis in aqueous solutions: the role of peroxynitrite, nitrogen dioxide, and hydroxyl radical. *J. Am. Chem. Soc.* 2007, *129*, 10597–10601.

140 Papée, H.M. and Petriconi, G.L. Formation and decomposition of alkaline "peroxynitrite." *Nature* 1964, *204*, 142–144.

141 Shuali, U., Ottolenghi, M., Rabani, J., and Yelin, Z. On the photochemistry of aqueous nitrate solutions excited in the 195-nm band. *J. Phys. Chem.* 1969, *73*, 3445–3451.

142 Barat, F., Gilles, L., Hickel, B., and Sutton, J. Flash photolysis of the nitrate ion in aqueous solution: excitation at 200 nm. *J. Chem. Soc. A. Inorg. Phys. Theor. Chem.* 1970, 1982–1986.

143 Goldstein, S., Lind, J., and Merényi, G. Chemistry of peroxynitrites as compared to peroxynitrates. *Chem. Rev.* 2005, *105*, 2457–2470.

144 Løgager, T. and Sehested, K. Formation and decay of peroxynitrous acid: a pulse radiolysis study. *J. Phys. Chem.* 1993, *97*, 6664–6669.

145 Beckman, J.S., Beckman, T.W., Chen, J., Marshall, P.A., and Freeman, B.A. Apparent hydroxyl radical production by peroxynitrite: implications for endothelial injury from nitric oxide and superoxide. *Proc. Natl. Acad. Sci. U.S.A.* 1990, *87*, 1620–1624.

146 Blough, N.V. and Zafiriou, O.C. Reaction of superoxide with nitric oxide to form peroxonitrite in alkaline aqueous solution. *Inorg. Chem.* 1985, *24*, 3502–3504.

147 Huie, R.E. and Padmaja, S. The reaction of NO with superoxide. *Free Radic. Res. Commun.* 1993, *18*, 195–199.

148 Nauser, T. and Koppenol, W.H. The rate constant of the reaction of superoxide with nitrogen monoxide: approaching the diffusion limit. *J. Phys. Chem. A* 2002, *106*, 4084–4086.

149 Kissner, R., Nauser, T., Bugnon, P., Lye, P.G., and Koppenol, W.H. Formation and properties of peroxynitrite as studied by laser flash photolysis, high-pressure stopped-flow technique, pulse radiolysis. *Chem. Res. Toxicol.* 1997, *10*, 1285–1292.

150 Botti, H., Moller, M.N., Steinmann, D., Nauser, T., Koppenol, W.H., Denicola, A., and Radi, R. Distance-dependent diffusion-controlled reaction of $^{\bullet}NO$ and $O_2^{\bullet-}$ at chemical equilibrium with $ONOO^-$. *J. Phys. Chem. B.* 2011, *114*, 16584–16593.

151 Hughes, M.N. and Nicklin, H.G. Autoxidation of hydroxylamine in alkaline solutions. *J. Chem. Soc. A. Inorg. Phys. Theor. Chem.* 1971, 164–168.

152 Koppenol, W.H., Kissner, R., and Beckman, J.S. Syntheses of peroxynitrite: to go with the flow or on solid grounds? *Method Enzymol.* 1996, *269*, 296–302.

153 Leis, J.R., Peña, M.E., and Ríos, A. A novel route to peroxynitrite anion. *J. Chem. Soc. Chem. Commun.* 1993, 1298–1299.

154 Reed, J.W., Ho, H.H., and Jolly, W.L. Chemical syntheses with a quenched flow reactor. Hydroxytrihydroborate and peroxynitrite [24]. *J. Am. Chem. Soc.* 1974, *96*, 1248–1249.

155 Sturzbecher-Höhne, M., Kissner, R., Nauser, T., and Koppenol, W.H. Preparation and properties of lithium and sodium peroxynitrite. *Chem. Res. Toxicol.* 2008, *21*, 2257–2259.

156 Bohle, D.S., Hansert, B., Paulson, S.C., and Smith, B.D. Biomimetic synthesis of the putative cytotoxin peroxynitrite, $ONOO^-$, and its characterization as a tetramethylammonium salt. *J. Am. Chem. Soc.* 1994, *116*, 7423–7424.

157 Goldstein, S. and Czapski, G. The reaction of NO· with $O_2^{\bullet-}$ and $HO_2\cdot$: a pulse radiolysis study. *Free Radic. Biol. Med.* 1995, *19*, 505–510.

158 Pryor, W.A. and Squadrito, G.L. The chemistry of peroxynitrite: a product from the reaction of nitric oxide with superoxide. *Am. J. Physiol. -Lung Cell. Mol. Physiol.* 1995, *268*, L699–L722.

159 Merényi, G., Lind, J., Goldstein, S., and Czapski, G. Mechanism and thermochemistry of peroxynitrite decomposition in water. *J. Phys. Chem. A* 1999, *103*, 5685–5691.

160 Gerasimov, O.V. and Lymar, S.V. The yield of hydroxyl radical from the decomposition of peroxynitrous acid. *Inorg. Chem.* 1999, *38*, 4317–4321.

161 Hodges, G.R. and Ingold, K.U. Cage-escape of geminate radical pairs can produce peroxynitrate from peroxynitrite under a wide variety of experimental conditions. *J. Am. Chem. Soc.* 1999, *121*, 10695–10701.

162 Kirsch, M., Korth, H.-G., Wensing, A., Sustmann, R., and De Groot, H. Product formation and kinetic simulations in the pH range 1–14 account for a free-radical mechanism of peroxynitrite decomposition. *Arch. Biochem. Biophys.* 2003, *418*, 133–150.

163 Kissner, R., Thomas, C., Hamsa, M.S.A., Van Eldik, R., and Koppenol, W.H. Evaluation of activation volumes for the conversion of peroxynitrous to nitric acid. *J. Phys. Chem. A* 2003, *107*, 11261–11263.

164 Coddington, J.W., Wherland, S., and Hurst, J.K. Pressure dependence of peroxynitrite reactions. Support for a radical mechanism. *Inorg. Chem.* 2001, *40*, 528–532.

165 Goldstein, S., Meyerstein, D., Van Eldik, R., and Czapski, G. Spontaneous reactions and reduction by iodide of peroxynitrite and peroxynitrate: mechanistic insight from activation parameters. *J. Phys. Chem. A* 1997, *101*, 7114–7118.

166 Goldstein, S., Meyerstein, D., Van Eldik, R., and Czapski, G. Peroxynitrous acid decomposes via homolysis: evidence from high-pressure pulse radiolysis. *J. Phys. Chem. A* 1999, *103*, 6587–6590.

167 Gratzel, M., Henglein, A., Lilie, J., and Beck, G. Pulse radiolysis of some elementary oxidation-reduction processes of nitrite ion. *Berichte Der Bunsen-Gesellschaft Fur Physikalische Chemie* 1969, *73*, 646–653.

168 Merényi, G. and Lind, J. Free radical formation in the peroxynitrous acid (ONOOH)/peroxynitrite ($ONOO^-$) system. *Chem. Res. Toxicol.* 1998, *11*, 245–246.

169 Goldstein, S., Saha, A., Lymar, S.V., and Czapski, G. Oxidation of peroxynitrite by inorganic radicals: a pulse radiolysis study. *J. Am. Chem. Soc.* 1998, *120*, 5549–5554.

170 Goldstein, S., Czapski, G., Lind, J., and Merényi, G. Gibbs energy of formation of peroxynitrite—order restored. *Chem. Res. Toxicol.* 2001, *14*, 657–660.

171 Gratzel, M., Taniguch, S., and Henglein, A. Study with pulse radiolysis of NO-oxidation-reduction processes of nitrite ion. *Berichte Der Bunsen-Gesellschaft Fur Physikalische Chemie* 1970, *74*, 488–492.

172 Treinin, A. and Hayon, E. Absorption spectra and reaction kinetics of NO_2, N_2O_3, and N_2O_4 in aqueous solution. *J. Am. Chem. Soc.* 1970, *92*, 5821–5828.

173 Goldstein, S., Czapski, G., Lind, J., and Merényi, G. Effect of ·NO on the decomposition of peroxynitrite: reaction of N_2O_3 with $ONOO^-$. *Chem. Res. Toxicol.* 1999, *12*, 132–136.

174 Gunaydin, H. and Houk, K.N. Molecular dynamics simulation of the HOONO decomposition and the $HO^\bullet/NO_2^\bullet$ caged radical pair in water. *J. Am. Chem. Soc.* 2008, *130*, 10036–10037.

175 Kissner, R. and Koppenol, W.H. Product distribution of peroxynitrite decay as a function of pH, temperature, and concentration. *J. Am. Chem. Soc.* 2002, *124*, 234–239.

176 Lymar, S.V., Khairutdinov, R.F., and Hurst, J.K. Hydroxyl radical formation by O-O bond homolysis in peroxynitrous acid. *Inorg. Chem.* 2003, *42*, 5259–5266.

177 Pfeiffer, S., Gorren, A.C.F., Schmidt, K., Werner, E.R., Hansert, B., Bohle, D.S., and Mayer, B. Metabolic fate of peroxynitrite in aqueous solution. Reaction with nitric oxide and pH-dependent decomposition to nitrite and oxygen in a 2:1 stoichiometry. *J. Biol. Chem.* 1997, *272*, 3465–3470.

178 Gupta, D., Harish, B., Kissner, R., and Koppenol, W.H. Peroxynitrate is formed rapidly during decomposition of peroxynitrite at neutral pH. *Dalton Trans.* 2009, 5730–5736.

179 Miyamoto, S., Ronsein, G.E., Corrêa, T.C., Martinez, G.R., Medeiros, M.H.G., and Di Mascio, P. Direct evidence of singlet molecular oxygen generation from peroxynitrate, a decomposition product of peroxynitrite. *Dalton Trans.* 2009, 5720–5729.

180 Liu, Y.D. and Zhong, R.-G. Theoretical studies on the isomerization of peroxynitrite to nitrate mediated by peroxynitrous acid. *Jiegou Huaxue* 2008, *27*, 171–176.

181 Sturzbecher-Höhne, M., Nauser, T., Kissner, R., and Koppenol, W.H. Photon-initiated homolysis of peroxynitrous acid. *Inorg. Chem.* 2009, *48*, 7307–7312.

182 Keith, W.G. and Powell, R.E. Kinetics of decomposition of peroxynitrous acid. *J. Chem. Soc. A. Inorg. Phys. Theor. Chem.* 1969, 90–99.

183 Radi, R., Cosgrove, T.P., Beckman, J.S., and Freeman, B.A. Peroxynitrite-induced luminol chemiluminescence. *Biochem. J.* 1993, *290*, 51–57.

184 Zhu, L., Gunn, C., and Beckman, J.S. Bactericidal activity of peroxynitrite. *Arch. Biochem. Biophys.* 1992, *298*, 452–457.

185 Squadrito, G.L. and Pryor, W.A. Mapping the reaction of peroxynitrite with CO_2: energetics, reactive species, and biological implications. *Chem. Res. Toxicol.* 2002, *15*, 885–895.

186 Lymar, S.V. and Hurst, J.K. Rapid reaction between peroxonitrite ion and carbon dioxide: implications for biological activity. *J. Am. Chem. Soc.* 1995, *117*, 8867–8868.

187 Alfassi, Z.B., Dhanasekaran, T., Huie, R.E., and Neta, P. On the reactions of CO_3^- radicals with NO(x) radicals. *Radiat. Phys. Chem.* 1999, *56*, 475–482.

188 Meli, R., Nauser, T., and Koppenol, W.H. Direct observation of intermediates in the reaction of peroxynitrite with carbon dioxide. *Helv. Chim. Acta* 1999, *82*, 722–725.

189 Goldstein, S. and Czapski, G. Formation of peroxynitrate from the reaction of peroxynitrite with CO_2: evidence for carbonate radical production. *J. Am. Chem. Soc.* 1998, *120*, 3458–3463.

190 Uppu, R.M. and Pryor, W.A. The reactions of ozone with proteins and unsaturated fatty Acids in reverse micelles. *Chem. Res. Toxicol.* 1994, *7*, 47–55.

191 Lymar, S.V. and Hurst, J.K. CO_2-catalyzed one-electron oxidations by peroxynitrite: properties of the reactive intermediate. *Inorg. Chem.* 1998, *37*, 294–301.

192 Lymar, S.V., Jiang, Q., and Hurst, J.K. Mechanism of carbon dioxide-catalyzed oxidation of tyrosine by peroxynitrite. *Biochemistry* 1996, *35*, 7855–7861.

193 Goldstein, S. and Czapski, G. The effect of bicarbonate on oxidation by peroxynitrite: implication for its biological activity. *Inorg. Chem.* 1997, *36*, 5113–5117.

194 Goldstein, S., Czapski, G., Lind, J., and Merényi, G. Carbonate radical ion is the only observable intermediate in the reaction of peroxynitrite with CO_2. *Chem. Res. Toxicol.* 2001, *14*, 1273–1276.

195 Uppu, R.M., Squadrito, G.L., Bolzan, R.M., and Pryor, W.A. Nitration and nitrosation by peroxynitrite: role of CO_2 and evidence for common intermediates. *J. Am. Chem. Soc.* 2000, *122*, 6911–6916.

196 Goldstein, S. and Czapski, G. Viscosity effects on the reaction of peroxynitrite with CO_2: evidence for radical formation in a solvent cage. *J. Am. Chem. Soc.* 1999, *121*, 2444–2447.

197 Houk, K.N., Condroski, K.R., and Pryor, W.A. Radical and concerted mechanisms in oxidations of amines, sulfides, and alkenes by peroxynitrite, peroxynitrous acid, and the peroxynitrite-CO_2 adduct: density functional theory transition structures and energetics. *J. Am. Chem. Soc.* 1996, *118*, 13002–13006.

198 Goldstein, S., Lind, J., and Merényi, G. The reaction of $ONOO^-$ with carbonyls: estimation of the half-lives of $ONOOC(O)O^-$ and $O_2NOOC(O)O^-$. *Dalton Trans.* 2002, 808–810.

199 Beckman, J.S., Ischiropoulos, H., Zhu, L., Van Der Woerd, M., Smith, C., Chen, J., Harrison, J., Martin, J.C., and Tsai, M. Kinetics of superoxide dismutase- and iron-catalyzed nitration of phenolics by peroxynitrite. *Arch. Biochem. Biophys.* 1992, *298*, 438–445.

200 Al-Ajlouni, A.M. and Gould, E.S. Electron Transfer. 132. Oxidations with Peroxynitrite 1. *Inorg. Chem.* 1996, *35*, 7892–7896.

201 Goldstein, S. and Czapski, G. Direct and indirect oxidations by peroxynitrite. *Inorg. Chem.* 1995, *34*, 4041–4048.

202 Martínez, E., Mucientes, A.E., Poblete, F.J., and Rodríguez, A. Oxidation of hexacyanoferrate(II) by peroxynitrite. A mechanistic and kinetic study. *Pol. J. Chem.* 2004, *78*, 261–271.

203 Herrero, H., Rodado, G.M., and Mucientes, A.E. Computational techniques applied to the study of the oxidation kinetics of iron and molybdenum cyanocomplexes by peroxynitrous acid. *J. Chemometrics* 2008, *22*, 556–562.

204 Radi, R., Beckman, J.S., Bush, K.M., and Freeman, B.A. Peroxynitrite-induced membrane lipid peroxidation: the cytotoxic potential of superoxide and nitric oxide. *Arch. Biochem. Biophys.* 1991, *288*, 481–487.

205 Radi, R., Beckman, J.S., Bush, K.M., and Freeman, B.A. Peroxynitrite oxidation of sulfhydryls: the cytotoxic potential of superoxide and nitric oxide. *J. Biol. Chem.* 1991, *266*, 4244–4250.

206 Pryor, W.A., Jin, X., and Squadrito, G.L. One- and two-electron oxidations of methionine by peroxynitrite. *Proc. Natl. Acad. Sci. U.S.A.* 1994, *91*, 11173–11177.

207 Ischiropoulos, H., Zhu, L., Chen, J., Tsai, M., Martin, J.C., Smith, C.D., and Beckman, J.S. Peroxynitrite-mediated tyrosine nitration catalyzed by superoxide dismutase. *Arch. Biochem. Biophys.* 1992, *298*, 431–437.

208 Geletii, Y.V., Musaev, D.G., Khavrutskii, L., and Hill, C.L. Peroxynitrite reactions with dimethylsulfide and dimethylselenide: an experimental study. *J. Phys. Chem. A* 2004, *108*, 289–294.

209 Padmaja, S., Squadrito, G.L., Lemercier, J.N., Cueto, R., and Pryor, W.A. Rapid oxidation of DL-selenomethionine by peroxynitrite. *Free Radic. Biol. Med.* 1996, *21*, 317–322.

210 Padmaja, S., Squadrito, G.L., Lemercier, J.N., Cueto, R., and Pryor, W.A. Peroxynitrite-mediated oxidation of D,L-selenomethionine: kinetics, mechanism and the role of carbon dioxide. *Free Radic. Biol. Med.* 1997, *23*, 917–926.

211 Padmaja, S., Squadrito, G.L., and Pryor, W.A. Inactivation of glutathione peroxidase by peroxynitrite. *Arch. Biochem. Biophys.* 1998, *349*, 1–6.

212 Ramezanian, M.S., Padmaja, S., and Koppenol, W.H. Nitration and hydroxylation of phenolic compounds by peroxynitrite. *Chem. Res. Toxicol.* 1996, *9*, 232–240.

213 Karoui, H., Hogg, N., Fréjaville, C., Tordo, P., and Kalyanaraman, B. Characterization of sulfur-centered radical intermediates formed during the oxidation of thiols and sulfite by peroxynitrite: ESR-spin trapping and oxygen uptake studies. *J. Biol. Chem.* 1996, *271*, 6000–6009.

214 Fontana, M., Giovannitti, F., and Pecci, L. The protective effect of hypotaurine and cysteine sulphinic acid on peroxynitrite-mediated oxidative reactions. *Free Radical Res.* 2008, *42*, 320–330.

215 Fontana, M., Duprè, S., and Pecci, L. The reactivity of hypotaurine and cysteine sulfinic acid with peroxynitrite. *Adv. Exp. Med. Biol.* 2006, *583*, 15–24.

216 Fontana, M., Amendola, D., Orsini, E., Boffi, A., and Pecci, L. Oxidation of hypotaurine and cysteine sulphinic acid by peroxynitrite. *Biochem. J.* 2005, *389*, 233–240.

217 Balcerczyk, A., Sowa, K., Bartosz, G. Metal chelators react also with reactive oxygen and nitrogen species. *Biochem. Biophys. Res. Commun.* 2007, *352*, 522–525.

218 Musaev, D.G., Geletii, Y.V., and Hill, C.L. Theoretical studies of the reaction mechanisms of dimethylsulfide and dimethylselenide with peroxynitrite. *J. Phys. Chem. A* 2003, *107*, 5862–5873.

219 Samuni, A. and Goldstein, S. One-electron oxidation of acetohydroxamic acid: the intermediacy of nitroxyl and peroxynitrite. *J. Phys. Chem. A* 2011, *115*, 3022–3028.

220 Jencks, W.P. When is an intermediate not an intermediate? Enforced mechanisms of general acid-base catalyzed, carbocation, carbanion, and ligand exchange reactions. *Acc. Chem. Res.* 1980, *13*, 161–169.

221 Yang, D., Tang, Y.C., Chen, J., Wang, X.C., Bartberger, M.D., Houk, K.N., and Olson, L. Ketone-catalyzed decomposition of peroxynitrite via dioxirane intermediates. *J. Am. Chem. Soc.* 1999, *121*, 11976–11983.

222 Tibi, S. and Koppenol, W.H. Reactions of peroxynitrite with phenolic and carbonyl compounds: flavonoids are not scavengers of peroxynitrite. *Helv. Chim. Acta* 2000, *83*, 2412–2424.

223 Merényi, G., Lind, J., and Goldstein, S. The rate of homolysis of adducts of peroxynitrite to the C=O double bond. *J. Am. Chem. Soc.* 2002, *124*, 40–48.

224 Sikora, A., Zielonka, J., Lopez, M., Joseph, J., and Kalyanaraman, B. Direct oxidation of boronates by peroxynitrite: mechanism and implications in fluorescence imaging of peroxynitrite. *Free Radic. Biol. Med* 2009, *47*, 1401–1407.

225 Sikora, A., Zielonka, J., Lopez, M., Dybala-Defratyka, A., Joseph, J., Marcinek, A., and Kalyanaraman, B. Reaction between peroxynitrite and boronates: EPR spin-trapping, HPLC analyses, and quantum mechanical study of the free radical pathway. *Chem. Res. Toxicol.* 2011, *24*, 687–697.

226 Ferrer-Sueta, G., Batinić-Haberle, I., Spasojević, I., Fridovich, I., and Radi, R. Catalytic scavenging of peroxynitrite by isomeric Mn(III) N-methylpyridyl porphyrins in the presence of reductants. *Chem. Res. Toxicol.* 1999, *12*, 442–449.

227 Stern, M.K., Jensen, M.P., and Kramer, K. Peroxynitrite decomposition catalysts. *J. Am. Chem. Soc.* 1996, *118*, 8735–8736.

228 Furtmüller, P.G., Jantschko, W., Zederbauer, M., Schwanninger, M., Jakopitsch, C., Herold, S., Koppenol, W.H., and Obinger, C. Peroxynitrite efficiently mediates the interconversion of redox intermediates of myeloperoxidase. *Biochem.* 2005, *337*, 944–954.

229 Quijano, C., Hernandez-Saavedra, D., Castro, L., McCord, J.M., Freeman, B.A., and Radi, R. Reaction of peroxynitrite with Mn-superoxide dismutase. Role of the metal center in decomposition kinetics and nitration. *J. Biol. Chem.* 2001, *276*, 11631–11638.

230 Floris, R., Piersma, S.R., Yang, G., Jones, P., and Wever, R. Interaction of myeloperoxidase with peroxynitrite. A comparison with lactoperoxidase, horseradish peroxidase and catalase. *Eur. J. Biochem.* 1993, *215*, 767–775.

231 Crow, J.P., Beckman, J.S., and McCord, J.M. Sensitivity of the essential zinc-thiolate moiety of yeast alcohol dehydrogenase to hypochlorite and peroxynitrite. *Biochemistry* 1995, *34*, 3544–3552.

232 Castro, L., Rodriguez, M., and Radi, R. Aconitase is readily inactivated by peroxynitrite, but not by its precursor, nitric oxide. *J. Biol. Chem.* 1994, *269*, 29409–29415.

233 Thomson, L., Trujillo, M., Telleri, R., and Radi, R. Kinetics of cytochrome c2+ oxidation by peroxynitrite implications for superoxide measurements in nitric oxide-producing biological systems. *Arch. Biochem. Biophys.* 1995, *319*, 491–497.

234 Bourassa, J.L., Ives, E.P., Marqueling, A.L., Shimanovich, R., and Groves, J.T. Myoglobin catalyzes its own nitration. *J. Am. Chem. Soc.* 2001, *123*, 5142–5143.

235 Herold, S., Kalinga, S., Matsui, T., and Watanabe, Y. Mechanistic studies of the isomerization of peroxynitrite to nitrate catalyzed by distal histidine metmyoglobin mutants. *J. Am. Chem. Soc.* 2004, *126*, 6945–6955.

236 Herold, S. and Shivashankar, K. Metmyoglobin and methemoglobin catalyze the isomerization of peroxynitrite to nitrate. *Biochemistry* 2003, *42*, 14036–14046.

237 Denicola, A., Freeman, B.A., Trujillo, M., and Radi, R. Peroxynitrite reaction with carbon dioxide/bicarbonate: kinetics and influence on peroxynitrite-mediated oxidations. *Arch. Biochem. Biophys.* 1996, *333*, 49–58.

238 Alayash, A.I., Brockner Ryan, B.A., and Cashon, R.E. Peroxynitrite-mediated heme oxidation and protein modification of native and chemically modified hemoglobins. *Arch. Biochem. Biophys.* 1998, *349*, 65–73.

239 Surmeli, N.B., Literman, N.K., Miller, A., and Groves, J.T. Peroxynitrite mediates active sites tyrosine nitration in manganese superoxide dismutase. Evidence of a role for the carbonate radical anion. *J. Am. Chem. Soc.* 2010, *132*, 17174–17185.

240 Alvarez, B., Demicheli, V., Durán, R., Trujillo, M., Cerveñansky, C., Freeman, B.A., and Radi, R. Inactivation of human Cu,Zn superoxide dismutase by peroxynitrite and formation of histidinyl radical. *Free Radic. Biol. Med.* 2004, *37*, 813–822.

241 Blanchard-Fillion, B., Souza, J.M., Friel, T., Jiang, G.C.T., Vrana, K., Sharov, V., Barrón, L., Schöneich, C., Quijano, C., Alvarez, B., Radi, R., Przedborski, S., Fernando, G.S., Horwitz, J., and Ischiropoulos, H. Nitration and inactivation of tyrosine hydroxylase by peroxynitrite. *J. Biol. Chem.* 2001, *276*, 46017–46023.

242 Floris, R., Piersma, S.R., Yang, G., Jones, P., and Wever, A. Erratum: interaction of myeloperoxidase with peroxynitrite. A comparison with lactoperoxidase, horseradish peroxidase and catalase (Eur. J. Biochem., Volume 215, No. 3, page 774). *Eur. J. Biochem.* 1993, *216*, 881.

243 Herold, S., Matsui, T., and Watanabe, Y. Peroxynitrite isomerization catalyzed by His64 myoglobin mutants. *J. Am. Chem. Soc.* 2001, *123*, 4085–4086.

244 Lee, J., Hunt, J.A., and Groves, J.T. Mechanism of iron porphyrin reactions with peroxynitrite. *J. Am. Chem. Soc.* 1998, *120*, 7493–7501.

245 Shimanovich, R. and Groves, J.T. Mechanisms of peroxynitrite decomposition catalyzed by FeTMPS, a bioactive sulfonated iron porphyrin. *Arch. Biochem. Biophys.* 2001, *387*, 307–317.

246 Jensen, M.P. and Riley, D.P. Peroxynitrite decomposition activity of iron porphyrin complexes. *Inorg. Chem.* 2002, *41*, 4788–4797.

247 Crow, J.P. Manganese and iron porphyrins catalyze peroxynitrite decomposition and simultaneously increase nitration and oxidant yield: implications for their use as peroxynitrite scavengers in vivo. *Arch. Biochem. Biophys.* 1999, *371*, 41–52.

248 Alvarez, B. and Radi, R. Peroxynitrite reactivity with amino acids and proteins. *Amino Acids* 2003, *25*, 295–311.

249 Herold, S. The outer-sphere oxidation of nitrosyliron(II)hemoglobin by peroxynitrite leads to the release of nitrogen monoxide. *Inorg. Chem.* 2004, *43*, 3783–3785.

250 Herold, S. and Boccini, F.N. $O^{\bullet}$ release from MbFe(II)NO and HbFe(II)NO after oxidation by peroxynitrite. *Inorg. Chem.* 2006, *45*, 6933–6943.

251 Herold, S., Fago, A., Weber, R.E., Dewilde, S., and Moens, L. Reactivity studies of the Fe(III) and Fe(II)NO forms of human neuroglobin reveal a potential role against oxidative stress. *J. Biol. Chem.* 2004, *279*, 22841–22847.

252 Herold, S. and Koppenol, W. H. Peroxynitritometal complexes. *Coord. Chem. Rev.* 2005, *249*, 499–506.

253 Herold, S. and Puppo, A. Oxyleghemoglobin scavenges nitrogen monoxide and peroxynitrite: a possible role in functioning nodules? *J. Biol. Inorg. Chem.* 2005, *10*, 935–945.

254 Herold, S. and Puppo, A. Kinetics and mechanistic studies of the reactions of metleghemoglobin, ferrylleghemoglobin, and nitrosylleghemoglobin with reactive nitrogen species. *J. Biol. Inorg. Chem.* 2005, *10*, 946–957.

255 Herold, S. and Rehmann, F.-K. Kinetics of the reactions of nitrogen monoxide and nitrite with ferryl hemoglobin. *Free Radic. Biol. Med.* 2003, *34*, 531–545.

256 Daiber, A., Herold, S., Schöneich, C., Namgaladze, D., Peterson, J.A., and Ullrich, V. Nitration and inactivation of cytochrome P450(BM-3) by peroxynitrite: stopped-flow measurements prove ferryl intermediates. *Eur. J. Biochem.* 2000, *267*, 6729–6739.

257 Sharma, V.K., Yngard, R.A., Homonnay, Z., Dey, A., and He, C. The kinetics of the interaction between iron(III)-ethylenediaminetetraacetate and peroxynitrite. *Aquat. Geochem.* 2010, *16*, 483–490.

258 Sharma, V.K., Szilágyi, P.A., Homonnay, Z., Kuzmann, E., and Vértes, A. Mössbauer investigation of peroxo species in Fe(III)-EDTA-H_2O_2 system. *Eur. J. Inorg. Chem.* 2005, *2005*, 4393–4400.

259 Brausam, A., Maigut, J., Meier, R., Szilágyi, P.A., Buschmann, H.-J., Massa, W., Homonnay, Z., and Van Eldik, R. Detailed spectroscopic, thermodynamic, and kinetic studies on the protolytic equilibria of Fe^{III}cydta and the activation of hydrogen peroxide. *Inorg. Chem.* 2009, *48*, 7864–7884.

260 Brausam, A. and Van Eldik, R. Further mechanistic information on the reaction between Fe^{III}(edta) and hydrogen peroxide: observation of a second reaction step and importance of pH. *Inorg. Chem.* 2004, *43*, 5351–5359.

261 Alvarez, B., Ferrer-Sueta, G., Freeman, B.A., and Radi, R. Kinetics of peroxynitrite reaction with amino acids and human serum albumin. *J. Biol. Chem.* 1999, *274*, 842–848.

262 Briviba, K., Kissner, R., Koppenol, W.H., and Sies, H. Kinetic study of the reaction of glutathione peroxidase with peroxynitrite. *Chem. Res. Toxicol.* 1998, *11*, 1398–1401.

263 Bryk, R., Griffin, P., and Nathan, C. Peroxynitrite reductase activity of bacterial peroxiredoxins. *Nature* 2000, *407*, 211–215.

264 Takakura, K., Beckman, J.S., MacMillan-Crow, L.A., and Crow, J.P. Rapid and irreversible inactivation of protein tyrosine phosphatases PTP1B, CD45, and LAR by peroxynitrite. *Arch. Biochem. Biophys.* 1999, *369*, 197–207.

265 Konorev, E.A., Hogg, N., and Kalyanaraman, B. Rapid and irreversible inhibition of creatine kinase by peroxynitrite. *FEBS Lett.* 1998, *427*, 171–174.

266 Souza, J.M. and Radi, R. Glyceraldehyde-3-phosphate dehydrogenase inactivation by peroxynitrite. *Arch. Biochem. Biophys.* 1998, *360*, 187–194.

267 Trujillo, M. and Radi, R. Peroxynitrite reaction with the reduced and the oxidized forms of lipoic acid: new insights into the reaction of peroxynitrite with thiols. *Arch. Biochem. Biophys.* 2002, *397*, 91–98.

268 Koppenol, W.H., Moreno, J.J., Pryor, W.A., Ischiropoulos, H., and Beckman, J.S. Peroxynitrite, a cloaked oxidant formed by nitric oxide and superoxide. *Chem. Res. Toxicol.* 1992, *5*, 834–842.

269 Perrin, D. and Koppenol, W.H. The quantitative oxidation of methionine to methionine sulfoxide by peroxynitrite. *Arch. Biochem. Biophys.* 2000, *377*, 266–272.

270 Carballal, S., Alvarez, B., Turell, L., Botti, H., Freeman, B.A., and Radi, R. Sulfenic acid in human serum albumin. *Amino Acids* 2007, *32*, 543–551.

271 Balazy, M., Kaminski, P.M., Mao, K., Tan, J., and Wolin, M.S. S-nitroglutathione, a product of the reaction between peroxynitrite and glutathione that generates nitric oxide. *J. Biol. Chem.* 1999, *273*, 32009–32015.

272 Masumoto, H., Kissner, R., Koppenol, W.H., and Sies, H. Kinetic study of the reaction of ebselen with peroxynitrite. *FEBS Lett.* 1996, *398*, 179–182.

273 Jensen, J.L., Miller, B.L., Zhang, X., Hug, G.L., and Schöneich, C. Oxidation of threonylmethionine by peroxynitrite. Quantification of the one-electron transfer pathway by comparison to one-electron photooxidation. *J. Am. Chem. Soc.* 1997, *119*, 4749–4757.

274 Alvarez, B., Rubbo, H., Kirk, M., Barnes, S., Freeman, B.A., and Radi, R. Peroxynitrite-dependent tryptophan nitration. *Chem. Res. Toxicol.* 1996, *9*, 390–396.

275 Van Der Vliet, A., Chr.'t Hoen, P.A., Wong, P.S., Bast, A., and Cross, C.E. Formation of S-nitrosothiols via direct nucleophilic nitrosation of thiols by peroxynitrite with elimination of hydrogen peroxide. *J. Biol. Chem.* 1998, *273*, 30255–30262.

276 Augusto, O., Gatti, R.M., and Radi, R. Spin-trapping studies of peroxynitrite decomposition and of 3-morpholinosydnonimine N-ethylcarbamide autooxidation: direct evidence for metal-independent formation of free radical intermediates. *Arch. Biochem. Biophys.* 1994, *310*, 118–125.

277 Kato, Y., Kawakishi, S., Aoki, T., Itakura, K., and Osawa, T. Oxidative modification of tryptophan residues exposed to peroxynitrite. *Biochem. Biophys. Res. Commun.* 1997, *234*, 82–84.

278 Pietraforte, D. and Minetti, M. One-electron oxidation pathway of peroxynitrite decomposition in human blood plasma: evidence for the formation of protein tryptophan-centred radicals. *Biochem. J.* 1997, *321*, 743–750.

279 Van Der Vliet, A., O'Neill, C.A., Halliwell, B., Cross, C.E., and Kaur, H. Aromatic hydroxylation and nitration of phenylalanine and tyrosine by peroxynitrite: evidence for hydroxyl radical production from peroxynitrite. *FEBS Lett.* 1994, *339*, 89–92.

280 Van Der Vliet, A., Eiserich, J.P., O'Neill, C.A., Halliwell, B., and Cross, C.E. Tyrosine modification by reactive nitrogen species: a closer look. *Arch. Biochem. Biophys.* 1995, *319*, 341–349.

281 Yamakura, F., Matsumoto, T., Fujimura, T., Taka, H., Murayama, K., Imai, T., and Uchida, K. Modification of a single tryptophan residue in human Cu,Zn-superoxide dismutase by peroxynitrite in the presence of bicarbonate. *Biochim. Biophys. Acta: Proteins Struct. Milecular Enzymol.* 2001, *1548*, 38–46.

282 Nuriel, T., Hansler, A., and Gross, S.S. Protein nitrotryptophan: formation, significance and identification. *J. Proteo.* 2011, *74*, 2300–2312.

283 Seeley, K.W. and Stevens, S.M. Investigation of local primary structure effects on peroxynitrite-mediated tyrosine nitration using targeted mass spectrometry. *J. Proteomics* 2012, *75*, 1691–1700.

284 Goldstein, S. and Czapski, G. Reactivity of peroxynitrite versus simultaneous generation of $^{\cdot}NO$ and $O_2^{\cdot-}$ toward NADH. *Chem. Res. Toxicol.* 2000, *13*, 736–741.

285 Santos, C.X.C., Bonini, M.G., and Augusto, O. Role of the carbonate radical anion in tyrosine nitration and hydroxylation by peroxynitrite. *Arch. Biochem. Biophys.* 2000, *377*, 146–152.

286 Reiter, C.D., Teng, R.-J., and Beckman, J.S. Superoxide reacts with nitric oxide to nitrate tyrosine at physiological pH via peroxynitrite. *J. Biol. Chem.* 2000, *275*, 32460–32466.

287 Pietraforte, D., Salzano, A.M., Marino, G., and Minetti, M. Peroxynitrite-dependent modifications of tyrosine residues in hemoglobin. Formation of tyrosyl radical(s) and 3-nitrotyrosine. *Amino Acids* 2003, *25*, 341–350.

288 Gunaydin, H. and Houk, K.N. Mechanisms of peroxynitrite-mediated nitration of tyrosine. *Chem. Res. Toxicol.* 2009, *22*, 894–898.

289 Castro, L., Demicheli, V., Tórtora, V., and Radi, R. Mitochondrial protein tyrosine nitration. *Free Radic. Res.* 2011, *45*, 37–52.

290 Bhabak, K.P., Satheeshkumar, K., Jayavelu, S., and Mugesh, G. Inhibition of peroxynitrite- and peroxidase-mediated protein tyrosine nitration by imidazole-based thiourea and selenourea derivatives. *Org. Biomol. Chem.* 2011, *9*, 7343–7350.

291 Massari, J., Fujiy, D.E., Dutra, F., Vaz, S.M., Costa, A.C.O., Micke, G.A., Tavares, M.F.M., Tokikawa, R., Assunção, N.A., and Bechara, E.J.H. Radical acetylation of 2′-deoxyguanosine and L-histidine coupled to the reaction of diacetyl with peroxynitrite in aerated medium. *Chem. Res. Toxicol.* 2008, *21*, 879–887.

292 Massari, J., Tokikawa, R., Zanolli, L., Tavares, M.F.M., Assunção, N.A., and Bechara, E.J.H. Acetyl radical production by the methylglyoxal-peroxynitrite system: a possible route for L-lysine acetylation. *Chem. Res. Toxicol.* 2010, *23*, 1762–1770.

293 Piñeyro, M.D., Arcari, T., Robello, C., Radi, R., and Trujillo, M. Tryparedoxin peroxidases from *Trypanosoma cruzi*: high efficiency in the catalytic elimination of hydrogen peroxide and peroxynitrite. *Arch. Biochem. Biophys.* 2011, *507*, 287–295.

294 Castro, H., Budde, H., Flohé, L., Hofmann, B., Lünsdorf, H., Wissing, J., and Toms, A.M. Specificity and kinetics of a mitochondrial peroxiredoxin of Leishmania infantum. *Free Radic. Biol. Med.* 2002, *33*, 1563–1574.

295 Budde, H., Flohé, L., Hecht, H.J., Hofmann, B., Stehr, M., Wissing, J., and Lünsdorf, H. Kinetics and redox-sensitive oligomerisation reveal negative subunit cooperativity in tryparedoxin peroxidase of trypanosoma brucei brucei. *Biol. Chem.* 2003, *384*, 619–633.

296 Chae, H.Z., Kim, H.J., Kang, S.W., and Rhee, S.G. Characterization of three isoforms of mammalian peroxiredoxin that reduce peroxides in the presence of thioredoxin. *Diabetes Res. Clin. Pract.* 1999, *45*, 101–112.

297 Dubuisson, M., Vander Stricht, D., Clippe, A., Etienne, F., Nauser, T., Kissner, R., Koppenol, W.H., Rees, J.F., and Knoops, B. Human peroxiredoxin 5 is a peroxynitrite reductase. *FEBS Lett.* 2004, *571*, 161–165.

298 Guerrero, S.A., Lopez, J.A., Steinert, P., Montemartini, M., Kalisz, H.M., Colli, W., Singh, M., Alves, M.J.M., and Flohé, L. His-tagged tryparedoxin peroxidase of *Trypanosoma cruzi* as a tool for drug screening. *Appl. Microbiol. Biotechnol.* 2000, *53*, 410–414.

299 Trujillo, M., Clippe, A., Manta, B., Ferrer-Sueta, G., Smeets, A., Declercq, J.P., Knoops, B., and Radi, R. Pre-steady state kinetic characterization of human peroxiredoxin 5: taking advantage of Trp84 fluorescence increase upon oxidation. *Arch. Biochem. Biophys.* 2007, *467*, 95–106.

300 Trajillo, M., Budde, H., Piñeyro, M.D., Stehr, M., Robello, C., Flohé, L., and Radi, R. Trypanosoma brucei and *Trypanosoma cruzi* tryparedoxin peroxidases catalytically detoxify peroxynitrite via oxidation of fast reacting thiols. *J. Biol. Chem.* 2004, *279*, 34175–34182.

301 Nogoceke, E., Gommel, D.U., Kieß, M., Kalisz, H.M., and Flohé, L. A unique cascade of oxidoreductases catalyses trypanothione-mediated peroxide metabolism in *Crithidia fasciculata. Biol. Chem.* 1997, *378*, 827–836.

302 Ogusucu, R., Rettori, D., Munhoz, D.C., Soares Netto, L.E., and Augusto, O. Reactions of yeast thioredoxin peroxidases I and II with hydrogen peroxide and

peroxynitrite: rate constants by competitive kinetics. *Free Radic. Biol. Med.* 2007, *42*, 326–334.

303 Manta, B., Hugo, M., Ortiz, C., Ferrer-Sueta, G., Trujillo, M., and Denicola, A. The peroxidase and peroxynitrite reductase activity of human erythrocyte peroxiredoxin 2. *Arch. Biochem. Biophys.* 2009, *484*, 146–154.

304 Selles, B., Hugo, M., Trujillo, M., Srivastava, V., Wingsle, G., Jacquot, J.P., Radi, R., and Rouhier, N. Hydroperoxide and peroxynitrite reductase activity of poplar thioredoxin-dependent glutathione peroxidase 5: kinetics, catalytic mechanism and oxidative inactivation. *Biochem. J.* 2012, *442*, 369–380.

305 Toledo, J.C., Jr. Audi, R., Ogusucu, R., Monteiro, G., Netto, L.E.S., and Augusto, O. Horseradish peroxidase compound I as a tool to investigate reactive protein-cysteine residues: from quantification to kinetics. *Free Radic. Biol. Med.* 2011, *50*, 1032–1038.

306 Baty, J.W., Hampton, M.B., and Winterbourn, C.C. Detection of oxidant sensitive thiol proteins by fluorescence labeling and two-dimensional electrophoresis. *Proteomics* 2002, *2*, 1261–1266.

307 Eaton, P. Protein thiol oxidation in health and disease: techniques for measuring disulfides and related modifications in complex protein mixtures. *Free Radic. Biol. Med.* 2006, *40*, 1889–1899.

308 Radi, R., Rodriguez, M., Castro, L., and Telleri, R. Inhibition of mitochondrial electron transport by peroxynitrite. *Arch. Biochem. Biophys.* 1994, *308*, 89–95.

309 Rubbo, H., Denicola, A., and Radi, R. Peroxynitrite inactivates thiol-containing enzymes of *Trypanosoma cruzi* energetic metabolism and inhibits cell respiration. *Arch. Biochem. Biophys.* 1994, *308*, 96–102.

310 Roberts, E.S., Lin, H., Crowley, J.R., Vuletich, J.L., Osawa, Y., and Hollenberg, P.F. Peroxynitrite-mediated nitration of tyrosine and inactivation of the catalytic activity of cytochrome P450 2B1. *Chem. Res. Toxicol.* 1998, *11*, 1067–1074.

311 Zou, M.H. and Ullrich, V. Peroxynitrite formed by simultaneous generation of nitric oxide and superoxide selectively inhibits bovine aortic prostacyclin synthase. *FEBS Lett.* 1996, *382*, 101–104.

312 Zou, M., Martin, C., and Ullrich, V. Tyrosine nitration as a mechanism of selective inactivation of prostacyclin synthase by peroxynitrite. *Biol. Chem.* 1997, *378*, 707–713.

313 Hühmer, A.F.R., Nishida, C.R., Ortiz De Montellano, P.R., and Schöneich, C. Inactivation of the inducible nitric oxide synthase by peroxynitrite. *Chem. Res. Toxicol.* 1997, *10*, 618–626.

314 Asahi, M., Fujii, J., Takao, T., Kuzuya, T., Hori, M., Shimonishi, Y., and Taniguchi, N. The oxidation of selenocysteine is involved in the inactivation of glutathione peroxidase by nitric oxide donor. *J. Biol. Chem.* 1997, *272*, 19152–19157.

315 Fu, Y., Porres, J.M., and Lei, X.G. Comparative impacts of glutathione peroxidase-1 gene knockout on oxidative stress induced by reactive oxygen and nitrogen species in mouse hepatocytes. *Biochem. J.* 2001, *359*, 687–695.

316 Keyer, K. and Imlay, J.A. Inactivation of dehydratase [4Fe-4S] clusters and disruption of iron homeostasis upon cell exposure to peroxynitrite. *J. Biol. Chem.* 1997, *272*, 27652–27659.

317 Castro, L.A., Robalinho, R.L., Cayota, A., Meneghini, R., and Radi, R. Nitric oxide and peroxynitrite-dependent aconitase inactivation and iron-regulatory

protein-1 activation in mammalian fibroblasts. *Arch. Biochem. Biophys.* 1998, *359*, 215–224.

318 Hausladen, A. and Fridovich, I. Superoxide and peroxynitrite inactivate aconitases, but nitric oxide does not. *J. Biol. Chem.* 1994, *269*, 29405–29408.

319 Berlett, B.S., Levine, R.L., and Stadtman, E.R. Carbon dioxide stimulates peroxynitrite-mediated nitration of tyrosine residues and inhibits oxidation of methionine residues of glutamine synthetase: both modifications mimic effects of adenylylation. *Proc. Natl. Acad. Sci. U.S.A.* 1998, *95*, 2784–2789.

320 Marcondes, S., Turko, I.V., and Murad, F. Nitration of succinyl-CoA:3-oxoacid CoA-transferase in rats after endotoxin administration. *Proc. Natl. Acad. Sci. U.S.A.* 2001, *98*, 7146–7151.

321 Macmillan-Crow, L.A., Crow, J.P., Kerby, J.D., Beckman, J.S., and Thompson, J.A. Nitration and inactivation of manganese superoxide dismutase in chronic rejection of human renal allografts. *Proc. Natl. Acad. Sci. U.S.A.* 1996, *93*, 11853–11858.

322 Ara, J., Przedborski, S., Naini, A.B., Jackson-Lewis, V., Trifiletti, R.R., Horwitz, J., and Ischiropoulos, H. Inactivation of tyrosine hydroxylase by nitration following exposure to peroxynitrite and 1-methyl-4-phenyl-1,2,3,6-tetrahydropyridine (MPTP). *Proc. Natl. Acad. Sci. U.S.A.* 1998, *95*, 7659–7663.

323 Kuhn, D.M. and Geddes, T.J. Peroxynitrite inactivates tryptophan hydroxylase via sulfhydryl oxidation. Coincident nitration of enzyme tyrosyl residues has minimal impact on catalytic activity. *J. Biol. Chem.* 1999, *274*, 29726–29732.

324 Klebl, B.M., Ayoub, A.T., and Pette, D. Protein oxidation, tyrosine nitration, and inactivation of sarcoplasmic reticulum Ca^{2+}-ATPase in low-frequency stimulated rabbit muscle. *FEBS Lett.* 1998, *422*, 381–384.

325 Viner, R.I., Hühmer, A.F.R., Bigelow, D.J., and Schoneich, C. The oxidative inactivation of sarcoplasmic reticulum Ca^{2+}-ATPaSe by peroxynitrite. *Free Radical Res.* 1996, *24*, 243–259.

326 Haendeler, J., Weiland, U., Zeiher, A.M., and Dimmeler, S. Effects of redox-related congeners of NO on apoptosis and caspase-3 activity. *Nitric Oxide—Biol. Chem.* 1997, *1*, 282–293.

327 Forsmark-Andrée, P., Persson, B., Radi, R., Dallner, G., and Ernster, L. Oxidative modification of nicotinamide nucleotide transhydrogenase in submitochondrial particles: effect of endogenous ubiquinol. *Arch. Biochem. Biophys.* 1996, *336*, 113–120.

328 Guittet, O., Ducastel, B., Salem, J.S., Henry, Y., Rubin, H., Lemaire, G., and Lepoivre, M. Differential sensitivity of the tyrosyl radical of mouse ribonucleotide reductase to nitric oxide and peroxynitrite. *J. Biol. Chem.* 1998, *273*, 22136–22144.

329 Sok, D.E. Active site of brain Zn^{2+}-glycerophosphocholine cholinephosphodiesterase and regulation of enzyme activity. *Neurochem. Res.* 1998, *23*, 1061–1067.

330 Sergeeva, S.V., Slepneva, I.A., and Khramtsov, V.V. Effect of selenolipoic acid on peroxynitrite-dependent inactivation of NADPH-cytochrome P450 reductase. *Free Radical Res.* 2001, *35*, 491–497.

331 Francescutti, D., Baldwin, J., Lee, L., and Mutus, B. Peroxynitrite modification of glutathione reductase: modeling studies and kinetic evidence suggest the

modification of tyrosines at the glutathione disulfide binding site. *Protein Eng.* 1996, *9*, 189–194.

332 Savvides, S.N., Scheiwein, M., Böhme, C.C., Arteel, G.E., Andrew Karplus, P., Becker, K., and Heiner Schirmer, R. Crystal structure of the antioxidant enzyme glutathione reductase inactivated by peroxynitrite. *J. Biol. Chem.* 2002, *277*, 2779–2784.

333 Wong, P.S., Eiserich, J.P., Reddy, S., Lopez, C.L., Cross, C.E., and Van Der Vliet, A. Inactivation of glutathione S-transferases by nitric oxide-derived oxidants: exploring a role for tyrosine nitration. *Arch. Biochem. Biophys.* 2001, *394*, 216–228.

334 Aykac-Toker, G., Bulgurcuoglu, S., and Kocak-Toker, N. Effect of peroxynitrite on glutaredoxin. *Hum. Exp. Toxicol.* 2001, *20*, 373–376.

335 Knapp, L.T., Kanterewicz, B.I., Hayes, E.L., and Klann, E. Peroxynitrite-induced tyrosine nitration and inhibition of protein kinase C. *Biochem. Biophys. Res. Commun.* 2001, *286*, 764–770.

336 Seidel, E.R., Ragan, V., and Liu, L. Peroxynitrite inhibits the activity of ornithine decarboxylase. *Life Sci.* 2001, *68*, 1477–1483.

337 Houston, M., Chumley, P., Radi, R., Rubbo, H., and Freeman, B.A. Xanthine oxidase reaction with nitric oxide and peroxynitrite. *Arch. Biochem. Biophys.* 1998, *355*, 1–8.

338 Lee, C.I., Liu, X., and Zweier, J.L. Regulation of xanthine oxidase by nitric oxide and peroxynitrite. *J. Biol. Chem.* 2000, *275*, 9369–9376.

339 Celano, L., Gil, M., Carballal, S., Durán, R., Denicola, A., Banerjee, R., and Alvarez, B. Inactivation of cystathionine β-synthase with peroxynitrite. *Arch. Biochem. Biophys.* 2009, *491*, 96–105.

340 Osoata, G.O., Yamamura, S., Ito, M., Vuppusetty, C., Adcock, I.M., Barnes, P.J., and Ito, K. Nitration of distinct tyrosine residues causes inactivation of histone deacetylase 2. *Biochem. Biophys. Res. Commun.* 2009, *384*, 366–371.

341 Stanyer, L., Jorgensen, W., Hori, O., Clark, J.B., and Heales, S.J.R. Inactivation of brain mitochondrial Lon protease by peroxynitrite precedes electron transport chain dysfunction. *Neurochem. Int.* 2008, *53*, 95–101.

342 Abd-Allah, G.M. and Mariee, A.D. Nitrite-mediated inactivation of human plasma paraoxonase-1: possible beneficial effect of aromatic amino acids. *Appl. Biochem. Biotechnol.* 2008, *150*, 281–288.

343 Donnini, S., Monti, M., Roncone, R., Morbidelli, L., Rocchigiani, M., Oliviero, S., Casella, L., Giachetti, A., Schulz, R., and Ziche, M. Peroxynitrite inactivates human-tissue inhibitor of metalloproteinase-4. *FEBS Lett.* 2008, *582*, 1135–1140.

344 Dairou, J., Atmane, N., Rodrigues-Lima, F., and Dupret, J.M. Peroxynitrite irreversibly inactivates the human xenobiotic-metabolizing enzyme arylamine N-acetyltransferase 1 (NAT1) in human breast cancer cells: a cellular and mechanistic study. *J. Biol. Chem.* 2004, *279*, 7708–7714.

345 Görg, B., Qvartskhava, N., Voss, P., Grune, T., Häussinger, D., and Schliess, F. Reversible inhibition of mammalian glutamine synthetase by tyrosine nitration. *FEBS Lett.* 2007, *581*, 84–90.

346 Ji, Y., Neverova, I., Van Eyk, J.E., and Bennett, B.M. Nitration of tyrosine 92 mediates the activation of rat microsomal glutathione S-transferase by peroxynitrite. *J. Biol. Chem.* 2006, *281*, 1986–1991.

347 Neta, P., Huie, R.E., and Ross, A.B. Rate constants for reaction of inorganic radicals in aqueous solution. *J. Phys. Chem. Ref. Data* 1988, *17*, 1027–1294.

348 Ranguelova, K., Rice, A.B., Khajo, A., Triquigneaux, M., Garantziotis, S., Magliozzo, R.S., and Mason, R.P. Formation of reactive sulfite-derived free radicals by the activation of human neutrophils: an ESR study. *Free Radic. Biol. Med.* 2012, *52*, 1264–1271.

349 Redpath, J.L. and Willson, R.L. Chain reactions and radiosensitization: model enzyme studies. *Int. J. Radiat. Biol.* 1975, *27*, 389–398.

350 Rustgi, S.N. and Riesz, P. An e.s.r. and spin-trapping study of the reactions of the SO_4^- radical with protein and nucleic acid constituents. *Int. J. Appl. Radiat. Isot.* 1978, *34*, 301–316.

351 Gau, B.C., Chen, H., Zhang, Y., and Gross, M.L. Sulfate radical anion as a new reagent for fast photochemical oxidation of proteins. *Anal. Chem.* 2010, *82*, 7821–7827.

352 Neta, P. and Huie, R.E. Free-radical chemistry of sulfite. *Environ. Health Perspect.* 1985, *64*, 209–217.

353 Rickman, K.A. and Mezyk, S.P. Kinetics and mechanisms of sulfate radical oxidation of β-lactam antibiotics in water. *Chemosphere* 2010, *81*, 359–365.

354 Shukla, P., Fatimah, I., Wang, S., Ang, H.M., and Tadé, M.O. Photocatalytic generation of sulphate and hydroxyl radicals using zinc oxide under low-power UV to oxidise phenolic contaminants in wastewater. *Catal. Today* 2010, *157*, 410–414.

355 Anipsitakis, G.P., Dionysiou, D.D., and Gonzalez, M.A. Cobalt-mediated activation of peroxymonosulfate and sulfate radical attack on phenolic compounds. Implications of chloride ions. *Environ. Sci. Technol.* 2006, *40*, 1000–1007.

356 Choi, H., Al-Abed, S.R., Dionysiou, D.D., Stathatos, E., and Lianos, P. Chapter 8 TiO_2-based advanced oxidation nanotechnologies for water purification and reuse. *In Anonymous* 2010, *2*, 229–254.

357 Liang, C. and Su, H.W. Identification of sulfate and hydroxyl radicals in thermally activated persulfate. *Ind. Eng. Chem. Res.* 2009, *48*, 5558–5562.

358 Yang, Q., Choi, H., Al-Abed, S.R., and Dionysiou, D.D. Iron-cobalt mixed oxide nanocatalysts: heterogeneous peroxymonosulfate activation, cobalt leaching, and ferromagnetic properties for environmental applications. *Appl. Catal. B: Environ.* 2009, *88*, 462–469.

359 Zalibera, M., Rapta, P., Staško, A., Brindzová, L., and Brezová, V. Thermal generation of stable SO_4^- spin trap adducts with super-hyperfine structure in their EPR spectra: an alternative EPR spin trapping assay for radical scavenging capacity determination in dimethylsulphoxide. *Free Radical Res.* 2009, *43*, 457–469.

360 Mora, V.C., Rosso, J.A., Carrillo Le Roux, G., Mártire, D.O., and Gonzalez, M.C. Thermally activated peroxydisulfate in the presence of additives: a clean method for the degradation of pollutants. *Chemosphere* 2009, *75*, 1405–1409.

361 Bao, Z.C. and Barker, J.R. Temperature and ionic strength effects on some reactions involving sulfate radical [SO_4^-(aq)]. *J. Phys. Chem.* 1996, *100*, 9780–9787.

362 McElroy, W.J. and Waygood, S.J. Kinetics of the reactions of the SO_4 radical with SO_4, $S_2O_8^{2-}$, H_2O, and Fe^{2+}. *J. Chem. Soc. Faraday Trans.* 1990, *86*, 2557–2564.

363 Maruthamuthu, P. and Neta, P. Reactions of phosphate radicals with organic compounds. *J. Phys. Chem.* 1977, *81*, 1622–1625.

364 Das, T.N. Reactivity and role of $SO_5^{\bullet-}$ radical in aqueous medium chain oxidation of sulfite to sulfate and atmospheric sulfuric acid generation. *J. Phys. Chem. A* 2001, *105*, 9142–9155.

365 Sharma, V.K. and Cabelli, D.E. Reduction of oxyiron(V) by sulfite and thiosulfate. *Environ. Sci. Technol.* 2009, *113*, 8901–8906.

366 Nfodzo, P. and Choi, H. Sulfate radicals destroy pharmaceuticals and personal care products. *Environ. Eng. Sci.* 2011, *28*, 605–609.

367 Dell'Arciprete, M.L., Cobos, C.J., Furlong, J.P., Mártire, D.O., and Gonzalez, M.C. Reactions of sulphate radicals with substituted pyridines: a structure-reactivity correlation analysis. *ChemPhysChem* 2007, *8*, 2498–2505.

368 Caregnato, P., David Gara, P.M., Bosio, G.N., Gonzalez, M.C., Russo, N., Michelini, M.D.C., and Mártire, D.O. Theoretical and experimental investigation on the oxidation of gallic acid by sulfate radical anions. *J. Phys. Chem. A* 2008, *112*, 1188–1194.

369 Chan, T.W., Graham, N.J.D., and Chu, W. Degradation of iopromide by combined UV irradiation and peroxydisulfate. *J. Hazard. Mater.* 2010, *181*, 508–513.

370 Antoniou, M.G., De La Cruz, A.A., and Dionysiou, D.D. Intermediates and reaction pathways from the degradation of microcystin-LR with sulfate radicals. *Environ. Sci. Technol.* 2010, *44*, 7238–7244.

371 Gara, P.M.D., Bosio, G.N., Gonzalez, M.C., Russo, N., Del Carmen Michelini, M., Diez, R.P., and Mártire, D.O. A combined theoretical and experimental study on the oxidation of fulvic acid by the sulfate radical anion. *Photochem. Photobiol. Sci.* 2009, *8*, 992–997.

372 Gara, P.M.D., Bosio, G.N., Gonzalez, M.C., and Mártire, D.O. Kinetics of the sulfate radical-mediated photo-oxidation of humic substances. *Int. J. Chem. Kinet.* 2008, *40*, 19–24.

373 Zhu, L., Michael Nicovich, J., and Wine, P.H. Temperature-dependent kinetics studies of aqueous phase reactions of hydroxyl radicals with dimethylsulfoxide, dimethylsulfone, and methanesulfonate. *Aquat. Sci.* 2003, *65*, 425–435.

374 Lomoth, R., Naumov, S., and Brede, O. Transients of the oxidation of pyrimidines with $SO_4^{\bullet-}$: structure and reactivity of the resulting radicals. *J. Phys. Chem. A* 1999, *103*, 6571–6579.

375 Méndez-Diaz, J., Sánchez-Polo, M., Rivera-Utrilla, J., Canonica, S., and von Gunten, U. Advanced oxidation of the surfactant SDBS by means of hydroxyl and sulphate radicals. *Chem. Eng. J.* 2010, *163*, 300–306.

376 Kishore, K. and Asmus, K.D. Radical cations from one-electron oxidation of aliphatic sulphoxides in aqueous solution. A radiation chemical study. *J. Chem. Soc. Perkin Trans.* 1989, *2*, 2079–2084.

377 Bosio, G., Criado, S., Massad, W., Nieto, F.J.R., Gonzalez, M.C., García, N.A., and Mártire, D.O. Kinetics of the interaction of sulfate and hydrogen phosphate radicals with small peptides of glycine, alanine, tyrosine and tryptophan. *Photochem. Photobiol. Sci.* 2005, *4*, 840–846.

378 Criado, S., Marioli, J.M., Allegretti, P.E., Furlong, J., Rodríguez Nieto, F.J., Mártire, D.O., and García, N.A. Oxidation of di- and tripeptides of tyrosine and valine mediated by singlet molecular oxygen, phosphate radicals and sulfate radicals. *J. Photochem. Photobiol. B: Biol.* 2002, *65*, 74–84.

379 Ito, T., Morimoto, S., Fujita, S., Kobayashi, K., Tagawa, S., and Nishimoto, S. Carbon- and nitrogen-centered radicals produced from l-lysine by radiation-induced oxidation: a pulse radiolysis study. *Chem. Phys. Lett.* 2008, *462*, 116–120.

380 Ito, T., Morimoto, S., Fujita, S., and Nishimoto, S. Radical intermediates generated in the reactions of l-arginine with hydroxyl radical and sulfate radical anion: a pulse radiolysis study. *Radiat. Phys. Chem.* 2009, *78*, 256–260.

381 Morimoto, S., Ito, T., Fujita, S., and Nishimoto, S. A pulse radiolysis study on the reactions of hydroxyl radical and sulfate radical anion with guanidine derivatives in aqueous solution. *Chem. Phys. Lett.* 2008, *461*, 300–304.

382 Villata, L.S., Gonzalez, M.C., and Mártire, D.O. A kinetic study of the reactions of sulfate and dihydrogen phosphate radicals with epicatechin, epicatechingallate, and epigalocatechingallate. *Int. J. Chem. Kinet.* 2010, *42*, 391–396.

383 Rayshell, M., Ross, J., and Werbin, H. Evidence that *N*-acetoxy-*N*-acetyl-2-aminofluorene cross-links DNA to protein by a free radical mechanism. *Carcinogenesis* 1983, *4*, 501–507.

384 Fancy, D.A. and Kodadek, T. Chemistry for the analysis of protein-protein interactions: rapid and efficient cross-linking triggered by long wavelength light. *Proc. Natl. Acad. Sci. U.S.A.* 1999, *96*, 6020–6024.

385 Rosso, J.A., Caregnato, P., Mora, V.C., Gonzalez, M.C., and Mártire, D.O. Reactions of phosphate radicals with monosubstituted benzenes. A mechanistic investigation. *Helv. Chim. Acta* 2003, *86*, 2509–2524.

386 Maruthamuthu, P. and Neta, P. Phosphate radicals. Spectra, acid-base equilibria, and reactions with inorganic compounds. *J. Phys. Chem.* 1978, *82*, 710–713.

387 Mártire, D.O. and Gonzalez, M.C. Aqueous phase kinetic studies involving intermediates of environmental interest: phosphate radicals and their reactions with substituted benzenes. *Prog. React. Kinet. Mechan* 2001, *26*, 201–218.

388 Huber, J.R. and Hayon, E. Flash photolysis in the vacuum ultraviolet region of the phosphate anions $H_2PO_4^-$, HPO_4^{2-}, and $P_2O_7^{4-}$ in aqueous solutions. *J. Phys. Chem.* 1968, *72*, 3820–3827.

389 Rosso, J.A., Rodríguez Nieto, F.J., Gonzalez, M.C., and Mártire, D.O. Reactions of phosphate radicals with substituted benzenes. *J. Photochem. Photobiol. A.* 1998, *116*, 21–25.

390 Rosso, J.A., Criado, S., Bertolotti, S.G., Allegretti, P.E., Furlong, J., García, N.A., Gonzalez, M.C., and Mártire, D.O. Kinetic study of the oxidation of phenolic derivatives of α,α,α-trifluorotoluene by singlet molecular oxygen [$O_2(^1\Delta_g)$] and hydrogen phosphate radicals. *Photochem. Photobiol. Sci.* 2003, *2*, 882–887.

391 Hansch, C., Hoekman, D., and Gao, H. Comparative QSAR: toward a deeper understanding of chemicobiological interactions. *Chem. Rev.* 1996, *96*, 1045–1075.

392 Neta, P., Madhavan, V., Zemel, H., and Fessenden, R.W. Rate constants and mechanism of reaction of SO_4^- with aromatic compounds. *J. Am. Chem. Soc.* 1977, *99*, 163–164.

393 Mohan, H., Mudaliar, M., Aravindakumar, C.T., Rao, B.S.M., and Mittal, J.P. Studies on structure-reactivity in the reaction of OH radicals with substituted halobenzenes in aqueous solutions. *J. Chem. Soc. Perkin Trans.* 1991, *2*, 1387–1392.

394 Peskin, A.V. and Winterbourn, C.C. Kinetics of the reactions of hypochlorous acid and amino acid chloramines with thiols, methionine, and ascorbate. *Free Radic. Biol. Med.* 2001, *30*, 572–579.

395 Ross, A.B. and Neta, P. Rate constants for reactions of inorganic radicals in aqueous solution. *NBS Natl. Std. Ref. Ser.* 1979, *65*, 1–55.

396 Das, T.N., Huie, R.E., Neta, P., and Padmaja, S. Reduction Potential of the sulfhydryl radical: pulse radiolysis and laser flash photolysis studies of the formation and reactions of ·SH and $HSSH^-$ in aqueous solutions. *J. Phys. Chem. A* 1999, *103*, X1–5226.

397 Carballal, S., Trujillo, M., Cuevasanta, E., Bartesaghi, S., Möller, M.N., Folkes, L.K., García-Bereguiaín, M.A., Gutiérrez-Merino, C., Wardman, P., Denicola, A., Radi, R., and Alvarez, B. Reactivity of hydrogen sulfide with peroxynitrite and other oxidants of biological interest. *Free Radic. Biol. Med.* 2011, *50*, 196–205.

398 Folkes, L.K., Candeias, L.P., and Wardman, P. Kinetics and mechanisms of hypochlorous acid reactions. *Arch. Biochem. Biophys.* 1995, *323*, 120–126.

399 Pattison, D.I. and Davies, M.J. Absolute rate constants for the reaction of hypochlorous acid with protein side chains and peptide bonds. *Chem. Res. Toxicol.* 2001, *14*, 1453–1464.

400 Winterbourn, C.C. and Metodiewa, D. Reactivity of biologically important thiol compounds with superoxide and hydrogen peroxide. *Free Radic. Biol. Med.* 1999, *27*, 322–328.

401 Filipovic, M.R., Miljkovic, J., Allgäuer, A., Chaurio, R., Shubina, T., Herrmann, M., and Ivanovic-Burmazovic, I. Biochemical insight into physiological effects of H 2S: reaction with peroxynitrite and formation of a new nitric oxide donor, sulfinyl nitrite. *Biochem. J.* 2012, *441*, 609–621.

402 Su, J. and Groves, J.T. Mechanisms of peroxynitrite interactions with heme proteins. *Inorg. Chem.* 2010, *49*, 6317–6329.

403 Spasojević, I. Free radicals and antioxidants at a glance using EPR spectroscopy. *Crit. Rev. Clin. Lab. Sci.* 2011, *48*, 114–142.

404 Davis, M.F., Vigil, D., and Campbell, S.L. Regulation of Ras proteins by reactive nitrogen species. *Free Radic. Biol. Med.* 2011, *51*, 565–575.

405 Padmaja, S., Ramazenian, M.S., Bounds, P.L., and Koppenol, W.H. Reaction of peroxynitrite with L-tryptophan. *Redox Report* 1996, *2*, 173–177.

406 Yi, D., Smythe, G.A., Blount, B.C., and Duncan, M.W. Peroxynitrite-mediated nitration of peptides: characterization of the products by electrospray and combined gas chromatography-mass spectrometry. *Arch. Biochem. Biophys.* 1997, *344*, 253–259.

6

HIGH-VALENT Cr, Mn, AND Fe SPECIES

High-valent metal-oxo species are involved in chemical, industrial, biological, and environmental oxidations [1–5]. Compounds of Cr in the pentavalent oxidation state with the formula M_3CrO_8 ($M = Na^+$, K^+, Cs^+, Rb^+, and NH_4^+) have shown magnetic and ferroelectric properties for their possible applications in data storage and manipulations [6]. Compounds of Cr(VI) have useful properties, which include tensile strength and malleability and, hence, are used in various industries such as stainless steel welding, chrome plating, and production of dyes. The oxidative degradation of organic pollutants using chromate-induced activation of hydrogen peroxide has also been proposed as a potential advanced oxidation process [7]. Applications of Cr compounds may lead to environmental hazards due to their carcinogenicity and mammalian toxicity [8, 9]. High-valent complexes of Mn play a role in synthetic chemistry and in biometric oxidation reactions. For example, high-valent manganese-oxo species have been proposed in the oxygen evolving complex (OEC) of photosystem II (PSII) [5, 10, 11]. High-valent-iron-oxo species have been invoked in oxidation reactions of heme and nonheme iron enzymes [12–16]. Ferrates ($Fe^VO_4^{3-}$ and $Fe^{VI}O_4^{2-}$) have shown properties of disinfection and oxidation to purify water [17–22].

In this chapter, the basic chemistry of high-valent compounds of Cr, Mn, and Fe is presented, followed by a discussion on their reactivity with biological compounds including amino acids, peptides, and proteins.

Oxidation of Amino Acids, Peptides, and Proteins: Kinetics and Mechanism, First Edition. Virender K. Sharma.

6.1 CHROMIUM

The oxidation states of chromium range from –2 to +6, which vary in colors and geometries of their compounds [23]. The electronic configuration of elemental chromium is $3d^54s^1$. The most common and stable oxidation states are +3 and +6 with electronic configurations of $3d^34s^0$ and $3d^04s^0$, respectively. Chromium(III) is amphoteric, capable of forming complexes with both acids and bases, which are, to some extent, kinetically substitution-inert [24]. Compounds of Cr(III) are mainly octahedral [25]. Oxo compounds of Cr(VI) in aqueous solutions are tetrahedral in their geometries. Compounds of other oxidation states, particularly +2, +4, and +5, are unstable and are easily converted to stable oxidation states of Cr.

Chromium is considered an essential micronutrient for humans, but a recent review on this topic disputes this conception [26]. The results of the nutritional biochemistry of Cr(III) within the last two decades are considered, concluding there is still a lack of understanding of the identification of the molecular-level mechanism(s) by which administration of Cr to humans occurs, and further research is necessary to confirm if Cr is an essential element.

The potentials of redox pairs of chromium are presented in Figure 6.1, which suggests the +2 oxidation state is a good reductant, while the +6 oxidation state is a powerful oxidant [24]. In acidic solutions, Cr(VI) is unstable and proceeds to the most stable state, Cr(III). The estimated potentials of Cr(VI)/

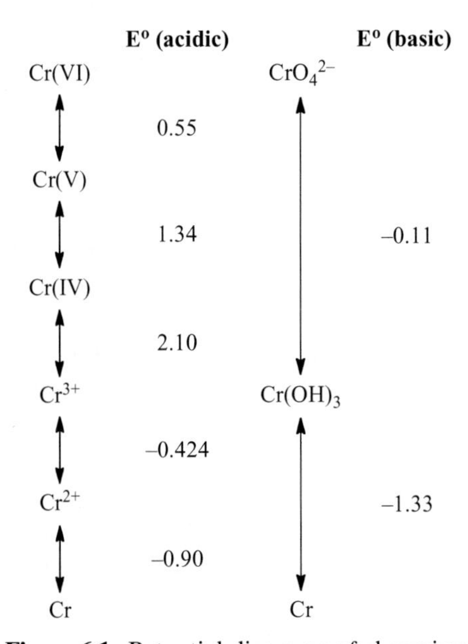

Figure 6.1. Potential diagrams of chromium.

Cr(V), Cr(V)/Cr(IV), and Cr(IV)/Cr(III) redox pairs are also given in Figure 6.1. In acidic solutions, the one-electron redox potentials increase in the order Cr(VI) < Cr(V) < Cr(IV). Cr(III) and Cr(VI) show different chemical behaviors, toxicities, and bioavailabilities [27]. The following sections first convey the chemistry of stable oxo species of Cr(III) and Cr(VI) in aqueous solutions, followed by their reactivity with biologically important molecules to understand their toxicity. The generation and involvement of unstable Cr(III), Cr(IV), and Cr(V) species are also summarized.

6.1.1 Aqueous Chemistry of Oxo-Cr Compounds

6.1.1.1 Cr(III). In aqueous solution, Cr(III) exists as a hexaqua ion, $[(Cr(H_2O)_6]^{3+}$, which is acidic ($pK_a = 4$) and is in dynamic equilibrium with its monomeric hydrolysis product. These species dimerize to form hydroxo-bridged species, which further polymerize slowly to trimers, teramers, hexamers, and so on [28]. The main aqueous species of Cr(III) are Cr^{3+}, $Cr(OH)^{2+}$, $Cr(OH)_3^o$, and $Cr(OH)_4^-$ [29]. In an acidic medium, Cr^{3+} is the main species, while $Cr(OH)_3^o$ and $Cr(OH)_4^-$ predominantly exist in the alkaline medium. The spectra of Cr(III) were obtained by mixing Cr(III) with either 0.005 M (pH 11.7) or 0.5 M NaOH (pH 13.7) solutions [30]. The spectra of a monomer and a dimer had a maxima at 230 nm. The molar absorptivities at 230 nm were determined as $\varepsilon_{monomer} = 245$/M/cm and $\varepsilon_{dimer} = 830$/M/cm at pH 11.7, and $\varepsilon_{monomer} = 348$/M/cm and $\varepsilon_{dimer} = 965$/M/cm at pH 13.7.

The kinetics of the monomer and dimer were studied at 230 nm as a function of pH. The kinetics curve was explained by the following set of reactions:

$$Cr^{III}(H_2O)_3(OH)_3 \rightleftharpoons H^+ + Cr^{III}(H_2O)_2(OH)_4^- \quad pK_a(\text{calc.}) = 12.82 \tag{6.1}$$

$$Cr^{III}(OH)_3 + Cr^{III}(OH)_3 \rightleftharpoons [(Cr)_2^{III,III}]_x \tag{6.2}$$

$$Cr^{III}(OH)_3 + Cr^{III}(OH)_4^- \rightleftharpoons [(Cr)_2^{III,III}]_y \tag{6.3}$$

$$Cr^{III}(OH)_4^- + Cr^{III}(OH)_4^- \rightleftharpoons [(Cr)_2^{III,III}]_z. \tag{6.4}$$

Cr(III) behaves as a typical "Lewis acid" and forms complexes with inorganic and organic ligands, which is a slow process due to substitution of the coordinated water molecules with a ligand. Interestingly, the frequency of dehydration of water from Cr(III) is $10^{-6.3}$/s while the frequencies for Al(III) and Fe(III) are $10^{-0.8}$ and $10^{-3.5}$/s [31].

6.1.1.2 Cr(VI). Oxo complexes of Cr(VI) have been studied experimentally and theoretically over the last several years. The chemical bonds in the CrO_4^{2-} ion are composed of d orbitals of Cr atoms and p orbitals of O atoms. There are two stable forms of Cr(VI) in the solid state: monomeric, CrO_4^{2-}, and dimeric, $Cr_2O_7^{2-}$. In aqueous solution, additional species, $HCrO_4^-$, $H_2CrO_4^{2-}$, and $HCr_2O_7^-$, are also involved in the following equilibria [32]:

$$H_2CrO_4 \rightleftharpoons H^+ + HCrO_4^- \quad pK_{a1} = 0.74 \text{ (ionic strength} = 0.16, 25°C) \tag{6.5}$$

$$HCrO_4^- \rightleftharpoons H^+ + CrO_4^{2-} \quad pK_{a2} = 5.84 \text{ (ionic strength} = 0.5, 25°C) \tag{6.6}$$

$$2HCrO_4^- \rightleftharpoons Cr_2O_7^{2-} + H_2O \quad pK_{a3} = -1.87 \text{ (ionic strength} = 0.5, 25°C). \tag{6.7}$$

Overall, the acidity and total concentration of Cr(VI) determine the participation of the species. The equilibria and the rate constants of reaction (6.7) depend on the type of buffer and the concentrations of H^+ and OH^- ions [32]. In dilute solution ($<10^{-3}$ M), the monomeric form predominates and $HCrO_4^-$ is the principal species in the pH range of 1–5. The dimeric form is not present in any significant concentration in biological and environmental conditions. At pH > 8, the yellow ion, CrO_4^{2-}, is the major species. The relative proportion of the species depends on both the pH and the total Cr(VI) concentration [33].

6.1.1.3 Cr(III) Superoxo and Hydroperoxo Complexes. The chemistry of chromium(III) superoxo ($Cr_{aq}OO^{2+}$) and hydroperoxo ($Cr_{aq}OOH^{2+}$) complexes have been studied in detail due to their importance in oxygen activation in industrial and biological processes [34–36]. These complexes were synthesized from the reaction of Cr(II) with molecular oxygen [37]. The decomposition of $Cr_{aq}OO^{2+}$ in acidic solution depends on the pH, and either hemolytic or heterolytic steps are responsible for the decomposition of $Cr_{aq}OO^{2+}$. This is given as a scheme in Figure 6.2. A homolysis Cr–O_2 bond is the major step in acidic solution, which is followed by the reaction between Cr_{aq}^{2+} and $Cr_{aq}OO^{2+}$ to ultimately produce Cr^{3+} and Cr^{6+}. Under the conditions that the pH is changed, the conjugate base, $(OH)Cr_{aq}OO^{2+}$, dissociates to Cr_{aq}^{3+}, and the free superoxide disproportionates to O_2 and H_2O_2 (Fig. 6.2). The products O_2, Cr_{aq}^{3+}, and Cr(VI) are also generated by the parallel reactions between $Cr_{aq}OO^{2+}$ and (OH) $Cr_{aq}OO^{2+}$ (Fig. 6.2).

The decay kinetics of $Cr_{aq}OOH^{2+}$ depends on the concentrations of H^+ and the hydroperoxo complex. Similar to superoxo complexes, the decay reaction generates mixtures of O_2, H_2O_2, Cr_{aq}^{3+}, and Cr^{6+} [35]. The multiple steps with the involvement of Cr(IV) and Cr(V) as intermediates of the reactions were suggested. In recent years, substitution of the hydroperoxo group in $Cr_{aq}OOH^{2+}$ with nitrate was carried out to form a $Cr_{aq}NO_2^{2+}$ complex, which may be important transients in the biological system and in aqueous atmospheric photochemistry [38].

6.1.1.4 Aqueous Cr(IV) Ion.

Acid Medium. In aqueous solution, Cr^{IV} was formed from the oxidation of Cr^{III} by $^{\bullet}OH$ and $SO_4^{\bullet-}$, produced by pulse radiolysis at pH 3.0 and 3.7 [39]. Initially, a precursor complex is formed, followed by an electron transfer from Cr^{III} to the radical to generate Cr^{IV} (reaction 6.8):

$$Cr^{III} + {}^{\bullet}OH/SO_4^{\bullet-} \rightleftharpoons Cr^{III}({}^{\bullet}OH)/SO_4^{\bullet-} \rightarrow Cr^{IV}. \tag{6.8}$$

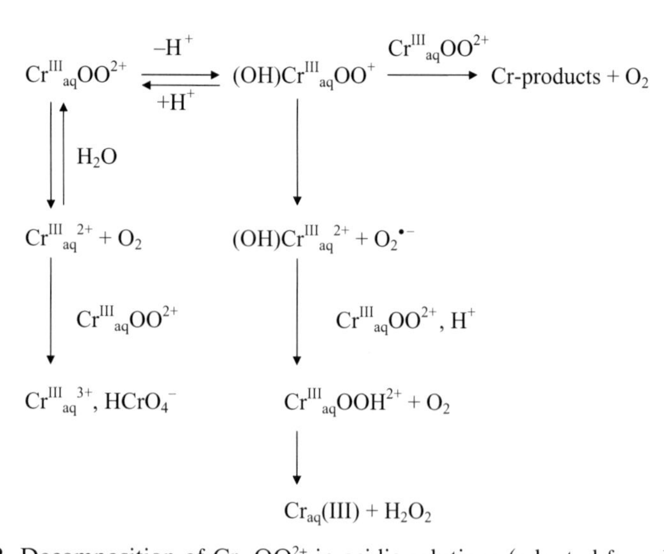

Figure 6.2. Decomposition of $Cr_{aq}OO^{2+}$ in acidic solutions (adapted from Bakac [35] with permission from Elsevier Inc.).

The rate constants of the electron transfer steps are of the order of 10^6 and 10^4/s for $^{\bullet}OH$ and $SO_4^{\bullet -}$, respectively. The spectrum of Cr^{IV} showed a weak rising band between 420 and 250 nm, which had an increase in the molar absorption coefficient from 4.3 to 48.0/M/cm [39]. The final product was Cr^{VI}. A mechanism was proposed in which the first step was the formation of Cr^{III} and Cr^{V} (reaction 6.9). This step was followed by the rearrangement of the coordination shell of Cr^{V} from octahedral to tetrahedral (reaction 6.10). The reaction between Cr^{V} and Cr^{IV} yielded Cr^{VI} (reaction 6.11):

$$Cr_{oct}^{IV} + Cr_{oct}^{IV} \rightarrow Cr_{oct}^{III} + Cr_{oct}^{V} \quad \text{slow} \tag{6.9}$$

$$Cr_{oct}^{V} \rightleftharpoons Cr_{tet}^{V} \quad \text{fast} \tag{6.10}$$

$$Cr_{tet}^{V} + Cr_{oct}^{IV} \rightarrow Cr_{tet}^{VI} + Cr_{oct}^{III} \quad \text{fast.} \tag{6.11}$$

The aquachromyl(IV) ion, $Cr_{aq}O^{2+}$, had also been produced from the reaction of Cr_{aq}^{2+} with oxygen in dilute perchloric acid [40]. This study suggested a formation of a precursor before the formation of Cr^{III} and Cr^{V}. The cleavage of the O–H bond to yield the product was confirmed by a large solvent kinetic effect ($k_H/k_D = 6.9$). The additional steps were the same as proposed earlier (reactions 6.10 and 6.11) [39].

Alkaline Medium. Cr^{IV} was obtained in the pulse radiolysis of solution of Cr^{III} in 0.5 M NaOH [30]. Aging of Cr^{III} solution formed both monomeric and dimeric species (reactions 6.1–6.4), and the pulse radiolysis of these species results in Cr^{IV} species (reactions 6.12 and 6.13):

$$\mathrm{Cr^{III}(OH)_4^- + O^- \rightleftharpoons Cr^{IV}O(OH)_4^{2-}} \quad (6.12)$$

$$\mathrm{[(Cr)_2^{III,III}]_z + O^- \rightleftharpoons [(Cr)_2^{III,IV}]_z}. \quad (6.13)$$

The decay of Cr^{IV} ultimately yielded Cr^{VI} species (reactions 6.14 and 6.15). Significantly, no evidence of Cr^{V} as an intermediate was observed. This is in contrast to the decay of Cr^{IV} in acidic solution (see reactions 6.9 and 6.10):

$$\mathrm{2Cr^{IV} \rightarrow (Cr)_2^{IV,IV}} \quad (6.14)$$

$$\mathrm{(Cr)_2^{IV,IV} + }n\mathrm{(Cr^{III}) \rightarrow Cr^{VI}(\text{-O-}Cr^{III})}_n\mathrm{ + Cr^{IV} + 2Cr^{III}}. \quad (6.15)$$

6.1.1.5 Cr(V). The pulse radiolysis of N_2O-saturated solution containing Cr^{VI} and formate yields Cr^{V} [41]. The formate radical produced in the system reacts rapidly with Cr^{VI} to form Cr^{V} ($Cr^{VI} + {}^{\bullet}CO_2 \rightarrow Cr^{V}$; $k \approx 10^8$/M/s). The spectra obtained at different pH values are given in Figure 6.3. The spectrum of Cr^{V} did not vary much within the pH range of 1.75–4.75 [41]. However, in the alkaline pH range, the spectrum had a higher molar absorptivity than in the acidic pH range (Fig. 6.3) [41]. The spectrum of Cr^{V} at pH 8 and 13.7 were

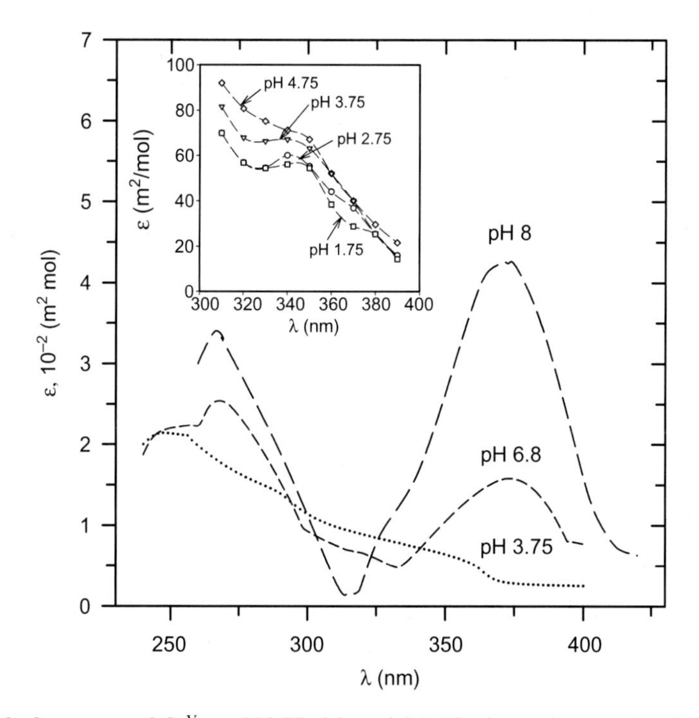

Figure 6.3. Spectrum of Cr^{V} at pH 3.75, 6.8, and 8.0. The inset shows a detail spectrum at pH 1.75, 2.75, 3.75, and 4.75 (redrawn from Buxton and Djouider [41]).

similar though [41]. Using the disproportionation rate of Cr^V as a function of pH as well as the conductivity measurements, the following sequence of reactions were proposed for the decay of Cr^V (reactions 6.16–6.20):

$$HCr^{VI}O^-_{4(tet)} + CO_2^{\bullet -} \rightarrow HCr^{V}O^{2-}_{4(tet)} + CO_2 \tag{6.16}$$

$$HCr^{V}O^{2-}_{4(tet)} + H^+ \rightleftharpoons H_2Cr^{V}O_4^- \quad K_{17} = 6.0 \times 10^3 \tag{6.17}$$

$$H_2Cr^{V}O_4^- + H^+ \rightleftharpoons H_3Cr^{V}O_{4(oct)} \quad K_{18} = 5.6 \times 10^2 \tag{6.18}$$

$$HCr^{V}O^{2-}_{4(tet)} + H_3Cr^{V}O_{4(oct)} \rightarrow HCr^{VI}O^-_{4(tet)} + H_3Cr^{V}O^-_{4(oct)} \tag{6.19}$$

$$HCr^{V}O^{2-}_{4(tet)} \rightleftharpoons H^+ + CrO_4^{3-} \quad K_{20} = 1.0 \times 10^{-7}. \tag{6.20}$$

The speciation of Cr^V as a function of pH in dilute aqueous solution using the constants of protonic equilibria of reactions (6.17), (6.18), and (6.20) is shown in Figure 6.4. At a neutral pH, both monoprotonated and deprotonated species of Cr^V coexist.

6.1.2 Chromium(VI, V, and IV) Complexes

In addition to the oxo complexes of Cr(VI), there are other simple complexes such as CrO_3X ($X = ClO_4^-$, HSO_4^-, and Hal^- [halogenation]) [25, 42]. The reactions between Cr(VI) and glutathione (GSH) and smaller thiolate reductants form complex species [43]. For example, the reaction between K_2CrO_7 and

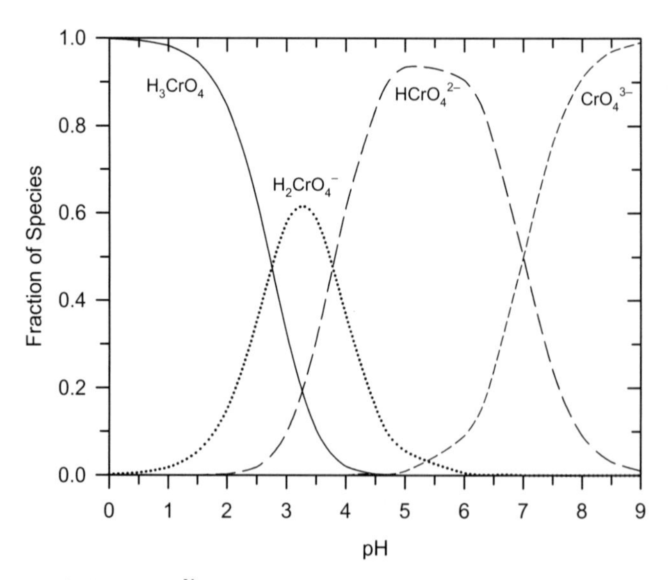

Figure 6.4. Speciation of Cr^V as a function of pH in dilute aqueous solution (redrawn from Buxton and Djouider [41]).

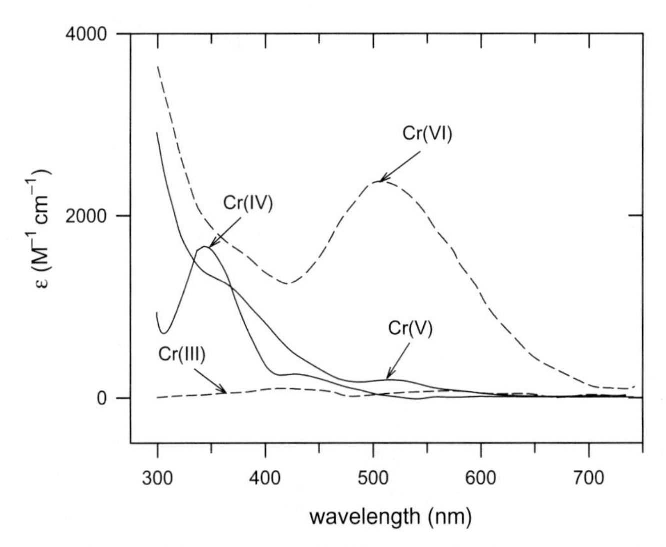

Figure 6.5. Typical UV–visible spectra of different oxidation states of Cr in 250 mM $ehbaH_2$/ehba/H buffer, pH 3.5. Main absorbing species (>90%) are Cr(VI) = $[HCrO_4^-]$; Cr(V) = $[CrO(ehba)_2]^-$; Cr(IV) = $[CrO(ehbaH)_2]^o$; Cr(III) = $[Cr(ehbaH)_2(OH_2)_2]^+$ (adapted from Codd et al. [44] with permission from Elsevier Inc.).

GSH at pH≤6 yields Cr(VI)–thioester. Complexes between cysteine, N-acetylcysteine, and γ-glutamylcysteine have also been characterized [43]. There have been several reports on the formation of Cr(V) complexes in the reaction of Cr(VI) with reductants such as hydroxyl acids, sugars, diols, ascorbic acids, heterocycles, amino acids, peptides, and proteins. The chemistry and their spectroscopic characterization have been reviewed [25, 44–46]. Cr(V)–porphyrin and Cr(V)–nitride complexes have also been prepared [1, 47, 48].

There are only a few examples of Cr(IV) complexes. A relatively stable Cr(IV)-2-hydroxy acid complex has been produced in aqueous solutions at pH 2–4 [49, 50]. The spectra of Cr(VI/V/IV, III) species with the 2-ethyl-2-hydroxobutanoato (ehba) ligand are shown in Figure 6.5 [44]. The Cr(IV) complex showed absorbance in the wavelength range from 400 to 600 nm ($\varepsilon_{max} = (2-4) \times 10^3$/M/cm. The spectrum of Cr(IV) was useful in studying the reactions of Cr(IV) with DNA and other molecules [44].

6.1.3 Reduction of Cr(VI/V/IV) by Substrates

Numerous studies have reviewed the reduction of Cr(VI) by inorganic and organic substrates (S) [51–58]. Most of the work was conducted under strong acidic conditions (pH≤2). One-electron and two-electron transfer steps have been proposed in the mechanisms of Cr(VI) reductions (reactions 6.21–6.23):

$$Cr(VI) + S \rightleftharpoons Cr(VI)\text{-}S \tag{6.21}$$

$$Cr(VI) - S \rightarrow Cr(V) + S^{\bullet} \tag{6.22}$$

$$Cr(VI) - S \rightarrow Cr(IV) + P. \tag{6.23}$$

The first step is commonly the formation of the Cr(VI)–S complex, followed by either a one-electron reduction to form Cr(V) and the substrate radical ($S^{\bullet}$) or a two-electron reduction to form Cr(IV) and the product (P). The following set of subsequent reactions can occur (reactions 6.24–6.31):

$$2Cr(IV) \rightarrow Cr(V) + Cr(III) \tag{6.24}$$

$$Cr(IV) + Cr(VI) \rightarrow 2Cr(V) \tag{6.25}$$

$$Cr(IV) + S \rightarrow Cr(III) + R^{\bullet} \tag{6.26}$$

$$Cr(VI) + R^{\bullet} \rightarrow Cr(V) + P \tag{6.27}$$

$$Cr(IV) + S \rightarrow Cr(II) + P \tag{6.28}$$

$$Cr(VI) + Cr(II) \rightarrow Cr(V) + Cr(III) \tag{6.29}$$

$$Cr(V) + S \rightarrow Cr(IV) + R^{\bullet} \tag{6.30}$$

$$Cr(V) + S \rightarrow Cr(III) + P. \tag{6.31}$$

The reductions of Cr(VI) by substrates differ depending on the involvement of Cr(V) or Cr(IV) in the steps of the mechanisms. Cr(V) can be formed by (1) disproportionation of the Cr(IV) (reaction 6.24), (2) the reaction of Cr(IV) with Cr(VI) (reaction 6.25), and (3) the one-electron transfer from Cr(VI) by either a radical (Eq. 6.27) or Cr(II) (reaction 6.29). Cr(IV) can oxidize a substrate by a one-electron pathway to form a radical (reaction 6.26) and by a two-electron pathway to yield the oxidized product of the substrate (reaction 6.28). The reduction of Cr(V) can proceed by either a one-electron or a two-electron transfer pathway (reactions 6.30 and 6.31). Therefore, different mechanisms occur depending on the reaction conditions as well as the concentrations of the chromium species and the substrate. The stabilization of the intermediate Cr(V/IV) species by substrates results in complex species, which can also influence the reaction mechanisms [59]. These complex species may also react with the substrate and the components of the reaction medium [59, 60].

6.1.4 Reactivity of Cr Species

In this section, the reduction of Cr(VI) with reductants of biological importance is discussed. The importance of these reactions will be discussed in the next section.

6.1.4.1 Fe(II) Complexes. The reduction of Cr(VI) by biological Fe(II) complexes (e.g., heme proteins) may result in Cr(V) complexes. The reaction between *trans*-bis(2-ethyl-2-hydroxybutanoato(2-))oxochromate(V) and cytochrome c^{II} has been studied [61]. The Cr(V) complex has several negatively charged donor centers, which could interact with the positively charged surfaces of the cytochrome c^{II} and resulted in the precursor complex. A proposed mechanism involves the outer-sphere one-electron transfer from Fe(II) to Cr(V), which proceeds by a precursor complex.

6.1.4.2 Hydrogen Peroxide. The Cr(VI)/H_2O_2 reaction has been extensively reviewed [62]. This reaction has an important role in the reduction of Cr(VI) by biomolecules [63]. Several peroxo species such as tetraperoxochromate(V) and oxodiperoxochromate(VI) complexes have been reported [64]. Recently, the kinetics of the reaction between Cr(VI) and H_2O_2 was examined in acetate and phosphate buffer solutions as a function of pH (4.6–7.3) [65]. Cr(VI) reduces to Cr(III) and acts as a catalyst for the dismutation of H_2O_2. Two diperoxochromium(V) complexes were observed, which play an important role in the conversion of H_2O_2 to O_2. Later work on Cr(VI)-H_2O_2 spectroscopically identified four species depending on the pH of the solution and the concentration of the reactants [63].

6.1.4.3 Carbohydrates. A series of reactions between Cr(VI) and carbohydrates of biological importance under highly acidic conditions have been examined [66–71]. The mechanisms undergo $Cr^{VI} \rightarrow Cr^{IV} \rightarrow Cr^{III}$ and $Cr^{VI} \rightarrow Cr^{V} \rightarrow Cr^{III}$ pathways [72–75]. Carbohydrates stabilize Cr^{V} with the formation of five- and six-coordinate oxo-Cr^{V} complexes [76]. Oxo-Cr(V)-carbohydrate complexes have also been characterized [77]. Some of the general conclusions drawn using absorption and electron paramagnetic resonance (EPR) spectroscopies were (1) the electron transfers occur in slow steps having numerous acid and nonacid catalyzed parallel pathways and the total number of electron transfers is always two; (2) transfer of electrons occurs intramolecularly within the redox precursor complex and the formation of the complex is not acid catalyzed; and (3) in slow redox steps, the highly reactive Cr^{IV} is formed, which reacts rapidly with carbohydrates [78].

6.1.4.4 Hydroxy Acids. Reduction of chromic acid by α-hydroxy acids has been extensively studied using kinetics and spectral measurements [74, 75]. Hydroxycarboxylato functional sites may stabilize $Cr^{VI,V}$, and such studies may unravel the role of chromium in biological systems. Products of the reactions were formed by the cleavage of either C–C or C–H [79], depending on the particular molecular characteristics. Importantly, the Cr(V) intermediate was observed. Cr(V) complexed with hydroxamic acid ligands has been independently synthesized and characterized by UV–visible (UV–vis) absorption, EPR, and X-ray absorption techniques [80]. The Cr(V) complexes with the naturally occurring *tert*-2-hydroxy acids, quinic acid (1R,3R,4R,5R-1,3,4,5-tetrahydroxycyclohexanecarboxylic acid, qaH_3), have also been obtained [81]. Quinic acid contains both 2-hydroxy acids and 1,2-diol groups, and the coordination of Cr(V) to these molecules depends on the pH. The 1,2-diols stabilize Cr(V) at pH 7.4, while 2-hydroxy acid ligands stabilize at pH ~4. Recently, the interaction of Cr(VI) with an amino sugar (2-amino-2-deoxy-D-glucopyranose, Nglc) has been investigated [82]. The kinetics traces and EPR measurements suggest the formation of Cr^{VI} esters, oxo-Cr^{V} complexes, superoxo-Cr(III) (CrO_2^{2+}), and free radicals as intermediates of the reaction. The substitution of C_2-OH group by C_2-NH_2 in Nglc species inhibited the oxidation of Nglc by $Cr^{VI/V}$. This indicates the role of $-NH_2$ group present at C_2 of Nglc in the oxidation reaction [82].

6.1.4.5 Ascorbic Acid. Several studies have performed the reaction of K_2CrO_7 with ascorbate [83–89]. The rate law for the decay kinetics of Cr(VI) is described as

$$-d[Cr(VI)]/dt = k_{RH}[Cr(VI)][RH^-], \tag{6.32}$$

where k_{RH} is the second-order rate constant of the reaction. [Cr(VI)] and $[RH^-]$ are the concentrations of Cr(VI) and ascorbate, respectively. The values of k_{RH} are pH dependent [90]. Cr(V), $CO_2^{\cdot-}$, the ascorbate radical, and the carbon-based spin-trap radical adducts were identified. The yields of the products depended on the ratio of ascorbate to dichromate. The formation of Cr(IV) was indirectly observed by the reaction with Mn(II) [83]. Later work on the reaction of Cr(VI) with ascorbic was carried out over a wide pH range [91]. Both the Cr(V)/ascorbate complex and the ascorbate radical were observed. The Cr(V)/ascorbate complex was most stable at a ratio of 1:1 ([Cr(VI)]:[ascorbate]). The reaction was influenced by O_2 in the solution, which produced Cr(V)/ascorbate/peroxo complexes with a maximum intensity in the EPR signal at a ratio of 1:2. Both complexes were not stable at pH > 10 and reduced rapidly to Cr(III). In the mechanism, reactions (6.21), (6.22), (6.25), (6.26), and (6.30) were proposed in the reduction of Cr(VI) by ascorbate in neutral aqueous solution [83]. Formation of Cr(V) in the reaction may have the relevance in genotoxicty of Cr. Significantly, excess concentration of ascorbate may rapidly eliminate high-valent Cr species and hence diminish the potential of carcinogenicity. Workers in the industry of chromate may thus be better off by taking high vitamin C.

6.1.4.6 Catecholamines. Catechol compounds are present in neurotrasmitters and their precursors, for example, are dopamine, dihydroxyphenylalanine (DOPA), hormones (e.g., adrenaline and noradrenaline), and melanin. The reactions of Cr(VI) with various catecholamines were studied using the EPR technique [92]. The reduction of Cr(VI) by catecholamines resulted in EPR signal of Cr(V) species and organic radicals [92]. The primary species formed from the complexation of Cr^{V} with catechol-derived ligands with an assigned octahedral structure ($[Cr^{V}(catechol)_2(catechol)]^{+}$). The complexation chemistry was affected by the reactant ratios and the pH. These results indicate the possibility of reduction of Cr(VI) by catechol moieties in catecholamines *in vivo*. Furthermore, generated organic radicals may participate in further radical reactions.

6.1.4.7 Thiols. The reactivity of $CrOO^{2+}$ with thiols (L-Cys, GSH, and DL-pencillamine) has been studied in aqueous perchloric acid [93]. The oxidation of thiols by $CrOO^{2+}$ resulted in sulfinic and sulfonic derivatives. A proposed mechanism involves the participation of the $CrOO^{2+}$/thiol complex, $CrOOH^{2+}$, and CrO^{2+} as intermediates of reactions. Several studies describe the kinetics and mechanism of the reduction of Cr(VI) by Cys, GSH, and related thiols [85, 86, 89, 94–98]. The techniques to perform measurements include kinetic studies, UV–vis spectroscopy, electrospray mass spectrometry (ESMS), X-ray absorption fine structure (XAFS), and X-ray absorption near-edge structure (XANES) spectroscopy. The reactivity of Cr(VI) with various compounds at pH 7.4 and 25°C is shown in Figure 6.6. Ascorbate, cysteine, GSH, and pencillamine were the only reactive compounds, while other compounds had no significant reactivity. A number of studies demonstrated the formation of Cr(VI) complexes with reacted thiols (cysteine, GSH, N-acetyl-2-mercaptoethylamine, and bromobenzoenethiol) [84, 94, 97, 99–101]. The formation of the Cr(V) species was seen in the reaction of Cr(VI) with biological thiols [84, 102]. The reduction of Cr(VI) in the glutathione-D-glucose solution resulted in a stable Cr(V)–carbohydrate species, which was more stable than the Cr(V)–thiol species [91]. Based on the stated observation above, reactions (6.33)–(6.35) were proposed in the mechanism of the reduction of Cr(VI) by thiols (RSH):

$$\mathrm{Cr(VI)} + \mathrm{RSH} \underset{k_{-33}}{\overset{k_{33}}{\rightleftharpoons}} \mathrm{Cr(VI)\text{-}SR} \tag{6.33}$$

$$\mathrm{Cr(VI)\text{-}SR} \xrightarrow{k_{34}} \mathrm{Cr(V)} + \mathrm{RS}^{\bullet} \tag{6.34}$$

$$\mathrm{Cr(VI)\text{-}SR} \xrightarrow{k_{35}} \mathrm{Cr(IV)} + \mathrm{RSSR}. \tag{6.35}$$

The following rate law (Eq. 6.36) considers the Cr(VI)–thiol complex under steady-state concentrations:

$$-\mathrm{d[Cr(VI)]/dt} = (k_{34}k_{33}[\mathrm{RSH}]_{\mathrm{t}}^{2} + k_{35}k_{33}[\mathrm{RSH}])[\mathrm{Cr(VI)}]/(k_{-33} + k_{35+}k_{34}[\mathrm{RSH}]_{\mathrm{t}}). \tag{6.36}$$

Correlations of the rates with dissociation constants of thiols suggest an important role of the $-NH_3^+$ group to determine the kinetics of the formation of Cr(VI)–thioester [87].

The reaction between a water-soluble and stable aqua ethylene diaminebis(peroxo)chromium(IV) with GSH have been carried out at a neutral pH [103]. The EPR technique and mass spectrometry/mass spectrometry (MS/MS) characterization were applied to determine the intermediates of the reactions. The GSH molecule coordinates to the Cr(IV)-peroxo complex,

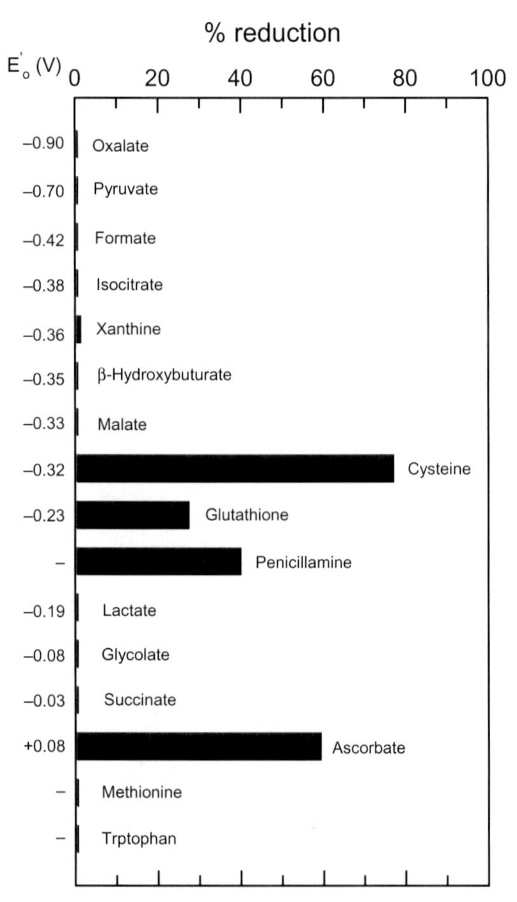

Figure 6.6. The extent of reaction of chromium with various compounds after 1 minute in 1 M Tris-HCl, pH 7.4. Reactions were carried out at 25°C for all compounds except methionine, tryptophan, glycolate, and xanthine, which were at 37°C. Initial [Cr(VI)] = 2.7×10^{-4} M and [compound] = 2.5×10^{-2} M, except [xanthine] = 6.3×10^{-3} M (redrawn from Connett and Wetterhahn [87]).

which triggers the reduction of Cr(IV) via metastable Cr(V) intermediates [103]. The intermediates of Cr(V) and Cr(IV) as mono- and bisglutathionato complexes with or without a coordinated peroxo moiety were characterized.

Recently, the reductions of $[(salen)Cr^{V}{=}O]^{+}$ (salen = bis(salicylidene)ethylenediamine) complexes by methionine and substituted methionine in aqueous acetonitrile have been conducted [104]. The rate law was determined to be first order with respect to each reactant. The reaction involved the transfer of oxygen from the complex to methionine to form methionine sulfoxide and a stable μ-oxo dimer of Cr. Kinetics of the reactions of Cr^{VI} and Cr^{V} complexes with dialkyl sulfides showed the rate constants for the reduction of Cr^{V} were two orders of magnitude higher than that of Cr^{VI} [105].

6.1.4.8 Non-Sulfur-Containing Amino Acids and Peptides. The reduction of Cr(VI) by Gly, Ala, and 2-amino-2-methylpropanoic acid did not occur either in dark or light [106]. However, when reactions were performed in methanol, Cr(V) complexes were formed. The mixture of Cr(VI)/glycine/methanol formed a complex, which is similar to the Cr(V)–methanol complex. The reaction performed under light increased the yield of the complex. Therefore, light was needed for the reduction to proceed at an appreciable rate. The Cr(VI)–methanol reaction in the presence of alanine resulted in the Cr(V)–alanine complex under both dark and light conditions. Several spectroscopic techniques were applied to characterize this complex as a Cr(V)–alanine dimer.

The Cr(V)–peptide complexes without thiol residues have also been characterized by the EPR technique [106, 107]. The reaction of Cr(VI) with methanol yielded two Cr(V)–methanol intermediates. However, methanol and glycine peptides (triglycine, tetraglycine, and pentaglycine) resulted in the reduction of Cr(VI) to form Cr(V)–methanol and Cr(V)–peptide complexes. However, the reaction of Cr(VI) and alanine peptides (trialanine, tetraalanine, and pentaalanine) formed only Cr(V)–peptide complexes. Other Cr(V)–peptide complexes containing the C-terminus of the proteins were also prepared [108]. The peptides were *N,N*-dimethylurea derivatives of tripeptides, Aib_3, AibLAlaAib-DMF, and AibSAlaAib-DMF (Aib-2-amino-2-methylpropanoic acid; Ala, alanine; *N,N*-dimethylformamide [DMF]). Results suggest the possible formation of Cr(V) species from the Cr(III) peptides *in vivo*. However, the biological activity of such complexes needs to be fully examined in further work.

6.1.4.9 Proteins. The chemistry of Cr(VI/V/IV) with a model thiolato complex $[Zn(SR)_2]$ (RSH=*O*-ethyl-L-cysteine) was studied [109]. The thiolato complex represents the tetrahedral (2*S*,2*N*) Zn(II) binding site in zinc-finger proteins. The reaction occurs through sequences of two- and one-electron transfer steps with the formation of Cr(V/IV) intermediates. An intramolecular formation of the disulfide bond was proposed. The reaction of Cr(VI) with human saliva has been studied to model the formation of Cr species in the human respiratory tract after inhaling Cr(VI). The EPR spectroscopy identified different Cr(V)-sialic (neuraminic) acid species in the reaction [110]. The EPR characteristics of the Cr(V) species were similar to those obtained in the mixtures of Cr(VI) and isolated components of salivary glycoproteins [110]. Cr(V) species may damage the cell through cleavage of DNA and/or oxidation of other biomolecules. A mechanism of biological damage caused by such Cr(V) species still requires further examination.

A number of studies on the interaction of Cr(VI) with thioredoxins (Trx1 and Trx2) and enzymes (peroxyredoxins, Prx1 and Prx3) have been performed [9, 111, 112]. The oxidation of Prx1 and Prx3 by Cr(VI) was observed in which a significant oxidation of thioredoxins also occurred [9]. Cytosolic Trx1 was less susceptible to Cr(VI) than mitochondrial Trx2. Ascorbate did not alter the effects on the thioredoxin reductase (TrxR)/Trx/Prx systems. The study performed on human bronchial epithelial cells (BEAS-2B) showed that the treatment of Cr(VI) resulted in inhibition of TrxR and aconitase, and a number of reducing equivalents in TrxR determined the oxidation of the thioredoxins in the BEAS-2B cells [112].

6.1.4.10 NADPH/NADP. The reduction of Cr(VI) by nicotinamide adenine dinculeotides (NAD(P)H/NAD(P)$^+$) results in Cr(V), evidenced by EPR measurements [113, 114]. The stabilization of Cr(V) at pH$\geq$7 occurs through the *cis*-1,2-diolato moiety of the ribose ring [114]. Details of the mechanism are still unknown. However, the studies with the NAD(P)H model compound, 10-methyl-9,10-dihydroacridine, under acidic conditions demonstrated the formation of Cr(V), Cr(IV), and organic radical intermediates [115]. The stabilization of Cr(V) could not be achieved due to the acidic conditions of the experiments.

6.1.4.11 Nitric Oxide Synthase (NOS). The formation of the Cr(V) species from the Cr(VI)/NOS has been examined by the EPR technique [116]. This reaction may thus be involved in the cytotoxic and vascular effects of Cr(VI) pollution. Significantly, the formation of NO from NOS did not preclude the reduction of Cr(VI) by the enzyme. The reduction of Cr(VI) was independent of calcium/calmodulin at the reductase domain.

6.1.5 Mechanism

The mechanism of chromium carcinogenicity is not fully understood, but numerous studies support the genotoxicity and the mutagenicity of Cr(VI) *in vitro* and *in vivo* [44, 83, 86, 87, 92, 96, 98, 117–130]. Figure 6.7 shows the potential steps of molecular mechanisms that may occur in Cr(VI) carcinogenesis. Cr(VI) can penetrate cell walls and undergoes intracellular reduction processes to produce Cr(V), Cr(IV), and Cr(III) species. Because of the structural similarity of CrO_4^{2-} to physiological SO_4^{2-} and phosphate ions, Cr(VI) can enter into cells via nonspecific anion channels where it can metabolically reduce to Cr(V), Cr(IV), and Cr(III) to induce a wide range of genomic DNA damage, which ultimately leads to the inhibition of DNA replication. As discussed in the previous section, the reduced substrates include Cys, lipoic acid, GSH, ascorbate, fructose, ribose, NAD(P)H, and hydrogen peroxide. Some redox proteins are also active in the reduction of Cr(VI). These include heme proteins (cytochrome P450 and hemoglobin) and NADPH-dependent flavoenzymes (GSH reductase and NADPH-cytochrome P450 reductase).

In addition to Cr(V) and Cr(IV) species, superoxide anions, hydroxyl radicals, and free radicals have also been suggested in the metabolism of Cr(VI). It is possible the hydroxyl radical is produced from the Fenton-like reaction of Cr(IV) and Cr(V) [131, 132]. Electron spin resonance (ESR) techniques in conjunction with spin trapping agents, 5,5-dimethyl-1-pyrroline-N-oxide

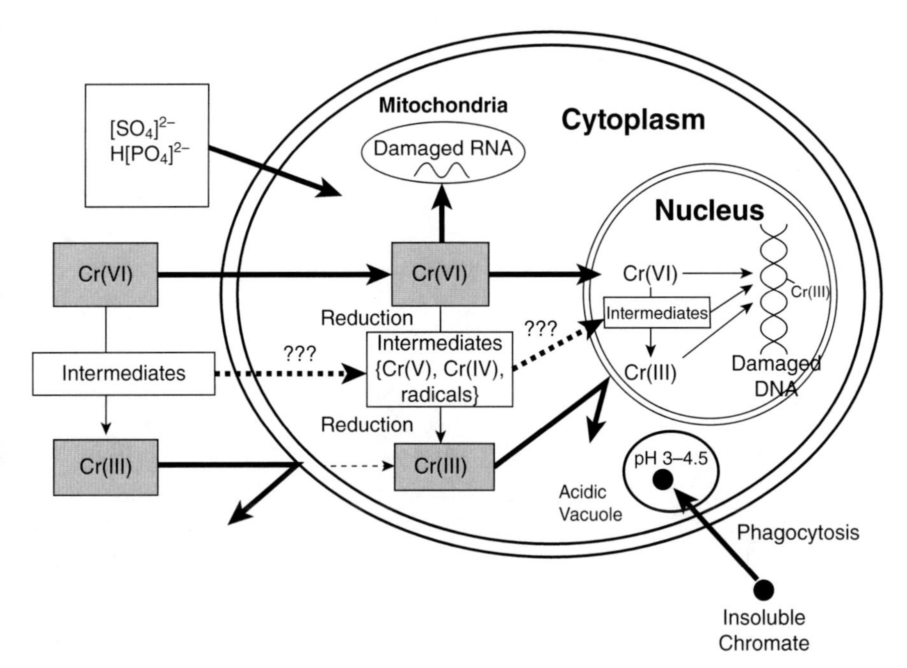

Figure 6.7. An overview of Cr redox chemistry in biological systems (adapted from Codd et al. [44] with the permission of Elsevier Inc.).

(DMPO) and α-(4-pyridyl-1-oxide)-N-t-butylnitrone (POBN), have been applied extensively to establish the formation of Cr(V) and $^{\bullet}OH$ radicals in the reductions of Cr(VI) using reductants such as ascorbate, cysteine, pencillamine, GSH, and NAD(P)H-dependent flavoenzymes (e.g., GSH reductase and lipoyl dehydrogenase) [27, 122, 129, 131–137]. A close relationship between the formation of organic radicals and the species of Cr(V)/Cr(IV) in the reduction of Cr(VI) has been determined [96]. The reactive species such as free radicals O_2^- and $^{\bullet}OH$ may effectively be produced from the reaction of Cr(VI) with biological reductants [131, 132]. The possible reactions to produce such species are as follows:

$$Cr(VI) + RH \rightarrow Cr(V) + R^{\bullet} \quad (6.37)$$

$$Cr(VI) + RH \rightarrow Cr(IV) + R^{\bullet} \quad (6.38)$$

$$R^{\bullet} + O_2 \rightarrow R + O_2^- \quad (6.39)$$

$$2O_2^- + 2H^+ \rightarrow H_2O_2 + O_2. \quad (6.40)$$

Cr(V) and Cr(IV) produced in reactions (6.37) and (6.38) may be further involved in Fenton-like reactions to yield $^{\bullet}OH$ radicals (reactions 6.41 and 6.42):

$$Cr(V) + H_2O_2 \rightarrow Cr(VI) + {}^{\bullet}OH \tag{6.41}$$

$$Cr(IV) + H_2O_2 \rightarrow Cr(V) + {}^{\bullet}OH. \tag{6.42}$$

In aerated solutions, there is consumption of oxygen in the reaction of Cr(VI/V/IV). The possible reactions for such observations can be explained by reactions (6.43) and (6.44):

$$R^{\bullet} + O_2 \rightarrow R + O_2^{-} \tag{6.43}$$

$$R_{ox}^{\bullet} + O_2 \rightarrow ROO^{\bullet} \rightarrow \text{oxidized products.} \tag{6.44}$$

Earlier studies have explored the formation of the chromate intermediate and mode of reactive species *in vitro* and *in vivo* [44, 83, 86, 87, 92, 96, 98, 117–123, 125, 127–129, 137–139]. The primary reducing substrate(s) in the reduction of Cr(VI) are determined by their cellular availabilities and reaction rates with Cr(VI). In previous studies, the importance of both GSH and ascorbate has been emphasized [125, 140, 141]. GSH is abundantly intracellular and its concentration has been determined at millimolar levels. The levels of ascorbate are also at millimolar *in vivo*. Most of the results have shown that ascorbate is more dominant and a kinetically favored biological reductant causing ~90% of *in vivo* metabolism of Cr(VI) [125, 140]. GSH has been identified as a modulator of cell stress in Cr(VI) cytotoxicity. When GSH is depleted, overexpression of the heme-oxygenase 1 gene expression in human dermal fibroblasts by Cr(VI) has been detected [141]. Significantly, this can be considered a marker in Cr(VI)-induced cell stress and cytotoxicity [141]. The stable Cr(III) species cannot cross the cellular membrane, and hence, it detoxifies the cellular membrane. However, if the reduction of Cr(VI) to Cr(III) occurs in the cell, the weak permeability of Cr(III) intracellularly traps the molecule. This results in the formation of a stable complex of Cr(III) with nucleic acid and proteins that lead to DNA damage.

Several studies support the role of ROS in Cr(VI)-induced oxidative stress [125, 137, 142–144], but a direct relationship between DNA–ROS and Cr(VI)-induced DNA damage is not fully understood and is heavily debated [25, 27, 44, 98, 129, 138, 144–150]. One of the debate issues includes the identity of the specific Cr(V) species responsible for the generation of ${}^{\bullet}OH$ [117, 137]. A tetraperoxochromate(V) species was suggested in the reduction of Cr(VI) and H_2O_2, leading to the formation of ${}^{\bullet}OH$ and singlet oxygen and resulting in damage to DNA [117]. However, an independent study using the ESR technique has shown that the reaction of Cr(VI) and H_2O_2 did not result in the generation of ${}^{\bullet}OH$. Furthermore, the reduction of Cr(VI) by ascorbate, GSH, GSH reductase, and vitamin B_2 did not yield the formation of tetraperoxochromate(V) [137]. Another example of the complexity of Cr(VI)-mediated oxidative stress and damage is when Cr(VI) was reduced to Cr(V) by vitamin B_2 (riboflavin) in Chinese hamster V79 cells, which resulted in an increase in hydroxyl radicals, chromosomal aberrations, and mutations at the

HGPRT locus [145, 146]. However, pretreatment with vitamin E, a free radical scavenger, showed a significant decrease in Cr(VI)-induced single-strand breaks and cytotoxicity [151]. Importantly, there was no difference observed in Cr(VI) uptake, Cr(VI)-induced DNA–protein cross-links (DPCs), and the levels of GSH and GSH reductase activity [151]. An indication of the dependence of antioxidant regulation on the concentration of Cr(VI) has also been suggested by *in vitro* studies. Considering the relative ratio of the concentration of available intracellular reductants and the total dosage of Cr(VI) over time as well as the dose rate may clarify the species involved in Cr(VI)-induced oxidative stress [9, 129].

6.1.6 Carcinogenesis

Mutagenicity is considered an initiation step in Cr(VI)-induced carcinogenesis [152]. A number of Cr(V)–amino acid and peptide complexes with ligands (Ala, Ala_3, Gly, and Gly_3) were found to be mutagenic against *Salmonella typhimurium* with an order of reactivity as Cr(V)-Ala_3 > Cr(V)-Ala > Cr(V)-Gly > Cr(V)-Gly_3 [106]. Complexes of Cr(V) with Gly_4 and Gly_5 were not mutagenic to the species. Comparatively, complexes of Cr(III) with AibH (α-aminoisobutanoic acid), Ala_3, Gly_3, Gly_4, and Gly_5 were also not mutagenic to *S. typhimurium* [106, 153].

Cr(VI) produced structural genetic lesions such as DPCs, DNA inter- and intrastrand cross-links, DNA-strand breaks, DNA adducts, and oxidized bases. High levels of DPC have been determined in peripheral blood lymphocytes among chrome platers, welders, and leather tanners [154]. A good correlation of DPCs with the levels of Cr in red blood cells has been observed [139]. The formation of DPCs *in vitro* has been studied in detail [154, 155]. The mechanism of DPC first involves the reduction of Cr(VI) to Cr(III), followed by the formation of DNA–Cr(III) adducts, which subsequently capture proteins (reaction 6.45):

$$Cr^{VI} \rightarrow Cr^{III} \rightarrow Cr^{III}\text{-DNA} \rightarrow \text{protein-}Cr^{III}. \qquad (6.45)$$

The role of ascorbic acid and GSH as reductants of Cr(VI) in the formation of DPC has been examined in A549 cells (Fig. 6.8) [154]. The GSH inhibitor, L-buthionine-[S,R]-sulfoxime (BSO), was applied during preincubation to reduce the level of GSH from $4.2 \pm 0.6\,mM$ to $0.15 \pm 0.2\,mM$ ($n = 3$). As shown in Figure 6.8A, this resulted in lower levels of DPC immediately and 18 hours after exposure to Cr(VI). When the cellular reducing capacity was restored using physiological ascorbic levels ($0.9 \pm 0.1\,mM$) prior to treatment with Cr(VI), the yield of DPC was increased (Fig 6.8B). However, no significant changes in the uptake of Cr(VI) in addition to ascorbic were observed (Fig 6.8C). It appears ionic interactions between negatively charged phosphate groups of DNA and positively charged Arg or Lys side chain groups are

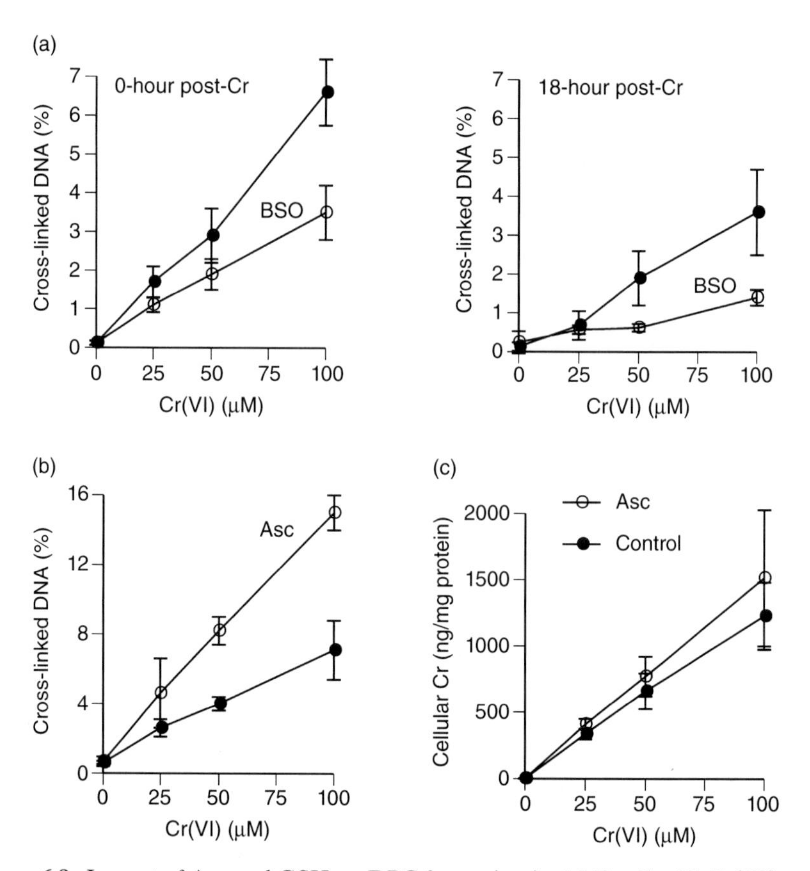

Figure 6.8. Impact of Asc and GSH on DPC formation in A549 cells. All Cr(VI) exposures were for 3 hours in a serum-free medium. Data are means ± SDs. (a) DPC levels in cells with and without pretreatment with 0.1 mM BSO for 24 hours. DPCs were measured either immediately (left panel) or 18 hours after Cr(VI) treatments (right panel). (b) Formation of DPC in control and Asc-restored cells. (c) Cr accumulation by A549 cells with and without Asc preloading (adapted from MacFie et al. [154] with the permission of the American Chemical Society).

involved in stable protein–DNA association. It is likely that Cr(III) binds favorably to His, Cys, and –COOH groups of Glu and Asp, which may be in contact with the duplex in DNA-binding proteins [154]. Recently, the level of DPCs in conjunction with levels of protein oxidation and lipid peroxidation in Cr(VI) exposure to MOLT4 cells was determined [155].

The formation of protein carbonyls and DPCs are shown in Figure 6.9. Tiron or α-tocopherol treated cells prior to the addition of Cr(VI) inhibited the formation of the protein carbonyl by 63% and 56% and DPCs by 57% and 52%, respectively, compared to cells treated with Cr(VI) alone (Fig. 6.9). The generation of H_2O_2 and ultimately $^{\bullet}OH$ radicals were provoked to explain the

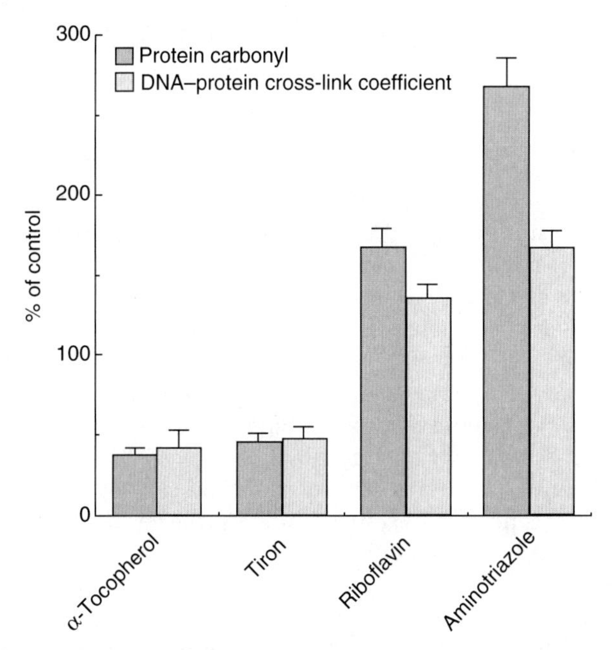

Figure 6.9. Effect of antioxidants and cellular chromate reductants on potassium chromate-induced protein oxidation and DNA–protein cross-linking. Cells were retreated with the respective agents in a complete medium for 16 hours, washed, and then treated with 200 μM chromate for 2 hours in salts-glucose medium (adapted from Mattagajasingh et al. [155] with the permission of Wiley Inc.).

results of Figure 6.9. This is supported by increased levels of Cr(VI)-induced DPCs and protein carbonyls in riboflavin pretreated cells. Furthermore, it is also supported by the use of a known catalase inhibitor, aminotriazole, in the pretreatment of cells in which an increase in Cr(VI)-induced DPCs and protein carbonyls by 67% and 66%, respectively, were observed (Fig. 6.9). A separate study has also shown that Cr(V) species produced from Cr(VI) can also oxidize human orosomucoid (α_1-acid glycoprotein) [156].

A primary cause of genetic lesions induced by Cr(VI) in mammalian cells was suggested due to the formation of the chromium–DNA adduct [124, 140, 157, 158]. Other types of adducts in the Cr–DNA are between Cr(III)–ligand–DNA complexes, which can be either binary or ternary. Among several adducts, ternary adducts are more relevant due to their abundance and importance in toxicology [124, 140, 157, 158]. Cr(III)–histidine–, Cr(III)–ascorbate–, Cr(III)–cysteine–, and Cr(III)–GSH–DNA adducts are examples of ternary adducts that are present after exposure and intracellular reduction of Cr(VI). DNA–chromium–DNA cross-links and DNA–chromium–GSH cross-links were suggested to cause significant damage to DNA [159] The role of Cr(V) in the oxidation of DNA has been examined by studying the reactions of Cr(V) complexes with DNA [160–162]. The analysis of oxidized products suggests the possibility of oxidation at the sugar moiety in an addition to the base.

6.1.7 Genotoxicity and Cytotoxicity

Solubility plays an important role in carcinogenicity, which has recently been demonstrated in human bronchial cells by performing a comparative study with four Cr(VI) compounds: sodium chromate, zinc chromate, barium chromate, and lead chromate [163, 164]. Sodium chromate is a soluble compound, while the others are particulate Cr(VI) compounds. The cytotoxic effect of the Cr(VI) compounds is shown in Figure 6.10A. It is clear that zinc chromate and barium chromate were more cytotoxic than sodium chromate and lead chromate. The results of genotoxicity experiments are presented in Figure 6.10B. The formation of γ-H2AX foci suggests the levels of DNA damage were similar for all three particulate chromates. Zinc chromate was more clastogenic than all the other Cr(VI) compounds. Furthermore, there was no difference in the induction of DNA double-strand breaks for any of the compounds. It is possible that the zinc ion may be involved in the repairing of DNA, resulting in a difference in the carcinogenic potency of zinc chromate over other chromium compounds [163]. An EPR study was also conducted to understand the mechanism of cytotoxicity of different chromates using human lung epithelian cells (BEAS-2B). Two Cr(V) EPR signals were observed in the incubation of cells with sodium chromate. Of the two signals, only one signal was thio dependent, and both signals were largely NAD(P)H dependent. Both EPR signals were also seen with zinc chromates. The use of lead chromate did not give any EPR signal. The most sensitive cells in the clonogenic assays were with sodium chromate and zinc chromate, while lead chromate was much less sensitive. A scheme was given to explain the signals in the EPR experiments (Fig. 6.11). This scheme is consistent with the role of reducing substrates in the generation of Cr(V) complexes as intermediates before converting to Cr(III) complexes, which were also detected in the cells.

Cellular resistance took place in the Cr(VI)-induced early stage of carcinogenesis, although the mechanism remains unclear. The involvement of aberrant DNA repair mechanisms, the dysregulation of critical survival signaling pathways and transcriptional repatterning have been suggested and are shown in Figure 6.12. In an attempt to repair damaged DNA, a normal cell may go through a transient checkpoint at the relevant doses of Cr(VI). Although some cells have DNA repair mechanisms, they lack repair genes such as *MLH1*, *MSH6*, *ATM*, and *PMS2* and cause the development of genomic instability (Table 6.1) [165–167]. Some cells that are unable to repair the damage may undergo terminal growth arrest or apoptosis (replicative death). In this process, some cells may survive by acquiring an intrinsic mechanism(s) of death resistance.

A summary of death resistance after Cr(VI) exposure is given in Table 6.1 [165–172]. It appears that dysregulated DNA repair mechanisms and/or

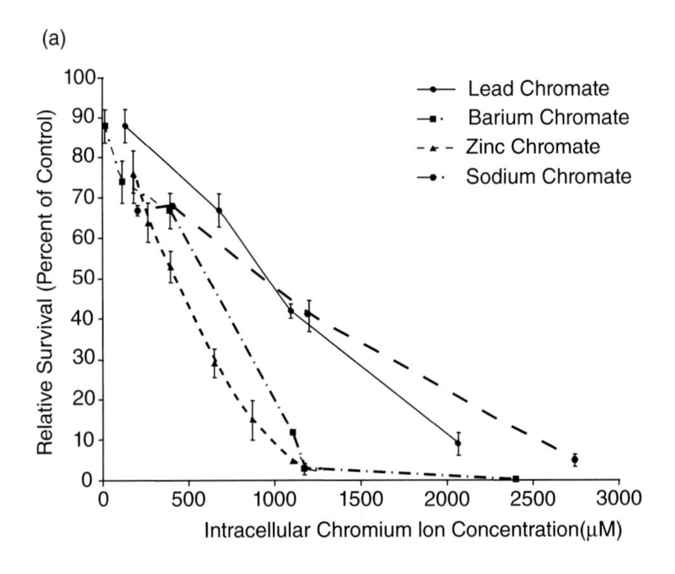

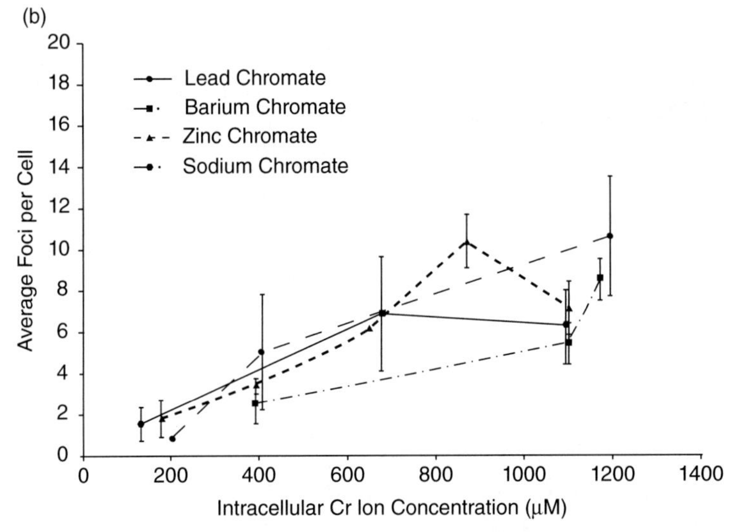

Figure 6.10. (a) The cytotoxic effect of all four compounds corrected for intracellular chromium ion concentration. Barium chromate and zinc chromate were significantly more cytotoxic than lead chromate and sodium chromate ($p < 0.005$); there was no significant difference between barium chromate and zinc chromate or between lead chromate and sodium chromate. (b) All compounds induced similar levels of DNA double-strand breaks at similar intracellular concentrations, for example, 500 μM intracellular Cr induced three, four, five, and six average foci per cell after exposure to barium chromate, zinc chromate, lead chromate, and sodium chromate, respectively (adapted from Wise et al. [163] with the permission of the American Chemical Society).

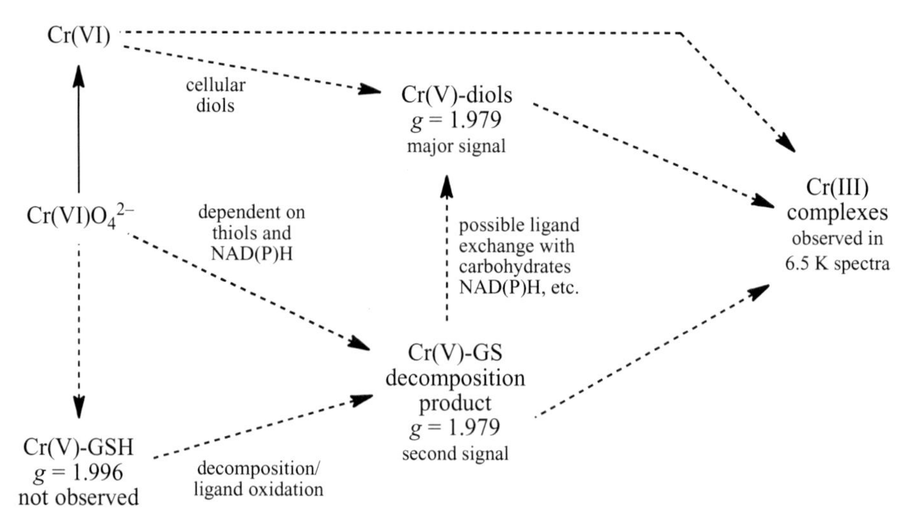

Figure 6.11. General scheme for the formation of the Cr species responsible for the ESR signals observed in BEAS-2B cells. The species associated with each signal are in boldface type. The dashed arrows indicate possible pathways to the formation of certain species, but the exact pathways are not known. The diagram is not meant to be inclusive of all possible pathways or Cr species but rather is intended to relate the various species observed in these cells (adapted from Borthiry et al. [164] with the permission of Elsevier Inc.).

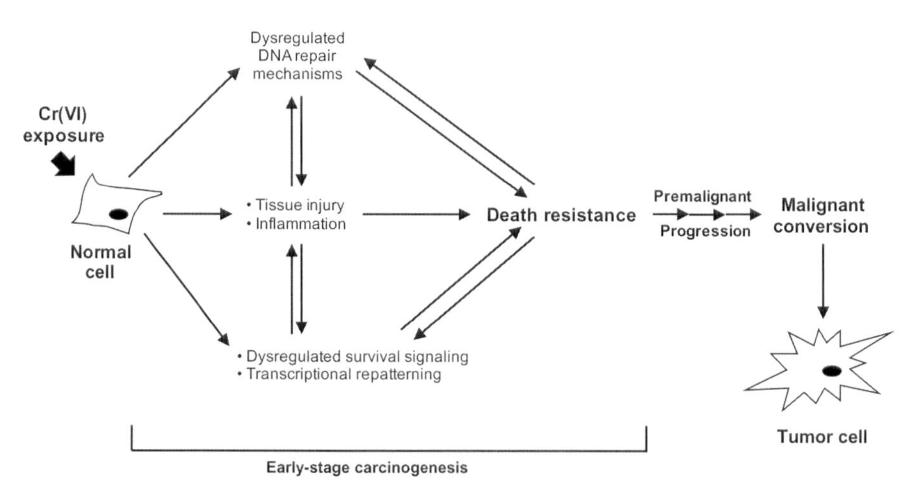

Figure 6.12. Cellular resistance to Cr(VI)-induced death and early-stage carcinogenesis (adapted from Nickens et al. [129] with the permission of Elsevier Inc.).

TABLE 6.1. Death Resistance after Cr(VI) Exposure

Mode of Inducing Cr(VI) Resistance	Model System	Cr(VI) Transport Status	Resistant Phenotype	Reference
Single exposure selection	Chinese hamster mutant cells	Dependent	Reduced uptake of labeled sulfate	[168]
Chronic exposure	Chinese hamster ovary cells	Dependent	Enhanced long-term survival Cr(VI)-specific resistance	[169]
Chronic exposure	293 embryonic kidney cells	Dependent	Sensitive to density-dependent growth inhibition. Grew slower than wild-type (WT) cells. Cr(VI)-specific resistance	[172]
Single exposure selection	BJ-hTERT human foreskin fibroblasts	Independent	Enhanced clonogenic survival p53 independent, increased Bcl-X gene expression, increased XPF gene expression; phenotype not Cr(VI)-specific. Cells also display resistance to H_2O_2	[170]
Genetic deletion of ATM	Human dermal fibroblasts	Independent	Resistant to Cr(VI)-induced apoptosis, display increased sensitivity to Cr(VI)-induced clonogenic lethality	[166]
Genetic deletion of MMR genes	Human colon HCT116 and DLD1 cells , and mouse embryonic fibroblasts	Independent	Enhanced clonogenic survival, resistant to Cr(VI)-induced apoptosis, p53-independent Cr(VI)-specific resistance	[165]
Chronic exposure	Human bronchial epithelial cell line, BEAS-2B	Data not provided	Decreased clonogenic efficiency, altered cellular morphology and growth pattern, upregulated expression of DNA repair genes, Upregulated expression of survival genes, Aneuploidy	[171]

Adapted from Nickens et al. [129] with the permission of Elsevier Inc.

dysregulated survival signaling and transcriptional repatterning are involved in the mechanisms of death resistance. Studies have demonstrated in the Cr(VI) exposure to the cells that the phosphorylation of proteins, which regulates the dysregulation processes, provides a critical molecular switch for rapid control of signaling pathways [173]. Signaling pathways include serine/threonine protein kinases that sense DNA damage and activate p53 and proliferating protein tyrosine kinase cascades, and cell death/cell growth-arresting tyrosine phosphatases. Tissue injury and severe inflammatory responses may result from direct exposure of Cr(VI) to the cells (Fig. 6.12). Therefore, there may be a further contribution to the microenvironment potentiating the death resistance phenotype to participate in early-stage carcinogenesis. This phenomenon over time may lead to the malignant conversion of the predisposed precursor cells to tumor cells [129]. In a recent study, it was demonstrated that Cr(VI) could induce genes [128]. It was postulated that the exposure of Cr(VI) stimulates gene *Fyn* to initiate innate immune gene induction in human airway epithelial cells through an innate immune-like signal transducer and activator of transcription 1 (STAT1)-dependent pathway [128].

6.1.8 Conclusions

Compounds of high-valent species of Cr have roles in developing new materials and in causing carcinogenic and toxic effects. Formations of Cr(V) an Cr(IV) in the reduction of Cr(VI) by substrates of biological importance have been suggested and, in some cases, direct spectroscopic evidences were given to suggest the formation of Cr(V) and Cr(IV) complexes. Reactive oxygen species $O_2^{\bullet-}$ and $^{\bullet}OH$ have also been proposed in the metabolism of Cr(VI). A multistage-multipath mechanism may be involved in causing cancer by Cr(VI). The oxidation of thioredoxin and peroxiredoxins by Cr(VI) is feasible and has implication in biological systems. Progress has been made to understand the genotoxicity and mutagenicity of Cr(VI) *in vitro* and *in vivo*; however, additional studies are needed to comprehend the mechanism of chromium carcinogenicity. The identification of high-valent Cr species, formed *in vivo*, in future studies may include the application of EPR and rapid synchrotron X-ray fluorescence mapping with XANES. Lastly, the Cr(VI)/H_2O_2 system may be applied to the oxidation of organic pollutants in water.

6.2 MANGANESE

Manganese compounds with oxidation states varying from −1 to +7 have been synthesized. Examples of Mn(−1) and Mn(O) compounds are the $Mn(CO)_5^-$ ion and $Mn_2(CO)_{10}$, respectively [174]. The Mn^+ ion does not exist in aqueous solution, but the +1 oxidation state is present in the $[Mn(CN)_6]^{5-}$ ion. Mn^{2+} is stable over a wide range of potentials in acidic solutions. However, Mn^{2+}

hydrolyzes to $Mn(OH)_2$ in alkaline solutions, where it easily oxidizes to Mn(III) and Mn(IV) species. The Mn(III) species could be stabilized by forming their complexes. In alkaline solutions, Mn_2O_3 and MnOOH are known compounds of Mn(III). Examples of Mn(IV) compounds include MnO_2, $Mn(SO_4)_2$, and MnF_4. MnO_2 is the only species stable in water. Hypomanganate (MnO_4^{3-}), manganate (MnO_4^{2-}), and permanganate (MnO_4^-) are oxo compounds of Mn(V), Mn(VI), and Mn(VII), respectively.

In the solid state, Mn(V) compounds in the tetrahedral oxo coordination in host compounds $Ca_2(MO_4)Cl$ (M=As(V), V(V), and P(V)) have been synthesized to obtain their single-crystal EPR spectra [175]. The EPR spectra of Mn(VI) have also been determined by stabilizing the +6 oxidation state of Mn in numerous host compounds possessing the β-K_2SO_4 structure. The crystalline oxo compounds of Mn(V) and Mn(VI) with low doping concentrations could be distinguished using their EPR spectra. However, at high concentrations of Mn, it was difficult to distinguish between the two oxidation states [175]. The low-temperature absorption and luminescence spectra of MnO_4^{2-}-doped crystals of CsBr, CsI, $SrCrO_4$, and Cs_2CrO_4 are also known [176].

In aqueous solution, various oxidation states of manganese have distinct colors: MnO_4^-—purple, MnO_4^{2-}—dark green, MnO_4^{3-}—light blue, and finally MnO_2—brown precipitate. The UV–vis spectra of the different species are shown in Figure 6.13. The spectra of Mn(V) and Mn(VI), shown in Figure 6.13,

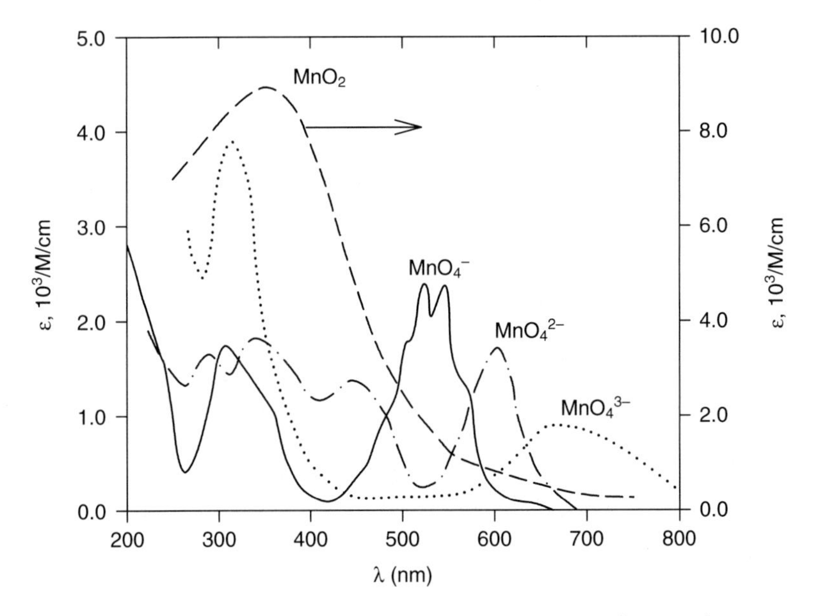

Figure 6.13. Ultraviolet and visible spectra of the MnO_4^-, MnO_4^{2-}, MnO_4^{3-}, and MnO_2. Data were taken from References 179 and 216.

are in strong alkaline solutions. All of the species, except MnO_2, absorb strongly in the visible region. There are distinct absorption maxima for Mn(VII) and Mn(VI) ions with high molar extinction coefficients at 522 and 426 nm, respectively. The spectral differences were used to study the photochemistry of aqueous Mn(VII) in acidic and basic solutions [177]. The final product was determined to be MnO_2^- under neutral conditions. In the case of basic solution, the product was identified to be MnO_4^{2-}. The suggested mechanism involved a formation of the Mn(V)-peroxo complex as an intermediate in the light-induced decomposition of MnO_4^-.

The spectra of manganese porphyrin complexes, MnTM-2PyP (TM-2PyP = chlorotetra(N-Me-2-pyridyl)porphyrin), are shown in Figure 6.14. Mn(V) had a strong band at 434 nm. The oxoMn(IV) porphyrin complex had a lower absorbance with a maximum at 426 nm. The oxo Mn(III) porphyrin complex had a characteristic band at 450 nm.

The potential diagrams for Mn are shown in Figure 6.15. A discussion on the reduction potentials of different species and their impact on the oxidation/reduction (redox) reactions have been thoroughly reviewed [174]. The redox potential values can be used to determine the stability and redox reactions of the different oxidation states of Mn. Of the several species, permanganate is the most important compound. The MnO_4^- ion is thermodynamically unstable in aqueous solution and its potential is greater than the evolution of oxygen over the entire pH range. In acidic solutions, it is reduced to the Mn^{2+} ion. MnO_2 is the reduced species of the MnO_4^- ion in alkaline solutions.

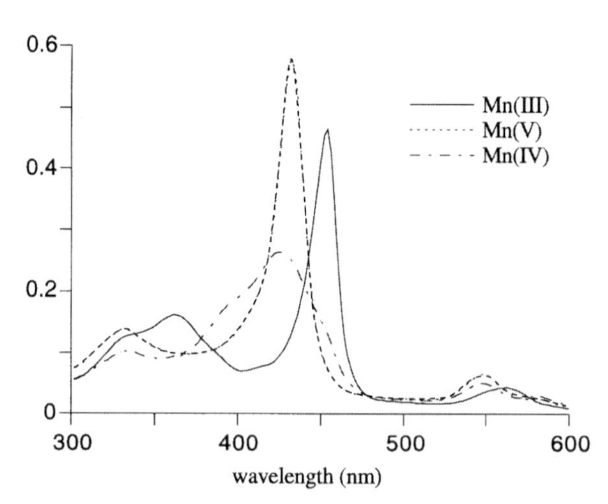

Figure 6.14. UV–vis spectra MnTM-2-PyP porphyrins. All three are 5 μm in a 50 mM pH 7.4 phosphate buffer. OxoMn(V)TM-2-PyP (λ_{max} = 434 nm) was prepared by oxidation of Mn(III)TM-2-PyP (λ_{max} = 454 nm) by 1 equiv of oxone. *In situ* reduction of Mn(V) by $NaNO_2$ (10 equiv) produced oxoMn(IV)TM-2-PyP (λ_{max} = 426 nm) (adapted from Jin and Groves [195] with permission of the American Chemical Society).

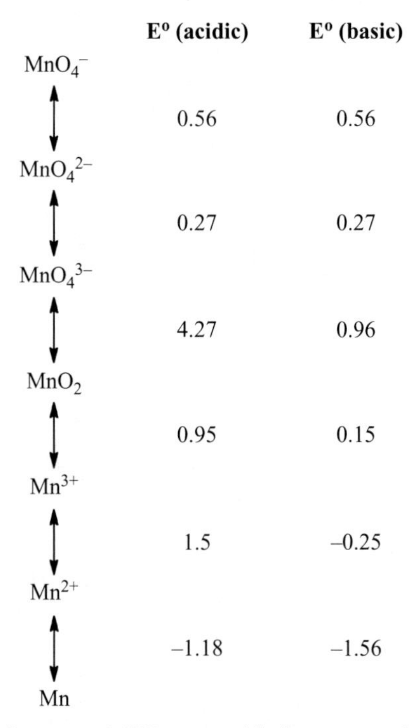

Figure 6.15. Potential diagram of different oxidation states of oxo compounds of Mn.

6.2.1 Aqueous Chemistry of Oxo-Mn Compounds

6.2.1.1 Mn(III). The oxidation of Mn^{2+} by O_3 in acidic solution resulted in Mn(III) (Fig. 6.16) [178]. The absorption maximum is at 220 nm with a molar absorptivity of ~5000/M/cm. A small difference in $HClO_4$ and H_2SO_4 solutions was not significant. A formation of Mn(II)/Mn(III)-SO_4^{2-} complexes in H_2SO_4 solutions was suggested. The proposed mechanism is presented in Equations (6.46) and (6.47). The formation of $^{\bullet}OH$ in the reaction steps was ruled out experimentally. Initially, the manganyl ion (MnO^{2+}) was formed, which rapidly reacted with excess Mn^{2+} to result in only Mn(III) as the final product:

$$Mn^{2+} + O_3 \rightarrow MnO^{2+} + O_2 \tag{6.46}$$

$$MnO^{2+} + Mn^{2+} + 2H^+ \rightarrow Mn(III) + H_2O. \tag{6.47}$$

In a recent detailed mechanistic study of the O_3-treated Mn^{2+}-containing waters also yielded MnO_2 and MnO_4^- ions [179]. In neutral solution, aqua-Mn^{2+} was mainly oxidized to colloidal MnO_2 by O_3. Comparatively, Mn(III) was formed in acidic solution (pH 0) when an excess [Mn^{2+}] over [O_3] was present. However, with low concentrations of Mn^{2+} and a large excess of [O_3] under

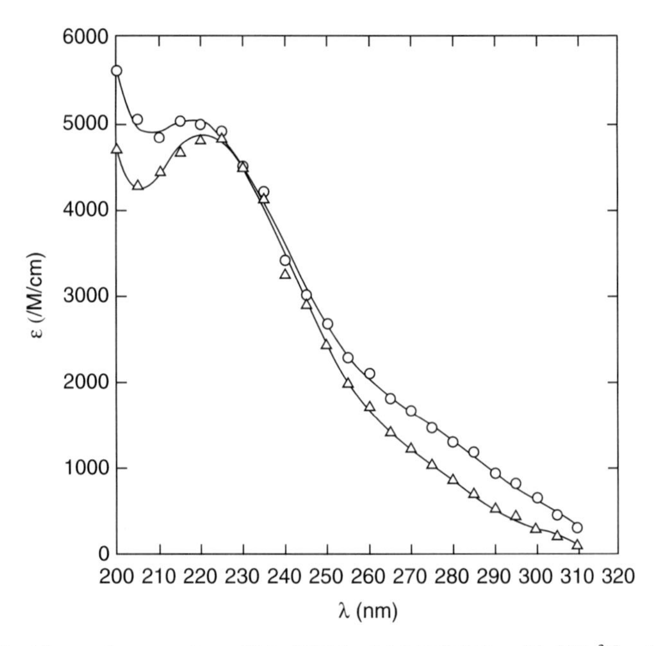

Figure 6.16. Absorption spectra of Mn(III) in 0.1 M $HClO_4$ with $[Mn^{2+}] = 2.5 \times 10^{-4}$ M and $[O_3] = 6.5 \times 10^{-5}$ M at 25°C. (Δ—$HClO_4$ and O—H_2SO_4) (adapted from Jacobsen et al. [178] with the permission of Wiley, Inc.).

highly acidic conditions, the MnO_4^- ion was the dominant species [179]. Fe^{2+} in Mn^{2+}-containing waters induced the formation of MnO_4^- ion. This reaction may be playing a significant role in generating MnO_4^- ion because Fe^{2+} reacts 650 times faster with O_3 than Mn^{2+} does. In neutral solutions, the reactions of O_3 with oxalate, bicarbonate, phosphate, and pyrophosphate complexes of Mn^{2+} formed the MnO_4^- ion [179]. Moreover, natural organic matter present in water may also influence the formation of MnO_4^- ion from Mn^{2+}. Overall, the results showed that the formation of MnO_4^- ion was more influenced by the Fe^{2+} ion rather than inorganic ligands or organics in water [179].

The hydrolysis constant of Mn(III) has been estimated (reaction 6.48) [178]:

$$Mn^{3+} + H_2O \rightarrow Mn(OH)^{2+} + H^+ \quad pK_{48} = 0.2. \tag{6.48}$$

$Mn(OH)^{2+}$ is the dominant species of Mn(III) in the pH range of 0–2. The disproportionation of Mn(III) corresponds to reaction (6.49). The final product of Mn(IV) was determined as MnO_2 (reaction 6.50), which was observed as an increase in absorption >300 nm:

$$2Mn(III) \rightleftharpoons Mn(II) + Mn(IV) \tag{6.49}$$

$$Mn(IV) \rightarrow MnO_2. \tag{6.50}$$

The rate of nucleation of Mn(IV) was dependent on $[Mn^{2+}]$ relative to [Mn(III)] and also increased with an increase in pH. At pH > 3 in $HClO_4$ solutions, reactions (6.46) and (6.47) were not kinetically separable. The activation energy of reaction (6.46) was determined as 39.5 kJ/mol [178].

The rate of the reaction between Mn(III) and H_2O_2 has also been studied in the acidic medium (pH 0–2) [178], which followed mixed first- and second-order kinetics. The mechanism involved the formation of the Mn(II)-superoxide complex (reaction 6.51) as well as the reaction between Mn(IV) and H_2O_2 (reaction 6.52). Mn(IV) was from the disproportionation of Mn(III) (reaction 6.49):

$$Mn(III) + H_2O_2 \rightarrow (MnO_2^+ + 2H^+) \rightarrow Mn^{2+} + HO_2 + H^+ \tag{6.51}$$

$$Mn(IV) + H_2O_2 \rightarrow Mn^{2+} + O_2 + H^+. \tag{6.52}$$

The rate constants of reaction (6.51) varied from $0.25–11 \times 10^4$/M/s in the pH range of 0–5.2 [178]. The estimated rate constant for reaction (6.52) was $\geq 10^6$/M/s.

The formation of Mn(III) has also been observed in the reaction of Mn(II) with the SO_5^- radical. This reaction was studied in the presence of excess sulfite at pH 3.0 with a 0.01 M ionic strength [180]. Under these conditions, $[Mn(II)]_{total}$ existed as $Mn^{2+}(aq)$, $[Mn(HSO_3)]^+$, and $[Mn(SO_3)Mn]^{2+}$, represented in Equations (6.53)–(6.55):

$$Mn^{2+} + SO_5^- + H^+ \rightarrow Mn^{3+} + HSO_5^- \tag{6.53}$$

$$[Mn(HSO_3)]^+ + SO_5^- + H^+ \rightarrow [Mn(HSO_3)]^{2+} + HSO_5^- \tag{6.54}$$

$$[Mn(SO_3)Mn]^{2+} + SO_5^- + H^+ \rightarrow [Mn(SO_3)Mn]^{3+} + HSO_5^-. \tag{6.55}$$

The overall second-order rate constant for the reaction of Mn(II) and the SO_5^- radical ranged from 2×10^8/M/s to 2×10^{10}/M/s, which was dependent on the kind of Mn(II) species under the experimental conditions.

The reaction of $Mn^{2+}(aq)$ with acylperoxyl radicals ($CH_3OO^\bullet$) and alkylperoxyl radicals ($ROO^\bullet$) in acidic aqueous solutions and in 95% acetic acid also produced Mn(III) [181]. The proposed scheme is represented by Equations (6.56) and (6.57):

$$Mn(II) + ROO^\bullet \rightleftharpoons Mn(III)OOR \tag{6.56}$$

$$Mn(III)OOR + H^+ \rightarrow Mn(III) + ROOH. \tag{6.57}$$

The initial formation of Mn(III)OOR was first order for each reactant. The second-order rate constants were $(0.5–1.6) \times 10^6$/M/s and $(0.5–5.0) \times 10^5$/M/s for acylperoxyl radicals and alkylperoxyl radicals, respectively. Mn(II) catalyzed the dissociation of Mn(III)OOR back to Mn(II) and $ROO^\bullet$ (reaction 6.58) [181]:

$$\mathrm{Mn(II)+Mn(III)OOR \rightarrow 2Mn(II)+ROO^{\bullet}.} \tag{6.58}$$

The reaction of Mn(III) with benzyl radicals proceeded at a fast rate in aqueous solution (reaction 6.59):

$$\mathrm{Mn(III)+PhCH_2^{\bullet}+H_2O \rightarrow Mn(II)+PhCH_2OH+H^+} \quad k_{59}=1\times10^7\,\mathrm{/M/s.} \tag{6.59}$$

The rate constants for reaction (6.59) were 2.3×10^8/M/s and 3.7×10^8/M/s in glacial acetic acid and 95% acetic acid, respectively.

6.2.1.2 Mn(IV). Mn(IV) in aqueous acetic acid has recently been prepared [182]. UV–vis spectroscopy was used to characterized Mn(IV) acetate. Magnetic susceptibility data of Mn(IV) acetate (μ = 3.57 BM) helped to assign the +4 oxidation state of Mn. The reactivity of Mn(IV) acetate was studied with Br^-, which gave first-order dependence on $[Br^-]$. The reaction produced 1 equiv of Br_2/Br^- [182].

Mn(IV) has also generated in alkaline solution from the reduction of Mn(V) with e_{aq}^- and $^{\bullet}CO_2$ in an argon-saturated 10 M NaOH solution containing formate (reaction 6.60) [183]:

$$\mathrm{Mn(V)+e_{aq}^- \rightarrow Mn(IV)} \quad k_{60}=5.0\times10^9\,\mathrm{/M/s.} \tag{6.60}$$

The spectrum of Mn(IV) is presented in Figure 6.17, which suggests that Mn(IV) would most likely appear as a blue or blue-green color. As shown in the inset of Figure 6.17, a similar spectrum was also obtained when dissolving pyrolusite (MnO_2) in concentrated base [184]. It is likely that the Mn(IV) species in the solution mixture were polymeric. This was supported by the ESR-silent property of dimeric or polymeric species. For example, Mn(IV) oxalate complexes were ESR silent due to the possible presence of μ-oxo-bridged Mn(IV) dimers [185]. Generally, the yellow/brown precipitates obtained in the reduction of permanganate in various reactions are highly polymeric.

6.2.1.3 Mn(V) and Mn(VI). Mn(V) and Mn(VI) have been observed in the reactions of Mn(VII) with As(III) and propane-1,2-diol, respectively, in acidic solutions [186–189]. In alkaline solutions, tetraoxyanions of Mn(VI) and Mn(V) were conveniently generated by pulse radiolysis in which Mn(VII) and Mn(VI) ions were reduced by hydrated electrons and radicals, respectively (Table 6.2) [183, 190–192]. Both Mn(VII) and Mn(VI) are strong oxidizing agents and react with the hydrated electron at the diffusion-controlled rates (Table 6.2). However, the rate constant for the reduction of Mn(VI) by O^- was somewhat slower than the rate of diffusion-controlled reactions. The reaction of Mn(VI) with O_2^- was relatively slow (reaction 6.61) (see Table 6.2):

$$\mathrm{MnO_4^{2-}+O_2^- \rightarrow MnO_4^{3-}.} \tag{6.61}$$

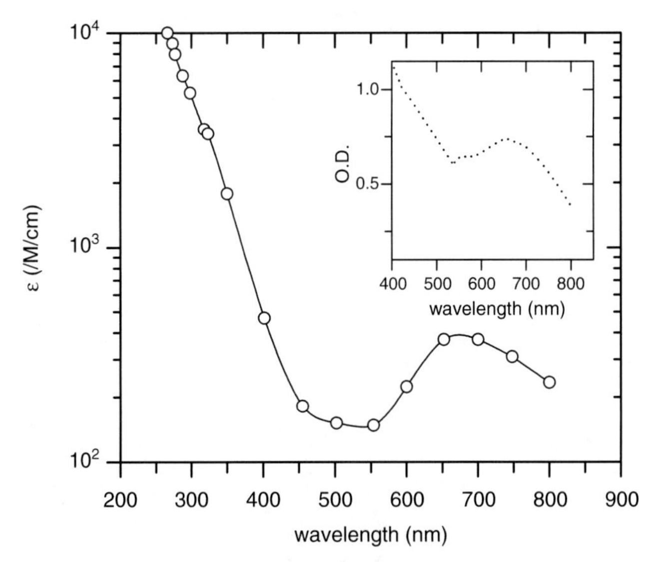

Figure 6.17. Point-by-point absorption spectrum (O) of Mn(IV) in 10M NaOH obtained by pulse-radiolytic reduction of MnO_4^{3-} with $^{\bullet}CO_2$ radicals and e_{aq}^-. Measurements of absorption differences were made ≈5 μs after pulse. Inset: spectrum of MnO_2 in concentrated KOH (adapted from Rush and Bielski [183] with the permission of the American Chemical Society).

TABLE 6.2. Rate Constants of Reactions of Radicals with MnO_4^- and MnO_4^{2-}

Radical	$k(MnO_4^-)$ (/M/s)	$k(MnO_4^{2-})$ (/M/s)	References
e_{aq}^-	2.6×10^{10}	2.0×10^{10}	[190–192]
O^-	–	8.0×10^8	[192]
$^{\bullet}O_2^-$	9.50×10^{5a}	1.8×10^{3a}	[183]
$^{\bullet}CO_2^-$	6.90×10^{9a}	8.1×10^{8a}	[183]
$^{\bullet}CH_2O^-$	–	1.9×10^{9b}	[183]
$CH_3C^{\bullet}HO^-$	3.30×10^{9c}	1.8×10^{9b}	[183]
$CH_3C^{\bullet}(O^-)CH_3$	7.20×10^{9b}	2.0×10^{9b}	[183]
$^{\bullet}CH_2C(OH)(CH_3)_2$	1.85×10^{9d}	3.3×10^{8b}	[183]

[a] 1.0M NaOH at 25°C.
[b] The observed rate constants are at an average of ≥8 runs at 25°C in 0.1 M NaOH.
[c] From Rush and Bielski [389] at pH 10.4 (alcohol radicals are in the protonated forms).
[d] Premix pulse radiolysis at pH 9.4, 0.05 M Na_2HPO_4/borate buffer at 23 ± 1°C.

The reactions of formate and alcohol radicals were slightly lower than reactions of e_{aq}^- with Mn(VII) and Mn(VI) (Table 6.2).

The spectra of Mn(VI) obtained by the reduction of Mn(VII) by ethanol radicals and e_{aq}^- in the pH range from 4 to 9 are shown in Figure 6.18 [183]. The molar extinction coefficient as a function of pH at 610 nm is presented in the inset of Figure 6.18:

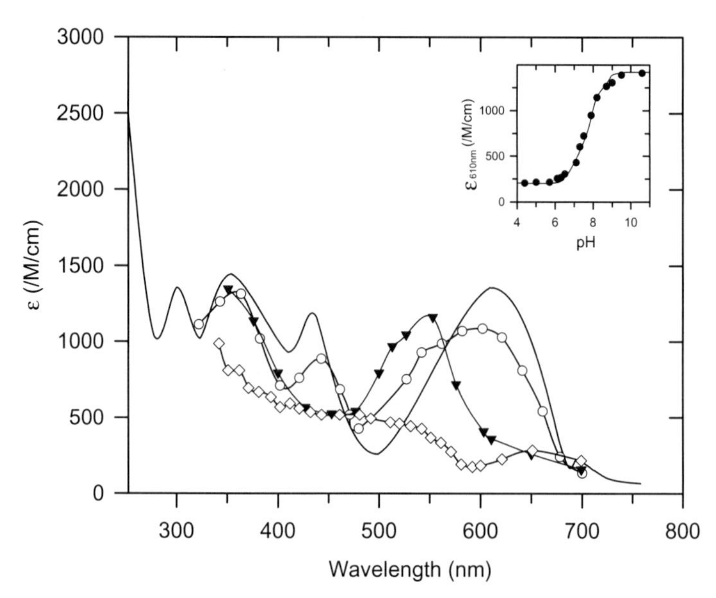

Figure 6.18. Spectrum of $O_3Mn(OH)^-$ (□) in argon-purged 0.025 M phosphate, 0.1 M ethanol, pH 4, spectrum of MnO_4^{2-} (solid curve) in 0.1 M NaOH, and Mn(VI) adduct spectra of the $^{\bullet}CO_2^-$ radical (○) in N_2O-saturated 2 mM formate and of the *tert*-butyl alcohol radical (▼) in N_2O-saturated 0.1 M *tert*-butyl alcohol solution at pH 9.4 (≈1 mM borate buffer). These spectra recorded 2–10 μs after the pulse. The inset shows the pH dependence of the Mn(VI) extinction coefficient at 610 nm in 0.025 M phosphate buffer, 0.1 M ethanol (adapted from Rush and Bielski [183] with the permission of the American Chemical Society).

$$O_3Mn(OH)^- \rightleftharpoons H^+ + MnO_4^{2-}. \tag{6.62}$$

The solid line in the inset of Figure 6.18 was calculated using the pK_{62} value of 7.4 [183]. The kinetics of the disproportionation of the MnO_4^{2-} ion in acidic solution has been performed [193]. The decay of the MnO_4^{2-} ion followed a pseudo-first-order rate law, which resulted in the formation of MnO_4^- ion. The formation of the permanganate ion obeyed second-order kinetics.

The spectra of Mn(V) under different conditions are presented in Figure 6.19 [183]. It is clear from Figure 6.19a that the spectra were sensitive to pH in the range of 0.01–10.0 M NaOH. This sensitivity was utilized to estimate the acid dissociation constant of Mn(V):

$$O_3Mn(OH)^{2-} \rightleftharpoons H^+ + MnO_4^{3-} \quad pK_{63} \approx 13.7. \tag{6.63}$$

The spectrum of the Mn(V) ester was also obtained by reducing Mn(VI) by *tert*-butyl radicals (Fig. 6.19b) [183]. In the reduction process, an intermediate (**A**) was initially formed, which subsequently decayed by first order to the final product of Mn(V) (**B**):

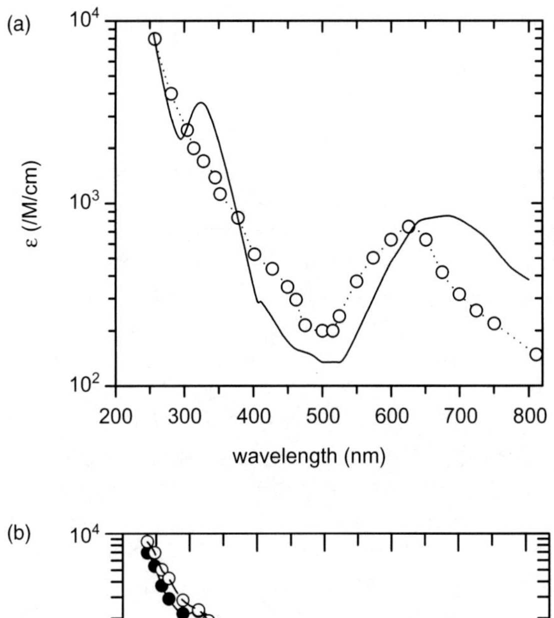

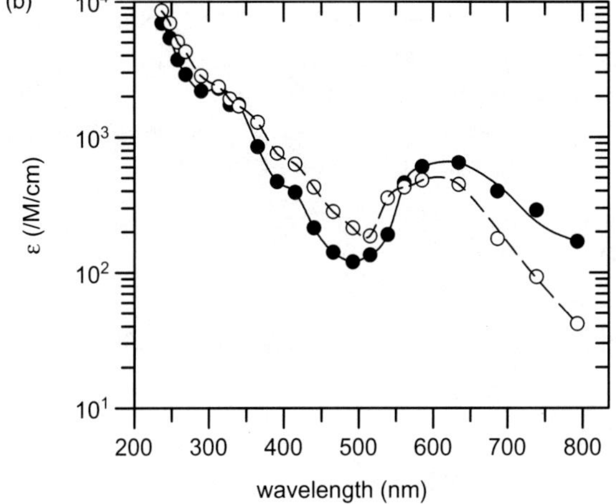

Figure 6.19. (a) Scanned (solid line) spectrum of MnO_4^{3-} in 10 M NaOH compared with the point-by-point (○) spectrum of $O_3Mn(OH)^{2-}$ obtained by the pulse-radiolytic reduction of MnO_4^{2-} in 0.01 M NaOH. (b) Point-by-point spectrum (○), measured 50 μs after the pulse, of the Mn(V) ester formed by the reaction of *tert*-butyl alcohol radicals with MnO_4^{2-} in 1 M NaOH. The spectrum of the intermediate (a) in reaction (6.64), decays to the final spectrum (●) of (b) ($MnO_4^{3-}/O_3Mn(OH)^{2-}$); 1.0 M NaOH and 25°C (adapted from Rush and Bielski [183] with the permission of the American Chemical Society).

$$O_3MnO^{2-} + {}^{\bullet}CH_2C(OH)(CH_3)_2 \rightarrow [O_3Mn(OC_4H_8OH)]^{2-} + OH^-/H_2O \rightarrow \mathbf{A}$$
$$MnO_4^{3-}/O_3Mn(OH)^{2-} + OHCH_2C(OH)(CH_3)_2 \rightarrow \mathbf{B}. \qquad (6.64)$$

Oxo-Mn(V) porphyrins have also been generated by the oxidation of Mn(III) porphyrins in aqueous solutions using peroxides and hypobromite [194, 195]. Mn^V-oxo complexes $(TBP_8Cz)Mn^V(O)$; TBP_8Cz = octa-*tert*-butylphenylcorrolazinato) have also been isolated in order to examine high-valent manganese reactions with one-electron reductants [196]. This complex was recently converted to $Mn^V(O)$ π-cation radical porphyrinoid complex, which was successfully characterized by UV–vis, EPR, and laser deposition and ionization mass spectrometry (MDI-MS) [197]. The $Mn^V(O)$ π-cation radical porphyrinoid complex was found to be more reactive than the $TBP_8Cz)$ $Mn^V(O)$ complex. The Mn(V)-imido complex $((TBP_8Cz)Mn^V(NMes)$ has also been synthesized [198]. The complex, $(PPh_4)_2[Mn^V(N)(CN)_4]$, has been shown to catalyze epoxidation of alkenes and oxidation of alcohols by H_2O_2 [199].

The Mn(VI)-ester was observed in the interaction of Mn(VII) with the *tert*-butyl alcohol radicals at pH 9.4 [183]. Initially, a spectrum, which appeared to be similar to MnO_4^{2-}, shown in Figure 6.19b, was observed (reaction 6.65). This species decayed by first order to produce a second transient (reaction 6.66). The decay of this second species finally yielded MnO_4^{2-} and an oxidized product of *tert*-butyl alcohol (reaction 6.67):

$$MnO_4^- + {}^{\bullet}CH_2C(OH)(CH_3)_2 \rightarrow [O_3Mn^{VI}OCH_2C(OH)(CH_3)_2]^- \qquad (6.65)$$
$$k_{65} = 1.85 \pm 0.2\,0 \times 10^9\,/M/s$$
$$[O_3Mn^{VI}OCH_2C(OH)(CH_3)_2]^- \rightarrow [O_2Mn(O_2C_4H_8)](chelate) + OH^- \qquad (6.66)$$
$$k_{66} = 5.5 \pm 1.0 \times 10^1\,/s$$
$$[O_2Mn(O_2C_4H_8)](chelate) \rightarrow MnO_4^{2-} + OHCH_2C(OH)(CH_3)_2 \qquad (6.67)$$
$$k_{67} = 0.8 \pm 0.2/s.$$

The formation of Mn(V) and Mn(VI) has also been seen in the spectral changes during the reactions of sulfite with Mn(VII) in alkaline solutions [200–202].

6.2.2 Reactivity of Complexes of Hypervalent Mn

In a recent study, the activation of a water molecule by electrochemical oxidation of the Mn-aquo complex has been reported [203]. The oxidation of a non-porphyrinic six-coordinated $Mn(II)(OH)_2$ complex resulted in a Mn(O) complex, which involved the sequential (2 × 1 electron/a proton) and direct (2 electron/2 proton) proton-coupled transfers. The intermediate Mn(III) $(OH)_2$ and Mn(III)OH complexes were analyzed [203]. Complexes of salen, porphyrine, and carrole with Mn(III), Mn(V), and Mn(VI) have been synthesized and characterized to understand their role in the generation of reactive intermediates in catalytic atom- or group-transfer reactions for several biological and synthetic systems [204–208]. For example, Mn(III)-corrole protects rat pancreatic beta cells against intracellular nitration by peroxynitrite, which causes subsequent cell death. The structure of corrole is in the form of the corrin and the positive charged Mn(III)-corrole rapidly decomposes peroxynitrite through a mechanism that does not involve the usual nitrating reaction intermediates [209]. Recently, evidence for the simultaneous generation of Mn^{III}-OOC(O)R, Mn^{IV}=O, and Mn^{V}=O as active oxidant species in olefin epoxidation by Mn(III) complexes have been reported [210]. A potential role of the oxidant-Mn-oxo(imido) intermediate has also been suggested in the epoxidation of alkenes with a series of iodosylarenes as oxidants catalyzed by Mn(V) oxo and imido complexes [198].

The OEC within PSII, containing Mn and Ca ions as well as amino acids in the structure, catalyzes the oxidation of water to form oxygen [211]. PSII contains two redoxactive tyrosines, YD and YZ, which performed different roles in catalysis [212]. An essential role of the redox-active tyrosines, YZ, for oxygen evolution has been explained. Biophysical and inorganic chemistry analyses, X-ray crystallography, and theoretical calculations were performed to comprehend the structures of the Mn and Ca ions, the redox-active tyrosine, and the surrounding amino acids that are present in the OEC [213, 214]. The manganese-catalytic site rapidly reduced the YZ. The PSII is useful in studying proton-coupled electron transfer reactions [11].

6.2.3 Oxidation by Mn(VII)

Permanganate has been used widely in the synthesis of organic compounds [215–221]. The Mn(VII) ion has also demonstrated its ability to be a versatile industrial oxidant in the preparation of many organic compounds [215]. Permanganate is now considered a green oxidant because of recent success in the recycling of its byproduct, MnO_2, back to permanganate.

Examples of selective oxidation carried out by the permanganate ion are presented in Table 6.3 [215]. Selectivity is generally defined by the conditions under which oxidations were carried out. Table 6.3 suggests the use of permanganate in the synthesis of organic compounds. The application of $KMnO_4$ adsorbed onto a solid support as a heterogeneous reagent or under solvent-free conditions has made significant advances in organic synthesis [215]. Other applications of permanganate include its use in the oxidation of organic contaminants in water purification [222–224]. Permanganate generally converts organic molecules into carbon dioxide and water. The oxidation of polycyclic aromatic hydrocarbons by permanganate is an example of a wide range of applications in the remediation of contaminants [225]. Oxidation of the aromatic ring is usually much slower than that of the side chains in hydrocarbons by the permanganate ion in slightly acidic, neutral, or basic conditions.

TABLE 6.3. Typical Reactions of Permanganate

Reductant	Process	Product
RCH=CHR	*cis*-Dihydroxylation	OH OH (RCH(OH)-CH(OH)R): RCH-CHR
RCH=CHR	Oxidative cleavage	RCH(=O) or RCOH(=O)
$RCH=CR_2$	Ketol formation	$RC(=O)-C(OH)R_2$
$R_2C=CR_2$	Oxidative cleavage	$R_2C=O$
$ArCH_3$	Oxidation	ArCOOH
$ARCH_2R$	Oxidation	ArC(=O)R
RSH	Oxidation	$ArSO_3H$
ArSH	Oxidation	RSSR
R_2CHNH_2	Oxidation	$R_2C=O$
$ArNH_2$	Oxidative coupling	ArN=Nar

Adapted from Singh and Lee [215] with the permission of the American Chemical Society.

Advances have also been made to identify the oxidation states of Mn formed in *in situ* chemical oxidation of the organic compound by $KMnO_4$ and in solid surfaces of filtration media samples from drinking water treatment [226, 227].

The reactivity of the permanganate ion with several inorganic and organic compounds has been reviewed [215, 216, 221, 228–230]. The results of the oxidation kinetics and mechanism indicate the possibility of several mechanisms, depending on the nature of the substrate, the reaction conditions, and the nature of the reactive species of manganese [221]. Recent oxidation procedures include the use of solid $KMnO_4$ adsorbed onto a solid support as a heterogeneous reagent or under solvent-free conditions [215]. The oxidation of alkanes and arylalkanes by the MnO_4^- ion may be accelerated with the use of an acetonitrile–BF_3 reaction mixture [231, 232]. Density functional theory (DFT) has been utilized in an effort to understand the reaction mechanisms [231, 233]. Below is a summary of the reactivity of permanganate with amino acids and amino polycarboxylates.

6.2.3.1 Amino Acids. The oxidation of amino acids by permanganate in acidic and neutral media has been extensively studied [234–236, 236–244]. The oxidation of amino acids resulted in aldehydes, ammonia, and CO_2 as products of the reactions (e.g., Eqs. 6.68–6.70) [234, 239]:

$$\text{Gly: } 2MnO_4^- + 3NH_3^+CH_2COO^- + 2H^+ \rightarrow 2MnO_2 + 3NH_3 + 3HCHO + 3CO_2 + H_2O \quad (6.68)$$

$$\text{Ala: } 2MnO_4^- + 5CH_3CH(NH_2)COOH + 11H^+ \rightarrow 2Mn^{2+} + 5CH_3CHO + 5NH_4^+ + 5CO_2 + 3H_2O \quad (6.69)$$

$$\text{Ser: } 2MnO_4^- + 3OHCH_2CH(NH_2)COOH + 2H^+ \rightarrow 2MnO_2 + 3OHCH_2CHO + 3NH_3 + 3CO_2 + H_2O. \quad (6.70)$$

The kinetics results of the reactions between Mn(VII) and amino acids showed an autocatalytic pattern. The formation of either the Mn(II) or Mn(III) complex involving the amino acid as a ligand was proposed to understand the autocatalytic behavior of the reactions in the acidic medium [239, 245]. However, in a neutral solution, a heterogeneous reaction pathway occurring on the colloidal MnO_2, the reduced product of Mn(VII), was proposed to explain the kinetics of the reaction [240]. The soluble form of colloidal MnO_2 can either be recognized as the reaction product or as the long-lived intermediate, which may constitute molecular aggregates with varying nanoparticle sizes [246].

Recently, a detailed role of MnO_2 colloidal particles in the autocatalytic reaction pathway in the oxidation of Gly by permanganate in neutral aqueous solution has been demonstrated [234]. Essentially, the kinetics of the reaction in phosphate solutions under various conditions was monitored at 526 and 418 nm for MnO_4^- and colloidal MnO_2 species, respectively, to gain information on different steps of the mechanisms. Significantly, long-lived intermediates such as Mn(VI) and Mn(V) were not observed, although it is a possibility they were present in insignificant steady-state concentrations. The rate law was deduced from the plot of the belled-shaped rate (v) versus time (t) (Eq. 6.71):

$$v = k_1c + k_2c(c_o - c). \quad (6.71)$$

where k_1 and k_2 are pseudo-first-order and pseudo-second-order rate constants of the noncatalytic and autocatalytic reaction pathways, respectively, in which Gly was present in excess. The rate law assumed both the noncatalytic and the autocatalytic reaction pathways. Reactions were first order with respect to $[MnO_4^-]$ and first order with respect to the colloidal MnO_2 surface for the noncatalytic and the autocatalytic reaction pathways, respectively. Values of k_1 and k_2 were determined using integrated (Eq. 6.72) and differential (Eq. 6.73) methods:

$$\ln[(k_1 + k_2(c_o - c))/c] = \ln(k_1/c_o) + (k_1 + k_2c_o)t \quad (6.72)$$

$$v/c = (k_1 + k_2c_o) - k_2c. \quad (6.73)$$

Most studies on the permanganate oxidation of amino acids and amines used integration to evaluate k_1 and k_2. Linear plots of (v/c) versus c of the differential method were applied to determine rate constants for the autocatalytic reactions of permanganate with formic acid, dimethylamine, trimethylamine, Gly, L-Ala, and L-Thr.

The kinetics of the permanganate–Gly reaction displayed a curvature in the plot. The curvature was dependent on the concentration of phosphate in the reaction mixture. Upward and downward curvatures were observed at low and high phosphate concentrations. A downward curvature increased with an increase in the initial concentration of permanganate. Effects of other parameters of the reaction were also examined. The downward curvature decreased with an increase in the concentration of the phosphate buffer, pH, and temperature, and decreased under the use of either cationic surfactant (benzyltriethylammonium chloride) or polymeric protective colloidal (gum arabic). The phosphate concentration dependence demonstrated that phosphate inhibited the autocatalytic reaction pathway. This was supported by the plotted results of the bell-shaped rate versus time, which was greatly affected by the concentration of phosphate (Fig. 6.20a). Importantly, the influence of phosphate was only observed when the differential method was applied (Fig. 6.20B). The integration method (Eq. 6.72) had almost no concentration dependence for k_1 and k_2 results (Fig. 6.20b). In contrast, the rate constants obtained by the differential method (Eq. 6.73) had a strong variation depending on the concentration of phosphate (Fig. 6.20b). This is an indication of the involvement of phosphate in the autocatalytic pathway of the mechanism.

The reaction mechanism is expressed in Equations (6.74)–(6.82), in which the $NH_2CH_2COO^-$ species was postulated as the reactive form of Gly to react with permanganate:

$$NH_3^+CH_2COO^- \rightleftharpoons H^+ + NH_2CH_2COO^- \quad K_{74} \tag{6.74}$$

$$MnO_4^- + NH_2CH_2COO^- \rightarrow HMnO_4^- + NH^{\bullet}CH_2COO^- \quad k_{75} \tag{6.75}$$

$$MnO_4^- + NH^{\bullet}CH_2COO^- \rightarrow MnO_4^{2-} + NH{=}CH_2 + CO_2 \quad k_{76} \tag{6.76}$$

Figure 6.20. (a) Rate versus time plots for the reaction of $KMnO_4$ (5.12×10^{-4} M) with Gly (0.160 M) in KH_2PO_4 (circles, 0.032 M; triangles, 0.080 M)–K_2HPO_4 (circles, 0.032 M; triangles, 0.080 M) buffer, at ionic strength 0.320 M (KCl), pH 6.64, and 25.0°C. (b) Dependence of the apparent rate constants for the noncatalytic (bottom) and autocatalytic (top) reaction pathways, obtained either by the integrated method (empty circles) or by the differential method with extrapolation at $t = 0$ (filled circles), on the total phosphate concentration during the reaction of $KMnO_4$ (5.12×10^{-4} M) with Gly (0.160 M) in KH_2PO_4–K_2HPO_4 buffer ($[KH_2PO_4] = [K_2HPO_4] = [\text{phosphate}]_T/2$), at ionic strength 0.320 M (KCl), pH 6.64, and 25.0°C (adapted from Perez-Benito [234] with the permission of the American Chemical Society).

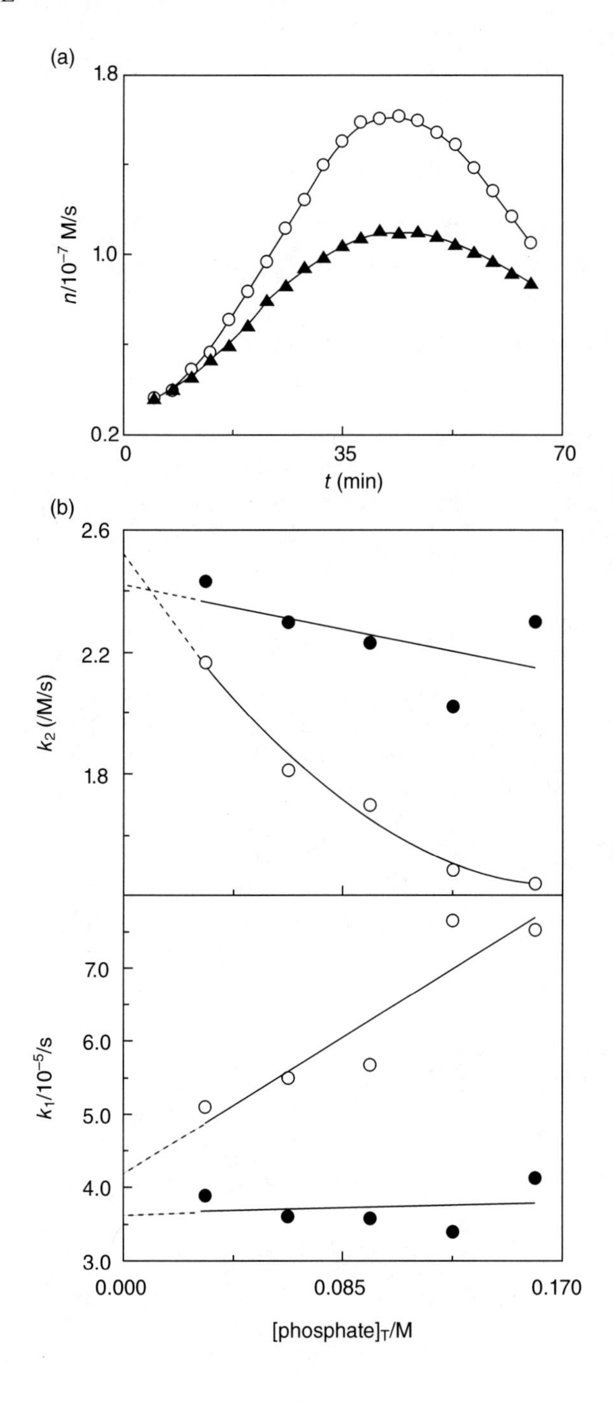
(a)
1.8
1.0
0.2
$n/10^{-7}$ M/s
0
35
70
t (min)
(b)
2.6
2.2
1.8
k_2 (/M/s)
7.0
6.0
5.0
4.0
3.0
$k_1/10^{-5}$/s
0.000
0.085
0.170
[phosphate]$_T$/M

$$HMnO_4^- + NH_2CH_2COO^- \rightarrow H_2MnO_4^{2-} + NH{=}CH_2 + CO_2 \quad k_{77} \tag{6.77}$$

$$H_2MnO_4^{2-} + H^+ \rightarrow MnO_2 + H_2O + OH^- \quad k_{78} \tag{6.78}$$

$$NH{=}CH_2 + H_2O \rightarrow NH_3 + HCHO \quad k_{79} \tag{6.79}$$

$$MnO_2 + NH_2CH_2COO^- \rightleftharpoons MnO_2\text{-}NH_2CH_2COO^- \quad K_{80} \tag{6.80}$$

$$MnO_2\text{-}NH_2CH_2COO^- + H_2PO_4^- \rightleftharpoons MnO_2\text{-}H_2PO_4^- + NH_2CH_2COO^- \quad K_{81} \tag{6.81}$$

$$MnO_2\text{-}NH_2CH_2COO^- + MnO_4^- \rightarrow MnO_2 + HMnO_4^- + NH^{\bullet}CH_2COO^- \quad k_{82}. \tag{6.82}$$

In the noncatalytic reaction, a one-electron (1 – e$^-$) transfer step was the rate-determining step (Eq. 6.75), which leads to the formation of Mn(VI) and the glycine radical. A radical formation was observed in the reaction mixture of permanganate-L-serine [234, 247]. The products of the reaction (Eqs. 6.77 and 6.78) were in agreement with the experimental stoichiometric reaction (Eq. 6.68).

In the autocatalytic pathway, adsorption of Gly onto MnO_2 colloidal particles occurred (Eq. 6.80). Competition of the $H_2PO_4^-$ ion for the same site displaced Gly from the MnO_2 colloidal surface (Eq. 6.81). It is a possibility that the remaining glycine molecules reacted with Mn(VII) from the solution surrounding the colloidal particles (Eq. 6.82). The Mn(VI) compound and a glycine radical, formed in reaction (6.82), could then participate in the reactions suggested for the noncatalytic pathways (Eqs. 6.74–6.79). The mechanism shows the $H_2PO_4^-$ species was used to stabilize the MnO_2 colloidal particles rather than the HPO_4^{2-} species of the phosphate ion because the $H_2PO_4^-$ species has a higher tendency to adsorb onto the colloidal MnO_2 surface than that of the HPO_4^- species. The stabilization processes suggest that the formation of the hydrogen bond between the hydrogen atom of the $H_2PO_4^-$ species and the oxygen atom of MnO_2 occurred. Furthermore, the MnO_2 particle size also participated in the stabilization process. The size of the colloidal MnO_2 surface increased with an increase in the molar ratio of either [MnO_2]/[phosphate] or [MnO_2/[ammonium ion]. Using Equations (6.74)–(6.82) and assuming Equation (6.82) was the rate-determining step, the values of k_1 and k_2 for the autocatalytic reaction were expressed as

$$k_1 = K_{74}k_{75}[NH_3^+CH_2COO^-]_T/(K_{74} + [H^+]) \tag{6.83}$$

$$k_2 = K_{74}K_{80}k_{82}[NH_3^+CH_2COO^-]_T/\{K_{74}K_{80}[NH_3^+CH_2COO^-]_T + (K_{74} + [H^+])(1 + K_{80}K_{81}[H_2PO_4^-]\}, \tag{6.84}$$

where $[NH_3^+CH_2CO_2^-]_T = [NH_3^+CH_2CO_2^-] + [NH_2CH_2CO_2^-]$ represents the total concentration of Gly (in large excess with respect to permanganate). Overall, the differential method (Eq. 6.73) has some advantages over the integration method (Eq. 6.72) in determining the correct values of noncatalytic (k_1) and autocatalytic (k_2) rate constants, and in understanding the oxidation of amino acids by permanganate in aqueous solutions.

Studies on the effects of buffers and colloidal MnO_2 have also been studied on other α-amino acids [248]. Similar to Gly, both noncatalytic and autocatalytic pathways contributed to rate constants for the oxidation of Ala, Glu, Leu, Ile, and Val by permanganate. Activation parameters of these two pathways helped to propose mechanisms, which agreed with experimental results.

Recently, the kinetics of the Tyr-MnO_4^- system in the presence of cetyltrimethylammonium bromide (CTAB) under acidic conditions has been carried out to learn the effect of the pre- and postcritical micellar concentration (CMC) of the CTAB on the reaction rates [249]. The rates of the reaction decreased if the concentration of CTAB was below the CMC. However, if the concentration was above the CMC, the rates increased. The –OH group in Tyr was suggested to be responsible for the reactivity of Tyr with MnO_4^-. The tyrosine radical and Mn(VI) compound as intermediates and dityrosine as the product of the reaction were suggested. This is similar to the oxidation of Tyr by MnO_4^- and H_2O_2 [250, 251]. Recently, a similar study has also been carried out for the MnO_4^-–Val system [252]. This study demonstrated the role of the hydrophobic chain in the oxidation mechanism in the presence of CTAB.

The kinetics and mechanistic study on the oxidation of L-Met by enneamolybdomangnate(IV) in perchloric acid has also been carried out [253]. The orders of the reaction were first order with respect to each reactant. An increase in $[H^+]$ increased the reaction rate, which was related to the protonation of the oxidant. The product of the reaction was methionine sulfoxide. The postulated mechanism of the reaction involved a direct two-electron ($2 - e^-$) transfer step without the formation of a free radical.

Some reactions involving the oxidation of amino acids by permanganate have also been performed in alkaline solutions. An example is the oxidation of L-Phe by Mn(VII) in an alkaline medium [254]. The major oxidation products were identified as ammonia and aldehyde, and Mn(VI) was the reduced product of Mn(VII). This is in contrast to the reaction studied under neutral conditions in which colloidal MnO_2 was the reduced species (Eq. 6.78). This indicates that Mn(VI), which formed initially in the alkaline medium (e.g., Eq. 6.85), did not further react with the reactant and products of the reaction. The stoichiometry of the reaction was expressed in Equation (6.85):

$$\begin{aligned}&2MnO_4^- + C_6H_5CH_2\text{-}CH(NH_2)COOH + 2OH^- \\ &\rightarrow 2MnO_4^{2-} + NH_3 + CO_2 + C_6H_5CH_2CHO + H_2O.\end{aligned} \tag{6.85}$$

6.2.3.2 Aminopolycarboxylates (APCs). Oxidation of APCs by permanganate in high alkaline solutions (pH 12–14) has been performed in detail due to the importance of the reactions in the oxidation of wastes containing chelating agents such as nitriloacetate (NTA) and ethylenediaminetetraacetate

(EDTA) [255]. The oxidation of ethylenediamine (EN) was also studied to understand the reaction mechanism of permanganate with APCs. The yields and breakdown products of the reactions varied with pH and the concentration of permanganate. For example, the molar yield of CO_2 for the oxidation of EDTA decreased with an increase in pH, and the highest doses of permanganate were 5.0, 3.4, and 1.9 for pH 12, 13, and 14, respectively.

In the oxidation of EN by Mn(VII), mineralization of the molecules to CO_2 was more significant at pH 12 than at pH 14. High yields of ammonia and oxalate at all pH values were obtained. Glyoxal was also detected in small amounts. This is an indication that EN was deaminated by Mn(VII), and possible pathways A and B of the mechanism are expressed in Figure 6.21. The mechanistic steps include sequential $1 - e^-$ transfer steps, which resulted in the formation of transient imines, their deprotonation, and hydrolysis. Glyoxal formed as a transient product, possibly in a more oxidized state than oxalate and CO_2. The formation of oxalate rather than formate indicates the transfers of both electrons were not occurring at the same sites of the EN molecule. Following the transfer of the first electron, oxidation of the *N*′-nitrogen atom by the second electron formed the transient, *N*,*N*′-diimine, which broke down and released ammonia and glyoxal, a precursor of oxalate (pathway B in Fig. 6.21). The degradation of NTA by permanganate also formed CO_2, ammonia, and oxalate with an additional contribution of iminodiacetate (IDA). This suggests *N*-dealkylation pathways in the reaction between NTA and Mn(VII).

The oxidation of EDTA by Mn(VII) provided a complicated outcome as the dealkylayion reaction products of EDTA and ethylediaminediacetate

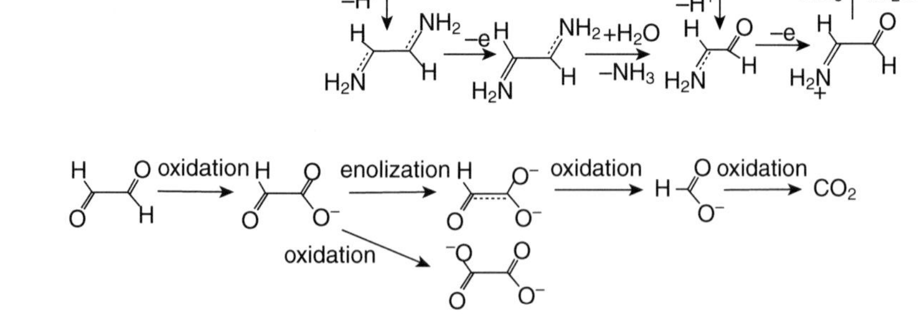

Figure 6.21. Oxidation of EN by Mn(VII) in alkaline solutions (adapted from Chang et al. [255] with the permission of the American Chemical Society).

(EDDA) were not predominating. However, a relatively high concentration of IDA occurred while EDDA was absent at pH 12–14 [255]. This suggests the attack was on the ethylene group rather than on the acetate groups of EDTA. The formation of IDA and other products, glycolate and oxalate, from the oxidation of EDTA and EDDA by Mn(VII) could not be ruled out completely. The postulated mechanism of the formation of IDA without the involvement of EDTA is displayed in Figure 6.22. The mechanism involved two pathways, A and B, which represent the electron transfer steps in the breakdown of EDTA. In pathway A, the electron transfer occurred at the same nitrogen atom while the formation of *N,N′*-diimine proceeded through a mechanism given for the oxidation of EN (see Fig. 6.21). The reduction of Mn(VII) resulted in Mn(VI), which was observed spectroscopically. The oxidized products were formed through oxygen transfer from the water molecules (Fig. 6.22).

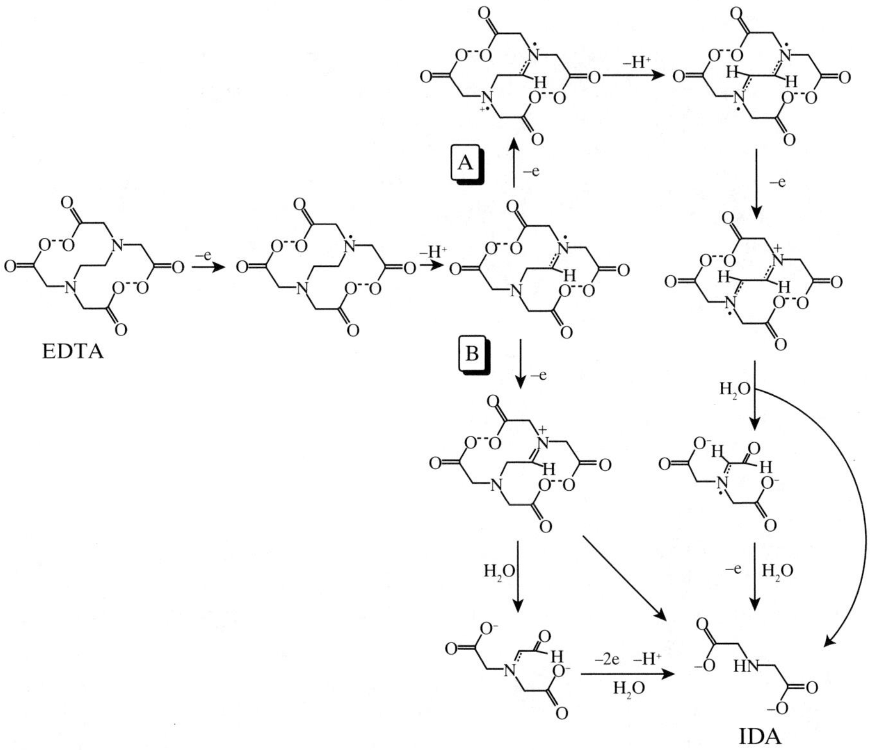

Figure 6.22. Suggested pathway of oxidation of EDTA by permanganate to result in IDA at high pH via oxidative pathway on the ethylene group (adapted from Chang et al. [255] with the permission of the American Chemical Society).

6.2.4 Conclusions

The high-valent manganese species undergo disproportionation and comproportionation reactions. The nature of transient species in the reaction of high-valent manganese species can be understood by their absorption spectra. A relatively simple and rapid pulse radiolysis technique may be applied to identify the intermediate Mn species. Moreover, this technique is useful in cases where the concentrations of species are low (micromolar) and the conventional techniques are not suitable. The permanganate ion has been extensively used in the oxidation of organic compounds in aqueous and organic phases, and current focus is on heterogeneous and solvent-free conditions (i.e., solid support). The reaction mechanisms under these conditions need to be explored further, which will provide the importance of high-valent Mn species in industrial and biochemical reactions.

The oxidation of organic substrates by Mn(VII) under alkaline conditions occurred through Mn(VI) and Mn(V) as intermediates, and Mn(IV) and Mn(II) were not involved in the oxidation mechanism. However, MnO_2 was involved as the autocatalytic reaction pathway in the oxidation by permanganate under acidic and neutral conditions. The oxidation of α-amino acids by Mn(VII) was autocatalyzed by MnO_2 colloidal particles. When reactions were conducted in phosphate-buffered solutions, the concentration of phosphate ions influenced the rates of oxidation of α-amino acids by the permanganate ion. Studies on the effects of MnO_2 colloidal particles and phosphate ions may also be extended for other amino acids to understand how side chains and aromaticity control the rates of oxidation of amino acids by Mn(VII). Moreover, studies on high-valent manganese compounds may also provide information for the role of high-valent iron species in oxidation reactions of various enzymes, which is discussed in the next section. High-valent Mn species and amino acids are involved in the water oxidation of PSII. The water oxidation mechanism needs detailed understanding and requires further experimental and theoretical elucidation.

6.3 IRON

The activation of O_2 and H_2O_2 by iron complexes in industrial, environmental, and biological redox processes has been of great interest [256–258]. Several reviews on this subject has been reported [258–272]. In the biological environment, Fe(IV)- and Fe(V)-oxo species have been suggested as the active oxidants of numerous heme and nonheme enzymes [260, 263, 273]. A general scheme showing the formation of oxo species of Fe(IV) and Fe(V) is shown in Figure 6.23 [260]. Iron(IV)-oxo complexes have the capability to hydroxylate C–H bonds of substrates efficiently. The heme-based enzyme P450 is involved in key biochemical reactions in the body such as the biosynthesis of hormones and drug metabolism as well as the detoxification of the liver. The

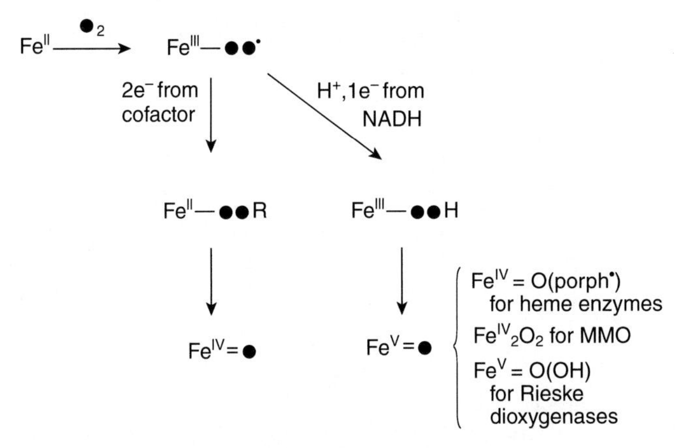

Figure 6.23. Generalized mechanistic scheme for oxygen activation at biological iron sites (adapted from Shan and Que [260] with the permission of Elsevier Inc.).

indirect evidence suggests the Fe(IV)-oxo heme cation radical is the intermediate responsible for the monooxygenation of a substrate. DFT and quantum mechanics/molecular mechanics (QM/MM) calculations on the activity of P450 (models) also support the Fe(IV)-oxo cation as the intermediate in reactions catalyzed by these enzymes [264, 269, 271]. Epoxidation, sulfoxodation, aliphatic and aromatic hydroxylation, dehydrogenation, and N-dealkylation are some of the reactions catalyzed by iron oxygenases [261, 274–277]. In a recent study, the potential roles of amino acids residues (Arg172 and Arg38) have been demonstrated in the transfer of protons from the ascorbate to the ferryl oxygen in the heme enzyme ascorbate peroxidase (APX) [278]. Both Arg172 and Arg38 had dual roles of forming ferryl species and of binding ascorbate and also facilitating proton transfer between ferryl and ascorbate.

Mononuclear iron-containing enzymes are also involved in the activation of O_2 to perform many metabolically important functions including the hydroxylation of arene, the oxidation of aliphatic C–H bonds, and the *cis*-dihydroxylation of arene double bonds [257, 260, 267, 279, 280]. Numerous oxo-iron(IV) model complexes, exhibiting intermediate $S = 1$ spin states, have been synthesized to understand the reaction mechanisms regarding the catalytic cycles of many nonheme enzymes. Comparatively, known $S = 2$ oxo-iron(IV) complexes are limited [12, 281, 282]. As suggested by the proposed reaction mechanism (Fig. 6.23), enzymes in need of cofactors, such as tetrahydrobiopterin or 2-oxoacids (iron(IV)-oxo or iron(II)-peroxo species), are formed through the simultaneous delivery of two electrons. Another example is the generation of iron(IV)-oxo species in the catalytic cycles of four 2-oxoglutarate-dependent enzymes, in which high-valent iron species act as the oxidant for the cleavage of the key C–H bond in many reactions [283]. The use of NADH in methane monooxygenase or Rieske dioxygenases usually delivered two electrons, one at a time, at different points to result in iron(III)-peroxo and iron(V)-oxo species (Fig. 6.23).

In the last few years, efforts have been made to understand the role of Fe^V=O in enzymatic reactions by carrying out oxidation of organic substrates by the mixture of iron complexes and H_2O_2 [284–287]. The generation of an Fe(V)=O species in nonheme Fe(II) complex/H_2O_2 has been demonstrated using variable-temperature mass spectrometry (VT-MS) [288]. This system catalyzed the oxidation of olefins, which are otherwise difficult to achieve conventionally. Oxygen atoms from both H_2O_2 and H_2O were shown involved in the *cis*-dihydroxylation reaction of olefins using isotopic labeling experiments. The Fe(V)-nitrido complex has also been synthesized and characterized spectroscopically [289]. Under reducing conditions, the reaction of this complex with water produced ammonia with a final iron product as Fe(II). These results may have implications in the chemistry of nitrogenase.

Trace amounts of the iron-tetraamidomacrocyclic ligand (Fe-TAML) catalysts were able to activate hydrogen peroxide to generate intermediates, Fe^{IV}=O and Fe^V=O [290–296]. This Fe-TAML-H_2O_2 system demonstrates peroxidase-like activities and longevities. The Fe^{III}-TAML activation has also been applied to demonstrate degradation of various pollutants [297–299]. Examples include degradation of estrogens, bisphenols, pharmaceuticals, and Orange II dye and inactivation of bacterial spores [297, 298, 300–302]. Products of the oxidation reaction were found to be nontoxic [297, 298]. Other examples are the desulfurization of heavy oil and the remediation of pulp and paper industry effluent [300, 303].

High-valent iron-based compounds (ferrates) are emerging disinfectants and oxidants in treating water [18, 19, 224, 304–311]. Ferrates are environmentally friendly and can address the concerns associated with the common treatment approaches. For example, ferrate(VI) ($Fe^{VI}O_4^{2-}$, Fe(VI)) does not produce bromate ion because of its nonreactivity with bromide ion [19]. Comparatively, ozone, a commonly used treatment oxidant, forms carcinogenic bromate ion. Disinfection tests of sodium ferrate(VI) on spore-forming bacteria showed that aerobic spore formers are reduced up to 3-log units while sulfite-reducing clostridia are effectively killed by ferrate(VI) [308]. Both aerobic spore formers and sulfite-reducing clostridia resist chlorine treatment. The multifunctional properties of ferrate(VI) can thus be utilized in a single dose for recycling and reuse of water and wastewater. Ferrate(VI) is a "green chemistry" chemical for coagulation and disinfection, and an oxidant for the multipurpose treatment of water and wastewater. A use of ferrate(VI) in developing a high charge-storage rechargeable battery, a "super iron battery" has also been demonstrated [312–318]. The rust generated from the discharge of the super iron battery is preferable to toxic manganese compounds currently used in commercial batteries.

The importance of ferrate species in the chemistry of natural water and atmospheric water droplets has also been demonstrated [259]. For example, the Fenton reaction plays an important role in atmospheric chemistry by contributing to the production of $^{\bullet}OH$ radicals. Experimental and model calculations have shown the production of ferryl ($Fe^{IV}O^{2+}$) as an active intermediate in the Fenton reaction [319–321]. Formation of ferryl species has also been suggested in the photoassisted Fenton reaction [322, 323].

The following sections present basic chemistry and reactivity of the high-valent iron species.

6.3.1 Iron(IV) and Iron(V)

6.3.1.1 Ferryl(IV) Ion. The ferryl ion, FeO^{2+}, has been generated by the oxidation of Fe^{2+} by ozone in aqueous acidic solution (pH 0–3) (Eq. 6.86):

$$Fe^{2+} + O_3 \rightarrow FeO^{2+} + O_2 \quad k_{96} = 8.2 \times 10^5 /M/s. \tag{6.86}$$

The spectrum of FeO^{2+} includes a small broad peak around 320 nm ($\varepsilon_{320nm} \approx 500/M/cm$) with a continuum that grows in the UV region (Fig. 6.24a) [321, 324]. The +4 oxidation state for the ferryl ion was confirmed by a Mössbauer spectroscopic technique [266, 325]. Experimental data and DFT calculations were consistent with the high spin ($S = 2$) of the ferryl species [325].

The absorption at 320 nm for the FeO^{2+} ion was used to study its decay as a function of pH in the acidic pH region [178]. An increase in the observed first-order rate constant, k (/s), with an increase in pH was observed. The decay rate increased approximately 50 times with an increase in pH. The results were interpreted based on the protolytic equilibrium between the two different hydrolytic forms of the ferryl ion with $pK_a \approx 2.0$.

The decay of FeO^{2+} is expressed by reaction (6.87):

$$4FeO^{2+} + 4H^+ \rightarrow 4Fe^{3+} + O_2 + 2H_2O. \tag{6.87}$$

The half-life of the ferryl ion is on the order of minutes in the pH range of 0–1 but decreases as the pH is increased. The following sequence of reactions explains the decay of the FeO^{2+} ion [326]:

$$FeO^{2+} + H_2O \rightarrow Fe^{3+} + {}^{\bullet}OH + OH^- \quad k_{98} = 1.3 \times 10^{-2} /s \tag{6.88}$$

$$FeO^{2+} + {}^{\bullet}OH + OH^- \rightarrow Fe^{3+} + H_2O_2 \quad k_{99} = 1.3 \times 10^7 /M/s \tag{6.89}$$

$$FeO^{2+} + H_2O_2 \rightarrow Fe^{3+} + HO_2 + OH^- \quad k_{100} = 1.0 \times 10^4 /M/s \tag{6.90}$$

$$FeO^{2+} + HO_2 \rightarrow Fe^{3+} + O_2 + OH^- \quad k_{101} = 2.0 \times 10^6 /M/s. \tag{6.91}$$

All of the reactions participate at low pH, but only reactions (6.88) and (6.89) dominate at micromolar concentrations of the ferryl ion. The rate-determining step of the decay is expressed by reaction (6.88). The activation energy of the decay of the ferryl ion at pH 0 was determined as 34.0 ± 0.3.0 kJ/mol [326].

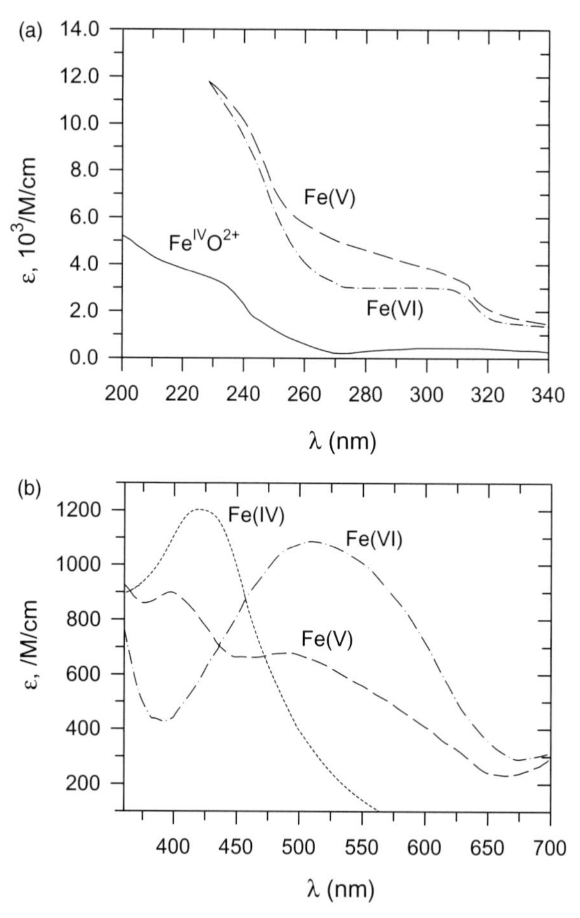

Figure 6.24. Spectra of iron(IV), iron(V), and iron(VI) in UV (a) and visible (b) wavelength regions.

The ferryl ion also reacts with hydrogen peroxide to first form HO_2, which in turn reacts with another ferryl ion to produce oxygen [321] (reactions 6.88 and 6.89). The formation of HO_2 by a $1 - e^-$ transfer is supported by the agreement between the experimental and the predicted values of the rate constants from the linear relationship between log k versus $E_o(H_2O_2/HO_2)$ (Fig. 6.25).

The rate constants for the reactions between FeO^{2+} and various inorganic ions are presented in Table 6.4. The rate constants ranged from 1×10^2/M/s to 1.8×10^5/M/s. A linear relationship between log k and the standard $1 - e^-$ redox potential, E_o, was observed (Fig. 6.25), suggesting electron transfer reactions [321] (e.g., reactions 6.92 and 6.93):

$$FeO^{2+} + HNO_2 \rightarrow Fe^{3+} + NO_2 + OH^- \quad (6.92)$$

$$FeO^{2+} + Mn^{2+} + 2H^+ \rightarrow Fe^{3+} + Mn^{3+} + H_2O. \quad (6.93)$$

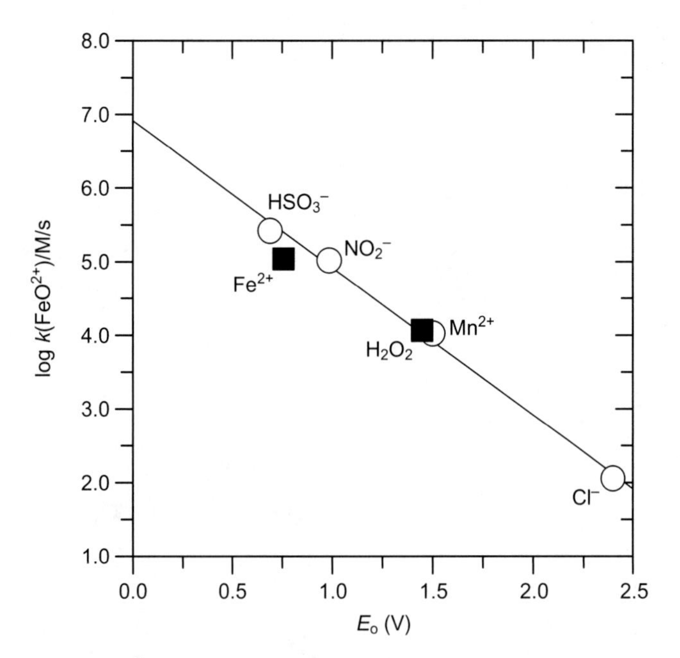

Figure 6.25. log $k(FeO^{2+})$ as a function of the standard $1 - e^-$ reduction potential (E_o) for the electron transfer reaction of FeO^{2+} with inorganic compounds. (●) Literature values (the point for NO_2^- also includes HNO_2) (adapted from Jacobsen et al. [321] with the permission of Wiley, Inc.).

TABLE 6.4. Reactivity of Ferryl Ion and Fe(IV) Pyrophosphate Complex with Inorganic Ions

Compound	k, /M/s 25°C	$\Delta H^‡$, kJ/mol	$\Delta S^‡$, J/mol	Metal(II)	k, /M/s 25°C
$Fe^{IV}O^{2+}$ (pH 1.0)				$[(P_2O_7)_2Fe^{IV}O]^{6-}$ (pH 10.0)	
HNO_2	1.1×10^4	32.0	−60.1	Mn^{2+}	1.2×10^6
NO_2^-	$\leq 10^5$	–		Fe^{2+}	1.6×10^6
Cl^-	1.0×10^2	–		Co^{2+}	5.5×10^5
HSO_3^-	2.5×10^5	–		Ni^{2+}	$<4.0 \times 10^2$
SO_2	4.5×10^5	–		Cu^{2+}	$<4.0 \times 10^6$
Mn^{2+}	1.0×10^4	18.8	−105.2		
Fe^{2+}	1.8×10^5	4.5	−136.8		

Data taken from References 321, 326, and 358).
‡Activation parameter.

The reaction between Fe^{2+} with the ferryl ion was studied at an initial ratio of $[Fe^{2+}]/[O_3] > 3$ (Eq. 6.94):

$$FeO^{2+} + Fe^{2+} + H_2O \rightarrow Fe^{3+} + 2OH^- \quad k_{104} = 8.5 \times 10^5 /M/s. \qquad (6.94)$$

In this reaction, an iron(III) dimer was observed as an intermediate; the yield of which was temperature dependent [326]. The results suggest reaction (6.94) can be branched into reactions (6.95a) and (6.95b):

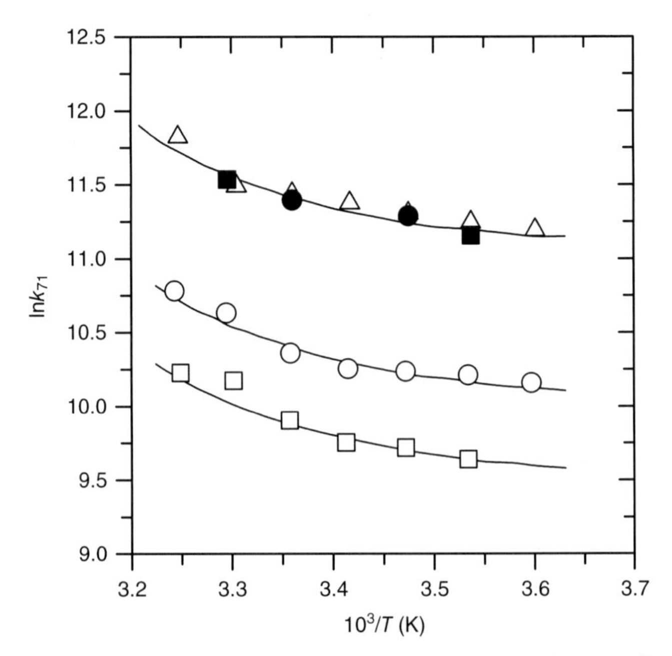

Figure 6.26. Arrhenius plot of the rate constant k_{71} for the reaction of Fe^{2+} with FeO^{2+}. Ionic strength adjusted with $NaClO_4$. (Δ) pH 0.0 (μ = 1.0); (○) pH 1.0 (μ = 0.1); (●) pH 1.0 (μ = 1.0); (■) pH 2.0 (μ = 0.01); (□) pH 2.0 (μ = 1.0) (adapted from Jacobsen et al. [326] with the permission of Wiley Inc.).

$$FeO^{2+} + Fe^{2+} + H_2O \rightarrow Fe^{3+} + 2OH^- \quad k_{105a} = 7.2 \times 10^4 \text{/M/s} \tag{6.95a}$$

$$FeO^{2+} + Fe^{2+} + H_2O \rightarrow Fe(OH)_2Fe^{4+} \quad k_{105b} = 1.8 \times 10^4 \text{/M/s.} \tag{6.95b}$$

The branching of reaction (6.95) was also supported by the curvature in the Arrhenius plot (Fig. 6.26). The yield of the dimer (reaction 6.95b) was dependent on the temperature, which may be related to a larger activation energy (E_a (6.95a) = 7.0 kJ/mol and E_a (6.95b) = 42.0 kJ/mol). The reactivity of the ferryl ion with H_2O_2 is about 200 times slower than with HO_2 [326].

The decay of the dimer followed first-order kinetics, which was independent of the initial Fe^{2+} and ozone concentrations. However, the decay was strongly dependent on the acidity (Fig. 6.27). The linear relationship of the first-order rate constant (k_{106}) with the proton concentration can be explained by reactions (6.96a) and (6.96b):

$$Fe(OH)_2Fe^{4+} \rightarrow 2Fe^{3+} + 2OH^- \quad k_{106a} = 0.63\text{/s (ionic strength} = 0.85) \tag{6.96a}$$

$$Fe(OH)_2Fe^{4+} + 2H^+ \rightarrow 2Fe^{3+} + 2H_2O \quad k_{106b} = 1.95\text{/M/s (ionic strength} = 0.85). \tag{6.96b}$$

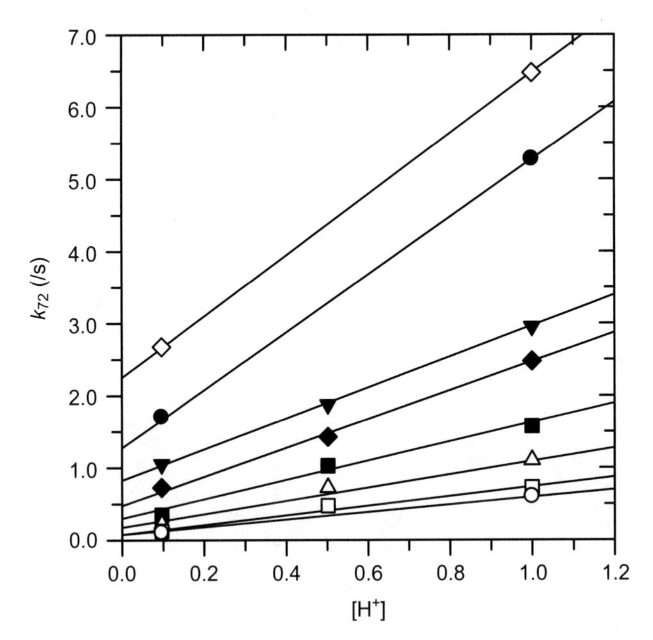

Figure 6.27. Apparent rate constant for the decay of the dimer $Fe(OH)_2Fe^{4+}$ measured at 320 nm as a function of $[H^+]$ at different temperatures, 5–40°C. Constant ionic strength at 1.0 M ClO_4^-. (○) 5°C, (□) 10°C, (△) 15°C, (■) 20°C, (◆) 25°C, (▼) 30°C, (●) 35°C, (◇) 40°C (adapted from Jacobsen et al. [326] with the permission of Wiley Inc.).

The activation energies for reaction (6.96a,b) were determined as 73.0 and 47.0 kJ/mol. The reaction of FeO^{2+} with H_2O_2 produced HO_2 by a one-electron transfer process (reaction 6.107). HO_2 further reacted with FeO_2^+ (reaction 6.98):

$$FeO^{2+} + H_2O_2 \rightarrow Fe^{3+} + HO_2 + OH^- \quad k_{107} = 1.0 \times 10^4 \text{/M/s} \qquad (6.97)$$

$$FeO^{2+} + HO_2 \rightarrow Fe^{3+} + O_2 + OH^- \quad k_{108} = 2 \times 10^4 \text{/M/s.} \qquad (6.98)$$

The rate constants for the FeO_2^+ reactions with H_2O_2 also followed a linear relationship shown in Figure 6.25, which also suggests the reactions with inorganic substrates proceed by a single electron transfer [326].

The reaction kinetics of the ferryl ion with phenol, nitrobenzene, o- and p-nitrophenol in 1 M $HClO_4$ was also studied [327]. The values of the second-order rate constants (k, /M/s) increased with an increase in the dissociation energy of the O–H bond of the studied molecules. A negative slope between log k and σ (NO_2) was determined. The rate constants decreased with an increase in the one-electron reduction potentials of the corresponding radical cations of the substrates. The mechanism involving the electrophilic addition and the electron transfer explained the observed trend in the kinetics of the reactions.

6.3.1.2 Iron(IV)- and Iron(V)-Oxo Complexes. The redox properties of nonheme Fe^{IV}=O complexes have been examined [279, 328–332]. The Fe^{IV}=O species generally undergoes a H-atom abstracting ability in the order Fe^{IV}=O >> Fe^{IV}-O-Fe > Fe^{IV}-OH [328]. Among the studied Fe^{IV}=O complexes, $[Fe^{IV}(O)(N4Py)]^{2+}$ (N4Oy = *N*,*N*-bis(2-pyridylmethyl)-bis(2-pyridyl)methylamine) had the ability to oxidize C–H bonds, even in cyclohexane ($D_{C\text{-}H}$ = 99.3 kcal/mol) [333]. The $[Fe^{IV}(O)(N4Py)]^{2+}$ has been synthesized using chemical, electrochemical, and photocatalytical methods [334–336]. The theoretical study supports the involvement of proton-coupled electron transfer in the O–O cleavage to result in oxoiron(IV) from the reaction of nonheme iron(II) complex with H_2O_2 [337].

Recently, the rate constants for the oxidation of dihydroanthracene (DHA) by $[Fe^{IV}(O)(N4Py)]^{2+}$ (k_2, M/s) as a function of $D_{O\text{-}H}$ were compared with other oxo complexes of Fe(IV), Ru(IV), and Mn to understand the transfer of a proton associated with the one-electron transfer process, responsible for the conversion of the Fe^{IV}=O unit to the Fe^{III}-OH species (Fig. 6.28) [329, 338–342]. The straight line in the plot of Figure 6.28 was constructed using appropriate values for $^tBuO^•$ and $^tBuOO^•$. This approach was used previously

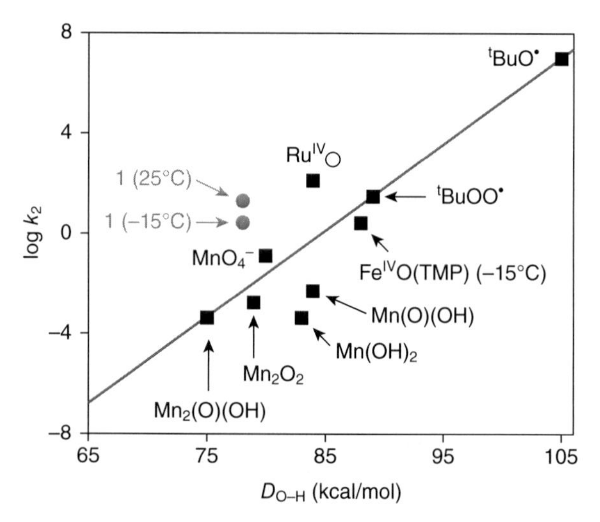

Figure 6.28. Plot of log k_2 of DHA oxidation and the strength of the O–H bond formed by the oxidants in CH_3CN at 25°C unless labeled otherwise. Data points shown in black filled squares were taken from References 32–37 and the blue filled circle was obtained in the present work of complex **1**. The red straight line was drawn through the points belonging to the two oxygen radicals following Mayer's precedent. Mn(O)(OH), $[Mn^{IV}(Me2EBC)(O)(OH)]^+$; $Mn(OH)_2$, $[Mn^{IV}(Me_2EBC)(OH)_2]^{2+}$; Mn_2O_2, $[(phen)_2Mn^{IV}(O)_2Mn^{III}\text{-}(phen)_2]^{3+}$; $Mn_2(O)(OH)$, $[(phen)_2Mn^{III}(O)(OH)Mn^{III}(phen)_2]^{3+}$; $Ru^{IV}O$, $[Ru^{IV}(O)(bpy)_2(py)]^{2+}$ (adapted from Wang et al. [329] with the permission of the American Chemical Society). See color insert.

[338]. Significantly, many of the Mn complexes have larger $D_{O\text{–}H}$ values than $[Fe^{IV}(O)(N4Py)]^{2+}$, but the ferryl complex reacted much faster with DHA than the Mn complexes. The unexpected larger relative reactivity of $[Fe^{IV}O(N4Py)]^{2+}$ was also observed in comparison to $^tBuOO^•$, $[Fe^{IV}(O)(TMP)]$ (TPMP = tetramesitylporphinate), and $Ru^{IV}(O)(bpy)_2(py)]^{2+}$ (Fig. 6.28). The results indicate the reactivity of $[Fe^{IV}(O)(N4Py)]^{2+}$ was not only driven by thermodynamics, but kinetic factors also participated in the enhancing ability of $[Fe^{IV}(O)(N4Py)]^{2+}$ to cleave the C–H bond in DHA. Furthermore, ligands present in the nonheme iron complexes may thus determine the efficiency of the Fe^{IV}=O units to perform H-atom abstractions.

In recent years, the reactivity of $[Fe^{IV}O(N4Py)]^{2+}$ with natural amino acids has been performed [334, 335]. A series of model compounds, *N*-acetyl, *tert*-butyl amide derivatives, of amino acids (Ac-AA-NHtBu) were synthesized (Fig. 6.29) [343]. The compounds represent individual residues within a polypeptide chain in which the N-terminus was acetylated and a *tert*-butyl amide was installed on the C-terminus [343]. The decomposition of $[Fe^{IV}(O)(N4Py)]^{2+}$ in the presence of 10 equiv of the model compounds was followed. The measured pseudo-first-order rate constants (k_{obs}) in a mixture of 1:1 H_2O/MeCN were determined (Table 6.5). The reactivity of $[Fe^{IV}(O)(N4Py)]^{2+}$ varied over five orders of magnitude. Comparatively, the kinetic rate constants for the reactions of $^•OH$ with amino acids varied over only three orders of magnitude [344]. This suggests $[Fe^{IV}(O)(N4Py)]^{2+}$ may be more selective than $^•OH$. The values of the k_{obs} and the relative rate constants (k_{rel}) in Table 6.5 demonstrate that Cys and Tyr had the highest reactivity with the ferryl species, followed by Trp and Met. The derivatives of Gly, His, and Ser showed intermediate reactivity (Table 6.5). The reactivities of other model compounds were similar to the control rate of the ferryl species.

Polypeptide

Residue Mimic

1 Ala	**8** Met	**15** Ser
2 Val	**9** Tyr	**16** Thr
3 Phe	**10** Trp	**17** His
4 Leu	**11** Gln	**18** Arg
5 Pro	**12** Lys	**19** Asn
6 Gly	**13** Glu	**20** Ile
7 Cys	**14** Asp	

Figure 6.29. Structure of Ac-AA-NHtBu substrates that mimic amino acid residues of polypeptides (adapted from Ekkati et al. [343] with the permission of Springer Americas).

TABLE 6.5. Pseudo-First-Order Rate Constants for Decomposition of $[Fe^{IV}(O)(N_4Py)]^{2+}$ by Amino Acid Substrates Ac-AA-NHtBu ([Fe] = 1 mM)[a,b]

Entry	Ac-AA-NHtBu	k_{obs} (/s)[c]	k_{rel}
1	Cys	1.7(1)	340,000
2	Tyr	$3.4(1) \times 10^{-1}$	68,000
3	Trp	$1.7(1) \times 10^{-2}$	3,400
4	Met	$3.2(1) \times 10^{-3}$	640
5	Gly	$5.8(2) \times 10^{-5}$	12
6	His	$3.8(2) \times 10^{-5}$	7.6
7	Ser	$3.3(7) \times 10^{-5}$	6.6
8	Ala	$2.5(2) \times 10^{-5}$	5.0
9	Asp	$1.7(3) \times 10^{-5}$	3.4
10	Gln	$1.6(2) \times 10^{-5}$	3.2
11	Thr	$1.4(6) \times 10^{-5}$	2.8
12	Phe	$1.4(4) \times 10^{-5}$	2.8
13	Asn	$1.4(1) \times 10^{-5}$	2.8
14	Arg·HCl[d]	$1.4(1) \times 10^{-5}$	2.8
15	Lys·HCl[d]	$1.4(2) \times 10^{-5}$	2.8
16	Glu	$1.3(1) \times 10^{-5}$	2.6
17	Val	$1.3(2) \times 10^{-5}$	2.6
18	Leu	$7.4(3) \times 10^{-6}$	1.5
19	Pro	$5.0(8) \times 10^{-6}$	1.0
20	Ile	$5.0(5) \times 10^{-6}$	1.0

[a] Reaction conditions: 1:1 H_2O/CH_3CN, 298 ± 2 K, 10 mM substrate (10 equiv).
[b] Background rate for decomposition of $[Fe^{IV}(O)(N4Py)]^{2+}$ in the absence of a substrate was $5.0(2) \times 10^6$/s.
[c] Number reported is the average of at least three runs; error as standard deviation is given in parentheses.
[d] HCl salts of the basic amino acids were used to avoid strongly basic conditions.
Adapted from Abouelatta et al. [334] with the permission of the American Chemical Society.

The kinetic isotope effect (KIE) and the analysis of products were performed to understand the mechanisms of the reactions of $[Fe^{IV}(O)(N4Py)]^{2+}$ with amino acids [334, 335]. Gly reacted 12 times faster than the control reaction (Table 6.5). In D_2O/CD_3CN, the rate was similar to the rate in H_2O/CH_3CN, which suggests the deprotonation of an amide proton of Ac-Gly-NhtBu did not participate in the rate-determining step of the decomposition of $[Fe^{IV}(O)(N4Py)]^{2+}$. However, when deuterated Gly at the α position (2,2-d_2) was used, the decomposition of the ferryl species was much slower ($k_{obs} = 1.2 \times 10^5$/s), resulting in a KIE of 4.8. The generation of a glycyl radical was proposed (Fig. 6.30). A hydrogen atom transfer (HAT) may also occur to yield a glycyl radical. In the case of other aliphatic amino acids, Ala, Val, Leu, Ile, and Pro, rapid decomposition of the ferryl species did not occur (Table 6.5). This observation suggests the abstraction of the α-H atom with

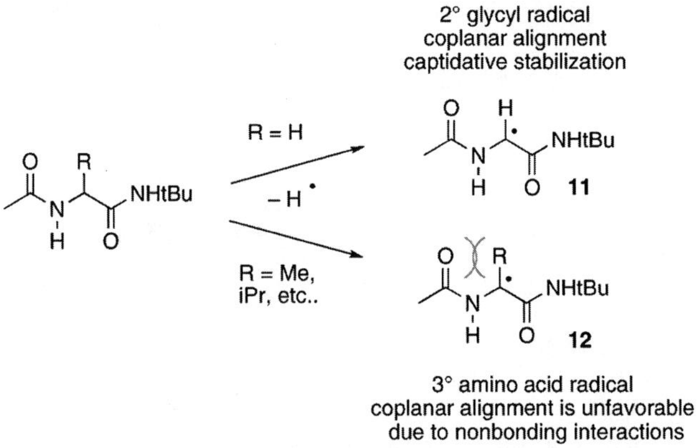

2° glycyl radical
coplanar alignment
captidative stabilization

R = H
−H•
R = Me, iPr, etc..

11

12

3° amino acid radical
coplanar alignment is unfavorable
due to nonbonding interactions

Figure 6.30. Carbon-centered radicals resulting from the abstraction of an α-hydrogen atom, stabilized in the case of Gly but destabilized with other amino acids (adapted from Abouelatta et al. [334] with the permission of the American Chemical Society). See color insert.

substituted amino acids was relatively slow in comparison with a compound derived from Gly. This difference in the reactivity of Gly from other aliphatic amino acids was explained by considering captodative stabilization of the radical formed (Fig. 6.30). For example, a nonbonding interaction between the amide carbonyl and the side chain (R) in Ala destabilizes the coplanar configuration in the radical, which does not have the full energetic favorability of captodative stabilization. In contrast, the glycine radical can obtain resonance stability through the coplanar alignment of six contiguous sp^2-hybridized atoms, necessary for maximization of the orbital overlap. Furthermore, steric effects may also be a factor in the relative unreactivity of secondary and tertiary amino acids [334].

The compounds derived from basic amino acids (Lys, Arg, and His) were not as reactive with the $[Fe^{IV}(O)(N4Py)]^{2+}$ species [334]. In contrast, the reactivity of anilines was significantly higher with the ferryl species. The results indicate a free base of amine may be required to participate in the electron transfer–proton transfer (ET-PT) mechanism for the reaction of amines with the $[Fe^{IV}(O)(N4Py)]^{2+}$ species [334]. Decomposition rates of the ferryl species for the compounds derived from Ser and Thr, containing alcohol functional groups, were only about three to seven times faster than the control reaction. Compounds derived from Asp, Gln, Glu, and Asn did not accelerate the decomposition of ferryl species under pseudo-first-order reaction conditions, having 10 equiv of the compound. Oxidation products of the compounds derived from basic and polar amino models were not observed by nuclear magnetic resonance (NMR) or ESMS analysis.

Sulfur-containing amino acids reacted very fast with the $[Fe^{IV}(O)(N4Py)]^{2+}$ species. The reactivities of Cys and Met were approximately five and two

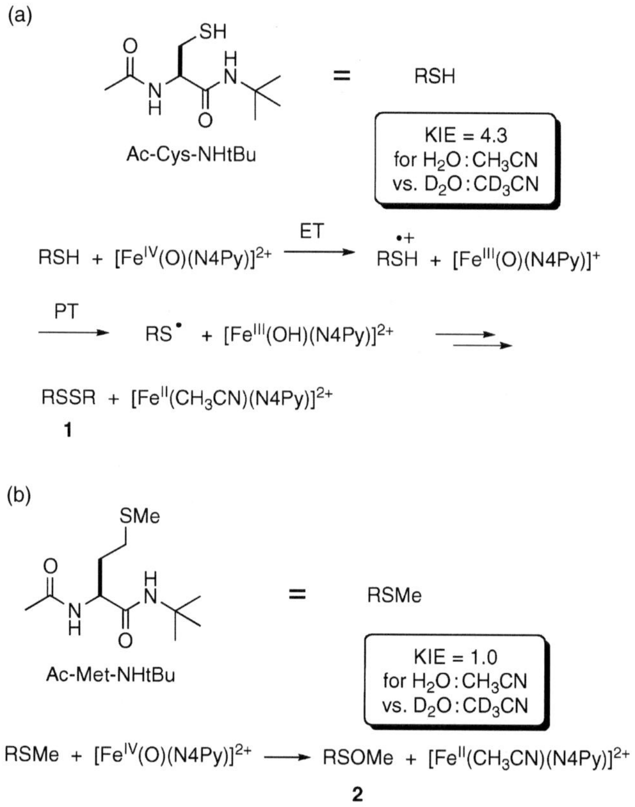

(a)

Ac-Cys-NHtBu = RSH

KIE = 4.3
for $H_2O:CH_3CN$
vs. $D_2O:CD_3CN$

$RSH + [Fe^{IV}(O)(N4Py)]^{2+} \xrightarrow{ET} RSH^{\bullet+} + [Fe^{III}(O)(N4Py)]^{+}$

$\xrightarrow{PT} RS^{\bullet} + [Fe^{III}(OH)(N4Py)]^{2+} \longrightarrow$

$RSSR + [Fe^{II}(CH_3CN)(N4Py)]^{2+}$
1

(b)

Ac-Met-NHtBu = RSMe

KIE = 1.0
for $H_2O:CH_3CN$
vs. $D_2O:CD_3CN$

$RSMe + [Fe^{IV}(O)(N4Py)]^{2+} \longrightarrow RSOMe + [Fe^{II}(CH_3CN)(N4Py)]^{2+}$
2

Figure 6.31. Proposed mechanism for the reaction of $[Fe^{IV}(O)(N4Py)]^{2+}$ with Ac-Cys-NHtBu (a) Ac-Met-NHtBu (b) (adapted from Abouelatta et al. [334] with the permission of the American Chemical Society).

orders of magnitude faster than the control reaction (Table 6.5). Second-order rate constants were determined as 1.9×10^2/M/s and 3.2×10^{-1}/M/s for Cys and Met, respectively. The reactions of Cys and Met with the ferryl species showed different KIE values (Fig. 6.31). The reactions performed under D_2O/CD_3CN solvent mixtures gave deuterium KIE (k_H/k_D) values of 4.3 and 1.0 for Cys and Met, respectively (Fig. 6.31). The results suggest a rate-limiting electron transfer, followed by a proton transfer in the reaction mechanism of the ferryl species with Ac-Cys-NHtBu (Fig. 6.31a). A concerted HAT was ruled out, which would have displayed a much larger KIE than 4.3. The electron transfer step is supported by the high $E_{1/2}$ value [345, 346] for the $Fe^{III/IV}$ couple of $[Fe^{IV}(O)(N4Py)]^{2+}$ and the reducing capability of thiols. The ET-PT steps formed the thiyl radical and the $[Fe^{III}(OH)(N4Py)]^{2+}$ species. The Fe^{III} species slowly decomposed to $[Fe^{II}(CH_3CN)(N4Py)]^{2+}$. Formation of disulfide **1** was confirmed by product analysis of the reaction mixture. Additionally, other possible products such as sulfenic and sulfinic acids derived from Cys were not observed. In the case of Met, no KIE was observed (Fig. 6.31b), which supports the rate-determining step did not involve the loss of a hydrogen atom from the compound. The sulfoxide product 2 from the reaction of the Ac-Met-NHt-Bu compound with the ferryl species was determined. This is in agreement with the results obtained in the reactions of the $[Fe^{IV}(O)(N4Py)]^{2+}$ species with aromatic sulfides [347]. Thus, an oxygen-atom transfer reaction mechanism was proposed (Fig. 6.31b).

The reactivity of $[Fe^{IV}(O)(N4Py)]^{2+}$ with model compounds derived from aromatic amino acids, Trp and Tyr, has also been studied [334]. The decomposition rates of the ferryl species were three and four orders of magnitude faster for Trp and Tyr, respectively, than the control reaction with the compounds. The second-order rate constants were determined as 1.64 and 4.2×10^1/M/s for Tyr and Trp, respectively. The decomposition of $[Fe^{IV}(O)(N4Py)]^{2+}$ by Ac-Trp-NHtBu in the D_2O/CD_3CN solvent produced a KIE of 5.2 (Fig. 6.32a). This deuterium KIE is similar to the KIE determined in the decomposition of $[Fe^{IV}(O)(N4Py)]^{2+}$ by the compounds derived from Gly and Cys (see Fig. 6.31). This is consistent with the ET-PT mechanism proposed in Figure 6.32a. Significantly, the KIE of 4.4 was determined for the oxidation of Trp in D_2O by the heme-containing iron enzyme tryptophan 2,3-dioxygenase [348]. In this study, the removal of the indole proton was suggested as partially rate-determining. The UV–vis spectroscopic results demonstrated a faster decomposition of the ferryl species than the regeneration of the Fe^{II} species (Fig. 6.32a). Products of the reaction mixture indicate the addition of a single oxygen atom to the Trp molecule, possibly located at position 3 of the indole ring. The results are similar to the products formed in the oxidation of Trp by ClO_2 [349].

The Ac-Tyr-NHtBu compound had a high KIE of 29 for the decomposition of $[Fe^{IV}(O)(N4Py)]^{2+}$ in the D_2O/CD_3CN solvent (Fig. 6.32b). The generation of a tyrosyl radical was proposed through a HAT rather than ET-PT (Fig. 6.32b). The proposed mechanism was supported by slower rate constants for the ferryl species in the presence of the protected Tyr compounds, Ac-Tyr(OAc)-NHtBu and Ac-Tyr(OMe)-NHtBu, having pseudo-first-order rate constants of 1.3×10^{-5}/s and 2.5×10^{-5}/s, respectively. The Ac-Phe-NHtBu compound also had a similar rate constant for the decomposition of the $[Fe^{IV}(O)(N4Py)]^{2+}$ species (see Table 6.5). The results clearly indicate the involvement of the functional hydroxyl group of Ac-Tyr-NHtBu rather than the electron-rich aromatic ring in the mechanism. The formation of the Tyr radical was also supported by the determination of a trace amount of the phenoxyl radical derived from Ac-Tyr-NHtBu using the EPR spectroscopy. Significantly, a phenoxyl radical was also observed in the treatment of 2,4,6,-tri-*tert*-butylphenol with $[Fe^{IV}(O)(N4Py)]^{2+}$. It appears the phenoxyl radicals polymerized because no major oxidation products were identified.

The mechanism for the oxidation of GSH by $[Fe^{IV}(O)(N4Py)]^{2+}$ has also been studied [350]. Initially, the reaction immediately produced an

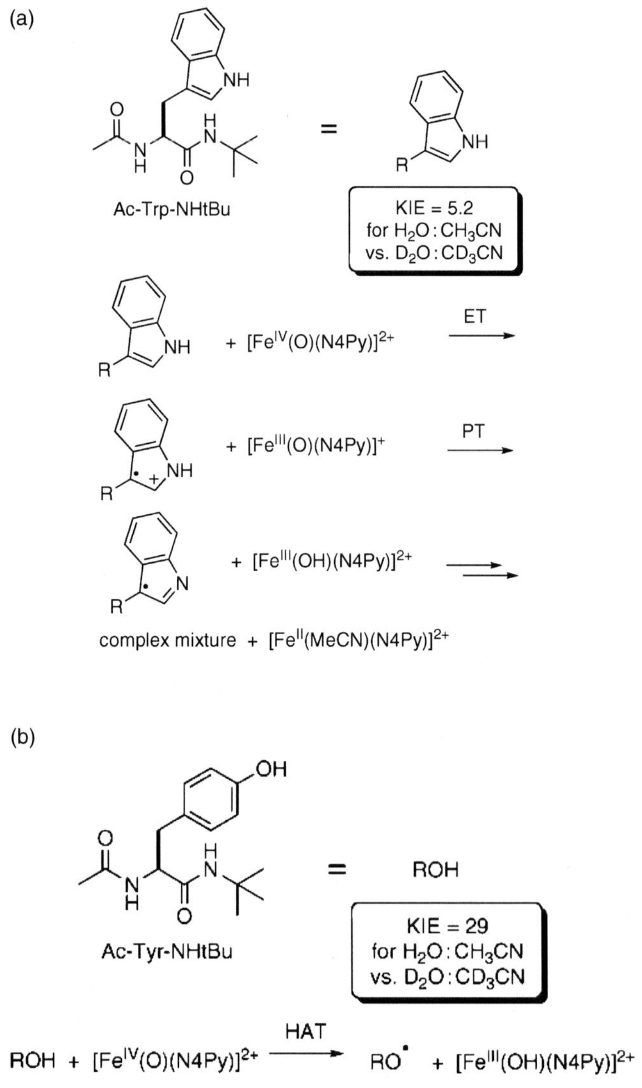

(a)

Ac-Trp-NHtBu

KIE = 5.2 for $H_2O:CH_3CN$ vs. $D_2O:CD_3CN$

$+ [Fe^{IV}(O)(N4Py)]^{2+}$ —ET→

$+ [Fe^{III}(O)(N4Py)]^{+}$ —PT→

$+ [Fe^{III}(OH)(N4Py)]^{2+}$ ⇀

complex mixture $+ [Fe^{II}(MeCN)(N4Py)]^{2+}$

(b)

Ac-Tyr-NHtBu = ROH

KIE = 29 for $H_2O:CH_3CN$ vs. $D_2O:CD_3CN$

$$ROH + [Fe^{IV}(O)(N4Py)]^{2+} \xrightarrow{HAT} RO^{\bullet} + [Fe^{III}(OH)(N4Py)]^{2+}$$

Figure 6.32. (A) Proposed mechanism for reaction of $[Fe^{IV}(O)(N4Py)]^{2+}$ with Ac-Trp-NHtBu (A) and Ac-Tyr-NHtBu (B) (adapted from Abouelatta et al. [334] with the permission of the American Chemical Society).

intermediate, $[Fe^{III}(OH)(N4Py)]^{2+}$, with a lower limit rate constant of 1×10^4/M/s (Eq. 6.99) [345]. The intermediate reacted with another molecule of GSH to yield a green intermediate, $[Fe^{III}(SG)(N4Py)]^{2+}$ (Eq. 6.100), which was characterized by UV–vis, EPR, and high-resolution time-of-flight mass spectrometry (HR-ToF-MS). The green intermediate could not be produced with other thiols. The value of $\varepsilon_{650} = 1870$/M/cm was calculated for the intermediate, and the proposed structure consists of a sulfur donor bound to the iron center of an N4Py complex. Reaction (6.100) followed a second-order process. The final step was the second-order decay of the green intermediate, homolysis of the Fe^{III}-SG bond, and dimerization of the two $GS^{\bullet}$ radicals to result in glutathione disulfide (GSSG) (Eq. 6.101):

$$[Fe^{IV}(O)(N4Py)]^{2+} + GSH \rightarrow [Fe^{III}(OH)(N4Py)]^{2+} + 1/2\ GSSG \qquad (6.99)$$

$$[Fe^{III}(OH)(N4Py)]^{2+} + GSH \rightarrow [Fe^{III}(SG)(N4Py)]^{2+} + H_2O\ k_{110} = 14.3/M/s \qquad (6.100)$$

$$2[Fe^{III}(SG)(N4Py)]^{2+} + H_2O \rightarrow [Fe^{III}(OH_2)(N4Py)]^{2+} + GSSG\ k_{111} = 1.68/M/s. \qquad (6.101)$$

Recently, the ferryl–peptide conjugate was synthesized using a ligand-dipeptide [351]. This conjugate was stable for more than 1 hour at room temperature. The decay of the ferryl–peptide conjugate, at a later time, was first order. The role of the functional group in the conjugated ferryl–peptide was also determined by the synthesis of several ester derivatives of the ferryl–peptide conjugate. Ester derivatives decayed at different rates, suggesting the role of a remote ester group to control the stability of the ferryl species. The mechanism of decomposition followed a hydrogen atom transfer pathway, which was supported by the KIE value of 4.5 and a slope (ρ) of −1.3 in the Hammett plot [351]. More recently, the study of oxidative inactivation of serine proteases trypsin and chymotrypsin by $[Fe^{IV}(O)(N4Py)]^{2+}$ and $[Fe^{IV}(O)(3CG\text{-}N4Py)]^{2+}$ (3CG-3-carbonguanidinium) has been performed [352]. The side chains residues those were involved in oxidation were Cys, Tyr, and Trp. The $[Fe^{IV}(O)(3CG\text{-}N4Py)]^{2+}$ was more effective in inactivating chymotrypsin than $[Fe^{IV}(O)(N4Py)]^{2+}$. A separate inactivating experiment using $[Fe^{II}(OH_2)(N4Py)]^{2+}$ or $[Fe^{II}(Cl)(3CG\text{-}N4Py)]^{2+}$ showed a role of ferryl species in inactiavting enzymes [352].

Recently, the role of metal ions in the oxidation carried out by a nonheme oxoiron(IV) complex, $[(TMC)Fe^{IV}(O)]^{2+}$ (TMC = 1,4,8,11-tetramethyl-1,4,8,11-tetraazacyclotetradecane), has been studied [353, 354]. The X-ray crystal structure characterization on the binding of the redox-inactive metal ion, Sc^{3+} to $[(TMC)Fe^{IV}(O)]^{2+}$ showed a structural distortion of the oxoiron(IV) moiety due to an oxo-Sc^{3+} interaction. This Lewis metal ion binding to the oxo atom can significantly alter the electron transfer behavior of a nonheme oxoiron(IV) complex [355]. For example, ferrocene reduces $[(TMC)Fe^{IV}(O)]^{2+}$ via a two-electron process in the presence of metal ions. Comparatively, only a one-electron reduction of $[(TMC)Fe^{IV}(O)]^{2+}$ by ferrocene occurs without the metal ions [206]. Thus, the binding of a positively charged metal ion to the oxo group of the nonheme oxoiron(IV) moiety facilitates further reduction. The results imply a role of such interactions in the oxygen-evolving center exists in the protein complex PSII [355]. The role of Sc^{3+} has also been studied in the sulfoxidation of thioanisoles by $[Fe^{IV}(O)(N4Py)]^{2+}$ [14]. The addition to Sc^{3+} enhanced the rate of sulfoxidation up to 10^2-fold and also switched the mechanism from the direct oxygen transfer to the Sc^{3+} ion-coupled electron transfer [14].

6.3.1.3 ***Ferrate(IV).*** In the solid phase, metalferrates(IV), $MFeO_3$ (alkali metals: Ba, Sr, Ag, and Pb), and orthoferrates(IV), M_2FeO_4 (alkali metals: Sr and Ba), have been synthesized by the thermal oxidation of two solid phases with an M:Fe ratio, corresponding to that in ferrate(IV) [356]. The crystals of Na_4FeO_4 are unstable under humid conditions and dissociate to Fe^{3+} and $Fe^{VI}O_4^{2-}$ in water (Eq. 6.102). The hydrolysis of Fe^{3+} then gives amorphous $Fe(OH)_3$ or FeO(OH) [357]:

$$3Na_4FeO_4 + 8H_2O \rightarrow 12Na^+ + Fe^{VI}O_4^{2-} + 2Fe^{3+} + 16OH^-. \quad (6.102)$$

Inorganic complexes of ferrate(IV) in basic solution have also been generated [358]. The Fe(IV) complexes with OH^- and $P_2O_7^{4-}$ in basic solution were produced by the oxidation of corresponding parent complexes with the OH/O^- radical (Eqs. 6.103 and 6.104):

$$Fe(OH)_4^- + OH/O \rightarrow FeO(OH)_n^{2-n} + H_2O/OH^- \quad k_{113} = 8.5\times10^7\,/M/s \quad (6.103)$$

$$[(P_2O_7)_2Fe^{3+}OH]^{6-} + OH \rightarrow [(P_2O_7)_2Fe^{IV}O]^{6-} + H_2O \quad k_{114} = 7.8\times10^7\,/M/s. \quad (6.104)$$

The spectrum of the ferrate(IV) complex is presented in Figure 6.24b. A peak at λ_{max} = 430 nm (ε = 1200/M/cm) was observed, similar to $FeO(OH)_n^{2-n}$ and $[(P_2O_7)_2Fe^{IV}O]^{6-}$ spectra. As the pH decreased, a blue shift in the 430-nm peak was observed. The $Fe(OH)_4^-$ species decayed mainly by a first-order process in 1 M NaOH ($k_{decay} \approx 2 \pm 1$/s at 25°C), and the pyrophosphate complex of iron(IV), formed at pH ≥ 10, was short lived ($t_{1/2}$ = 100–600 ms) [358]. The $[(P_2O_7)_2Fe^{IV}O]^{6-}$complex decayed by a second-order process with the formation of an Fe(III) pyrophosphate complex and molecular oxygen (Eq. 6.105) [358]:

$$2[(P_2O_7)_2Fe^{IV}O]^{6-} + 2H_2O \rightarrow 2[(P_2O_7)_2Fe^{III}OH]^{6-} + 1/2O_2. \quad (6.105)$$

Reactions of Fe(IV)–pyrophosphate complexes with divalent transition metal ions in pyrophosphate have been studied (reaction 6.116) (Table 6.4) [358]:

$$Fe(IV) + M(II) \rightarrow Fe(III) + M(III). \quad (6.106)$$

The Ni(II) and Cu(II) pyrophosphate complexes showed no reactivity, and hence, the upper limits of the rate constants are presented in Table 6.4. The rate constants are on the order of 10^6/M/s. The mechanism involves a rate-limiting inner-sphere association between M(II) pyrophosphate and $[(P_2O_7)_2\,Fe^{IV}O]^{6-}$ in addition to a rapid one-electron transfer [358].

6.3.1.4 Ferrate(V) and Ferrate(VI). Salts of sodium, potassium, and rubidium (Na_3FeO_4, K_3FeO_4, and Rb_3FeO_4) have been synthesized by heating alkali oxides [359, 360]. In later years, black crystals of K_3FeO_4 were synthesized by heating well-ground mixtures of K_2O_2 and $KFeO_2$ (1.78:1.00, Ag-tube) for several days at 350–470°C [361]. Heating a mixture of K_2FeO_4 and Na_2O at 450–600°C produced a mixture of Na_3FeO_4 and K_3FeO_4. The oxidation of a mixture of RbO_x or RbOH with Fe_2O_3 in a stream of O_2 at 600°C produced the salt with the composition of $Rb_3FeO_{4.4–4.7}$ [356, 362], where subsequent heating in a stream of N_2 at 350°C for a short time yielded Rb_3FeO_4. Comparatively, numerous salts of ferrate(VI) have been synthesized. Examples of these salts include Na_2FeO_4, K_2FeO_4, $BaFeO_4$, and $SrFeO_4$. Generally, three methods have been used to prepare ferrate(VI) compounds: (1) dry thermal synthesis, (2) wet chemical synthesis, and (3) electrochemical synthesis [363–366].

In solution, the reduction of ferrate(VI) by radicals generates ferrate(V). The radicals produced in pulse radiolysis are determined using the following scheme [367–369]:

$$H_2O \rightsquigarrow H(0.55), e^-_{aq}(2.65), OH(2.75), H_2O_2(0.72), H_2(0.45)I$$

$$N_2O + e^-_{aq} + H_2O \rightarrow OH + OH^- + N_2 \tag{6.107}$$

$$H + OH^- \rightarrow e^-_{aq} + H_2O \tag{6.108}$$

$$OH/O^- + ROH \rightarrow H_2O/OH^- + {}^{\bullet}ROH \tag{6.109}$$

$$Fe(VI) + {}^{\bullet}ROH \rightarrow Fe(V) + \text{product} \quad k_{120} = 9\times10^9\,/M/s\,(14). \tag{6.110}$$

In reaction (6.109), ROH is an alcohol (e.g., methanol, ethanol, isoproponol or *tert*-butanol) that reacts with an ${}^{\bullet}OH$ radical to form a simple carbon-centered radical, which reduces ferrate(VI) by a diffusion-controlled rate constant to produce ferrate(V). In a strong alkaline solution, ferrate(V) decays to a longer-lived transient ($t_{1/2} \approx$ seconds) via a first-order process. A relatively short lifetime ($k_{decay} = 4/s$) was obtained in 5M NaOH [367]. Ferrate(V) decayed by first-order kinetics in acidic media, while second-order kinetics followed in a moderately alkaline solution. The spectrum of ferrate(V) in alkaline solution is shown in Figure 6.24b. The spectrum has a maximum at 380nm ($\varepsilon_{380\,nm} = 1460/M/cm$), which undergoes a blue shift with decreasing pH. Ferrate(V) absorbs very strongly in the UV region ($\varepsilon_{270\,nm} \approx 5000/M/cm$) (Fig. 6.24B) [368]. Ferrate(VI) also absorbs strongly in the UV region with a shoulder between 275 and 320nm ($\varepsilon_{250\,nm} = 7000/M/cm$) (Fig. 6.24b) [368] in addition to an absorbance maximum at 510nm ($\varepsilon_{510\,nm} = 1150/M/cm$) (Fig. 6.24b). The spectra of ferrate(V) and ferrate(VI) are strongly influenced by pH.

6.3.2 Reactivity of Ferrate(V) and Ferrate(VI)

The reactivity of ferrate(VI) with a number of inorganic compounds has been performed to seek application of ferrate(VI) to remediate pollutants and to understand the mechanism of oxidation reactions [17, 19]. The correlations of the rate constants with one-electron and two-electron thermodynamic reduction potentials were able to distinguish one- and two-electron transfer steps [19]. A linear relationship was found between log k and the one-electron reduction potentials for iodide, cyanides, and superoxide, while oxy-compounds of nitrogen, sulfur, selenium, arsenic showed a linear relationship for two-electron reduction potentials [19]. Stoichiometric and products observed of oxidation reactions were consistent with steps obtained from the correlations [17]. Similar correlations were also performed with organosulfur compounds (sulfur-containing amino acids, aliphatic and aromatic thiols, and mercaptans) to understand oxygen transfer in oxidized products [20]. The ratios of ferrate(VI) to the various organosulfur compounds for the one-oxygen-atom transfer were 0.50 and 0.67 for Fe(II) and Fe(III) as final products, respectively. The oxidation of the compounds involved a $1-e^-$ transfer step from Fe(VI) to Fe(V), followed by $2-e^-$ transfer to Fe(III) as the reduced product (Fe(VI)$\rightarrow$Fe(V)$\rightarrow$Fe(III)). The $2-e^-$ transfer steps resulted in the formation of Fe(II) (Fe(VI)$\rightarrow$Fe(IV)$\rightarrow$Fe(II)). The schemes for oxygen transfer are presented in Equations (6.111)–(6.116):

Scheme 1

RS(O)H formation

$$HFeO_4^- + RSH \rightarrow HFeO_4^{2-} + RS^{\bullet+}H \qquad (6.111a)$$

$$HFeO_4^- + RS^{\bullet+}H + 2OH^- \rightarrow HFeO_4^{2-} + RS(O)H + H_2O \qquad (6.111b)$$

$$2HFeO_4^{2-} + 2RSH + 4H_2O \rightarrow 2Fe(OH)_3 + 2RS(O)H + 4OH^- \qquad (6.111c)$$

$RS(O_2)H$ formation

$$HFeO_4^- + RS(O)H \rightarrow HFeO_4^{2-} + RS^{\bullet+}(O)H \qquad (6.112a)$$

$$HFeO_4^- + RS^{\bullet+}(O)H + 2OH^- \rightarrow HFeO_4^{2-} + RS(O_2)H \qquad (6.112b)$$

$$2HFeO_4^{2-} + 2RS(O)H + 4H_2O \rightarrow 2Fe(OH)_3 + 2RS(O_2)H + 4OH^- \qquad (6.112c)$$

$RS(O_3)H$ formation

$$HFeO_4^- + RS(O_2)H \rightarrow HFeO_4^{2-} + RS^{\bullet+}(O_2)H \qquad (6.113a)$$

$$HFeO_4^- + RS^{\bullet+}(O_2)H + 2OH^- \rightarrow HFeO_4^{2-} + RS(O_3)H \qquad (6.113b)$$

$$2HFeO_4^{2-} + 2RS(O_2)H + 4H_2O \rightarrow 2Fe(OH)_3 + 2RS(O_3)H + 4OH^- \qquad (6.113c)$$

$$2HFeO_4^- + 3RNHCSO_3H + 4OH^- \rightarrow\rightarrow\rightarrow 2Fe(OH)_3 + 3SO_4^{2-} + 3RCONH_2 \qquad (6.114)$$

Scheme 2

RS(O)H formation

$$HFeO_4^- + RSH + 2OH^- \rightarrow HFeO_4^{3-} + RS(O)H + H_2O \quad (6.115a)$$

$$HFeO_4^{3-} + RSH + 2H_2O \rightarrow Fe(OH)_2 + RS(O)H + 3OH^- \quad (6.115b)$$

$RS(O_2)H$ formation

$$HFeO_4^- + RS(O)H + 2OH^- \rightarrow HFeO_4^{3-} + RS(O_2)H + H_2O \quad (6.116a)$$

$$HFeO_4^{3-} + RS(O)H + 2H_2O \rightarrow Fe(OH)_2 + RS(O_2)H + 3OH^-. \quad (6.116b)$$

Again, conclusions drawn from the correlations agreed with the experimentally determined stoichiometries and products of the reactions, for example, the oxidation of Cys and Met by ferrate(VI), which formed Fe(II) and Fe(III) as the reduced products of ferrate(VI), respectively (Eqs. 6.117 and 6.118):

$$HFeO_4^- + HSCH_2CH(NH_3^+)COO^- + H_2O$$
$$\rightarrow Fe(OH)_2 + HS(O_2)CH_2CH(NH_3^+)COO^- + OH^- \quad (6.117)$$

$$2HFeO_4^- + 3RSR' + 3H_2O \rightarrow 2Fe(OH)_3 + 3RS(O)R' + 2OH^-. \quad (6.118)$$

The oxidation of Met thus occurs through Scheme 1, while Scheme 2 is applied to the oxidation of Cys by ferrate(VI).

Besides inorganic and sulfur-containing compounds, ferrate(VI) also efficiently oxidizes emerging contaminants such as estrogens and pharmaceuticals [309, 370–373]. The next section describes the reactivity of ferrate(VI) with nitrogen-containing compounds including amino acids.

6.3.2.1 Amines. The reactions of ferrate(VI) with substrates are mostly second order, that is, first order in the total ferrate(VI) concentration ($[Fe(VI)]_{tot}$) and first order in the total concentration of the substrate ($[S]_{tot}$):

$$-d[Fe(VI)]/dt = k_{app}[Fe(VI)]_{tot}[S]_{tot}, \quad (6.119)$$

where k_{app} represents the apparent second-order rate constants. The values of k_{app} have been determined by several substrates including sulfur- and nitrogen-containing compounds and emerging contaminants in the environment [18, 18, 371]. Figure 6.33 shows the structure of organic amine substrates studied by ferrate(VI). The reactivity of substrates with ferrate(VI) at pH 7.0 is given in Table 6.6. Determination of k_{app} ranged from 9.5×10^{-1}/M/s to 3.3×10^2/M/s for the aliphatic amines. Dimethyldithiocarbamate (DMDC), a sulfur-containing substrate, had the highest reactivity (Table 6.6). The aromatic amine substrates showed relatively higher reactivity than most of the aliphatic amines ($k_{app} = 2.8 \times 10^2 - 1.6 \times 10^5$). The fate of carbon and nitrogen in the oxidation of monomethylamine (MMA), dimethylamine (DMA), and trimethylamine

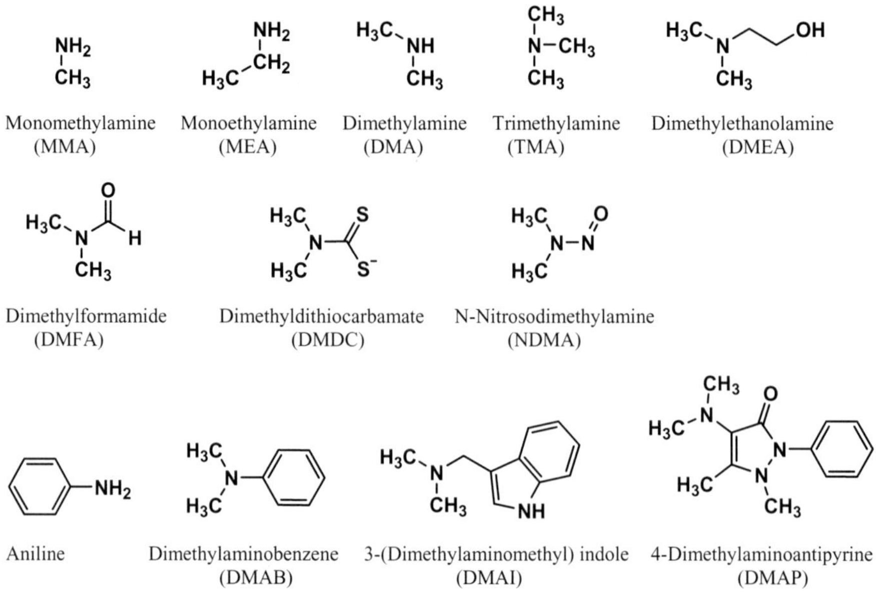

Monomethylamine (MMA) | Monoethylamine (MEA) | Dimethylamine (DMA) | Trimethylamine (TMA) | Dimethylethanolamine (DMEA)

Dimethylformamide (DMFA) | Dimethyldithiocarbamate (DMDC) | N-Nitrosodimethylamine (NDMA)

Aniline | Dimethylaminobenzene (DMAB) | 3-(Dimethylaminomethyl) indole (DMAI) | 4-Dimethylaminoantipyrine (DMAP)

Figure 6.33. Structures of amines.

(TMA) by ferrate(VI) has been examined in detail [305]. Bicarbonate, cyanate, and nitrogen were the products of the oxidation of MMA. The oxidation of DMA and TMA resulted in formic and nitrogen as the major products of the reactions. Fe(IV) was suggested as the reaction intermediate (Eqs. 6.120–6.122):

$$FeO_4^{2-} + CH_3NH_2 \rightarrow Fe(IV) + CH_2{=}NH + H_2O \tag{6.120}$$

$$FeO_4^{2-} + CH_2{=}NH \rightarrow Fe(IV) + HCO\text{-}NH_2 \tag{6.121}$$

$$FeO_4^{2-} + HCO\text{-}NH_2 \rightarrow Fe(IV) + NCO^- + H_2O. \tag{6.122}$$

The product studies on the oxidation of aniline have been performed [374–376]. In a previous study [375], both azobenzene and nitrobenzene products were demonstrated. When an excess molar amount of ferrate(VI) to aniline was used, nitrobenzene was the product of the reaction (Eq. 6.123), However, in excess aniline, azobenzene was formed (Eq. 6.124):

$$FeO_4^{2-} + C_5H_5NH_2 \rightarrow Fe(II) + C_5H_5NO_2 \tag{6.123}$$

$$FeO_4^{2-} + C_5H_5NH_2 \rightarrow Fe(II) + C_5H_5N{=}NC_5H_5. \tag{6.124}$$

In recent years, spin-trap EPR measurements were performed using excess amounts of aniline to ferrate(VI) [376]. A free radical mechanism was

TABLE 6.6. Second-Order Rate Constants, *k* (/M/s) of Amines at pH 7.0 and 25°C

Compound	k (/M/s)	Reference
MMA	4.0×10^{1}	[390]
DMA	3.3×10^{2}	[390]
DMA	8.9×10^{0}	[305]
TMA	1.6×10^{1}	[390]
DMAP	9.5×10^{-1}	[305]
DMEA	6.4×10^{0}	[305]
DMFA	5.5×10^{-1}	[305]
DMDC	$>1.5 \times 10^{6}$	[305]
NDMA	9.8×10^{-1}	[305]
Aniline	6.6×10^{3}	[306]
DMAB	4.4×10^{5}	[305]
DMAI	2.8×10^{2}	[305]
DMAP	1.6×10^{5}	[305]

proposed based on the EPR signal in the reaction between ferrate(VI) and aniline to produce azobenzene. Recently, a study under similar experimental conditions did not confirm the free radical-based mechanism, and the formation of an imidoiron(VI) intermediate before the formation of *cis*-azobenzene was suggested (Eq. 6.125) [376]:

$$FeO_4^{2-} \xrightarrow[\text{milliseconds-seconds}]{C_6H_5NH_2} \text{imidoiron(VI) intermediate} \xrightarrow[\text{seconds-minutes}]{C_6H_5NH_2} cis\text{-azobenzene} + Fe(II). \quad (6.125)$$

Similar steps have also been proposed in the oxidation of *para*-substituted anilines [376]. Significantly, the oxidation of *ortho*-substituted anilines proceeds via a one-electron step, involving a free radical mechanism [376].

6.3.2.2 Amino Acids. The oxidation of amino acids by ferrate(VI) and ferrate(V) under anaerobic and alkaline conditions has been studied [377]. The rates of the reactions were determined as a function of pH and ranged from 1.0×10^{1}/M/s to 1.5×10^{3}/M/s and 1.0×10^{5}/M/s to 1.5×10^{7}/M/s for the oxidation by ferrate(VI) and ferrate(V), respectively. Linear Taft type plots of the rate constants versus $\log(K/K_0)$ were constructed (Fig. 6.34) [377]. In the plots, K_0 is the ionization constant for the unsubstituted amino acid Gly ($H{-}CH(NH_3^+)COO^-$), and K's are the corresponding dissociation constants of amino acids, in which the α-H atom has been replaced by different R groups ($R{-}CH(NH_3^+)COO^-$). The plots suggest ferrate(VI) and ferrate(V) react preferentially with an α–$CH(NH_3^+)$ group of the amino acids. The attack of ferrate(V) on either nitrogen or the α-carbon atoms of the amino acids was further explored by performing the reactivity of ferrate(V) with a methyl

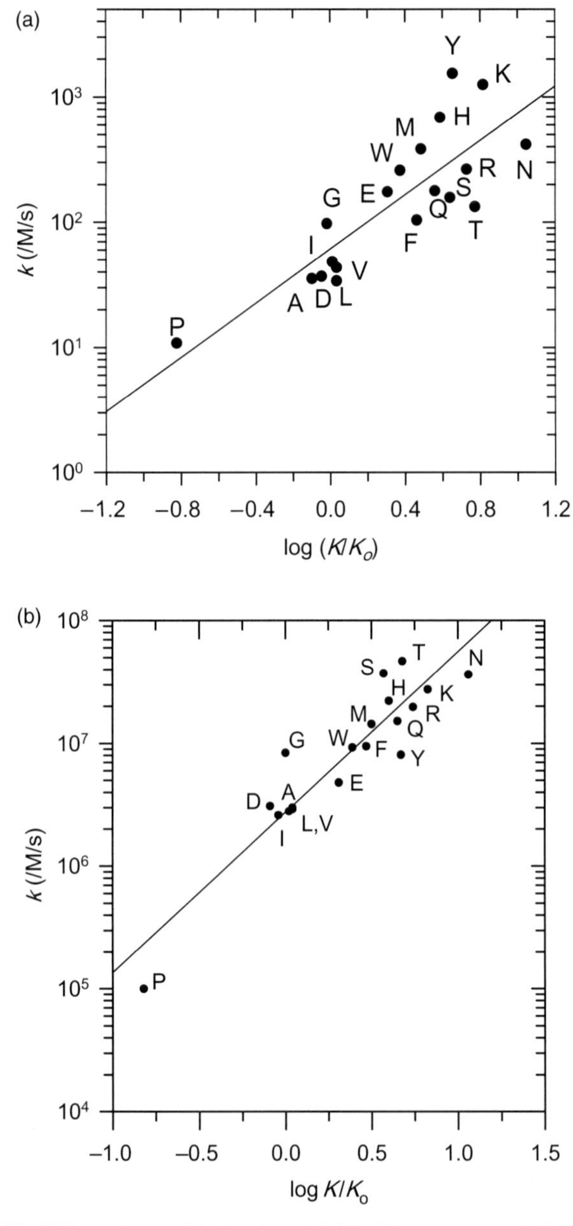

Figure 6.34. k(Fe(VI)+amino acids)) (a) and k(Fe(V)+amino acids)) (b) as a function of $\log(K/K_0)$. K_0 is the ionization constant for the unsubstituted amino acid Gly (H–CH(NH_3^+)COO^-), whereas K's are the corresponding dissociation constants of amino acids, in which the α–H atom has been replaced by different R groups (R–CH(NH_3^+)COO^-). A, Ala; R, Arg; N, Asn; D, Asp; E, Glu; Q, Gln; G, Gly; H, His; I, Ile; L, Leu; K, Lys' M, Met; F, Phe; P, Pro; S, Ser; T, Thr; W, Trp; Y, Tye; V, Val (adapted from Sharma and Bielski [377] with the permission of the American Chemical Society).

TABLE 6.7. Observed Rate Constants for the Reactions of Methyl-Substituted Amino Acids (10 mM) with Ferrate(V) at pH 8.8 and 23°C in 0.025 M Phosphate

Compound	Structure	k (/M/s)
Gly	H_2CNH_2COOH	$3.0 \pm 1.0 \times 10^5$
N-Methylglycine	$H_2CNH(CH_3)COOH$	$1.2 \pm 0.3 \times 10^5$
N-Dimethylglycine	$H_2CN(CH_3)_2COOH$	$<5.0 \times 10^3$ (estimated upper limit)
α-Aminobutyric	$(CH_3)_2CNH_2COOH$	$1.2 \pm 0.3 \times 10^5$
β-Aminobutyric	$HOOCCH_2CH(CH_3)NH_2$	$7.0 \pm 0.2 \times 10^4$

Adapted from Rush and Bielski [382] with the permission of Harwood Academic Publishers GmbH).

substitution at pH 8.8 (Table 6.7). Dimethylglycine showed no reactivity with ferrate(V). Other substituted molecules were only slightly less reactive than Gly. The comparative reactivity suggests that the attack of ferrate(V) was most likely at the nitrogen atom of the amino acids. A recent study showed the simultaneous formation of acetate, carbon dioxide, and ammonia for the oxidation of Gly by ferrate(VI). This indicates the possibility of an attack on nitrogen or the α-carbon atoms of Gly by the FeO_4^{2-} molecule to explain the observed products [378].

In the presence of excess ferrate(VI) and a ^{60}Co gamma ray source under steady-state conditions, a chain mechanism has been proposed for the oxidation of amino acids (AAs) (Eqs. 6.126–6.131) [377]:

Initiation and propagation steps

$$FeO_4^{2-} + AA^{\bullet} \rightarrow FeO_4^{3-} + \text{products} \tag{6.126}$$

$$FeO_4^{3-} + RCH(NH_3^+)COO^- \rightarrow FeO_4^{4-} + AA^{\bullet} \tag{6.127}$$

$$FeO_4^{4-} + RCH(NH_3^+)COO^- \rightarrow Fe^{3+} + AA^{\bullet} \tag{6.128}$$

Termination steps

$$FeO_4^{3-} + AA^{\bullet} \rightarrow FeO_4^{4-} + \text{products} \tag{6.129}$$

$$FeO_4^{4-} + AA^{\bullet} \rightarrow Fe^{3+} + \text{products} \tag{6.130}$$

$$Fe^{3+} + AA^{\bullet} \rightarrow Fe^{2+} + \text{products.} \tag{6.131}$$

Radical–radical reactions as well as reactions between similar ferrate species are other possible termination steps. The rates of the individual initiation step of the mechanism have been determined independently [379]. The second-order rate constants for the reactivity of ferrate(VI) with radicals are 1.4×10^9/M/s, 1.2×10^9/M/s, and 3.3×10^8/M/s at pH 12.4 and 24°C for Gly, Ala, and Asp, respectively. Ferrate(V) and ferrate(IV) have shown three to five orders of magnitude higher reactivity than ferrate(VI). The rate constants for ferrate(V) with radicals are thus expected to be much greater than ferrate(VI).

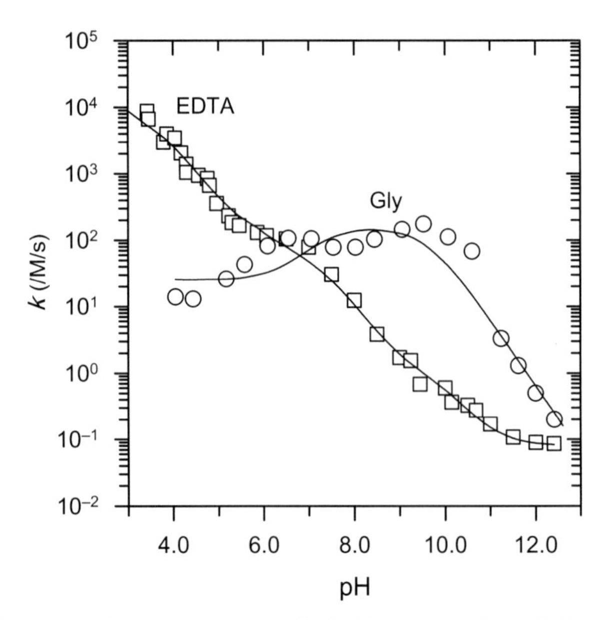

Figure 6.35. Second-order rate constants (k, /M/s) as a function of pH at 25°C for the oxidation of Gly and EDTA by Fe(VI). Solid lines represent best-fitted curves (adapted from Noorhasan et al. [378] with the permission of Elsvier, Inc.).

The reaction of ferrate(VI) with Gly in the acidic to basic pH range has been recently performed [378]. Generally, the rates decreased with a decrease in pH, particularly in alkaline media (Fig. 6.35). However, the reactivity with Gly showed only a small variation in the rates at pH values between 6.0 and 8.0. Also, an increase was observed with an increase in pH from 4.0 to 6.0. This dependence is unusual because ferrate(VI) is stronger upon protonation [370, 380]. For example, the oxidation of EDTA had a decrease in rate with an increase in pH (Fig. 6.35). Significantly, the oxidation rate of EDTA (3° amine, Fig. 6.36) was less than that of Gly in a strong alkaline medium, but the trend reversed in an acidic medium (Fig. 6.35). The pH dependence is usually related to the protonation of ferrate(VI) and substrates [381]. The equilibrium reactions of ferrate(VI) and Gly are given in Table 6.8. Based on the dissociation constants (Table 6.8), the calculated fractions of various ferrate(VI) and Gly as a function of pH are shown in Figures 6.37 and 6.38.

The pH dependence of k for the reaction of Fe(VI) with substrates can be quantitatively modeled by Equation (6.132):

$$k[\mathrm{Fe(VI)}]_{\mathrm{tot}}[\mathrm{S}]_{\mathrm{tot}} = \Sigma k_{\mathrm{ij}}\alpha_{\mathrm{ij}}\beta_{\mathrm{ij}}[\mathrm{Fe(VI)}]_{\mathrm{tot}}[\mathrm{S}]_{\mathrm{tot}}, \tag{6.132}$$

$$i = 1, 2, 3, 4$$

$$j = 1, 2, 3, 4, 5$$

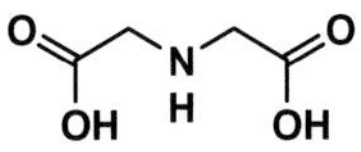

Iminodiacetate (IDA)

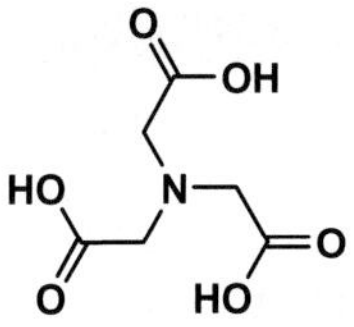

Nitrilotriacetate (NTA)

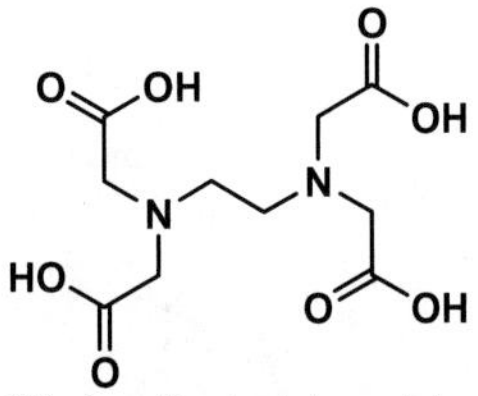

Ethylenediaminetetraacetate (EDTA)

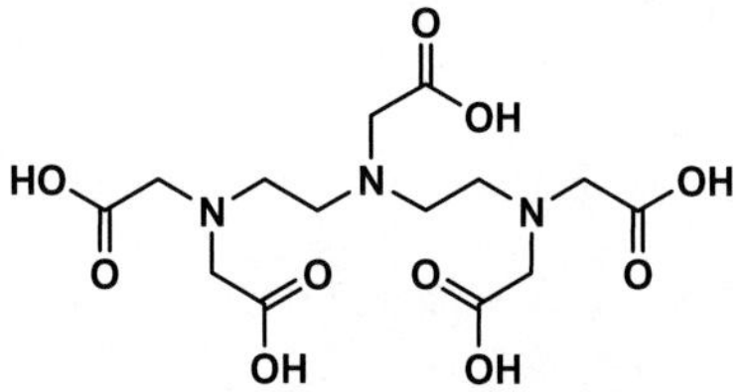

Diethylenetriaminepentaacetate (DTPA)

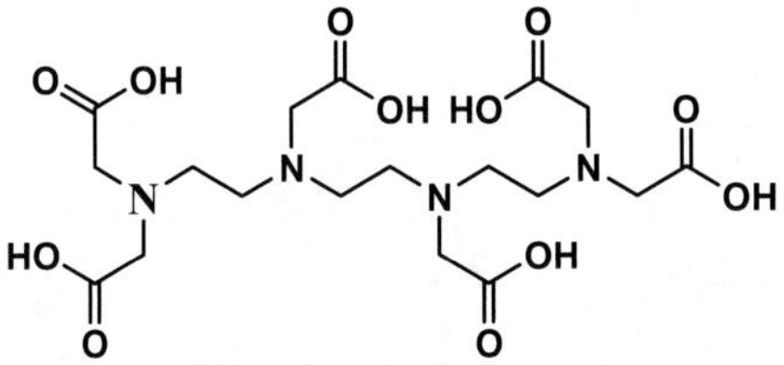

Triethylenetriaminehexaacetate (TTHA)

Figure 6.36. Structures of aminopolycarboxylates.

where $[Fe(VI)]_{tot} = [H_3FeO_4^+] + [H_2FeO_4] + [HFeO_4^-] + [FeO_4^{2-}]$; α_{ij} and β_{ij} represent the respective species distribution coefficients for Fe(VI) and the substrate; i and j represent each of the four species of Fe(VI) and each species of the substrate, and k_{ij} is the species-specific second-order rate constant for the reaction between the Fe(VI) species, i, with the substrate species, j. In the pH range studied for Gly, the four forms of Fe(VI) ($H_3FeO_4^+$, H_2FeO_4, $HFeO_4^-$, and FeO_4^{2-}) could potentially react with the three forms of Gly (H_2Gly^+, HGly, and Gly^-). Of the possible 12 reactions, it has been determined that only 3 reactions (Eqs. 6.133–6.135 [model I] or Eqs. 6.133, 6.134, 6.136 [model II]) were necessary to fit the experimental k values (see Fig. 6.35):

$$\begin{gathered} H_2FeO_4 + {}^+H_3N\text{-}CH_2\text{-}COO^- \rightarrow Fe(OH)_3 + \text{product(s)} \\ k_{143} = 3.0(\pm 0.6) \times 10^1\,/\text{M/s} \end{gathered} \tag{6.133}$$

$$\begin{gathered} HFeO_4^- + {}^+H_3N\text{-}CH_2\text{-}COO^- \rightarrow Fe(OH)_3 + \text{product(s)} \\ k_{144} = 2.5(\pm 0.3) \times 10^1\,/\text{M/s} \end{gathered} \tag{6.134}$$

TABLE 6.8. Dissociation Constants of Ferrate(VI), Ferrate(V), Gly, Glycylglycine, and EDTA at 25°C

Reaction	Constant
Ferrate(VI)	
$H_3FeO_4^+ + H_2O \rightleftarrows H^+ + H_2FeO_4$	$pK_{a1} = 1.6$ [391]
$H_2FeO_4 + H_2O \rightleftharpoons H^+ + HFeO_4^-$	$pK_{a2} = 3.5$ [392]
$HFeO_4^- + H_2O \rightleftharpoons H^+ + FeO_4^{2-}$	$pK_{a3} = 7.3$ [393]
Ferrate(V) [394]	
$H_3FeO_4 \rightleftharpoons H^+ + H_2FeO_4^-$	$5.5 \leq pK_{a1} \geq 6.5$
$H_2FeO_4^- \rightleftharpoons H^+ + HFeO_4^{2-}$	$pK_{a2} = 7.2$
$HFeO_4^{2-} \rightleftharpoons H^+ + FeO_4^{3-}$	$pK_{a3} = 10.1$
Glycine	
$^+H_3N\text{-}CH_2\text{-}COOH + H_2O \rightleftarrows {}^+H_3N\text{-}CH_2\text{-}COO^- + H_3O^+$ (H_2Gly^+) $(HGly)$	$pK_1 = 2.3$
$^+H_3N\text{-}CH_2\text{-}COO^- + H_2O \rightleftarrows H_2N\text{-}CH_2\text{-}COO^- + H_3O^+$ $(HGly)$ (Gly^-)	$pK_2 = 9.6$
Glycylglycine	
$^+H_3NCONHCH_2COOH + H_2O \rightleftarrows {}^+H_3NCONHCH_2COO^- + H_3O^+$	$pK_1 = 3.1$
$^+H_3NCONHCH_2COO^- + H_2O \rightleftarrows H_2NCONHCH_2COO^- + H_3O^+$	$pK_2 = 8.2$
EDTA	
$H_4Y + H_2O \rightleftarrows H_3O^+ + H_3Y^-$	$pK_3 = 2.0$
$H_3Y^- + H_2O \rightleftarrows H_3O^+ + H_2Y^{2-}$	$pK_4 = 2.7$
$H_2Y^{2-} + H_2O \rightleftarrows H_3O^+ + HY^{3-}$	$pK_5 = 6.2$
$HY^{3-} + H_2O \rightleftarrows H_3O^+ + Y^{4-}$	$pK_6 = 10.3$

$$HFeO_4^- + H_2N\text{-}CH_2\text{-}COO^- \rightarrow Fe(OH)_3 + \text{product(s)} \quad k_{145} = 3.0(\pm 0.1) \times 10^4\,/\text{Ms} \tag{6.135}$$

$$FeO_4^{2-} + {}^+H_3N\text{-}CH_2\text{-}COO^- \rightarrow Fe(OH)_3 + \text{product(s)} \quad k_{146} = 1.5(\pm 0.1) \times 10^2\,/\text{M/s}. \tag{6.136}$$

Reactions (6.135) and (6.136) introduce the proton ambiguity in the reactivity of ferrate(VI) with Gly. Thus, the experimental values of k fit reasonably well (a solid line in Fig. 6.35) by considering the set of reactions in model I or model II.

The reactivity of ferrate(VI) with the dipeptide, glycylglycine (Gly-Gly), has also been studied [378]. The rate law for the oxidation of Gly-Gly was first order with respect to each reactant in the acidic to alkaline pH range. The rate of the oxidation of Gly-Gly was slower than that of Gly at pH > 9.0 (Table 6.9), while the trend is reversed at pH < 9.0. The differences observed in the pK_a values of the two substrates may explain the differences in reactivity (see Table 6.8). The amine moiety of Gly-Gly has a pK_a of 8.2, 1.4 units lower than

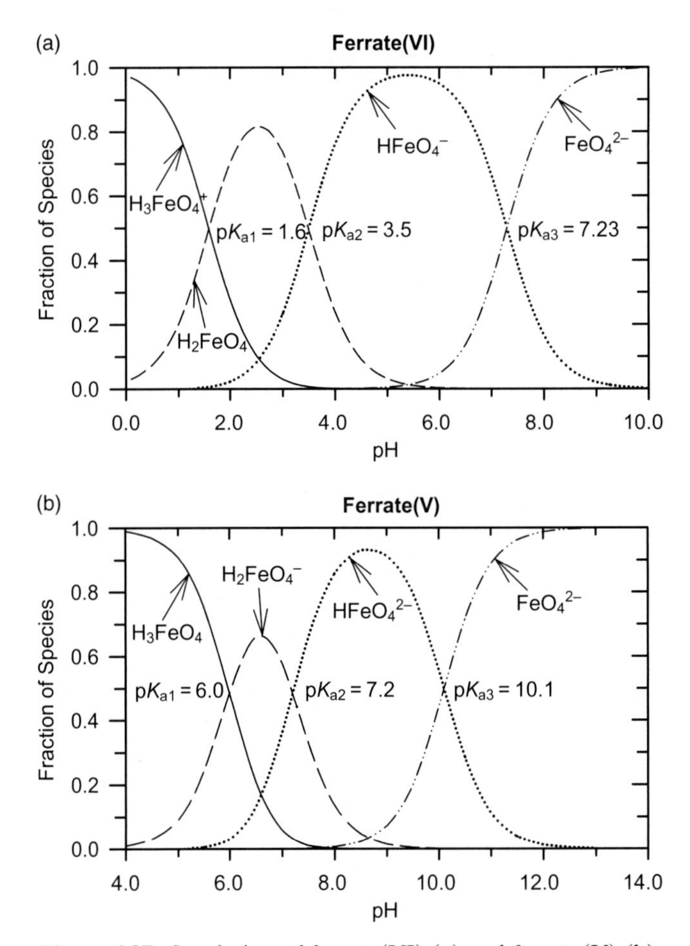

Figure 6.37. Speciation of ferrate(VI) (a) and ferrate(V) (b).

that of Gly. If the deprotonated amine moieties of Gly-Gly and Gly determine the reactivity with Fe(VI), the observed difference in reactivity would be expected. The maximum reactivity of Gly-Gly is higher than that of Gly with Fe(VI). The N-terminal of Gly-Gly is less electrophilic than the amino nitrogen of Gly, suggesting cause of the observed differences in reactivity.

The pH dependence of the reaction of ferrate(V) with amino acids has also been studied [382]. The rate-determining step was the two-electron reduction of ferrate(V) to Fe(III). The subsequent reaction yielded deamination products. The reaction rates as a function of pH showed a bell shape, which resulted in a maximum between pH 9 and 10 (Fig. 6.39). The maxima are in the protonation equilibria of both ferrate(V) and amino acids (Tables 6.8 and 6.10). The speciation of ferrate(V) is shown in Figure 6.37. Based on the species of

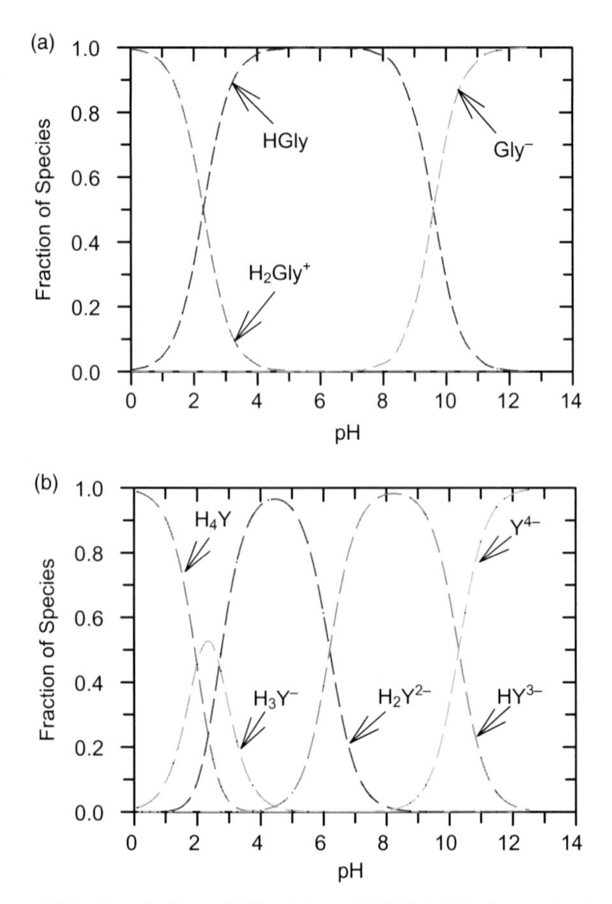

Figure 6.38. Speciation of Gly (a) and EDTA (b). See color insert.

ferrate(V) and the amino acids in the studied pH range, the following set of reactions (Eqs. 6.137–6.141) were considered to justify the pH dependence in Figure 6.39:

$$H_2FeO_4^- + RCH(NH_3^+)COO^- \rightarrow Fe(III) + \text{products} \quad (6.137)$$

$$HFeO_4^{2-} + RCH(NH_3^+)COO^- \rightarrow Fe(III) + \text{products} \quad (6.138)$$

$$FeO_4^{3-} + RCH(NH_3^+)COO^- \rightarrow Fe(III) + \text{products} \quad (6.139)$$

$$HFeO_4^{2-} + RCH(NH_2)COO^- \rightarrow Fe(III) + \text{products} \quad (6.140)$$

$$FeO_4^{3-} + RCH(NH_2)COO^- \rightarrow Fe(III) + \text{products.} \quad (6.141)$$

The contribution of reaction (6.139) was negligible to the overall rates of the oxidation of amino acids by ferrate(V) (Table 6.10). The order of reactivity

TABLE 6.9. Rate Constants for the Reaction of Fe(VI) and Fe(V) with APC at Different pH Values

	k (Fe^{VI}+APC) (/M/s)			k (Fe^{V}+APC) (/M/s)	$\frac{k(Fe(V)+APC)}{k(Fe(VI)+APC)}$
APC	pH 7.0	pH 9.0	pH 12.5	pH 12.5	pH 12.5
Gly	$1.00 \pm 0.12 \times 10^2$	$1.10 \pm 0.12 \times 10^2$	$1.6 \pm 0.1 \times 10^{-1}$	$(1.4 \pm 0.1) \times 10^4$	9.3×10^4
Gly-Gly	8.22×10^2 (3.0×10^1)[a]	1.51×10^2	–	–	–
IDA	2.3×10^2	$1.89 \pm 0.12 \times 10^1$	$3.8 \pm 0.5 \times 10^{-2}$	$4.0 \pm 0.3 \times 10^3$	1.1×10^3
NTA	7.00×10^0	$7.10 \pm 0.50 \times 10^{-1}$	$\leq 4.4 \times 10^{-2}$	$\leq 1.6 \times 10^2$	
EDTA	1.0×10^2	$1.72 \pm 0.08 \times 10^0$	$8.6 \pm 0.8 \times 10^{-2}$	$2.7 \pm 0.1 \times 10^2$	3.1×10^3
DTPA	–	$2.90 \pm 0.14 \times 10^0$	$1.7 \pm 0.1 \times 10^{-1}$	$2.6 \pm 0.2 \times 10^2$	1.5×10^3
TTHA	–	$3.60 \pm 0.12 \times 10^0$	$2.72 \pm 0.30 \times 10^{-1}$		

[a] pH 4.0.

Data were taken from References 378 and 385.

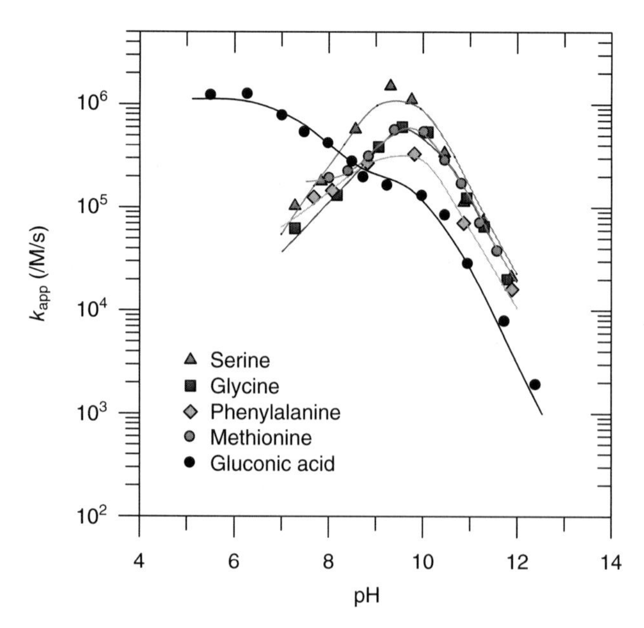

Figure 6.39. Second-order rate constants (k, /M/s) as a function of pH at 22°C for the oxidation of carboxylic acids by ferrate(V). See color insert.

TABLE 6.10. pK_a and Rate Constants (/M/s) for the Reaction of Ferrate(V) Species with Amino Acids in 0.025 M Phosphate at 23°C

	Gly	Ser	Met	Phe
pK_a	9.60	9.06	9.05	9.15
$k_{138}(H_2FeO_4^- + RCH(NH_3^+)COO^-)$	$\leq 3.0 \times 10^4$	$\leq 3.0 \times 10^4$	3.0×10^4	$\leq 3.0 \times 10^4$
$k_{139}(HFeO_4^{2-} + RCH(NH_3^+)COO^-)$	1.6×10^5	3.0×10^5	2.0×10^5	2.1×10^5
$k_{140}(FeO_4^{3-} + RCH(NH_3^+)COO^-)$	Estimated to be unimportant to the overall reaction			
$k_{141}(HFeO_4^{2-} + RCH(NH_2)COO^-)$	4.5×10^6	8.0×10^6	2.8×10^6	2.6×10^6
$k_{142}(FeO_4^{3-} + RCH(NH_2)COO^-)$	$<10^3$	$<10^3$	$<10^3$	$<10^3$

Data taken from Rush and Bielski [382].

of the ferrate(V) species was $H_2FeO_4^- < HFeO_4^{2-} > FeO_4^{3-}$ (Table 6.10). The higher reactivity of $HFeO_4^{2-}$ than FeO_4^{3-} could be explained by considering (1) the oxygen atoms of $HFeO_4^{2-}$ have a strong free radical character and (2) $HFeO_4^{2-}$ is substitutionally more labile than FeO_4^{3-}, allowing $HFeO_4^{2-}$ to expand its coordination sphere. The latter may occur when the amine group attack on the oxide ligand of ferrate(V) results in a ferrate(V)–AA complex. The complex formation may account for the two-electron oxidation of amino acids by ferrate(V). The second-order self-decay of the $HFeO_4^{2-}$ species, which has a rate constant of ~10^7/M/s, is similar to the rate constant of $HFeO_4^{2-}$ with the amino acid anion (Table 6.10). The lower reactivity of the zwitterion species, $RCH(NH_3^+)COO^-$, than the anionic species, $RCH(NH_2)COO^-$, is similar to the reactivity of $^{\bullet}OH$ with amino acids. The results indicate that the Fe^V=O system may abstract a hydrogen atom from the nitrogen within the amino acids. The pH dependence of the rate suggests the attack of ferrate(V) is also at the nitrogen site rather than at the sulfur site of Met. In contrast, the reactivity of $^{\bullet}OH$ with Met undergoes an attack at the sulfur site at the near diffusion-controlled rate [344].

In addition to amino acids, the reactivity of ferrate(V) with a number of carboxylic acids has been studied [382, 383]. The rates for the oxidation of carboxylic acids by Fe(V) depend on the nature of the substituent group at the α-carbon atom of the acids and decrease in the order α-C-NH_2 > α-C-OH > α-C-H [382, 383]. It has been suggested Fe(V) oxidizes amino acids and carboxylic acids by a two-electron step (Fe(V) + amino acids → Fe(III) + NH_3 + α-keto acids). For example, the oxidation of aspartic acid by Fe(V) yielded Fe(III), ammonia, and oxaloacetate. The steps of the mechanism are expressed in reactions (6.142) and (6.143):

$$HFe^VO_4^{2-} + {}^-OOCCH_2CH(NH_2)COO^- \rightarrow Fe^{III} + {}^-OOCCH_2C(=NH)COO^- \quad (6.142)$$

$${}^-OOCCH_2C(=NH)COO^- + H_2O \rightarrow {}^-OOCCH_2C(=O)COO^- + NH_3. \quad (6.143)$$

6.3.2.3 Aminopolycarboxylates. The kinetics of the oxidation of APCs by ferrate(VI) and ferrate(V) were conducted in detail [384, 385]. The structures of the studied APCs are given in Figure 6.36. Among the studied compounds, IDA is a secondary amine (2°), and NTA, EDTA, DTPA, and TTHA are tertiary amines (3°). The reactions followed a second-order rate law, first order with respect to the concentration of each reactant. The reactivity of ferrate(V) was three to five orders of magnitude more reactive than ferrate(VI) (Table 6.9). Both ferrate(VI) and ferrate(V) showed a decrease in the reaction rates with an increase in pH in the alkaline medium. Significantly, the rate constants demonstrated the primary amine had the highest rate than the secondary and tertiary amines, with the order of reactivity primary > secondary > primary (Table 6.9). The results suggest the oxidation occurred at the N–C bond because, if the oxidation occurred between the acetate carbons, a distinct separation in the reaction rates would not have been observed [386]. Among the tertiary amines, the rates increased with an increase in the $-CH_2COOH$ group, and thus, the rates increased in the order NTA < EDTA < DTPA < TTHA.

A spectral study using premix pulse radiolysis experiments was performed to determine whether Fe(IV) is formed in the oxidation of APC by Fe(V) [385]. The spectra obtained for IDA and EDTA are shown in Figure 6.40 [385].

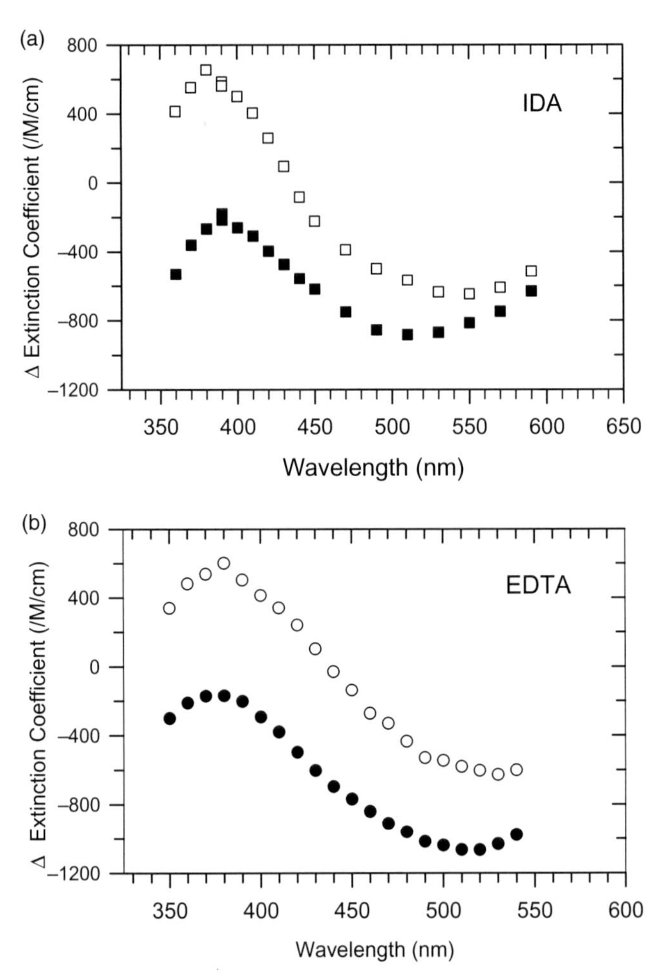

Figure 6.40. Δ Extinction coefficient versus wavelength measured at 0.01 ms (open circles: −Fe(VI) + Fe(V)) and 500 ms (filled circles: −Fe(VI)] (adapted from Noorhasan et al. [385] with the permission of Elsevier Inc.).

The characteristic spectrum of Fe(IV) (see Fig. 6.24) was not observed in these experiments, demonstrating the reaction of Fe(V) with APC proceeded via a concerted two-electron oxidation, which converted ferrate(V) to Fe(III) (Eq. 6.144):

$$\text{Fe(V)} + \text{APC} \rightarrow \text{Fe(III)} + \text{APC(oxidized)}. \quad (6.144)$$

The reactions of ferrate(VI) and ferrate(V) with EDTA as a function of pH have been studied (Figs. 6.35 and 6.41) [384, 385]. The dissociation constants of the EDTA species suggest there are five different species of EDTA

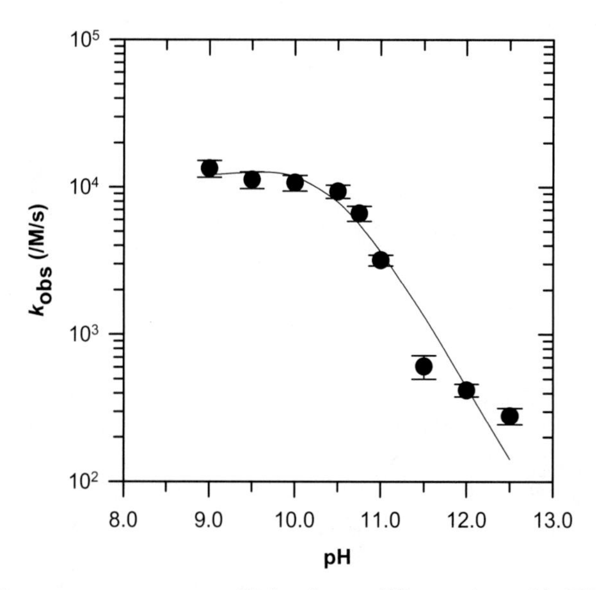

Figure 6.41. Rate constants versus pH for ferrate(V) reaction with EDTA (adapted from Noorhasan et al. [385] with permission of Elsvier Inc.).

in the studied pH range of the kinetic measurements (Table 6.8). The fractions of the species are presented in Figure 6.38. Of the possible 20 reactions between the species of ferrate(VI) and EDTA, only 6 reactions successfully fit the data over the entire pH range using the model in Equation (6.142) (a solid line in Fig. 6.35). The six reactions and their calculated individual rate constants are given in Equations (6.145)–(6.150):

$$H_3FeO_4^+ + H_3Y^- \rightarrow Fe(III) + products \quad k_{155} = 3.30 \pm 0.12 \times 10^5 /M/s \tag{6.145}$$

$$H_2FeO_4 + H_2Y^{2-} \rightarrow Fe(III) + products \quad k_{156} = 1.18 \pm 0.09 \times 10^4 /M/s \tag{6.146}$$

$$HFeO_4^- + H_2Y^{2-} \rightarrow Fe(III) + products \quad k_{157} = 2.73 \pm 0.50 \times 10^2 /M/s \tag{6.147}$$

$$HFeO_4^- + HY^{3-} \rightarrow Fe(III) + products \quad k_{158} = 6.71 \pm 0.17 \times 10^1 /M/s \tag{6.148}$$

$$FeO_4^{2-} + HY^{3-} \rightarrow Fe(III) + products \quad k_{159} = 6.88 \pm 0.19 \times 10^{-1} /M/s \tag{6.149}$$

$$FeO_4^{2-} + Y^{4-} \rightarrow Fe(III) + products \quad k_{160} = 8.00 \pm 0.50 \times 10^{-2} /M/s. \tag{6.150}$$

In the reaction of ferrate(V) with EDTA, $HFeO_4^{2-}$ and FeO_4^{3-} can react with two forms of EDTA in the studied pH range (Eqs. 6.151–6.154):

$$HFeO_4^{2-} + HEDTA^{3-} \rightarrow products \quad k_{161} = 1.18 \pm 0.12 \times 10^4 /M/s \tag{6.151}$$

$$FeO_4^{3-} + HEDTA^{3-} \rightarrow products \quad k_{162} = 2.85 \pm 0.29 \times 10^4 /M/s \tag{6.152}$$

$$HFeO_4^{2-} + EDTA^{4-} \rightarrow products \quad k_{163} = 3.58 \pm 0.36 \times 10^4 /M/s \tag{6.153}$$

$$FeO_4^{3-} + EDTA^{4-} \rightarrow \text{products.} \tag{6.154}$$

Reactions (6.143) and (6.144) involve proton ambiguity to justify the pH dependence of the reaction of ferrate(V) with EDTA. Thus, the experimental values of k fit reasonably well (a solid line in Fig. 6.41) by considering either reactions (6.151) and (6.152) (model I) or reactions (6.151) and (6.153) (model II). Reaction (6.154) was not imperative to fit the data.

6.3.3 Conclusions

Extensive research on the synthesis and reactivity of nonheme oxoiron(IV) complexes is being performed to understand oxidation reactions in enzymes. The C–H bond activation, oxygen-atom transfer reaction, and electron and hydride transfer in oxidations have been demonstrated. Recent focus on oxidation studies includes the influence of axial ligands, the topology of ligands, solvent pH, the spin state of the iron(IV) ion, and redox-inactive metal ions (e.g., Sc^{3+}). Future work may include the development of generality of such influences on oxidation reactions carried out by nonheme oxoiron(IV) complexes. Fe(V)-oxo complexes are synthesized and characterized to provide evidence of their generation in the enzyme-like C–H and C=C oxidation reactions. Direct evidence of the generation of Fe(V) is forthcoming, and a technique such as VT-MS would be useful in the investigation of reactive intermediates of the reactions. The role of Fe(V)-nitrido species in the biological production of ammonia has been suggested.

High-valent nonheme iron complexes are capable of modifying side chains of proteins by the oxidation of the amino acids. Cys, Try, and Trp residues of trypsin and chymotrypsin can be oxidized by high-valent iron complexes. However, the potential role of such complexes compared to reactive oxygen and nitrogen species in modification of proteins in biological systems still need to be established.

The reactivity of ferryl(IV) ion with inorganic substrates is generally understood. However, similar information of ferryl(IV) ion reactions with organic constituents of the atmospheric liquid phase (e.g., organic peroxides) is required to better evaluate the role of iron in the chemistry of the aqueous atmosphere. Moreover, product studies of known reactivity of the ferryl(IV) ion with organic constituents of the atmosphere (formic, formaldehyde, and acetone) are needed in order to understand the importance of such reactions.

The studies carried out by ferrate(VI) were mostly related to industrial and remediation processes. Oxidation reactions carried out by ferrate(VI) were completed in shorter time periods than oxidations carried out by permanganate and chromate [387, 388]. These results have implications in organic syntheses performed using transition metal oxidants. Iron, unlike chromium and manganese, is considered almost nontoxic; therefore, ferrate(VI) can make industrial processes more environmentally benign by achieving cleaner technologies for organic syntheses. Water treatment by ferrate(VI) may also be preferable over permanganate because of relatively faster rates of reactions. Studies on the applications of ferrate(VI) as a chemical treatment are being performed, and the current focus is on the oxidation of emerging contaminants such as pharmaceuticals and endocrine disruptors. Future work should include the identification and quantification of products from such oxidation reactions by ferrate(VI).

The studies on ferrate(IV) and ferrate(V) are still very limited. The oxidation of amino acids by ferrate(IV) with amino acids is not known. A few studies on the reactivity of ferrate(V) with amino acids were mostly carried out under alkaline pH conditions. Knowledge on the nature of products of oxidation by both ferrate(V) and ferrate(VI) is still largely missing. Studies on the reactions of ferrate species with amino acids and peptides under neutral pH conditions are needed to learn their relevance in biological and disinfection processes. A detailed analysis of the intermediate(s) and products are also required to learn the steps of the oxidation reaction mechanisms. Overall, the kinetics and mechanism of oxidation reactions of ferrates will help to understand the role of high-valent iron in biological and environmental reactions.

REFERENCES

1 Bendix, J., Anthon, C., Schau-Magnussen, M., Brock-Nannestad, T., Vibenholt, J., Rehman, M., and Sauer, S.P.A. Heterobimetallic nitride complexes from terminal chromium(V) nitride complexes: hyperfine coupling increases with distance. *Angew. Chem. Int. Ed.* 2011, *50*, 4480–4483.

2 Abu-Omar, M.M. High-valent iron and manganese complexes of corrole and porphyrin in atom transfer and dioxygen evolving catalysis. *Dalton Trans.* 2011, *40*, 3435–3444.

3 Blotevogel, J., Borch, T., Desyaterik, Y., Mayeno, A.N., and Sale, T.C. Quantum chemical prediction of redox reactivity and degradation pathways for aqueous phase contaminants: an example with HMPA. *Environ. Sci. Technol.* 2010, *44*, 5868–5874.

4 Chiu, A., Shi, X.L., Lee, W.K.P., Hill, R., Wakeman, T.P., Katz, A., Xu, B., Dalal, N.S., Robertson, J.D., Chen, C., Chiu, N., and Donehower, L. Review of chromium (VI) apoptosis, cell-cycle-arrest, and carcinogenesis. *J. Environ. Sci. Health C* 2010, *28*, 188–230.

5 Yiu, S., Man, W., Wang, X., Lam, W.W.Y., Ng, S., Kwong, H., Lau, K., and Lau, T. Oxygen evolution from BF_3/MnO_4^-. *Chem. Commun.* 2011, *47*, 4159–4161.

6 Samantaray, R., Clark, R.J., Choi, E.S., Zhou, H., and Dalal, N.S. M3-x(NH4)xCrO8 (M = Na, K, Rb, Cs): a new family of Cr^{5+}-based magnetic ferroelectrics. *J. Am. Chem. Soc.* 2011, *133*, 3792–3795.

7 Bokare, A.D. and Choi, W. Chromate-induced activation of hydrogen peroxide for oxidative degradation of aqueous organic pollutants. *Environ. Sci. Technol.* 2010, *44*, 7232–7237.

8 Vignati, D.A.L., Dominik, J., Beye, M.L., Pettine, M., and Ferrari, B.J.D. Chromium(VI) is more toxic than chromium(III) to freshwater algae: a paradigm to revise? *Ecotoxicol. Environ. Saf.* 2010, *73*, 743–749.

9 Myers, J.M. and Myers, C.R. The effects of hexavalent chromium on thioredoxin reductase and peroxiredoxins in human bronchial epithelial cells. *Free Radic. Biol. Med.* 2009, *47*, 1477–1485.

10 Kanady, J.S., Tsui, E.Y., Day, M.W., and Agapie, T. A synthetic model of the Mn3Ca subsite of the oxygen-evolving complex in photosystem II. *Science* 2011, *333*, 733–736.

11 Meyer, T.J., Huynh, M.H.V., and Thorp, H.H. The possible role of proton-coupled electron transfer (PCET) in water oxidation by photosystem II. *Angew. Chem. Int. Ed.* 2007, *46*, 5284–5304.

12 England, J., Guo, Y., Van Heuvelen, K.M., Cranswick, M.A., Rohde, G.T., Bominaar, E.L., Münck, E., and Que, L., Jr. A more reactive trigonal-bipyramidal high-spin oxoiron(IV) complex with a cis-labile site. *J. Am. Chem. Soc.* 2011, *133*, 11880–11883.

13 Yin, G. Active transition metal oxo and hydroxo moieties in nature's redox, enzymes and their synthetic models: structure and reactivity relationships. *Coord. Chem. Rev.* 2010, *254*, 1826–1842.

14 Park, J., Morimoto, Y., Lee, Y., Nam, W., and Fukuzumi, S. Metal ion effect on the switch of mechanism from direct oxygen transfer to metal ion-coupled electron transfer in the sulfoxidation of thioanisoles by a non-heme iron(IV)-oxo complex. *J. Am. Chem. Soc.* 2011, *133*, 5236–5239.

15 Morimoto, Y., Kotani, H., Park, J., Lee, Y., Nam, W., and Fukuzumi, S. Metal ion-coupled electron transfer of a nonheme oxoiron(IV) complex: remarkable enhancement of electron-transfer rates by Sc^{3+}. *J. Am. Chem. Soc.* 2011, *133*, 403–405.

16 Kundu, S., Thompson, J.V.K., Ryabov, A.D., and Collins, T.J. On the reactivity of mononuclear iron(V)oxo complexes. *J. Am. Chem. Soc.* 2011, *133*, 18546–18549.

17 Sharma, V.K. Oxidation of inorganic contaminants by ferrates(Fe(VI), Fe(V), and Fe(IV))—kinetics and mechanisms—a review. *J. Environ. Manage.* 2011, *92*, 1051–1073.

18 Sharma, V.K. Oxidation of nitrogen containing pollutants by novel ferrate(VI) technology: a review. *J. Environ. Sci. Health A Toxic/Hazard. Subst. Environ. Eng.* 2010, *45*, 645–667.

19 Sharma, V.K. Oxidation of inorganic compounds by ferrate(VI) and ferrate(V): one-electron and two-electron transfer steps. *Environ. Sci. Technol.* 2010, *45*, 5148–5152.

20 Sharma, V.K., Luther, G.W., III., and Millero, F.J. Mechanisms of oxidation of organosulfur compounds by ferrate(VI). *Chemosphere* 2011, *82*, 1083–1089.

21 Sharma, V.K. Disinfection performance of Fe(VI) in water and wastewater: a review. *Water Sci. Technol.* 2007, *55*, 225–232.

22 Filip, J., Yngard, R.A., Siskova, K., Marusak, Z., Ettler, V., Sajdl, P., Sharma, V.K., and Zboril, R. Mechanisms and efficiency of the simultaneous removal of metals and cyanides by using ferrate(VI): crucial roles of nanocrystalline iron(III) oxyhydroxides and metal carbonates. *Chem. Eur. J.* 2011, *17*, 10097–10105.

23 Shupack, S.I. The chemistry of chromium and some resulting analytical problems. *Environ. Health Perspect.* 1991, *92*, 7–11.

24 Niki, K. Chromium, molybdenum, and tungston. In *Standard Potentials in Aqueous Solution*, A.J. Bard, R. Parsons, and J. Jordan, eds. Int. Union Pure Appl. Chem., Marcel Dekker, New York, 1985, pp. 453–460.

25 Cieślak-Golonka, M. and Daszkiewicz, M. Coordination geometry of Cr(VI) species: structural and spectroscopic characteristics. *Coord. Chem. Rev.* 2005, *249*, 2391–2407.

26 Vincent, J.B. Chromium: celebrating 50 years as an essential element? *Dalton Trans.* 2010, *39*, 3787–3794.

27 Levina, A. and Lay, P.A. Mechanistic studies of relevance to the biological activities of chromium. *Coord. Chem. Rev.* 2005, *249*, 281–298.

28 Giusti, L. and Barakat, S. The monitoring of Cr(III) and Cr(VI) in natural water and synthetic solutions: an assessment of the performance of the DGT and DPC methods. *Water Air Soil Pollut.* 2005, *161*, 313–334.

29 Richard, F.C. and Bourg, A.C.M. Aqueous geochemistry of chromium: a review. *Water Res.* 1991, *25*, 807–816.

30 Zhao, Z., Rush, J.D., Holcman, J., and Bielski, B.H.J. The oxidation of chromium(III) by hydroxyl radical in alkaline solution. A stopped-flow and pre-mix pulse radiolysis study. *Radiat. Phys. Chem.* 1995, *45*, 257–263.

31 Christoffersen, M.R., Thyregod, H.C., and Christoffersen, J. Effects of aluminum(III), chromium(III), and iron(III) on the rate of dissolution of calcium hydroxyapatite crystals in the absence and presence of the chelating agent desferrioxamine. *Calcif. Tissue Int.* 1987, *41*, 27–30.

32 Tandon, R.K., Crisp, P.T., Ellis, J., and Baker, R.S. Effect of pH on chromium(VI) species in solution. *Talanta* 1984, *31*, 227–228.

33 Sharpe, P.H.G. and Sehested, K. The dichromate dosimeter: a pulse radiolysis study. *Radiat. Phys. Chem.* 1989, *34*, 763–768.

34 Bakac, A. Oxygen activation with transition-metal complexes in aqueous solution. *Inorg. Chem.* 2010, *49*, 3584–3593.

35 Bakac, A. Kinetic and mechanistic studies of the reactions of transition metal-activated oxygen with inorganic substrates. *Coord. Chem. Rev.* 2006, *250*, 2046–2058.

36 Song, W. and Bakac, A. Fast ligand substitution at a chromium(III) hydroperoxo complex. *Inorg. Chem.* 2010, *49*, 150–156.

37 Bakac, A. and Espenson, J.H. Chromium complexes derived from molecular oxygen. *Acc. Chem. Res.* 1993, *26*, X1–523.

38 Cheng, M. and Bakac, A. Photochemical oxidation of halide ions by a nitratochromium(III) complex. Kinetics, mechanism, and intermediates. *J. Am. Chem. Soc.* 2008, *130*, 5600–5605.

39 Buxton, G.V., Djouider, F., Alan Lynch, D., and Malone, T.N. Oxidation of Cr^{III} to Cr^{VI} initiated by $^{\cdot}OH$ and SO_4^{-} in acidic aqueous solution: a pulse radiolysis study. *J. Chem. Soc. Faraday Trans.* 1997, *93*, 4265–4268.

40 Nemes, A. and Bakac, A. Disproportionation of aquachromyl(IV) ion by hydrogen abstraction from coordinated water. *Inorg. Chem.* 2001, *40*, 2720–2724.

41 Buxton, G.V. and Djouider, F. Disproportionation of Cr^{V} generated by the radiation-induced reduction of Cr^{VI} in aqueous solution containing formate: a pulse radiolysis study. *J. Chem. Soc. Faraday Trans.* 1996, *92*, 4173–4176.

42 Cieślak-Golonka, M. Spectroscopy of chromium(VI) species. *Coord. Chem. Rev.* 1991, *109*, 223–249.

43 Brauer, S.L., Hneihen, A.S., McBride, J.S., and Wetterhahn, K.E. Chromium(VI) forms thiolate complexes with γ-glutamylcysteine, N-acetylcysteine, cysteine, and the methyl ester of N-acetylcysteine. *Inorg. Chem.* 1996, *35*, 373–381.

44 Codd, R., Dillon, C.T., Levina, A., and Lay, P.A. Studies on the genotoxicity of chromium: from the test tube to the cell. *Coord. Chem. Rev.* 2001, *216–217*, 537–582.

45 Nguyen, A., Mulyani, I., Levina, A., and Lay, P.A. Reactivity of chromium(III) nutritional supplements in biological media: an X-ray absorption spectroscopic study. *Inorg. Chem.* 2008, *47*, 4299–4309.

46 Nag, K. and Base, S.N. Chemistry of tetra- and pentavalent chromium. *Struct. Bond.* Anonymous 1985; *63*, 153–197.

47 Crestoni, M.E., Fornarini, S., Lanucara, F., Warren, J.J., and Mayer, J.M. Probing "spin-forbidden" oxygen-atom transfer: gas-phase reactions of chromium-porphyrin complexes. *J. Am. Chem. Soc.* 2010, *132*, 4336–4343.

48 Fujii, H., Yoshimura, T., and Kamada, H. ESR studies of oxochromium(V) porphyrin complexes: electronic structure of the Cr^{V}=O moiety. *Inorg. Chem.* 1997, *36*, 1122–1127.

49 Ghosh, M.C. and Gould, E.S. Electron transfer. 106. Stabilized aqueous chromium(IV), as prepared from the chromium(VI)-arsenic(III) reaction. *Inorg. Chem.* 1991, *30*, 491–494.

50 Levina, A., Foran, G.J., and Lay, P.A. X-ray absorption spectroscopic studies of the Cr(IV) 2-ethyl-2- hydroxybutanoato(1-) complex. *Chem. Commun.* 1999, 2339–2340.

51 Westheimer, F.H. The mechanisms of chromic acid oxidations. *Chem. Rev.* 1949, *45*, 419–451.

52 Espenson, J.H. Oxidation of transition metal complexes by chromium(VI). *Acc. Chem. Res.* 1970, *3*, 347–353.

53 Beattie, J.K. and Haight, J.G.P. Chrmoium(VI) oxidations of inorganic substrates. *Prog. Inorg. Chem.* 1972, *17*, 93–145.

54 Krumpolc, M., DeBoer, B.G., and Roček, J. A stable chromium(V) compound. Synthesis, properties, and crystal structure of potassium bis(2-hydroxy-2-methylbutyrato)-oxochromate(V) monohydrate. *J. Am. Chem. Soc.* 1978, *100*, 145–153.

55 Krumpolc, M. and Roček, J. Synthesis of stable chromium(V) complexes of tertiary hydroxy acids. *J. Am. Chem. Soc.* 1979, *101*, 3206–3209.

56 Mitewa, M. and Bontchev, P.R. Chromium(V) coordination chemistry. *Coord. Chem. Rev.* 1985, *61*, 241–272.

57 Krumpolc, M. and Roček, J. Chromium(V) oxidations of organic compounds. *Inorg. Chem.* 1985, *24*, 617–621.

58 Gould, E.S. Redox chemistry of chromium (IV) complexes. *Coord. Chem. Rev.* 1994, *135–136*, 651–684.

59 Cooper, J.N., Staudt, G.E., Smalser, M.L., Settzo, L.M., and Haight, G.P. Ligand capture in reductions of chromium(VI). *Inorg. Chem.* 1973, *12*, 2075–2079.

60 Scott, S.L., Bakac, A., and Espenson, J.H. Oxidation of alcohols, aldehydes, and carboxylates by the aquachromium(IV) ion. *J. Am. Chem. Soc.* 1992, *114*, 4205–4213.

61 Körner, M. and Van Eldik, R. Kinetics and mechanism of the outer-sphere oxidation of horse-heart cytochrome c by an anionic chromium(V) complex—kinetic evidence for precursor formation and a late electron-transfer transition state. *Eur. J. Inorg. Chem.* 1999, *1999*, 1805–1812.

62 Dickman, M.H. and Pope, M.T. Peroxo and superoxo complexes of chromium, molybdenum, and tungsten. *Chem. Rev.* 1994, *94*, 569–584.

63 Zhang, L. and Lay, P.A. EPR spectroscopic studies on the formation of chromium(V) peroxo complexes in the reaction of chromium(VI) with hydrogen peroxide. *Inorg. Chem.* 1998, *37*, 1729–1733.

64 Bartlett, B.L. and Quane, D. Tetraperoxychromate(V)-diperoxychromate(VI) equilibrium in basic aqueous hydrogen peroxide solutions. *Inorg. Chem.* 1973, *12*, 1925–1927.

65 Perez-Benito, J.F. and Arias, C. A kinetic study of the chromium(VI)-hydrogen peroxide reaction. Role of the diperoxochromate(VI) intermediates. *J. Phys. Chem. A* 1997, *101*, 4726–4733.

66 Sala, L.F., Palopoli, C., and Signorella, S. Oxidation of 2-acetamido-2-deoxy-d-glucose by Cr^{VI} in perchloric acid. *Polyhedron* 1995, *14*, 1725–1730.

67 Signorella, S., Daier, V., García, S., Cargnello, R., González, J.C., Rizzotto, M., and Sala, L.F. The relative ability of aldoses and deoxyaldoses to reduce Cr(VI) and Cr(V). A comparative kinetic and mechanistic study. *Carbohydr. Res.* 1999, *316*, 14–25.

68 Signorella, S., Frascaroli, M.I., Garcia, S., Santoro, M., Gonzalez, J.C., Palopoli, C., Daier, V., Casado, N., and Sala, L.F. Kinetics and mechanism of the chromium(VI) oxidation of methyl a-D-glucopyranoside and methyl a-D-mannopyranosidet. *J. Chem. Soc. Dalton Trans.* 2000, 1617–1623.

69 Signorella, S., Garcia, S., and Sala, L.F. Degradative oxidation of d-ribono-1,4-lactone by Cr^{VI} in perchloric acid. *Polyhedron* 1997, *16*, 701–706.

70 Signorella, S., Lafarga, R., Daier, V., and Sala, L.F. The reduction of Cr(VI) to Cr(III) by the α and β anomers of D-glucose in dimethyl sulfoxide. A comparative kinetic and mechanistic study. *Carbohydr. Res.* 2000, *324*, 127–135.

71 Signorella, S., Santoro, M., Palopoli, C., Brondino, C., Salas-Peregrin, J.M., Quiroz, M., and Sala, L.F. Kinetics and mechanism of the oxidation of D-galactono-1,4-lactone by Cr^{VI} and Cr^{V}. *Polyhedron* 1998, *17*, 2739–2749.

72 Mangiameli, M.F., González, J.C., García, S., Bellu, S., Santoro, M., Caffaratti, E., Frascaroli, M.I., Peregrín, J.M.S., Atria, A.M., and Sala, L.F. Redox, kinetics, and complexation chemistry of the $Cr^{VI}/Cr^{V}/Cr^{IV}$-D-glycero-D-gulo-heptono- 1,4-lactone System. *J. Phys. Org. Chem.* 2010, *23*, 960–971.

73 Mangiameli, M.F., González, J.C., García, S.I., Frascaroli, M.I., Van Doorslaer, S., Salas Peregrin, J.M., and Sala, L.F. Few insights on the mechanism of oxidation of D-galacturonic acid by hypervalent chromium. *Dalton Trans.* 2011, *40*, 7033–7045.

74 González, J.C., García, S., Bellú, S., Peregrín, J.M.S., Atria, A.M., Sala, L.F., and Signorella, S. Redox and complexation chemistry of the $Cr^{VI}/Cr^{V}/Cr^{IV}$-d-glucuronic acid system. *Dalton Trans.* 2010, *39*, 2204–2217.

75 González, J.C., García, S.I., Bellú, S., Atria, A.M., Pelegrín, J.M.S., Rockenbauer, A., Korecz, L., Signorella, S., and Sala, L.F. Oligo and polyuronic acids interactions with hypervalent chromium. *Polyhedron* 2009, *28*, 2719–2729.

76 Daier, V., Signorella, S., Rizzotto, M., Frascaroli, M.I., Palopoli, C., Brondino, C., Salas-Peregrin, J.M., and Sala, L.F. Kinetics and mechanism of the reduction of Cr(VI) to Cr(III) by D-ribose and 2-deoxy-D-ribose. *Can. J. Chem.* 1999, *77*, 57–64.

77 Codd, R., Irwin, J.A., and Lay, P.A. Sialoglycoprotein and carbohydrate complexes in chromium toxicity. *Curr. Opin. Chem. Biol.* 2003, *7*, 213–219.

78 Sala, L.F., Gonzalez, J.C., Garcia, S.I., Frascaroli, M.I., and Mangiomeli, M.F. Hypervalent chromium oxidation of carbohydrates of biological importance. *Global J. Inorg. Chem.* 2011, *2*, 18–38.

79 Signorella, S., Rizzotto, M., Daier, V., Frascaroli, M.I., Palopoli, C., Martino, D., Bousseksou, A., and Sala, L.F. Comparative study of oxidation by chromium(V) and chromium(VI). *J. Chem. Soc. Dalton Trans.* 1996, 1607–1611.

80 Gez, S., Luxenhofer, R., Levina, A., Codd, R., and Lay, P.A. Chromium(V) complexes of hydroxamic acids: formation, structures, and reactivities. *Inorg. Chem.* 2005, *44*, 2934–2943.

81 Codd, R. and Lay, P.A. Competition between 1,2-diol and 2-hydroxy acid coordination in Cr(V)- quinic acid complexes: implications for stabilization of Cr(V) intermediates of relevance to Cr(VI)-induced carcinogenesis. *J. Am. Chem. Soc.* 1999, *121*, 7864–7876.

82 Frascaroli, M.I., Signorella, S., González, J.C., Mangiameli, M.F., García, S., De Celis, E.R., Piehl, L., Sala, L.F., and Atria, A.M. Oxidation of 2-amino-2-deoxy-d-glucopyranose by hypervalent chromium: kinetics and mechanism. *Polyhedron* 2011, *30*, 1914–1921.

83 Stearns, D.M. and Wetterhahn, K.E. Reaction of chromium(VI) with ascorbate produces chromium(V), chromium (IV), and carbon-based radicals. *Chem. Res. Toxicol.* 1994, *7*, 219–230.

84 O'Brien, P. and Wang, G. A potentially significant one-electron pathway in the reduction of chromate by glutathione under physiological conditions. *J. Chem. Soc. Chem. Commun.* 1992, 690–692.

85 O'Brien, P. and Woodbridge, N. Evidence for epoxidation by the chromium(VI)-glutathione system. *Polyhedron* 1997, *16*, 2897–2899.

86 Connett, P.H. and Wetterhahn, K.E. Reaction of chromium(VI) with thiols: pH dependence of chromium(VI) thio ester formation. *J. Am. Chem. Soc.* 1986, *108*, 1842–1847.

87 Connett, P.H. and Wetterhahn, K.E. *In vitro* reaction of the carcinogen chromate with cellular thiols and carboxylic acids. *J. Am. Chem. Soc.* 1985, *107*, 4282–4288.

88 Connett, P.H. and Wetterhahn, K.E. Metabolism of the carcinogen chromate by cellular constituents. *Struct. Bond.* 1983, *54*, 94–122.

89 Dixon, D.A., Dasgupta, T.P., and Sadler, N.P. Mechanism of the oxidation of DL-penicillamine and glutathione by chromium(VI) in aqueous solution. *J. Chem. Soc. Dalton Trans.* 1995, 2267–2271.

90 Dixon, D.A., Sadler, N.P., and Dasgupta, T.P. Oxidation of biological substrates by chromium(VI). Part 1. Mechanism of the oxidation of L-ascorbic acid in aqueous solution. *J. Chem. Soc. Dalton Trans.* 1993, 3489–3495.

91 Zhang, L. and Lay, P.A. EPR spectroscopic studies of the reactions of Cr(VI) with L-ascorbic acid, L-dehydroascorbic acid, and 5,6-O-isopropylidene-L-ascorbic acid in water. Implications for chromium(VI) genotoxicity. *J. Am. Chem. Soc.* 1996, *118*, 12624–12637.

92 Pattison, D.I., Lay, P.A., and Davies, M.J. EPR studies of chromium(V) intermediates generated via reduction of chromium(VI) by DOPA and related catecholamines: potential role for oxidized amino acids in chromium-induced cancers. *Inorg. Chem.* 2000, *39*, 2729–2739.

93 Perez-Benito, J.F. and Arias, C. Kinetics and mechanism of the reactions of superoxochromium(III) ion with biological thiols. *J. Phys. Chem. A* 1998, *102*, 5837–5845.

94 Levina, A., Zhang, L., and Lay, P.A. Structure and reactivity of a chromium(V) glutathione complex. *Inorg. Chem.* 2003, *42*, 767–784.

95 Moghaddas, S., Gelerinter, E., and Bose, R.N. Mechanisms of formation and decomposition of hypervalent chromium metabolites in the glutathione-chromium(VI) reaction. *J. Inorg. Biochem.* 1995, *57*, 135–146.

96 Lay, P.A. and Levina, A. Activation of molecular oxygen during the reactions of chromium(VI/V/IV) with biological reductants: implications for chromium-induced genotoxicities. *J. Am. Chem. Soc.* 1998, *120*, 6704–6714.

97 Lay, P.A. and Levina, A. Kinetics and mechanism of chromium(III) reduction to chromium(VI) by L-cysteine in neutral aqueous solutions. *Inorg. Chem.* 1996, *35*, 7709–7717.

98 Cieślak-Golonka, M. Toxic and mutagenic effects of chromium(VI). A review. *Polyhedron* 1996, *15*, 3667–3689.

99 Levina, A. and Lay, P.A. Solution structures of chromium(VI) complexes with glutathione and model thiols. *Inorg. Chem.* 2004, *43*, 324–335.

100 Perez-Benito, J.F., Lamrhari, D., and Arias, C. Three rate constants from a single kinetic experiment: formation, decomposition, and reactivity of the chromium(VI)-glutathione thioester intermediate. *J. Phys. Chem.* 1994, *98*, 12621–12629.

101 Brauer, S.L. and Wetterhahn, K.E. Chromium(VI) forms a thiolate complex with glutathione. *J. Am. Chem. Soc.* 1991, *113*, 3001–3007.

102 Levina, A., Zhang, L., and Lay, P.A. Formation and reactivity of chromium(V)-thiolato complexes: a model for the intracellular reactions of carcinogenic chromium(VI) with biological thiols. *J. Am. Chem. Soc.* 2010, *132*, 8720–8731.

103 Marin, R., Ahuja, Y., and Bose, R.N. Potentially deadly carcinogenic chromium redox cycle involving peroxochromium(IV) and glutathione. *J. Am. Chem. Soc.* 2010, *132*, 10617–10619.

104 Venkataramanan, N.S., Rajagopal, S., and Vairamani, M. Oxidation of methionines by oxochromium(V) cations: a kinetic and spectral study. *J. Inorg. Biochem.* 2007, *101*, 274–282.

105 Ganesan, T.K., Rajagopal, S., and Bharathy, J.B. Comparative study of chromium(V) and chromium(VI) oxidation of dialkyl sulfides. *Tetrahedron* 2000, *56*, 5885–5892.

106 Headlam, H.A. and Lay, P.A. EPR spectroscopic studies of the reduction of Chromium(VI) by methanol in the presence of peptides. Formation of long-lived Chromium(V) peptide complexes. *Inorg. Chem.* 2001, *40*, 78–86.

107 Headlam, H.A., Weeks, C.L., Turner, P., Hambley, T.W., and Lay, P.A. Dinuclear chromium(V) amino acid complexes from the reduction of chromium(VI) in the presence of amino acid ligands: XAFS characterization of a chromium(V) amino acid complex. *Inorg. Chem.* 2001, *40*, 5097–5105.

108 Barnard, P.J., Levina, A., and Lay, P.A. Chromium(V) peptide complexes: synthesis and spectroscopic characterization. *Inorg. Chem.* 2005, *44*, 1044–1053.

109 Levina, A., Bailey, A.M., Champion, G., and Lay, P.A. Reactions of Chromium(VI/V/IV) with bis(O-ethyl-L- cysteinato-N,S)zinc(II): a model for the action of carcinogenic chromium on zinc-finger proteins. *J. Am. Chem. Soc.* 2000, *122*, 6208–6216.

110 Codd, R. and Lay, P.A. Chromium(V)-sialic (neuraminic) acid species are formed from mixtures of chromium(VI) and saliva. *J. Am. Chem. Soc.* 2001, *123*, 11799–11800.

111 Myers, J.M., Antholine, W.E., and Myers, C.R. Hexavalent chromium causes the oxidation of thioredoxin in human bronchial epithelial cells. *Toxicology* 2008, *246*, 222–233.

112 Myers, J.M., Antholine, W.E., and Myers, C.R. The intracellular redox stress caused by hexavalent chromium is selective for proteins that have key roles in cell survival and thiol redox control. *Toxicology* 2011, *281*, 37–47.

113 Garcia, J.D. and Wetterhahn Jennette, K. Electron-transport cytochrome P-450 system is involved in the microsomal metabolism of the carcinogen chromate. *J. Inorg. Biochem.* 1981, *14*, 281–295.

114 Wetterhahn Jennette, K. The role of metals in carcinogenesis: biochemistry and metabolism. *Environ. Health Perspect.* 1981, *40*, 233–252.

115 Pestovsky, O., Bakac, A., and Espenson, J.H. Kinetics and mechanism of the oxidation of 10-methyl-9,10- dihydroacridine by chromium(VI,V,IV): electron vs hydrogen atom vs hydride transfer. *J. Am. Chem. Soc.* 1998, *120*, 13422–13428.

116 Porter, R., Jáchymová, M., Martásek, P., Kalyanaraman, B., and Vásquez-Vivar, J. Reductive activation of Cr(VI) by nitric oxide synthase. *Chem. Res. Toxicol.* 2005, *18*, 834–843.

117 Kawanishi, S., Inoue, S., and Sano, S. Mechanism of DNA cleavage induced by sodium chromate(VI) in the presence of hydrogen peroxide. *J. Biol. Chem.* 1986, *261*, 5952–5958.

118 Farrell, R.P., Judd, R.J., Lay, P.A., Dixon, N.E., Baker, R.S.U., and Bonin, A.M. Chromium(V)-induced cleavage of DNA: are chromium(V) complexes the active carcinogens in chromium(VI)-induced cancers? *Chem. Res. Toxicol.* 1989, *2*, 227–229.

119 Borges, K.M., Boswell, J.S., Liebross, R.H., and Wetterhahn, K.E. Activation of chromium(VI) by thiols results in chromium(V) formation, chromium binding to DNA and altered DNA conformation. *Carcinogenesis* 1991, *12*, 551–556.

120 Standeven, A.M. and Wetterhahn, K.E. Possible role of glutathione in chromium(VI) metabolism and toxicity in rats. *Pharm. Toxicol.* 1991, *68*, 469–476.

121 Standeven, A.M. and Wetterhahn, K.E. Ascorbate is the principal reductant of chromium(VI) in rat liver and kidney ultrafiltrates. *Carcinogenesis* 1991, *12*, 1733–1737.

122 Shi, X., Leonard, S.S., Liu, K.J., Zang, L., Gannett, P.M., Rojanasakul, Y., Castranova, V., and Vallyathan, V. Cr(III)-mediated hydroxyl radical generation via Haber-Weiss cycle. *J. Inorg. Biochem.* 1998, *69*, 263–268.

123 Yuann, J.M.P., Liu, K.J., Hamilton, J.W., and Wetterhahn, K.E. *In vivo* effects of ascorbate and glutathione on the uptake of chromium, formation of chromium(V), chromium-DNA binding and 8-hydroxy-2′-deoxyguanosine in liver and kidney of osteogenic disorder Shionogi rats following treatment with chromium(VI). *Carcinogenesis* 1999, *20*, 1267–1275.

124 Zhitkovich, A., Song, Y., Quievryn, G., and Voitkun, V. Non-oxidative mechanisms are responsible for the induction of mutagenesis by reduction of Cr(VI) with cysteine: role of ternary DNA adducts in Cr(III)-dependent mutagenesis. *Biochemistry* 2001, *40*, 549–560.

125 Ding, M. and Shi, X. Molecular mechanisms of Cr(VI)-induced carcinogenesis. *Mol. Cell. Biochem.* 2002, *234–235*, 293–300.

126 Bagchi, D., Stohs, S.J., Downs, B.W., Bagchi, M., and Preuss, H.G. Cytotoxicity and oxidative mechanisms of different forms of chromium. *Toxicology* 2002, *180*, 5–22.

127 Levina, A., Harris, H.H., and Lay, P.A. Binding of chromium(VI) to histones: implications for chromium(VI)-induced genotoxicity. *J. Biol. Inorg. Chem.* 2006, *11*, 225–234.

128 Nemec, A.A., Zubritsky, L.M., and Barchowsky, A. Chromium(VI) stimulates fyn to initiate innate immune gene induction in human airway epithelial cells. *Chem. Res. Toxicol.* 2010, *23*, 396–404.

129 Nickens, K.P., Patierno, S.R., and Ceryak, S. Chromium genotoxicity: a double-edged sword. *Chem. Biol. Interact.* 2010, *188*, 276–288.

130 Asatiani, N., Kartvelishvili, T., Abuladze, M., Asanishvili, L., and Sapojnikova, N. Chromium(VI) can activate and impair antioxidant defense system. *Biol. Trace Elem. Res.* 2011, *142*, 388–397.

131 Jennette, K.W. Microsomal reduction of the carcinogen chromate produces chromium(V). *J. Am. Chem. Soc.* 1982, *104*, 874–875.

132 Sedman, R.M., Beaumont, J., McDonald, T.A., Reynolds, S., Krowech, G., and Howd, R. Review of the evidence regarding the carcinogenicity of hexavalent chromium in drinking water. *J. Environ. Sci. Health C* 2006, *24*, 155–182.

133 Hai, L., Lu, Y., Yan, M., Xianglin, S., and Dalal, N.S. Role of chromium(IV) in the chromium(VI)-related free radical formation, dG hydroxylation, and DNA damage. *J. Inorg. Biochem.* 1996, *64*, 25–35.

134 Liu, K.J., Shi, X., Jiang, J., Goda, F., Dalal, N., and Swartz, H.M. Low frequency electron paramagnetic resonance investigation on metabolism of chromium(VI) by whole live mice. *Ann. Clin. Lab. Sci.* 1996, *26*, 176–184.

135 Liu, K.J., Shi, X., and Dalal, N.S. Synthesis of Cr(IV)-GSH, its identification and its free hydroxyl radical generation: a model compound for Cr(VI) carcinogenicity. *Biochem. Biophys. Res. Commun.* 1997, *235*, 54–58.

136 Xianglin, S., Ding, M., Jianping, Y., Wang, S., Leonard, S.S., Zang, L., Castranova, V., Vallyathan, V., Chiu, A., Dalal, N., and Kejian, L. Cr(IV) causes activation of nuclear transcription factor-κB, DNA strand breaks and dG hydroxylation via free radical reactions. *J. Inorg. Biochem.* 1999, *75*, 37–44.

137 Shi, X. and Dalal, N.S. Generation of hydroxyl radical by chromate in biologically relevant systems: role of Cr(V) complexes versus tetraperoxochromate(V). *Environ. Health Perspect.* 1994, *102*, 231–236.

138 Bagchi, D., Bagchi, M., and Stohs, S.J. Chromium (VI)-induced oxidative stress, apoptotic cell death and modulation of p53 tumor suppressor gene. *Mol. Cell. Biochem.* 2001, *222*, 149–158.

139 Zhitkovich, A., Lukanova, A., Popov, T., Taioli, E., Cohen, H., Costa, M., and Toniolo, P. DNA-protein cross-links in peripheral lymphocytes of individuals exposed to hexavalent chromium compounds. *Biomarkers* 1996, *1*, 86–93.

140 Salnikow, K. and Zhitkovich, A. Genetic and epigenetic mechanisms in metal carcinogenesis and cocarcinogenesis: nickel, arsenic, and chromium. *Chem. Res. Toxicol.* 2008, *21*, 28–44.

141 Joseph, P., He, Q., and Umbright, C. Heme-oxygenase 1 gene expression is a marker for hexavalent chromium-induced stress and toxicity in human dermal fibroblasts. *Toxicol. Sci.* 2008, *103*, 325–334.

142 Patlolla, A.K., Barnes, C., Hackett, D., and Tchounwou, P.B. Potassium dichromate induced cytotoxicity, genotoxicity and oxidative stress in human liver carcinoma (HepG2) cells. *Int. J. Environ. Res. Public Health* 2009, *6*, 643–653.

143 Shi, X., Chiu, A., Chen, C.T., Halliwell, B., Castranova, V., and Vallyathan, V. Reduction of chromium(VI) and its relationship to carcinogenesis. *J. Toxicol. Environ. Health B Crit. Rev.* 1999, *2*, 87–104.

144 Azad, N., Iyer, A.K.V., Manosroi, A., Wang, L., and Rojanasakul, Y. Superoxide-mediated proteasomal degradation of Bcl-2 determines cell susceptibility to Cr(VI)-induced apoptosis. *Carcinogenesis* 2008, *29*, 1538–1545.

145 Sugiyama, M., Tsuzuki, K., Lin, X., and Costa, M. Potentiation of sodium chromate (VI)-induced chromosomal aberrations and mutation by vitamin B2 in Chinese hamster V79 cells. *Mut. Res. Mut. Res. Lett.* 1992, *283*, 211–214.

146 Sugiyama, M. Effects of vitamins on chromium(VI)-induced damage. *Environ. Health Perspect.* 1991, *92*, 63–70.

147 Ye, J. and Shi, X. Gene expression profile in response to chromium-induced cell stress in A549 cells. *Mol. Cell. Biochem.* 2001, *222*, 189–197.

148 Asatiani, N., Abuladze, M., Kartvelishvili, T., Kulikova, N., Asanishvili, L., Holman, H., and Sapojnikova, N. Response of antioxidant defense system to chromium(VI)-induced cytotoxicity in human diploid cells. *Biometals* 2010, *23*, 161–172.

149 Andrew, A.S., Warren, A.J., Barchowsky, A., Temple, K.A., Klei, L., Soucy, N.V., O'Hara, K.A., and Hamilton, J.W. Genomic and proteomic profiling of responses to toxic metals in human lung cells. *Environ. Health Perspect.* 2003, *111*, 825–838.

150 Martin, B.D., Schoenhard, J.A., and Sugden, K.D. Hypervalent chromium mimics reactive oxygen species as measured by the oxidant-sensitive dyes 2′,7′-dichloro-fluorescin and dihydrorhodamine. *Chem. Res. Toxicol.* 1998, *11*, 1402–1410.

151 Sugiyama, M. Effects of vitamin E and vitamin B2 on chromate-induced DNA lesions. *Biol. Trace Elem. Res.* 1989, *21*, 399–404.

152 McCarroll, N., Keshava, N., Chen, J., Akerman, G., Kligerman, A., and Rinde, E. An evaluation of the mode of action framework for mutagenic carcinogens case study II: chromium(VI). *Environ. Mol. Mutagen.* 2010, *51*, 89–111.

153 Shi, X., Dalal, N.S., and Vallyathan, V. One-electron reduction of carcinogen chromate by microsomes, ria, and *Escherichia coli*: identification of Cr(V) and ·OH radical. *Arch. Biochem. Biophys.* 1991, *290*, 381–386.

154 MacFie, A., Hagan, E., Zhitkovich, A., and Mechanism, D.N. A-protein cross-linking by chromium. *Chem. Res. Toxicol.* 2010, *23*, 341–347.

155 Mattagajasingh, S.N., Misra, B.R., and Misra, H.P. Carcinogenic chromium(VI)-induced protein oxidation and lipid peroxidation: implications in DNA-protein cross-linking. *J. Appl. Toxicol.* 2008, *28*, 987–997.

156 Shrivastava, H.Y. and Nair, B.U. Protein degradation by peroxide catalyzed by chromium(III): role of coordinated ligand. *Biochem. Biophys. Res. Commun.* 2000, *270*, 749–754.

157 Holmes, A.L., Wise, S.S., and Wise, J.P., Sr. Carcinogenicity of hexavalent chromium. *Indian J. Med. Res.* 2008, *128*, 353–372.

158 O'Brien, T.J., Ceryak, S., and Patierno, S.R. Complexities of chromium carcinogenesis: role of cellular response, repair and recovery mechanisms. *Mut. Res. Fund. Mol. Mechan. Mut.* 2003, *533*, 3–36.

159 O'Brien, T., Xu, J., and Patierno, S.R. Effects of glutathione on chromium-induced DNA cross-linking and DNA polymerase arrest. *Mol. Cell. Biochem.* 2001, *222*, 173–182.

160 Joudah, L., Moghaddas, S., and Bose, R.N. DNA oxidation by peroxo-chromium(V) species: oxidation of guanosine to guanidinohydantoin. *Chem. Commun.* 2002, 1742–1743.

161 Bose, R.N., Moghaddas, S., Mazzer, P.A., Dudones, L.P., Joudah, L., and Stroup, D. Oxidative damage of DNA by chromium(V) complexes: relative importance of base versus sugar oxidation. *Nucleic Acids Res.* 1999, *27*, 2219–2226.

162 Sugden, K.D. and Wetterhahn, K.E. Identification of the oxidized products formed upon reaction of Chromium(V) with thymidine nucleotides. *J. Am. Chem. Soc.* 1996, *118*, 10811–10818.

163 Wise, S.S., Holmes, A.L., Qin, Q., Xie, H., Katsifis, S.P., Douglas Thompson, W., and Wise, J.P., Sr. Comparative genotoxicity and cytotoxicity of four hexavalent chromium compounds in human bronchial cells. *Chem. Res. Toxicol.* 2010, *23*, 365–372.

164 Borthiry, G.R., Antholine, W.E., Myers, J.M., and Myers, C.R. Reductive activation of hexavalent chromium by human lung epithelial cells: generation of Cr(V) and Cr(V)-thiol species. *J. Inorg. Biochem.* 2008, *102*, 1449–1462.

165 Peterson-Roth, E., Reynolds, M., Quievryn, G., and Zhitkovich, A. Mismatch repair proteins are activators of toxic responses to chromium-DNA damage. *Mol. Cell. Biol.* 2005, *25*, 3596–3607.

166 Ha, L., Ceryak, S., and Patierno, S.R. Chromium(VI) activates ataxia telangiectasia mutated (ATM) protein. Requirement of ATM for both apoptosis and recovery from terminal growth arrest. *J. Biol. Chem.* 2003, *278*, 17885–17894.

167 Cogan, N., Baird, D.M., Phillips, R., Crompton, L.A., Caldwell, M.A., Rubio, M.A., Newson, R., Lyng, F., and Case, C.P. DNA damaging bystander signalling from stem cells, cancer cells and fibroblasts after Cr(VI) exposure and its dependence on telomerase. *Mutat. Res. Fund. Mol. Mech. Mut.* 2010, *683*, 1–8.

168 Campbell, C.E., Gravel, R.A., and Worton, R.G. Isolation and characterization of Chinese hamster cell mutants resistant to the cytotoxic effects of chromate. *Somat. Cell Genet.* 1981, *7*, 535–546.

169 Lu, Y.Y. and Yang, J.L. Long-term exposure to chromium(VI) oxide leads to defects in sulfate transport system in chinese hamster ovary cells. *J. Cell. Biochem.* 1995, *57*, 655–665.

170 Pritchard, D.E., Ceryak, S., Ramsey, K.E., O'Brien, T.J., Ha, L., Fornsaglio, J.L., Stephan, D.A., and Patierno, S.R. Resistance to apoptosis, increased growth potential, and altered gene expression in cells that survived genotoxic hexavalent chromium [Cr(VI)] exposure. *Mol. Cell. Biochem.* 2005, *279*, 169–181.

171 Rodrigues, C.F.D., Urbano, A.M., Matoso, E., Carreira, I., Almeida, A., Santos, P., Botelho, F., Carvalho, L., Alves, M., Monteiro, C., Costa, A.N., Moreno, V., and Alpoim, M.C. Human bronchial epithelial cells malignantly transformed by hexavalent chromium exhibit an aneuploid phenotype but no microsatellite instability. *Mut. Res. Fund. Mol. Mechan. Mut.* 2009, *670*, 42–52.

172 Son, K., Zhang, M., Rucobo, E., Nwaigwe, D., Montgomery, F., and Leffert, H.L. Derivation and study of human epithelial cell lines resistant to killing by chromium trioxide. *J. Environ. Sci. Health A Toxic/Hazard. Subst. Environ. Eng.* 2004, *67*, 1027–1049.

173 Bae, D., Camilli, T.C., Chun, G., Lal, M., Wright, K., O'Brien, T.J., Patierno, S.R., and Ceryak, S. Bypass of hexavalent chromium-induced growth arrest by a protein tyrosine phosphatase inhibitor: enhanced survival and mutagenesis. *Mutat. Res. Fundam. Mol. Mech. Mugag.* 2009, *660*, 40–46.

174 Hunter, J.C. and Kozawa, A. Manganese. In *Standard Potentials in Aqueous Solution*, A.J. Bard, R. Parsons, and J. Jordon, eds. Marcel Dekker, New York, 1985, pp. 429–439.

175 Lachwa, H. and Reinen, D. Color and electronic structure of manganese(V) and manganese(VI) in tetrahedral oxo coordination. A spectroscopic investigation. *Inorg. Chem.* 1989, *28*, 1044–1053.

176 Schenker, R.P., Brunold, T.C., and Güdel, H.U. Synthesis and optical spectroscopy of MnO_4^{2-}—doped crystals of Cs_2CrO_4, $SrCrO_4$, CsBr, and CsI. *Inorg. Chem.* 1998, *37*, 918–927.

177 Lee, D.G., Moylan, C.R., Hayashi, T., and Brauman, J.I. Photochemistry of aqueous permanganate ion. *J. Am. Chem. Soc.* 1987, *109*, 3003–3010.

178 Jacobsen, F., Holcman, J., and Sehested, K. Oxidation of manganese(II) by ozone and reduction manganese(III) by hydrogen peroxide in acidic solution. *Int. J. Chem. Kinet.* 1998, *30*, 207–214.

179 Reisz, E., Leitzke, A., Jarocki, A., Irmscher, R., and Von Sonntag, C. Permanganate formation in the reactions of ozone with Mn(II): a mechanistic study. *J. Water Suppl. Res. Technol.—AQUA* 2008, *57*, 451–454.

180 Berglund, J., Elding, L.I., Buxton, G.V., McGowan, S., and Salmon, G.A. Reaction of peroxomonosulfate radical with manganese(II) in acidic aqueous solution. a pulse radiolysis study. *J. Chem. Soc. Faraday Trans.* 1994, *90*, 3309–3313.

181 Jee, J. and Bakac, A. Reactions of Mn(II) and Mn(III) with alkyl, peroxyalkyl, and peroxyacyl radicals in water and acetic acid. *J. Phys. Chem. A* 2010, *114*, 2136–2141.

182 Jee, J., Pestovsky, O., and Bakac, A. Preparation and characterization of manganese(IV) in aqueous acetic acid. *Dalton Trans.* 2010, *39*, 11636–11642.

183 Rush, J.D. and Bielski, B.H.J. Studies of manganate(V), –(VI), and –(VII) tetra-oxyanions by pulse radiolysis. Optical spectra of protonated forms. *Inorg. Chem.* 1995, *34*, 5832–5838.

184 Lott, K.A.K. and Symons, M.C.R. Structure and reactivity of the oxyanions of transition metals. Part V. Quadrivalent manganese. *J. Chem. Soc.* 1959, 829–833.

185 Rush, J.D., Maskos, Z., and Koppenol, W.H. The superoxide dismutase activities of two higher-valent manganese complexes, Mn(IV) desferrioxamine and Mn(III) cyclam. *Arch. Biochem. Biophys.* 1991, *289*, 97–102.

186 Záhonyi-Budó, E. and Simándi, L.I. Oxidation of propane-1,2-diol by acidic manganese(V) and manganese(VI). *Inorg. Chim. Acta* 1996, *248*, 81–84.

187 Záhonyi-Budó, É. and Simándi, L.I. Oxidation of phosphorous acid and ethyl phosphonates by permanganate. *Inorg. Chim. Acta* 1993, *205*, 207–212.

188 Záhonyi-Budó, É. and Simándi, L.I. Induced oxidation of phosphorus(III) by a short-lived manganate(V) intermediate in the permanganate oxidation of arsenite(III). *Inorg. Chim. Acta* 1992, *191*, 1–2.

189 Záhonyi-Budó, E. and Simándi, L. Oxidations with unstable manganese(VI) in acidic solution. *Inorg. Chim. Acta* 1995, *237*, 173–175.

190 Baxendale, J.H., Fielden, E.M., and Keene, J.P. Absolute rate constants for the reactions of some metal ions with the hydrated electron. *Proc. Chem. Soc.* 1963, August, 242–243.

191 Fielden, E.M. and Hart, E.J. Primary radical yields in pulse-irradiated alkaline aqueous solution. *Radiat. Res.* 1967, *32*, 564–580.

192 Kirschenbaum, L.J. and Meyerstein, D. A pulse radiolysis study of the MnO_4^{2-} ion. The stability of Mn(V) in 0.1 M NaOH. *Inorg. Chim. Acta* 1981, *53*, L99–L100.

193 Sutter, J.H., Colquitt, K., and Sutter, J.R. Kinetics of the disproportionation of manganate in acid solution. *Inorg. Chem.* 1974, *13*, 1444–1446.

194 Jin, N., Ibrahim, M., Spiro, T.G., and Groves, J.T. Trans-dioxo manganese(V) porphyrins. *J. Am. Chem. Soc.* 2007, *129*, 12416–12417.

195 Jin, N. and Groves, J.T. Unusual kinetic stability of a ground-state singlet oxomanganese(V) porphyrin. Evidence for a spin state crossing effect. *J. Am. Chem. Soc.* 1999, *121*, 2923–2924.

196 Fukuzumi, S., Kotani, H., Prokop, K.A., and Goldberg, D.P. Electron- and hydride-transfer reactivity of an isolable manganese(V)-oxo complex. *J. Am. Chem. Soc.* 2011, *133*, 1859–1869.

197 Prokop, K.A., De Visser, S.P., and Goldberg, D.P. Unprecedented rate enhancements of hydrogen-atom transfer to a manganese(V)-oxo corrolazine complex. *Angew. Chem. Int. Ed.* 2010, *49*, 5091–5095.

198 Leeladee, P. and Goldberg, D.P. Epoxidations catalyzed by manganese(V) oxo and lmido complexes: role of the oxidant Mn oxo (imido) intermediate. *Inorg. Chem.* 2010, *49*, 3083–3085.

199 Kwong, H., Lo, P., Lau, K., and Lau, T. Epoxidation of alkenes and oxidation of alcohols with hydrogen peroxide catalyzed by a manganese(V) nitrido complex. *Chem. Commun.* 2011, *47*, 4273–4275.

200 Simándi, L.I., Jáky, M., and Schelly, Z.A. Short-lived manganate(VI) and manganate(V) intermediates in the permanganate oxidation of sulfite ion. *J. Am. Chem. Soc.* 1984, *106*, 6866–6867.

201 Simándi, L.I., Jáky, M., Savage, C.R., and Schelly, Z.A. Kinetics and mechanism of the permanganate ion oxidation of sulfite in alkaline solutions. The nature of short-lived intermediates. *J. Am. Chem. Soc.* 1985, *107*, 4220–4224.

202 Simándi, L.I. and Záhonyi-Budó, E. Relative reactivities of hydroxy compounds with short-lived manganese(V). *Inorg. Chim. Acta* 1998, *281*, 235–238.

203 Lassalle-Kaiser, B., Hureau, C., Pantazis, D.A., Pushkar, Y., Guillot, R., Yachandra, V.K., Yano, J., Neese, F., and Anxolabéhère-Mallart, E. Activation of a water molecule using a mononuclear Mn complex: Fsrom Mn-aquo, to Mn-hydroxo, to Mn-oxyl via charge compensation. *Energy Environ. Sci.* 2010, *3*, 924–938.

204 Lieb, D., Zahl, A., Shubina, T.E., and Ivanović-Burmazović, I. Water exchange on manganese(III) porphyrins. Mechanistic insights relevant for oxygen evolving complex and superoxide dismutation catalysis. *J. Am. Chem. Soc.* 2010, *132*, 7282–7284.

205 Goldsmith, C.R., Cole, A.P., and Stack, T.D.P. C-H activation by a mononuclear manganese(III) hydroxide complex: synthesis and characterization of a manganese-lipoxygenase mimic? *J. Am. Chem. Soc.* 2005, *127*, 9904–9912.

206 Fukuzumi, S., Fujioka, N., Kotani, H., Ohkubo, K., Lee, Y., and Nam, W. Mechanistic insights into hydride-transfer and electron-transfer reactions by a manganese(IV)-oxo porphyrin complex. *J. Am. Chem. Soc.* 2009, *131*, 17127–17134.

207 Kurahashi, T., Kikuchi, A., Shiro, Y., Hada, M., and Fujii, H. Unique properties and reactivity of high-valent manganese–oxo versus manganese–hydroxo in the salen platform. *Inorg. Chem.* 2010, *49*, 6664–6672.

208 Zhang, R., Horner, J.H., and Newcomb, M. Laser flash photolysis generation and kinetic studies of porphyrin-manganese-oxo intermediates. Rate constants for oxidations effected by porphyrin-Mn^{V}-oxo species and apparent disproportionation equilibrium constants for porphyrin-Mn^{IV}-oxo species. *J. Am. Chem. Soc.* 2005, *127*, 6573–6582.

209 Okun, Z., Kupershmidt, L., Amit, T., Mandel, S., Bar-Am, O., Youdim, M.B.H., and Gross, Z. Manganese corroles prevent intracellular nitration and subsequent death of insulin-producing cells. *ACS Chem. Biol.* 2009, *4*, 910–914.

210 Lee, S.H., Xu, L., Park, B.K., Mironov, Y.V., Kim, S.H., Song, Y.J., Kim, C., Kim, Y., and Kim, S. Efficient olefin epoxidation by robust re4 cluster-supported Mn^{III} complexes with peracids: evidence of simultaneous operation of multiple active oxidant species, Mn^{V}=O, Mn^{IV}=O, and Mn^{III}-OOC(O)R. *Chem. Eur. J.* 2010, *16*, 4678–4685.

211 Siegbahn, P.E.M. Structures and energetics for O_2 formation in photosystem II. *Acc. Chem. Res.* 2009, *42*, 1871–1880.

212 Keough, J.M., Jenson, D.L., Zuniga, A.N., and Barry, B.A. Proton coupled electron transfer and redox-active tyrosine Z in the photosynthetic oxygen-evolving complex. *J. Am. Chem. Soc.* 2011, *133*, 11084–11087.

213 Brudvig, G.W. Water oxidation chemistry of photosystem II. *Proc. R Soc. Lond B Biol. Sci.* 2008, *363*, 1211–1218.

214 Milikisiyants, S., Chatterjee, R., Weyers, A., Meenaghan, A., Coates, C., and Lakshmi, K.V. Ligand environment of the S2 state of photosystem II: a study of the hyperfine interactions of the tetranuclear manganese cluster by 2D 14N HYSCORE spectroscopy. *J. Phys. Chem. B* 2010, *114*, 10905–10911.

215 Singh, N. and Lee, D.G. Permanganate: a green and versatile industrial oxidant. *Org. Process Res. Dev.* 2001, *5*, 599–603.

216 Stewart, R. Oxidation by permanganate. In *Oxidation in Organic Chemistry*, K.B. Wiberg, ed. Academic Press, New York, 1973, pp. 1–70.

217 Shaabani, A., Bazgir, A., and Lee, D.G. Oxidation of organic compounds by potassium permanganate supported on montmorillonite K10. *Synth. Commun.* 2004, *34*, 3595–3607.

218 Shaabani, A., Mirzaei, P., Naderi, S., and Lee, D.G. Green oxidations. The use of potassium permanganate supported on manganese dioxide. *Tetrahedron* 2004, *60*, 11415–11420.

219 Shaabani, A., Rahmati, A., Sharifi, M., Rad, J.M., Aghaaliakbari, B., Farhangi, E., and Lee, D.G. Green oxidations. Manganese(II) sulfate aided oxidations of organic compounds by potassium permanganate. *Monatshefte fur Chemie* 2007, *138*, 649–651.

220 Shaabani, A., Tavasoli-Rad, F., and Lee, D.G. Potassium permanganate oxidation of organic compounds. *Synth. Commun.* 2005, *35*, 571–580.

221 Dash, S., Patel, S., and Mishra, B.K. Oxidation by permanganate: synthetic and mechanistic aspects. *Tetrahedron* 2009, *65*, 707–739.

222 Crimi, M., Quickel, M., and Ko, S. Enhanced permanganate *in situ* chemical oxidation through Mn_O2 particle stabilization: evaluation in 1-D transport systems. *J. Contam. Hydrol.* 2009, *105*, 69–79.

223 Waldemer, R.H. and Tratnyek, P.G. Kinetics of contaminant degradation by permanganate. *Environ. Sci. Technol.* 2006, *40*, 1055–1061.

224 Hu, L., Martin, H.M., Arce-Bulted, O., Sugihara, M.N., Keating, K.A., and Strathmann, T.J. Oxidation of carbamazepine by Mn(VII) and Fe(VI): reaction kinetics and mechanism. *Environ. Sci. Technol.* 2009, *43*, 509–515.

225 Forsey, S.P., Thomson, N.R., and Barker, J.F. Oxidation kinetics of polycyclic aromatic hydrocarbons by permanganate. *Chemosphere* 2010, *79*, 628–636.

226 Cerrato, J.M., Hochella, M.F., Jr., Knocke, W.R., Dietrich, A.M., and Cromer, T.F. Use of XPS to identify the oxidation state of Mn in solid surfaces of filtration media oxide samples from drinking water treatment plants. *Environ. Sci. Technol.* 2010, *44*, 5881–5886.

227 Loomer, D.B., Al, T.A., Banks, V.J., Parker, B.L., and Mayer, K.U. Manganese valence in oxides formed from *in situ* chemical oxidation of TCE by KMn_O4. *Environ. Sci. Technol.* 2010, *44*, 5934–5939.

228 Lee, D.G. and Chen, T. The oxidation of alcohols by permanganate. A comparison with other hexa-valent transition-metal oxidants. *J. Org. Chem.* 1991, *56*, 5341–5345.

229 Ladbury, J.W. and Cullis, C.F. Kinetics and mechanism of oxidation by permanganate. *Chem. Rev.* 1958, *58*, 403–438.

230 Soldatenkov, A.T., Temesgen, A.V., and Kolyadina, N.M. Oxidation of heterocyclic compounds by permanganate anion. (Review). *Chem. Heterocycl. Compd.* 2004, *40*, 537–560.

231 Lam, W.W.Y., Yiu, S., Lee, J.M.N., Yau, S.K.Y., Kwong, H., Lau, T., Liu, D., and Lin, Z. BF_3-activated oxidation of alkanes by MnO_4^-. *J. Am. Chem. Soc.* 2006, *128*, 2851–2858.

232 Du, H., Lo, P., Hu, Z., Liang, H., Lau, K., Wang, Y., Lam, W.W.Y., and Lau, T. Lewis acid-activated oxidation of alcohols by permanganate. *Chem. Commun.* 2011, *47*, 7143–7145.

233 Strassner, T. and Busold, M. Density functional theory study on the initial step of the permanganate oxidation of substituted alkynes. *J. Phys. Chem. A* 2004, *108*, 4455–4458.

234 Perez-Benito, J.F. Autocatalytic reaction pathway on manganese dioxide colloidal particles in the permanganate oxidation of glycine. *J. Phys. Chem. C* 2009, *113*, 15982–15991.

235 Khan, F.H. and Ahmad, F. Micellar effect on two-phase kinetics of oxidative degradation of lysine. *Oxidat. Commun.* 2004, *27*, 869–885.

236 Bahrami, H. and Zahedi, M. Delayed autocatalytic behavior of Mn(II) ions at a critical ratio: the effect of structural isomerism on permanganic oxidation of L-norleucine. *Int. J. Chem. Kinet.* 2006, *38*, 1–11.

237 De Andres, J., Brillas, E., Garrido, J.A., and Perez-Benito, J.F. Kinetics and mechanisms of the oxidation by permanganate of L-alanine. *J. Chem. Soc. Perkin Trans.* 1988, *2*, 107–112.

238 Hassan, R.M., Mousa, M.A., and Wahdan, M.H. Kinetics and mechanism of oxidation of β-phenylalanine by permanganate ion in aqueous perchloric acid. *J. Chem. Soc. Dalton Trans.* 1988, 605–609.

239 Hassan, R.M. Kinetics and mechanism of oxidation of DL-alanine by permanganate ion in acidic perchlorate media. *Can. J. Chem.* 1991, *69*, 2018–2023.

240 Perez-Benito, J.F., Brillas, E., and Pouplana, R. Identification of a soluble form of colloidal manganese(IV). *Inorg. Chem.* 1989, *28*, 390–392.

241 Perez-Benito, J.F., Arias, C., and Brillas, E. Kinetic study of the autocatalytic permanganate oxidation of formic acid. *Int. J. Chem. Kinet.* 1990, *22*, 261–287.

242 Insausti, M.J., Mata-Perez, F., and Alvarez-Macho, M.P. Kinetics and mechanics of the oxidation by permanganate of L-phenylalanine. *Int. J. Chem. Kinet.* 1991, *23*, 593–605.

243 Arrizabalaga, A., Andrés-Ordax, F.I., Fernández-Aránguiz, M.Y., and Peche, R. Kinetic studies on the permanganic oxidation of amino acids. Effect of the length of amino acid carbon chain. *Int. J. Chem. Kinet.* 1997, *29*, 181–185.

244 Bahrami, H. and Zahedi, M. Conclusive evidence for delayed autocatalytic behavior of Mn(II) ions at a critical concentration. *J. Iran Chem. Soc.* 2008, *5*, 535–545.

245 Verma, R.S., Reddy, M.J., and Shastry, V.R. Kinetic study of homogeneous acid-catalysed oxidation of certain amino-acids by potassium permanganate in moderately concentrated acidic media. *J. Chem. Soc. Perkin Trans.* 1976, *2*, 469–473.

246 Perez-Benito, J.F. and Arias, C. Occurrence of colloidal manganese dioxide in permanganate reactions. *J. Colloid Interface Sci.* 1992, *152*, 70–84.

247 Bajpai, U.D.N. and Bajpai, A.K. Aqueous polymerization of acrylamide initiated with the permanganate-L-serine redox system. *Macromolecules* 1985, *18*, 2113–2116.

248 Perez-Benito, J.F. Permanganate oxidation of α-amino acids: kinetic correlations for the nonautocatalytic and autocatalytic reaction pathways. *J. Phys. Chem. A* 2011, *115*, 9876–9885.

249 Malik, M.A., Basahel, S.N., Obaid, A.Y., and Khan, Z. Oxidation of tyrosine by permanganate in presence of cetyltrimethylammonium bromide. *Colloids Surf. B* 2010, *76*, 346–353.

250 Foppoli, C., De Marco, C., Blarzino, C., Coccia, R., Mosca, L., and Rosei, M.A. Dimers formation by cytochrome c-catalyzed oxidation of tyrosine and enkephalins. *Amino Acids* 1997, *13*, 273–280.

251 Giulivi, C. and Davies, K.J.A. Mechanism of the formation and proteolytic release of H_2O_2-induced dityrosine and tyrosine oxidation products in hemoglobin and red blood cells. *J. Biol. Chem.* 2001, *276*, 24129–24136.

252 Sheikh, R.A., Al-Nowaiser, F.M., Malik, M.A., Al-Youbi, A.O., and Khan, Z. Effect of cationic micelles of cetyltrimethylammonium bromide on the MnO_4^- oxidation of valine. *Colloids Surf. Physicochem. Eng. Aspects* 2010, *366*, 129–134.

253 Gurame, V.M., Supale, A.R., and Gokavi, G.S. Kinetic and mechanistic study of oxidation of L-methionine by Waugh-type enneamolybdomanganate(IV) in perchloric acid. *Amino Acids* 2010, *38*, 789–795.

254 Panari, R.G., Chougale, R.B., and Nandibewoor, S.T. Kinetics and mechanism of oxidation of L-phenylalanine by alkaline permanganate. *Pol. J. Chem.* 1998, *72*, 99–107.

255 Chang, H., Korshin, G.V., and Ferguson, J.F. Investigation of mechanisms of oxidation of EDTA and NTA by permanganate at high pH. *Environ. Sci. Technol.* 2006, *40*, 5089–5094.

256 Sychev, A.Y. and Isak, V.G. Iron compounds and the mechanisms of the homogeneous catalysis of the activation of O_2 and H_2O_2 and of the oxidation of organic substrates. *Russ. Chem. Rev.* 1995, *64*, 1105–1129.

257 Mukherjee, A., Cranswick, M.A., Chakrabarti, M., Paine, T.K., Fujisawa, K., Münck, E., and Que, L., Jr. Oxygen activation at mononuclear nonheme iron centers: a superoxo perspective. *Inorg. Chem.* 2010, *49*, 3618–3628.

258 Matsui, T., Iwasaki, M., Sugiyama, R., Unno, M., and Ikeda-Saito, M. Dioxygen activation for the self-degradation of heme: reaction mechanism and regulation of heme oxygenase. *Inorg. Chem.* 2010, *49*, 3602–3609.

259 Deguillaume, L., Leriche, M., Desboeufs, K., Maillhot, G., George, C., and Chaumerliac, N. Transition metals in atmospheric liquid phases: sources, reactivity, and sensitive parameters. *Chem. Rev.* 2005, *105*, 3388–3431.

260 Shan, X. and Que, J.L. High-valent nonheme iron-oxo species in biometric oxidations. *J. Inorg. Biochem.* 2006, *100*, 421–433.

261 Groves, J.T. High-valent iron in chemical and biological oxidations. *J. Inorg. Biochem.* 2006, *100*, 434–447.

262 Petrenko, T., George, S.D., Aliaga-Alcalde, N., Bill, E., Mienert, B., Xio, Y., Guo, Y., Sturham, W., Cramer, S.P., Wieghardt, K., and Neese, F. Characterization of a genuine iron(V)-nitrido species by nuclear resonant vibrational spectroscopy coupled to density functional calculations. *J. Am. Chem. Soc.* 2007, *109*, 11053–11060.

263 Nam, W. High-valent iron(IV)-oxo complexes of heme and non-heme ligands in oxygenation reactions. *Acc. Chem. Res.* 2007, *40*, 522–531.

264 Shaik, S., Hirao, H., and Kumar, D. Reactivity of high-valent iron-oxo species in enzymes and synthetic reagents: a tale of many states. *Acc. Chem. Res.* 2007, *40*, 532–542.

265 Oliveria, F.T.D., Chanda, A., Benerjee, D., Shan, X., Mondal, S., Que, J.L., Bominaar, E.L., Munck, E., and Collins, T.J. Chemical and spectroscopic evidence for an Fe^{V}-Oxo complex. *Science* 2007, *315*, 835–839.

266 Pestovsky, O. and Bakac, A. Identification and characterization of aqueous ferryl(IV) ion. *ACS Symp. Ser.* 2008, *985*(Ferrates), 167–176.

267 Solomon, E.I., Wong, S.D., Liu, L.V., Decker, A., and Chow, M.S. Peroxo and oxo intermediates in mononuclear nonheme iron enzymes and related active sites. *Curr. Opin. Chem. Biol.* 2009, *13*, 99–113.

268 Ryabov, A.D. and Collins, T.J. Mechanistic considerations on the reactivity of green Fe^{III}-TAML activators of peroxides. *Adv. Inorg. Chem.* 2009, *61*, 471–521.

269 Shaik, S., Cohen, S., Wang, Y., Chen, H., Kumar, D., and Thiel, W. P450 enzymes: their structure, reactivity, and selectivity—modeled by QM/MM calculations. *Chem. Rev.* 2010, *110*, 949–1017.

270 Rittle, J., Younker, J.M., and Green, M.T. Cytochrome P450: the active oxidant and its spectrum. *Inorg. Chem.* 2010, *49*, 3610–3617.

271 de Visser, S.P. Trends in substrate hydroxylation reactions by heme and nonheme iron(IV)-oxo oxidants give correlations between intrinsic properties of the oxidant with barrier height. *J. Amer. Chem. Soc.* 2010, *132*, 1088–1097.

272 Friedle, S., Reisner, E., and Lipard, S.J. Current challenges of modeling diiron enzyme active sites for dioxygen activation by biometric synthetic complexes. *Chem. Soc. Rev.* 2010, *39*, 2768–2779.

273 Que, L., Jr. The road to non-heme oxoferryls and beyond. *Acc. Chem. Res.* 2007, *40*, 493–500.

274 Sheng, X., Zhang, H., Im, S., Horner, J.H., Waskell, L., Hollenberg, P.F., and Newcomb, M. Kinetics of oxidation of benzphetamine by compounds I of cytochrome P450 2B4 and its mutants. *J. Am. Chem. Soc.* 2009, *131*, 2971–2976.

275 Grzyska, P.K., Appelman, E.H., Hausinger, R.P., and Proshlyakov, D.A. Insight into the mechanism of an iron dioxygenase by resolution of steps following the Fe^{IV}=O species. *Proc. Natl. Acad. Sci. U. S. A.* 2010, *107*, 3982–3987.

276 Lewis-Ballester, A., Batabyal, D., Egawa, T., Lu, C., Lin, Y., Marti, M.A., Capece, L., Estrin, D.A., and Yeh, S.R. Evidence for a ferryl intermediate in a heme-based dioxygenase. *Proc. Natl. Acad. Sci. U. S. A.* 2009, *106*, 17371–17376.

277 Ortiz De Montellano, P.R. Hydrocarbon hydroxylation by cytochrome P450 enzymes. *Chem. Rev.* 2010, *110*, 932–948.

278 Efimov, I., Badyal, S.K., Metcalfe, C.L., Macdonald, I., Gumiero, A., Raven, E.L., and Moody, P.C.E. Proton delivery of ferryl heme in a heme peroxidase:

enzymatic use of the Grotthuss mechanism. *J. Am. Chem. Soc.* 2011, *133*, 15376–15383.

279 Nehru, K., Jang, Y., Oh, S., Dallemer, F., Nam, W., and Kim, J. Oxidation of hydroquinones by a nonheme iron(IV)-oxo species. *Inorg. Chim. Acta* 2008, *361*, 2557–2561.

280 Yoon, J., Wilson, S.A., Jang, Y.K., Seo, M.S., Nehru, K., Hedman, B., Hodgson, K.O., Bill, E., Solomon, E.I., and Nam, W. Reactive intermediates in oxygenation reactions with mononuclear nonheme iron catalysts. *Angew. Chem. Int. Ed.* 2009, *48*, 1257–1260.

281 England, J., Martinho, M., Farquhar, E.R., Frisch, J.R., Bominaar, E.L., Munck, E., and Que, L., Jr. A synthetic high-spin oxoiron(IV) complex: generation, spectroscopic characterization, and reactivity. *Angew. Chem. Int. Ed.* 2009, *48*, 3622–3626, S3622/1–S3622/16.

282 Bigi, J.P., Harman, W.H., Lassalle-Kaiser, B., Robles, D.M., Stich, T.A., Yano, J., Britt, R.D., and Chang, C.J. A high-spin iron(IV)-oxo complex supported by a trigonal nonheme pyrrolide platform. *J. Am. Chem. Soc.* 2012, *134*, 1536–1542.

283 Krebs, C., Fujimori, D.G., Walsh, C.T., and Bollinger, J.M., Jr. Non-heme Fe(IV)-oxo intermediates. *Acc. Chem. Res.* 2007, *40*, 484–492.

284 Das, P. and Que, L., Jr. Iron catalyzed competitive olefin oxidation and ipso-hydroxylation of benzoic acids: further evidence for an Fe^{V}=O oxidant. *Inorg. Chem.* 2010, *49*, 9479–9485.

285 Feng, Y., Ke, C., Xue, G., and Que, L., Jr. Bio-inspired arene cis-dihydroxylation by a non-haem iron catalyst modeling the action of naphthalene dioxygenase. *Chem. Commun.* 2009, 50–52.

286 Feng, Y., England, J., and Que, L., Jr. Iron-catalyzed olefin epoxidation and cis-dihydroxylation by tetraalkylcyclam complexes: the importance of Cis-labile sites. *ACS Catal.* 2011, *1*, 1035–1042.

287 McDonald, A.R. and Que, L., Jr. Elusive iron(V) species identified. *Nat. Chem.* 2011, *3*, 761–762.

288 Prat, I., Mathieson, J.S., Guell, M., Ribas, X., Luis, J.M., Cronin, L., and Costas, M. Observation of Fe(V)=O using variable-temperature mass spectrometry and its enzyme-like C-H and C=C oxidation reactions. *Nat. Chem.* 2011, *3*, 788–793.

289 Scepaniak, J.J., Vogel, C.S., Khusniyarov, M.M., Heinemann, F.W., Meyer, K., and Smith, J.M. Synthesis, structure, and reactivity of an iron(V) nitride. *Science* 2011, *331*, 1049–1052.

290 Chanda, A., Shan, X., Chakrabarti, M., Ellis, W.C., Popescu, D.L., Oliveira, F.T.D., Wang, D., Que, J.L., Collins, T.J., Munck, E., and Bominaar, E.L. TAML)Fe^{IV}=O complex in aqueous solution: synthesis and spectroscopic and computational characterization. *Inorg. Chem.* 2008, *47*, 3669–3678.

291 Chanda, A., Popescu, D., Oliveira, D.T.D., Bominnar, E.L., Ryabov, A.D., Munck, E., and Collins, T.J. High-valent iron complexes with tetraamido macrocyclic ligands: structures, Mössbauer spectroscopy, and DFT calculations. *J. Inorg. Biochem.* 2006, *100*, 606–619.

292 Chanda, A., de Oliveira, F., Collins, T.J., Munck, E., and Bominaar, E.L. Density functional theory study of the structural, electronic, and magnetic properties of a μ-oxo bridged dinuclear Fe^{IV} complex based on a tetra-amido macrocyclic ligand. *Inorg. Chem.* 2008, *47*, 9372–9379.

293 Collins, T.J. TAML oxidant activators: a new approach to the activation of hydrogen peroxide for environmentally significant problems. *Acc. Chem. Res.* 2002, *35*, 782–790.

294 Collins, T.J., Khetan, S.K., and Ryabov, A.D. Chemistry and applications of iron-TAML catalysts in green oxidation processes based on hydrogen peroxide. *Handb. Green Chem. Green Catal.* 2009, *1*, 39–77.

295 Ghosh, A., Mitchell, D.A., Chanda, A., Ryabov, A.D., Popescu, D.L., Upham, E.C., Collins, G.J., and Collins, T.J. Catalase-peroxidase activity of iron(III)-TAML activators of hydrogen peroxide. *J. Am. Chem. Soc.* 2008, *130*, 15116–15126.

296 Tiago De Oliveira, F., Chanda, A., Banerjee, D., Shan, X., Mondal, S., Que, L., Jr., Bominaar, E.L., Muenck, E., and Collins, T.J. Chemical and spectroscopic evidence for an Fe^{IV}-Oxo Complex. *Science* 2007, *315*, 835–838.

297 Shappell, N.W., Vrabel, M.A., Madsen, P.J., Harrington, G., Billey, L.O., Hakk, H., Larsen, G.L., Beach, E.S., Horwitz, C.P., Ro, K., Hunt, P.G., and Collins, T.J. Destruction of estrogens using Fe-TAML/peroxide catalysis. *Environ. Sci. Technol.* 2008, *42*, 1296–1300.

298 Chahbane, N., Popescu, D., Mitchell, D.A., Chanda, A., Lenoir, D., Ryabov, A.D., Schramm, K., and Collins, T.J. Fe^{III}-TAML-catalyzed green oxidative degradation of the azo dye Orange II by H_2O_2 and organic peroxides: products, toxicity, kinetics, and mechanisms. *Green Chem.* 2007, *9*, 49–57.

299 Ellis, W.C., Tran, C.T., Roy, R., Rusten, M., Fischer, A., Ryabov, A.D., Blumberg, B., and Collins, T.J. Designing green oxidation catalysts for purifying environmental waters. *J. Am. Chem. Soc.* 2010, *132*, 9774–9781.

300 Horwitz, C.P., Collins, T.J., Spatz, J., Smith, H.J., Wright, L.J., Stuthridge, T.R., Wingate, K.G., and McGrouther, K. Iron-TAML catalysts in the pulp and paper industry. *ACS Symp. Ser.* 2006, *921*, 156–169.

301 Malecky, R.T., Beach, E., Horwitz, C.P., and Collins, T.J. Effective chemical remediation method for bisphenol A. *Abstracts of Papers, 232nd ACS National Meeting, San Francisco, CA, United States, Sept. 10–14, 2006* 2006, ENVR-167.

302 Shen, L.Q., Beach, E.S., Xiang, Y., Tshudy, D.J., Khanina, N., Horwitz, C.P., Bier, M.E., and Collins, T.J. Rapid, biomimetic degradation in water of the persistent drug sertraline by TAML catalysts and hydrogen peroxide. *Environ. Sci. Technol.* 2011, *45*, 7882–7887.

303 Mondal, S., Hangun-Balkir, Y., Alexandrova, L., Link, D., Howard, B., Zandhuis, P., Cugini, A., Horwitz, C.P., and Collins, T.J. Oxidation of sulfur components in diesel fuel using Fe-TAML catalysts and hydrogen peroxide. *Catal. Today* 2006, *116*, 554–561.

304 Lee, Y. and Gunten, U.V. Oxidative transformation of micropollutants during municipal wastewater treatment: comparison of kinetic aspects of selective (chlorine, chlorine dioxide, $ferrate^{VI}$, and ozone) and non-selective oxidants (hydroxyl radical). *Water Res.* 2010, *44*, 555–566.

305 Lee, C., Lee, Y., Schmidt, C., Yoon, J., and von Gunten, U. Oxidation of suspected N-nitrosodimethylamine (NDMA) precursors by ferrate(VI): kinetics and effect on the NDMA formation potential of natural waters. *Water Res.* 2008, *42*, 433–441.

306 Lee, Y., Zimmermann, S.G., Kieu, A.T., and Gunten, G.V. Ferrate (Fe(VI)) application for municipal wastewater treatment: a novel process for simultaneous

micropollutant oxidation and phosphate removal. *Environ. Sci. Technol.* 2009, *43*, 3831–3838.

307 Sharma, V.K., Graham, N.J.D., Li, X.Z., and Yuan, B.L. Ferrate(VI) enhanced photocatalytic oxidation of pollutants in aqueous TiO_2 suspensions. *Environ. Sci. Pollut. Res.* 2010, *17*(2), 453–461.

308 Sharma, V.K., Kazama, F., Jiangyong, H., and Ray, A.K. Ferrates as environmentally-friendly oxidants and disinfectants. *J. Water Health* 2005, *3*, 45–58.

309 Sharma, V.K., Li, X.Z., Graham, N., and Doong, R.A. Ferrate(VI) oxidation of endocrine disruptors and antimicrobials in water. *J. Water Suppl. Res. Technol.—AQUA* 2008, *57*, 419–426.

310 Sharma, V.K., Sohn, M., Anquandah, G., and Nesnas, N. Kinetics of the oxidation of sucralose and related carbohydrates by ferrate(VI). *Chemosphere* 2012, *87*, 644–648.

311 Anquandah, G.A.K., Sharma, V.K., Knight, D.A., Batchu, S.R., and Gardinali, P.R. Oxidation of trimethoprim by ferrate(VI): kinetics, products, and antibacterial activity. *Environ. Sci. Technol.* 2011, *45*, 10575–10581.

312 Licht, S. A high capacity Li-ion cathode: the Fe(III/VI) super-iron cathode. *Energies* 2010, *3*, 960–972.

313 Licht, S., Wang, B., Gosh, S., Li, J., and Naschitz, V. Insoluble Fe(VI) compounds: effects on the super-iron battery. *Electrochem. Commun.* 1999, *1*, 522–526.

314 Licht, S., Wang, Y., and Gourdin, G. Enhancement of reversible nonaqueous Fe(III)/Fe(VI) cathodic charge transfer. *J. Phys. Chem. C* 2009, *113*, 9884–9891.

315 Licht, S. and TelVered, R. Rechargeable Fe(III/VI) super-iron cathodes. *Chem. Commun.* 2004, *6*, 628–629.

316 Licht, S., Wang, B., and Ghosh, S. Energetic iron(VI) chemistry: the super-iron battery. *Science* 1999, *285*, 1039–1042.

317 Walz, K.A., Szczech, J.R., Suyama, A.N., Suyama, W.E., Stoiber, L.C., Zeltner, W.A., Armacanqui, M.E., and Anderson, M.A. Stabilization of iron(VI) ferrate cathode materials using nanoporous silica coatings. *J. Electrochem. Soc.* 2006, *153*, A1102–A1107.

318 Yu, X. and Licht, S. Recent advances in synthesis and analysis of Fe(VI) cathodes: solution phase and solid-state Fe(VI) syntheses, reversible thin-film Fe(VI) synthesis, coating-stabilized Fe(VI) synthesis, and Fe(VI) analytical methodologies. *J. Solid State Chem.* 2008, *12*, 1523–1540.

319 Verweij, H., Christianse, K., and Van Steveninck, J. Ozone-induced formation of O,O′-dityrosine cross-links in proteins. *Biochim. Biophys. Acta* 1982, *701*, 180–184.

320 Trehy, M.L. and Bieber, T.I. Detection, identification, and quantitative analysis of dihaloacetonitriles in chlorinated natural waters. In *Advances in the identification and Analysis of Organic Pollutants in Water,* L.H. Keoth, ed. Ann Arbor Science Publishers, Ann Arbor, MI, 1981, *2*, 941–975.

321 Jacobsen, F., Holcman, J., and Sehested, K. Reactions of the ferryl ion with some compounds found in cloud water. *Int. J. Chem. Kinet.* 1998, *30*, 215–221.

322 Pignatello, J.J., Liu, D., and Huston, P. Evidence for an additional oxidant in the photoassisted Fenton reaction. *Environ. Sci. Technol.* 1999, *33*, 1832–1836.

323 Pignatello, J.J., Oliveros, E., and MacKay, A. Advanced oxidation processes for organic contaminant destruction based on the Fenton reaction and related chemistry. *Crit. Rev. Environ. Sci. Technol.* 2006, *36*, 1–84.

324 Loegager, T., Holcman, J., Sehested, K., and Pedersen, T. Oxidation of ferrous ions by ozone in acidic solutions. *Inorg. Chem.* 1992, *31*, 3523–3529.

325 Pestovsky, O. and Bakac, A. Identification and characterization of aqueous ferryl(IV) ion. *Preprints of Extended Abstracts presented at the ACS National Meeting, American Chemical Society, Division of Environmental Chemistry* 2006, *46*(2), 573–577.

326 Jacobsen, F., Holcman, J., and Sehested, K. Activation parameters of ferryl ion reactions in aqueous acid solution. *Int. J. Chem. Kinetics* 1997, *29*, 17–24.

327 Martire, D.O., Careganato, P., Furlong, J., Allegretti, P., and Gonzalez, M.C. Kinetic study of the reactions of oxoiron(IV) with aromatic substrates in aqueous solutions. *Int. J. Chem. Kinet.* 2002, *34*, 488–494.

328 Fiedler, A.T. and Que, L., Jr. Reactivities of Fe(IV) complexes with oxo, hydroxo, and alkylperoxo ligands: an experimental and computational study. *Inorg. Chem.* 2009, *48*, 11038–11047.

329 Wang, W., Zhang, M., Buhlmann, P., and Que, J.L. Redox potential and C-H bond cleaving properties of a nonheme Fe^{IV}=O complex in aqueous solution. *J. Am. Chem. Soc.* 2010, *132*, 7638–7644.

330 Comba, P., Fukuzumi, S., Kotani, H., and Wunderlich, S. Electron-transfer properties of an efficient nonheme iron oxidation catalyst with a tetradentate bispidine ligand. *Angew. Chem. Int. Ed.* 2010, *49*, 2622–2625.

331 Hong, S., Lee, Y., Cho, K., Sundaravel, K., Cho, J., Kim, M.J., Shin, W., and Nam, W. Ligand topology effect on the reactivity of a mononuclear nonheme iron(IV)-oxo complex in oxygenation reactions. *J. Am. Chem. Soc.* 2011, *133*, 11876–11879.

332 Park, J., Morimoto, Y., Lee, Y., Nam, W., and Fukuzumi, S. Proton-promoted oxygen atom transfer vs proton-coupled electron transfer of a non-heme iron(IV)-oxo complex. *J. Am. Chem. Soc.* 2012, *134*, 3903–3911.

333 Klinker, E.J., Shaik, S., Hirao, H., and Que, L., Jr. A two-state reactivity model explains unusual kinetic isotope effect patterns in C-H bond cleavage by nonheme oxoiron(IV) complexes. *Angew. Chem. Int. Ed.* 2009, *48*, 1291–1295.

334 Abouelatta, A.I., Campanali, A.A., Ekkati, A.R., Shamoun, M., Kalapugama, S., and Kodanko, J.J. Oxidation of the natural amino acids by a ferryl complex: kinetic and mechanistic studies with peptide model compounds. *Inorg. Chem.* 2009, *48*, 7729–7739.

335 Ekkati, A.R. and Kodanko, J.J. Targeting peptides with an iron-based oxidant: cleavage of the amino acid backbone and oxidation of side chains. *J. Am. Chem. Soc.* 2007, *129*, 12390–12391.

336 Kotani, H., Suenobu, T., Lee, Y., Nam, W., and Fukuzumi, S. Photocatalytic generation of a non-heme oxoiron(IV) complex with water as an oxygen source. *J. Am. Chem. Soc.* 2011, *133*, 3249–3251.

337 Hirao, H., Li, F., Que, L., and Morokuma, K. Theoretical study of the mechanism of oxoiron(IV) formation from H_2O_2 and a nonheme iron(II) complex: O-O cleavage involving proton-coupled electron transfer. *Inorg. Chem.* 2011, *50*, 6637–6648.

338 Bryant, J.R. and Mayer, J.M. Oxidation of C-H bonds by $[(bpy)2(py)R^{uI}VO]^{2}+$ occurs by hydrogen atom abstraction. *J. Am. Chem. Soc.* 2003, *125*, 10351–10361.

339 Wang, K. and Mayer, J.M. Oxidation of hydrocarbons by $[(phen)_2Mn(\mu\text{-}O)_2Mn(phen)2]^{3+}$ via hydrogen atom abstraction. *J. Am. Chem. Soc.* 1997, *119*, 1470–1471.

340 Yin, G., Danby, A.M., Kitko, D., Carter, J.D., Scheper, W.M., and Busch, D.H. Oxidative reactivity difference among the metal oxo and metal hydroxo moieties: pH dependent hydrogen abstraction by a manganese(IV) complex having two hydroxide ligands. *J. Am. Chem. Soc.* 2008, *130*, 16245–16253.

341 Yin, G., Danby, A.M., Kitko, D., Carter, J.D., Scheper, W.M., and Busch, D.H. Understanding the selectivity of a moderate oxidation catalyst: hydrogen abstraction by a fully characterized, activated catalyst, the robust dihydroxo manganese(IV) complex of a bridged cyclam. *J. Am. Chem. Soc.* 2007, *129*, 1512–1513.

342 Gardner, K.A., Kuehnert, L.L., and Mayer, J.M. Hydrogen atom abstraction by permanganate: oxidations of arylalkanes in organic solvents. *Inorg. Chem.* 1997, *36*, 2069–2078.

343 Ekkati, A.R., Campanali, A.A., Abouelatta, A.I., Shamoun, M., Kalapugama, S., Kelley, M., and Kodanko, J.J. Preparation of N-acetyl, tert-butyl amide derivatives of the 20 natural amino acids. *Amino Acids* 2010, *38*, 747–751.

344 Buxton, G.V., Greenstock, C.L., Helman, W.P., and Ross, W.P. Critical review of rate constants for reactions of hydrated electrons, hydrogen atoms and hydroxyl radicals in aqueous solution. *J. Phys. Chem. Ref. Data* 1988, *17*, 513–886.

345 Collins, M.J., Ray, K., and Que, J.L. Electrochemical generation of a nonheme oxoiron(IV) complex. *Inorg. Chem.* 2006, *45*, 8009–8011.

346 Lee, Y.M., Kotani, H., Suenobu, T., Nam, W., and Fukuzumi, S. Fundamental electron-transfer properties of non-heme oxoiron(IV) complexes. *J. Am. Chem. Soc.* 2008, *130*, 434–435.

347 Sastri, C.V., Seo, M.S., Park, M.J., Kim, K.M., and Nam, W. Formation, stability, and reactivity of a mononuclear nonheme oxoiron(IV) complex in aqueous solution. *Chem. Commun.* 2005, 1405–1407.

348 Leeds, J.M., Brown, P.J., McGeehan, G.M., Brown, F.K., and Wiseman, J.S. Isotope effects and alternative substrate reactivities for tryptophan 2,3- dioxygenase. *J. Biol. Chem.* 1993, *268*, 17781–17786.

349 Stewart, D.J., Napolitano, M.J., Bakhmutova-Albert, E.V., and Margerum, D.W. Kinetics and mechanisms of chlorine dioxide oxidation of tryptophan. *Inorg. Chem.* 2008, *47*, 1639–1647.

350 Campanali, A.A., Kwiecien, T.D., Hryhorczuk, L., and Kodanko, J.J. Oxidation of glutathione by $[Fe^{IV}(O)(N_4Py)]^{2+}$: characterization of an $[Fe^{III}(SG)(N_4Py)]^{2+}$ intermediate. *Inorg. Chem.* 2010, *49*, 4759–4761.

351 Jabre, N.D., Hryhorczuk, L., and Kodanko, J.J. Stability of a ferryl-peptide conjugate is controlled by a remote substituent. *Inorg. Chem.* 2009, *48*, 8078–8080.

352 Prakash, J. and Kodanko, J.J. Selective inactivation of serine proteases by nonheme iron complexes. *Inorg. Chem.* 2011, *50*, 3934–3945.

353 Fukuzumi, S. Role of metal ion in controlling bioinspired electron-transfer systems. Metal ion coupled electron-transfer. *Prog. Inorg. Chem.* 2009, *56*, 49–153.

354 Fukuzumi, S., Morimoto, Y., Kotani, H., Naumov, P., Lee, Y., and Nam, W. Crystal structure of a metal ion-bound oxoiron(IV) complex and implications for biological electron transfer. *Nature Chem.* 2010, *2*, 756–759.

355 Karlin, K.D. Redox control of oxoiron(IV). *Nat. Chem.* 2010, *2*, 711–712.

356 Kokarovtseva, I.G., Belyaev, I.N., and Semenyakova, L.V. Oxygen compounds of iron (VI, V, IV). *Russ. Chem. Rev.* 1972, *41*, 929–937.

357 Jeannot, C., Malaman, B., Gerardin, R., and Oulladiaf, B. Synthesis, crystal, and magnetic structures of the sodium ferrate(IV) Na_4FeO_4 studied by neutron diffraction and Mössbauer techniques. *J. Solid State Chem.* 2002, *165*, 266–277.

358 Menton, J.D. and Bielski, B.H.J. Studies of the kinetics, spectral and chemical properties of Fe(IV) pyrophosphate by pulse radiolysis. *Radiat. Phys. Chem.* 1990, *36*, 725–733.

359 Klemm, W. New metallates with coordinated oxygen and fluorine. II. *Angew. Chem.* 1954, *66*, 468–474.

360 Wahl, K., Klemm, W., and Wehrmeyer, G. Several oxo complexes of transition elements. *Z. Anorg. Allg. Chem.* 1956, *285*, 322–326.

361 Hoppe, R. and Mader, K. On the constitution of $K_3[FeO_4]$. *Z. Anorg. Allg. Chem.* 1990, *586*, 115–124.

362 Scholder, R. Alkali oxometallates(V) of chromium, manganese, iron, and cobalt. *Bull. Soc. Chim. France* 1965, *4*, 1112–1114.

363 Perfiliev, Y.D. and Sharma, V.K. Higher oxidation states of iron in solid state: synthesis and their Mossbauer characterization. *ACS Symp. Ser.* 2008, *985*(Ferrates), 112–123.

364 Macova, Z., Bouzek, K., Hives, J., Sharma, V.K., Terryn, R.J., and Baum, J.C. Research progress in the electrochemical synthesis of ferrate(VI). *Electrochim. Acta* 2009, *54*, 2673–2683.

365 Perfiliev, Y.D., Benko, E.M., Pankratov, D.A., Sharma, V.K., and Dedushenko, S.K. Synthesis of ferrates(VI) by ozonation. *Abstracts of Papers, 232nd ACS National Meeting, San Francisco, CA, United States, Sept. 10–14, 2006* 2006, Envr-019.

366 Licht, S. and Yu, X. Recent advances in Fe(VI) synthesis. *ACS Symp. Ser.* 2008, *985*(Ferrate), 2–51.

367 Rush, J.D. and Bielski, B.H.J. Pulse radiolysis studies of alkaline Fe(III) and Fe(VI) solutions. Observation of transient iron complexes with intermediate oxidation states. *J. Am. Chem. Soc.* 1986, *108*, 523–525.

368 Bielski, B.H.J. and Thomas, M.J. Studies of hypervalent iron in aqueous solutions. 1. Radiation-induced reduction of iron(VI) to iron(V) by CO_2^-. *J. Am. Chem. Soc.* 1987, *109*, 7761–7764.

369 Bielski, B.H.J. Generation of iron(IV) and iron(V) complexes in aqueous solutions. *Method Enzym.* 1990, *186*, 108–113.

370 Sharma, V.K., Mishra, S.K., and Nesnas, N. Oxidation of sulfonamide antimicrobials by ferrate(VI) $[Fe^{VI}O_4^{2-}]$. *Environ. Sci. Technol.* 2006, *40*, 7222–7227.

371 Sharma, V.K. Oxidative transformations of environmental pharmaceuticals by Cl_2, ClO_2, O_3, and Fe(VI): kinetics assessment. *Chemosphere* 2008, *73*, 1379–1386.

372 Sharma, V.K., Anquandah, G.A.K., and Nesnas, N. Kinetics of the oxidation of endocrine disruptor nonylphenol by ferrate(VI). *Environ. Chem. Lett.* 2009, *7*, 115–119.

373 Sharma, V.K., Anquandah, G.A.K., Yngard, R.A., Kim, H., Fekete, J., Bouzek, K., Ray, A.K., and Golovko, D. Nonylphenol, octylphenol, and bisphenol-A in the aquatic environment: a review on occurrence, fate, and treatment. *J. Environ. Sci. Health A Toxic/Hazard. Subst. Environ. Eng.* 2009, *44*, 423–442.

374 Huang, H., Sommerfeld, D., Dunn, B.C., Lloyd, C.R., and Eyring, E.M. Ferrate(VI) oxidation of aniline. *J. Chem. Soc. Dalton Trans.* 2001, 1301–1305.

375 Johnson, M.D. and Hornstein, B.J. Unexpected selectivity in the oxidation of arylamine with ferrate-preliminary mechanistic considerations. *Chem. Commun.* 1996, 965–966.

376 Johnson, M.D., Hornstein, B.J., and Wischnewsky, J. Ferrate(VI) oxidation of nitrogenous compounds. *ACS Symp. Ser.* 2008, *985*(Ferrates), 177–188.

377 Sharma, V.K. and Bielski, B.H.J. Reactivity of ferrate(VI) and ferrate(V) with amino acids. *Inorg. Chem.* 1991, *30*, 4306–4311.

378 Noorhasan, N., Patel, B., and Sharma, V.K. Ferrate(VI) oxidation of glycine and glycylglycine: kinetics and products. *Water Res.* 2010, *44*, 927–937.

379 Bielski, B.H.J., Sharma, V.K., and Czapski, G. Reactivity of ferrate(V) with carboxylic acids: a pre-mix pulse radiolysis study. *Radiat. Phys. Chem.* 1994, *44*, 479–484.

380 Kamachi, T., Nakayama, T., and Yoshizawa, K. Mechanism and kinetics of cyanide decomposition by ferrate. *Bull. Chem. Soc. Jpn.* 2008, *81*, 1212–1218.

381 Sharma, V.K. Use of iron(VI) and iron(V) in water and wastewater treatment. *Water Sci. Technol.* 2004, *49*, 69–74.

382 Rush, J.D. and Bielski, B.H.J. The oxidation of amino acid by ferrate(V). A premix pulse radiolysis study. *Free Radic. Res.* 1995, *22*, 571–579.

383 Sharma, V.K. Ferrate(V) oxidation of pollutants: a premix pulse radiolysis. *Radiat. Phys. Chem.* 2002, *65*, 349–355.

384 Noorhasan, N. and Sharma, V.K. Kinetics of the reaction of aqueous iron(VI) with ethylediaminetetraacetic acid. *Dalton Trans.* 2008, 1883–1887.

385 Noorhasan, N.N., Sharma, V.K., and Cabelli, D. Reactivity of ferrate(V) ($F^eV_O^{43}$-) with aminopolycarboxylates in alkaline medium: a premix pulse radiolysis. *Inorg. Chim. Acta* 2008, *361*, 1041–1046.

386 Carr, J.D., Kelter, P.B., and Ericson, III, A.T. Ferrate(VI) oxidation of nitrilotriacetic acid. *Environ. Sci. Technol.* 1981, *15*, 184–187.

387 Delaude, L. and Laszlo, P. A novel oxidizing reagent based on potassium ferrate(VI). *J. Org. Chem.* 1996, *61*, 6360–6370.

388 Khalilzadeh, M.A., Hosseini, A., Sadeghifar, H., and Valipour, P. Rapid and efficient oxidation of Hantzsch 1,4-dihydropyridines with potassium ferrate under microwave irradiation. *Acta Chim. Slov.* 2007, *54*, 900–902.

389 Rush, J.D. and Bielski, B.H.J. Kinetics of ferrate(V) decay in aqueous solution. A pulse-radiolysis study. *Inorg. Chem.* 1989, *28*, 3947–3951.

390 Carr, J.D. Kinetics and product identification of oxidation by ferrate(VI) of water and aqueous nitrogen containing solutes. *ACS Symp. Ser.* 2008, *985*(Ferrates), 189–196.

391 Rush, J.D., Zhao, Z., and Bielski, B.H.J. Reaction of Ferrate(VI)/Ferrate(V) with hydrogen peroxide and superoxide anion- A stopped-flow and premix pulse radiolysis study. *Free Radic. Res.* 1996, *24*, 187–192.

392 Carr, J.D., Kelter, P.B., Tabatabai, A., Spichal, D., Erickson, L., and McLaughlin, C.W. Properties of ferrate(VI) in aqueous solution: an alternate oxidant in wastewater treatment. *Proceedings of the Conference on Water Chlorination and Chemical Environment Impact Health Effects* 1985, 1285–1298.

393 Sharma, V.K., Burnett, C.R., and Millero, F.J. Dissociation constants of monoprotic ferrate(VI) ions in NaCl media. *Phys. Chem. Chem. Phys.* 2001, *3*, 2059–2062.

394 Rush, J.D. and Bielski, B.H.J. Decay of ferrate(V) in neutral and acidic solutions. A premix pulse radiolysis study. *Inorg. Chem.* 1994, *33*, 5499–5502.

INDEX

Oxidation of Amino Acids, Peptides, and Proteins: Kinetics and Mechanism, First Edition. Virender K. Sharma.